T0231109

POWER AND ENERGY

PROCEEDINGS OF THE INTERNATIONAL CONFERENCE ON POWER AND ENERGY (CPE 2014), SHANGHAI, CHINA, 29–30 NOVEMBER 2014

Power and Energy

Editor

Richard Kong
Wuhan University, China

CRC Press is an imprint of the
Taylor & Francis Group, an **informa** business

A BALKEMA BOOK

CRC Press/Balkema is an imprint of the Taylor & Francis Group, an informa business

© 2015 Taylor & Francis Group, London, UK

Typeset by V Publishing Solutions Pvt Ltd., Chennai, India
Printed and bound in the UK and the US

Published by: CRC Press/Balkema
P.O. Box 11320, 2301 EH Leiden, The Netherlands
e-mail: Pub.NL@taylorandfrancis.com
www.crcpress.com – www.taylorandfrancis.com

ISBN: 978-1-138-02782-4 (Hbk)
ISBN: 978-1-315-68865-7 (eBook PDF)

Table of contents

Foreword

Conference proceedings CPE-14 is the companion volume of conference proceedings MSET-14, for the same purpose of promoting the creativity of Chinese nation but in the scope of Power and Energy. Power and Energy, as well and material science are always the companion of electric, electronic and energy.

The reasons to promote the subjects of science are extraordinarily the same as those for Power and Energy and we won't to repeat them in the foreword for CPE-14.

When we were collecting the papers for the Conference on Power and Energy, the interesting things were that a number of authors were quite keen to look for the same chance to make public the theoretical works of intellectual creativity or formal results of scientific research or practice, written by those who are their former class mates, campus fellows, friends, relatives, colleagues and cooperators. It is really a chance for us to organize a chance for our smart intellectuals to expose, exchange and confidently approve each other the value of their arduous work. In three months, we have collected 304 copies of papers in the scope the mentioned subjects and their close-related subjects, and 86 copies selected after peer-view. 29 copies are in the research of power system while 32 copies of energy, 25 copies are on the subjects of close-related to power and energy.

Energy is one of the most important matters of our life and with us as ever we have existed, unlike computer or information technology but unfortunately we have never known all enery we have used or around us. Nevertheless, we have been working very hard to discover new sources of enery we are in need of and striving non-stop for new synthetic stuffs or man-made matters. High demand has pushed our scholars, experts and professionals to continue the mission, not only for the materials themselves but non-perilous to the life and environment as well, causing the least hazard to the world. We appreciate our authors consciously to involve into that mission.

Power appeared in the antient time in human life. All tools we found or created beside we can simply use as part of our body such as our hands and feet, eyes and teeth, are mechanics, but modern power has extended its concept since steam and gasoline engine, electric motor were found. It has spread to all corners of life, without what, we can not survive in contemporary days. Consequently, there are countless subjects for our learned people to work at. As the organizer of the conference CPE-14, we have selected 86 copies of papers in power and engry which are comparatively at a high level of each research.

Conference on Power and Energy is scheduled to hold in Shanghai from Nov 29 to 30, 2014, experts and scholars, a group of the authors and other related people will attend the conference with apparent interest, we are expecting a full success of the conference.

We appreciate those who responded to our proposal and submitted their papers, especially those whose papers have been selected for the conference CPE-14, the sponsors who have provided their valuable and professional suggestions and instructions and the scholars and professors who have spent their efforts as peer reviewers. We expect the conference to be successful and we hope that our proposal 'China Made jumping to China Create' will result in many more conferences on various other scopes and subjects, nation-wide, encouraging more professionals, scholars and students to contribute their talents to improving our life.

Thank you!

Samson Yu
November 5, 2014

Organizing committee

GENERAL CHAIR

Prof. Samson Yu, *Shanghai Jiao Tong University, China*

TECHNICAL PROGRAM COMMITTEE

Prof. Junyong Liu, *Sichuan University, China*
Prof. Kaipei Liu, *Wuhan University, China*
Prof. Junichi Arai, *Kogakuin University, Japan*
Prof. Yasunori Mitani, *Kyushu Institute of Technology, Japan*
Prof. Liangzhong Yao, *AREVA T&D Technology Centre, UK*
Prof. Chia-Chi Chu, *National Tsing Hua University, Taiwan (China)*
Prof. Gary W. Chang, *Department of Electrical Engineering, National Chung Cheng University, Taiwan (China)*
Dr. Ming Xu, *Honeywell International Inc., USA*
Prof. Trillion Q. Zheng, *Beijing Jiaotong University, China*
Prof. Ahmed Faheem Zobaa, *Exeter University, UK*
Prof. San Shing Choi, *Electrical and Electronic Engineering, Nanyang Technological University, Singapore*
Prof. Henry Huang, *Pacific Northwest National Lab, USA*
Prof. Shumei Cui, *Harbin Institute of Technology, China*
Prof. Wei Jen Lee, *University of Texas at Arlington, USA*
Prof. Ryuichi Yokoyama, *Environment and Energy Engineering, Waseda University, Japan*
Prof. Fangxing(Fran) Li, *Tennessee at Knoxville University, China*
Prof. Hongbin Sun, *Tsinghua University, China*
Prof. Chengshan Wang, *Tianjin University, China*
Prof. Jingnan Liu, *Member of Chinese Academy of Engineering, China*
Prof. Xiaoxin Zhou, *China Electric Power Research Institute, China*
Prof. Marcelo Masera, *Ispra Joint Research Center—European Commission, Italy*
Prof. M. Ramamoorty, *R&D Advisor, Equipment Manufacturing Company, Thane, India*
Prof. Yinbiao Shu, *Director General of the State Grid Corporation of China, China*
Prof. Qiang Lu, *Academician of the Chinese Academy of Sciences, China*
Prof. Djoko Prasetijo, *Deputy Director for System Planning, PT PLN—The State Electricity Company, Jakarta, Djibouti*
Prof. C.C. Chan, *University of Hong Kong, Hong Kong (China)*
Prof. Yusheng Xue, *State Grid Electric Power Research Institute, Member of Chinese Academy of Science, China*

Power system

Power and Energy – Kong (Ed.)
© 2015 Taylor & Francis Group, London, ISBN 978-1-138-02782-4

The application of Trend Point Model in short-term wind power forecasting

Fei Lan, Xing Jiang, Meng Yang, Xiaohua He & Jin Hu
School of Electrical Engineering, Guangxi University, Nanning, Guangxi Autonomous Region, China

ABSTRACT: In this paper, a novel strategy named Trend Point Model (TPM) in wind power prediction is proposed. Aiming at extracting the available information at utmost from onefold historical data, TPM utilizes Pearson's Correlation Coefficient (PCC) to find the points next to the similar sequences, which could be called as the trend points. Then by introducing Ordinary Least Squares (OLS) Method to fit linear functions of the similar sequences, the forecasting values at a certain point could be calculated through the corresponding linear functions one by one. The final predicting value would be obtained by taking the average of the previous forecasting values. The concision of the form of TPM makes it simple, intuitive and easy to understand. In the end, experimental results exhibit the validity of TPM in wind power prediction.

1 INTRODUCTION

As the traditional fossil fuels decrease dramatically, wind energy has come and been playing an increasing role in the field of power. China, one of the largest countries of energy consumption, even has its wind generation capacity up to 44,733 MW by the end of 2010[1]. Sophisticated and changeable as the wind is, it's hardly to know the output of wind power turbines precisely in the future, let alone considering the probability of malfunctioning of wind power equipments[2]. Being one of the most pivotal impacts in wind power development, wind power prediction not only affect the stability of power system, but also associate with the energy costs in the market for wind generators[3–4].

Wind power prediction started from forming of the wind power plants more than thirty years ago[5]. So far, there has been various techniques, theories or tactics applied to the wind power forecasting. In the early 1990s, the meteorological basis for the assessment of wind energy resources was established in order to get the suitable data for evaluating the unseen wind energy by means of the European Wind Atlas[6]. Based on the artificial intelligence, Wind Power Prediction Tool (WPPT) could calibrate the observed situation automatically and be configured to predict the total wind power in a region or a single wind farm[7]. Instead of complicated mathematical formulations, Artificial Neural Networks (ANN) imitate human brains to set and adjust the weights of the neurons as well as their biases with gradient descent algorithm[8]. And ANN has proven itself as a simple, powerful and flexible tool for wind power prediction[9–10]. Taking the historical data and the random disturbance into consideration, time series models, proved to be practical for short-term forecast, have drawn a lot of attentions on wind speed and wind power forecasting [11–12]. Based on the structure minimization rather than empirical minimization, Support Vector Machine (SVM) has enjoyed a warmly welcome after its birth and the use of SVM to forecast the wind speed and wind power took efforts[13–14]. Since every single methodology has its own property in tackling data, the combinational forecasts of different methods usually outperform the individual cases[15]. Then, plenty of methodologies combined with two or more strategies emerged.

The typical two forms are the following:

1. Making some optimization to the parameters of the algorithms through other methods like Particle Swarm Optimization (PSO)[16–17]
2. Reassembling different prediction results together like entropy[18–19]

This paper, derived from data mining[20], presents a new simple strategy using PCC[21–22] of sub-sequences in the main-sequence that represents the historical wind power data to collect the trend points which could be mapped into the required values according to the relationship between the sub-sequences meeting the certain condition.

Extreme Learning Machine (ELM), a newborn single hidden layer feed-forward network that tends to have better and more stable performance than back propagation in wind power forecasting[23] and achieve similar generalization performance for regression than SVM[24], is adopted as a comparison to verify the feasibility of TPM.

2 METHODOLOGY

2.1 Trend points

A large number of historical data of wind power could be written into a main-sequence just as $X_m = [x_1, x_2, ..x_i..., x_t]$ which contains t values. Where x_i stands for the wind power value of time i. Let $X_1 = [x_{t-d+1}, x_{t-d+2}, ..., x_t]$ be the first sub-sequence with a length of d. Similarly, the second should be $X_2 = [x_{t-d}, x_{t-d+1}, ..., x_{t-1}]$. So there would be as much as $t - d + 1$ amount of sub-sequences totally, which could be intuitively demonstrated in Figure 1.

It takes no effort to compute the Pearson's r of X_1 and the other remaining $t - d$ severally. As following:

$$r(X_1, X_j) = \frac{\sum_{k=1}^{d}(x_{1k} - \bar{X}_1)(x_{jk} - \bar{X}_j)}{\sqrt{\sum_{k=1}^{d}(x_{1k} - \bar{X}_1)^2 \sum_{k=1}^{d}(x_{jk} - \bar{X}_j)^2}} \quad (1)$$

where $j = 2, ..., t - d + 1$; x_{1k} and $\bar{X}_1$ are the kth value and the mean value of X_1, respectively. And it's the same with x_{jk} and $\bar{X}_j$.

ε indicates the threshold between 0 and 1 which can be artificially set. If $r(X_1, X_j) \geq \varepsilon$, then the sequences X_1 and X_j would be called as similarity sequences. In other words, X_1 and X_j are similar to each other. What's more, the first value right after the sub-sequence X_j within the main-sequence becomes what we call trend point, which is also exactly as same as the last one in sub-sequence X_{j+1}. Take X_{t+d-1} in Figure 1 for example, if $r(X_1, X_{t+d-1}) \geq \varepsilon$, then X_{t+d-1} is the similarity consequence, and x_{d+1} becomes a trend point corresponding to X_{t+d-1}.

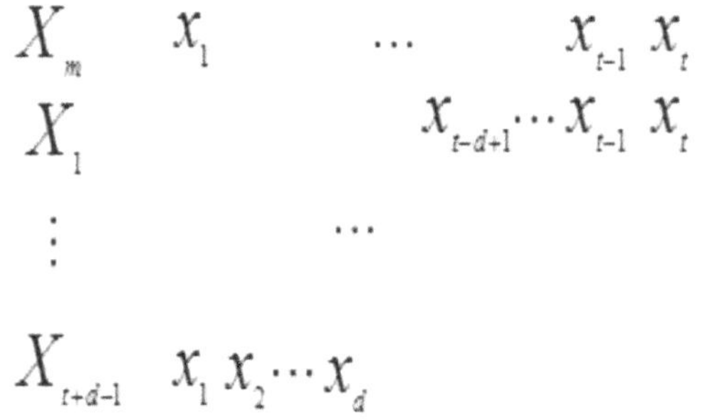

Figure 1. The diagram of main sequence and its corresponding sub-sequences.

Although the width of the sub-sequence d has a large number and ε is given a great figure close to 1. The similarity sequences may not be found occasionally. If so, either d or ε is supposed to be reduced to some extent. This essay choses ε to lessen.

And it is strongly suggested that if the sub-sequence X_j should be similar to X_1, then the corresponding trend point should be just the same similar to x_{t+1}, the real value of time t which is unknown at present.

2.2 Fitting

From equation (1), some conclusions can be drawn when ε is large enough close to 1:

a) X_j and X_1 get a linear relationship;
b) X_j has a tiny statistical fluctuation with some sort of linear sequence of X_1.

Apparently, b) is somewhat restating conclusion a) for the sake of manifestation of the importance of equation (1) in TPM.

When all the trend points are found, the next step is to find a way to fit the TPs to $\hat{x}_{t+1}$s, i.e. the forecasting value of time $t + 1$.

In statistics, OLS, which minimizes the sum of squared vertical distances between the observed responses in the dataset and the responses predicted by the linear approximation, is quite a simple and renowned fitting technique[25]. And the resulting estimator can be expressed by a simple formula, especially in the case of a single regressor on the right-hand side. In TPM, if X_{t+d-1} and X_1 satisfy equation (1), then X_1 may be well-fitting by OLS via formula $X_{t-d+1} \cdot a + b$, for instance. Consequently, $\hat{x}_{t+1}$ could be inferred through $\hat{x}_{t+1} = x_{d+1} \cdot a + b$.

With different trend point results in different set of (a, b), namely leading to different $\hat{x}_{t+1}$. Hence, there may be more than one predicting value at one point. Then, the final $\hat{x}_{t+1}$ can be gained by taking the average of different $\hat{x}_{t+1}$ corresponding to different trend point.

2.3 Adjustment

Wind power generated from wind farms has its physical limits that enforce its values to be non-negative. Along with many different $\hat{x}_{t+1}$s emerging at time $t + 1$, the forecast accuracy heightens when the negative $\hat{x}_{t+1}$s are deleted. Without the negative $\hat{x}_{t+1}$, however, the final $\hat{x}_{t+1}$ could not be obtained immediately owing to the possible underlying singularities.

In order to remove the probable outliers, we assume that the predicting values at one point except for the deleted negatives follow a normal distribution and therefore to set a confidence level turns necessary. Then the final $\hat{x}_{t+1}$ comes from the mean of the predicting values between the confidence interval.

Sometimes, there might be no pertinent $\hat{x}_{t+1}$ left because the $\hat{x}_{t+1}$ we compute may be entirely negative. In that case, reducing the size of ε will be able to help until we get $\hat{x}_{t+1}$ desirable.

2.4 Multistep prediction

Usually, the maximum predicting step could be longer than one. In TPM, the main-sequence updates itself by shorten the length of the main sequence, in other words, elongating the time

The prediction horizon	The main sequence X	Next prediction point
1	... x_{t-3} x_{t-2} x_{t-1} x_t	$\hat{x}_{t-1}$
2	... x_{t-6} x_{t-4} x_{t-2} x_t	$\hat{x}_{t-2}$
3	... x_{t-9} x_{t-6} x_{t-3} x_t	$\hat{x}_{t-3}$
...	... x_t	...

Figure 3. Change of the main-sequence with different prediction horizon.

Table 1. The average RMSE and MAPE comparisons of ELM and TPM in different maximum prediction horizons.

Max-step	RMSE/MW		MAPE	
	ELM	TPM	ELM	TPM
1-hour	4.5924	4.0600	0.0835	0.0743
2-hour	8.7070	6.1354	0.1312	0.0973
3-hour	15.4625	10.1803	0.3440	0.2215

span of the observed values. As is shown below in Table 1, the amount of the values in the main-sequence decreases from the original t to t/p when the length of prediction goes to p till p reaches the maximum prediction step. And all the sub-sequences would also make corresponding change in every step. Above all, a flow chart could substitute for the whole required steps of TPM in Figure 2.

Based on this deduction, the main sequence of wind power with 15-minute resolution would be $[..., x_{t-36}, x_{t-24}, x_{t-12}, x_t]$ when the next three hours becomes the prediction horizon, i.e., the prediction step is 12. Afterwards, the next forecasting value $\hat{x}_{t+12}$ could be calculated just the same way as $\hat{x}_{t+1}$. Another issue rise that the message of $[x_{t-11}, x_{t-10}, x_{t-9}, ..., x_{t-2}, x_{t-1}]$, which is from the current nearest known point x_t, is going to be abandoned. This as shown in Figure 3, in some way, may create a waste of effective information. Bring the lost data that may be useful into use needs some other algorithms to make a combination with TPM. And it will be discussed in our future work.

3 CASE ANALYSIS

3.1 Data and parameters

All data used in this paper are single history of wind power collected from several wind farms of a region located in Xijiang Province, northwest, China. The area has nominal wind power generation capacity of 1600 MW. Over 35,000 wind power data in the full year of 2012 with 15 minutes time interval are used in the case study.

All data have been normalized by equation (2) to be confined between 0 and 1 for the celerity and convenience of processing. TPM is trying to predict the wind power in one, two, and three hours to

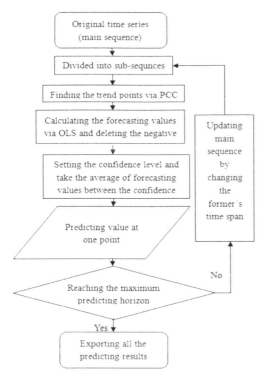

Figure 2. The flow chart of multistep prediction of TPM.

come, which means the maximum prediction step (P-max) are 4, 8, and 12 respectively. Then make a brief contrast with ELM forecasts. In this test, after the prediction step reaches the maximum horizon, we put the real values in lieu of the prediction values and roll forward.

$$x'_t = \frac{x_t - x_{t.\min}}{x_{t.\max} - x_{t.\min}} \qquad (2)$$

where $x_{t.\max}$ and $x_{t.\min}$ denote the maximum and minimum x_t respectively.

Due to the different wind power fluctuations in the intensity of different time series, neither a too great nor too small the threshold ε is suitable. The former increases the probability of losing similarity sequences. The latter may engender too many trend points making TPM meaningless. The determination of the length of sub-sequences d has the homologous puzzlement as threshold ε.

As we have not figured out an optimum way to get the eligible parameters so far, through exploratory analysis in this example, ε set to be 0.98 and the length of sub-sequences going to 6 would be a good pair. Every time when the similarity sequences do not exist, ε minus 0.01 of itself until the similarity sequences emerge. Then ε gets back to 0.98 for the next prediction step.

In addition, we choose 0.7 as the confidence level. Predicting values, computed by OLS, between the confidence interval at one point are considered to be reliable. By taking the mean of predicting values between the confidence, we can get the final predicting value at one point.

As for ELM, it is convenient that we can just need to determine the number of hidden layer nodes of the network and the hidden nodes number of ELM could be determined on the cross-validation approach[26-27]. In this experiment, ELM with a hidden nodes of 30 could be sufficient.

3.2 Evaluation standards

Two typical criteria for illustrating a prediction algorithm are taken in this paper: the Root Mean Squared Error (RMSE) and the Mean Absolutely Percentage Error (MAPE)[28].

If n is the maximum prediction horizon, $\hat{y}_i$ and y_i stand for the ith step prediction value and the true value of time ith, then

$$\text{RMSE} = \sqrt{\frac{1}{n}\sum_{i=1}^{n}(y_i - \hat{y}_i)^2}$$

$$\text{MAPE} = \frac{1}{n}\sum_{i=1}^{n}\left|\frac{y_i - \hat{y}_i}{y_i}\right|$$

3.3 Test results and illustration

The average RMSE and MAPE errors in ELM and TPM in the next one, two and three hours are revealed dividually in Table 1. The real and forecasting values in Table 2 have recovered from equation (2).

As is shown, both RMSE and MAPE errors increase as the time becomes longer. TPM gives little better performance than ELM in short-term wind power forecasting.

4 CONCLUSION

This paper introduced a fresh approach to predict the wind power in short terms, point by point. In a nutshell, several conclusions can be drawn:

1. The performance of TPM is probably not notable compared with ELM's, but it does, to a certain extent, improve the prediction accuracy when the wind speed, temperature, altitude, humidity, atmosphere and so forth factors fail to be available expect the past wind power.
2. Theoretically, the more the historical data of wind power is, the more beneficial for us to find the similarity sequences. While, this situation becomes paradoxical as the larger of the main-sequence spends the longer time on computing, which is not wanted.
3. TPM does not care how fiercely the wind power changes, linear or nonlinear. What it focus are just the little segments of wind power and the relationship between them, whether they are similar or not. That makes it concise and few parameters to be set.
4. Further work need to be done such as a considerable level of coordination between the length of sub-sequences and the size of the threshold.

ACKNOWLEDGEMENT

This research was supported by National Natural Science Foundation of China (NSFC) (51277034 & 51377027). Also, special thanks need go to the referees and the anonymous for their indispensable guidance and suggestions.

REFERENCES

1. http://www.chinapower.com.cn/article/1248/art1248844.asp.
2. Ekaterina Vladislavleva, Tobias Friedrich, et al. Predicting the energy output of wind farms based on weather data: important variables and their correlation. Renewable Energy 2013, 50: 236–243.

3. Bouffard F, Galiana F.D. Stochastic security for operations planning with significant wind power generation. IEEE Transaction on Power Systems 2008, 23(2): 306–316.

4. Fabbri A, Román T.G.S, etal. Assessment of the cost associated with wind generation prediction errors in a liberalized electricity market. IEEE Transactions on Power Systems 2005, 20(3): 1440–146.

5. Alexandre Costa, Antonio Crespo et al. A review on the young history of the wind power short-term prediction. Renewable & Sustainable Energy Reviews 2008 12(6): 1725–1744.

6. Troen I, Lundtang Petersen E. European wind atlas. Risø National Laboratory 1989.

7. Nielsen T.S, Madsen H. WPPT—a tool for wind power prediction. EWEA Special Topic Conference, Kassel 2000.

8. Anderson J.A, Rosenfeld E. Neurocomputing: foundations of research. Cambridge, MA:MIT Press 1989.

9. J.P.S. Catalao, H.M.I. Pousinho, V.M.F. Mendes. An arti—ficial neural network approach for short-term wind power forecasting in Portugal. 15th International Conference on Intelligent System Applications to Power Systems 2009: 1–5.

10. M.C. Mabel, E. Fernandez. Analysis of wind power gene—ration and prediction using ANN: a case study. Renew Energy 2008;33(5): 986–992.

11. J.L Torres, A. Garcia, M. De Blas, et al. Forecast of hourly average wind speed with ARMA models in Navarre (Spain). Solar Energy 2005, 79(1): 65–77.

12. L. Kamal, Y.Z. Jafri. Time series models to simulate and forecast hourly averaged wind speed in Quetta, Pakistan. Solar Energy 1997, 61(1): 23–32.

13. Mohandes M.A, Halawani T.O, Rehman S, Hussain A.A. Support vector machines for wind speed prediction. Renewable Energy 2004, 29(6): 939–947.

14. Shi J, Liu Yongqian, Yang Yongping, et al. The research and application of wavelet-support vector machine on short term wind power prediction.8th World Congress on Intelligent Control and Automation (WCICA) 2010, 4927–4931.

15. Bates J.M, Granger C.W.J. The Combination of Forecasts. Operation Research Quarterly 1969, 20(16): 451–468.

16. Sancho Salcedo-Sanz, et al. Short term wind speed predic-tion based on evolutionary support vector regression algorithms. Expert System with Applications 2011, 38(4): 4052–4057.

17. Wenyu Zhang, Jie Wu et al. Performance analysis of four modified approaches for wind speed forecasting. Applied Energy 2012;99: 324–333.

18. Shuang Han, Yongqian Liu. The study of wind power combination prediction. In Power and Energy Engineering Conference, Chengdu, China, 2010.

19. Shuang Cang, Hongnian Yu. A combination selection algorithm on forecasting. European Journal of Operational Research 2014;234(1): 127–139.

20. Kfeng, F. Meng, Q. Niu, Q. Yan. Time series predic—tion algorithm based on trends point state model. Application Research of Computers 2011; 28(12): 4510–4513 (In Chinese with English Abstract).

21. Jacob Benesty, Jingdong Chen, Yiteng Huang. On the importance of the pearson correlation coefficient in noise reduction. IEEE Trans. Audio, Speech, Language Processing, 2003, 16: 757–765.

22. AnlongChen, ChangjieTang, ChanganYuan, et al. Mining correlations between multi-streams based on haar wavelet. Advances in Computer Science-ASIAN 2005, 3818: 270–271

23. Cancelliere R, GossoA, Grosso A. Neural networks for wind power generation forecasting: A case study. Networking, Sensing and Control (ICNSC), 2013 10th IEEE International Conference on 2013: 666–671.

24. Guang-Bin Huang, Hongming Zhou, Xiaojian Ding and Rui Zhang. Extreme Learning Machine for Regression and Multiclass Classification, IEEE Transactions on system, man, and cybernetics Part B: Cybernetics, vol. 42, 2, April 2012.

25. Milne, G.W. Response of the ordinary least squares estimator (OLS) to deterministic and stochastic noise. IEEE AFRICON 4th, AFRICON, 1996(2): 1099–1104.

26. Can Wan, Zhao Xu, Pinson P, Zhao Yang Dong, and Kit Po Wong. Probabilistic forecasting of wind power generation using extreme learning machine. Power Systems, IEEE Transactions on. 2014, 29(3): 1033–1044.

27. G.B. Huang, Q.Y. Zhu, and C.K. Siew. Extreme learning machine: Theory and applications. Neurocomputing. 2006, 70(1–3): 489–501.

28. Madsen H, Pinson P, Kariniotakis G, Nielsen H, Nielsen T. Standardizing the performance evaluation of short-term wind power prediction models. Wind Eng 2005, 29(6): 475–89.

Power and Energy – Kong (Ed.)
© 2015 Taylor & Francis Group, London, ISBN 978-1-138-02782-4

Comprehensive analysis of multiple Hidden Failures in protection system with the context of N-k contingency

Guang Sun & Junyan Chen
Shanghai Electric Power Design Institute Co. Ltd., Shanghai, China

ABSTRACT: The Hidden Failures (HFs) in protection systems have been proved to be a critical reason in creating and spreading N-k contingencies. This paper presents a comprehensive investigation of HFs in protection system and proposes a multiple HFs mode. The method used to calculate probability of each HF and N-k contingencies is proposed based on a function group method and event tree method. Based on the multiple mode, probability analysis of N-k contingencies has been proposed and a method used to identify the N-k contingencies is established. Applying the method in a practice power grid in China, the results show that this method can improve the accuracy of the calculating of the probability of N-k contingencies and identify more N-k contingencies which can be used as guidance for system planners to handle HFs and prevent the cascading failure.

Keywords: Protection relay; Hidden Failure; vulnerability region; N-k contingency; function group

1 INTRODUCTION

The Hidden Failures (HFs) in protection system have been recognized as critical reasons in spreading the power system turbulence, and leading to high order contingencies called N-k (k > 1) contingencies. According to the North American Electric Reliability (NERC) report, 400 events of cascading collapses happened in the past 16 years from 1984 to 1999 are caused by the HFs in protection relays [1].

The hidden failure of a protection system is defined as a permanent defect in protection system, which remains dormant when everything is normal and manifest as abnormal operating conditions are reached, such as faults or overloads [2,3]. There are multiple models applied to describe HFs in protection system. Ref. [4–6] regards the occurrence of HF as the consequence of the increasing line flow, which focuses on the characteristic of zone 3 of distance or over-current protection essentially. Ref. [7] deems the HFs occurring in protection as a result of two phenomena: "a protection element functionality defect" and "the logical arrangement of devices with the functionality defect". The concept of the regions of vulnerability is proposed to describe how an HF can cause a false trip [7,8]. Ref. [9] focus on the analysis of the inappropriate relay settings which result in HFs by using the concept of vulnerability region whose scope and position are related to the settings of relays and the system operation scheme. Ref. [10] investigates the

impact of HFs in transducers and circuit breakers on bulk power system reliability while the properties of the HF in a breaker related to the breaker failure tripping schemes and bus configurations have not received much attention in the analysis of the HFs in protection system [11,12]. All of the work was mainly focused on one or two significant HF modes in protection system.

However, as a complex automatically controlled systems, HFs in protection systems should be multiple. One HF has an influence with other HFs and different HFs occur at different time points in the action of a protection relay. The multiple HFs in protection system have brought some new problems in the analysis of N-k contingencies. Different HFs may come to different N-k contingencies. The problems can be reduced to two aspects. One is the method to calculate the probability of N-k contingencies. The other one is the identifying of N-k contingencies in power system.

This paper proposes a multiple HF mode in protection system, which considers the relationship among typical HF modes in protection system. Several significant HF modes in protection system are discussed in section II. The calculating and identifying methods of the N-k contingencies based on the Functional Group are established in section III. Risk index and the flowchart of the risk assessment are given in section IV. In section V a practice power grid in China is used to apply the proposed method. The results show that the multiple HF mode can improve the accuracy of the calculating of the probability of N-k contingencies and identify more N-k contingencies, which is imperative for the system planners to deal with the HFs in protection system and mitigate the cascading failures from spreading to the entire power system.

2 TYPICAL HIDDEN FAILURES IN PROTECTION SYSTEM

The types of HFs in protection system are various and complex which are corresponding to different characteristic. In this paper, in order to simplify the analysis process and put more attention on the essence of the development of N-k contingency, we introduce two assumption: 1) line protection is investigated while the transformer protection and generator protection are neglected. 2) The initial faults set on lines are assumed to be three-phase permanent faults. Generally, transmission line protection consists of double primary protections with different protection principle, such as pilot distance protection and differential protection, and back up protection, such as grounded, phase distance and zero order current protection [13]. In this context, the ground and phase distance protection can be deal with as the distance protection and the zero order current protection which is often used to detect high impedance grounding faults are neglected.

In this section the typical HF modes in primary protection, such as pilot distance protection and differential protection, and backup protection including distance protection are investigated. The HFs of the communication channel which play a crucial role in the action of the primary relay and the relay settings of distance protection have been paid more attention. In addition, the HF of a breaker considering station configurations is also discussed in this section.

2.1 *The communication channel*

The communication channel in protection system is critical for a differential or pilot relay to act correctly, which is often affected by many factors, such as short circuits in lines, switch operations and so on. The HF of communication channel (called HF1) is revealed through the fault detector relay [7]. For example, when communication channel's failure takes place, the fault detector relay of the pilot distance protection cannot detect external faults. If a fault occurs at the vulnerability region on line B2–B4 illustrated in Figure 1, both of line B1–B2 and B2–B4 will be tripped in the absence of blocking carrier. It brings a disoperation and leads to an N-2 contingency from the initial fault.

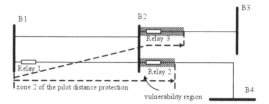

Figure 1. The vulnerability region for a pilot distance protection with HF of the communication channel.

The method to calculate the scope of vulnerability region is showed in section III.

Similarly, the differential current protection will fail to operate when a fault occurs on the internal zone of the relay if HF1 occurs. The relay that fails to act may cause the associated relays to act to isolate the faults which lead to the expansion of the fault area.

2.2 *The inappropriate relay settings*

The relay settings are usually obtained from off-line calculation, which are sometimes inappropriate especially when the real-time operation scheme of power system changes [9]. The inappropriate settings include two situations: "the inappropriate coordination relation among different relay settings" and "can not to evade the load impedance".

The inappropriate coordination relationship among different relay settings are correlatively from the settings of zone 1 and zone 2 of distance protection, which result in HFs of the backup protection. Figure 2 shows a protection scope of zone I, which cannot meet the selectivity (called HF2). A fault occurring in the vulnerability region on B2–B3 will trip line B1–B2 and B2–B3.

Figure 3 shows the situation that the settings of zone 2 cannot meet the selectivity (called HF3). This is a common situation when a short line needs to coordinate with a long one. The regions on different lines are different from each other which are related with impedance of the line and system operation. On line B2–B4, the scope of the zone 2 of Relay 1 is exceeding the scope of zone 1 of Relay 2 and coming with a vulnerability zone while the same situation is not appearing on line B2–B3. The vulnerability region associated with inappropriate settings is changing and uncertain as the system operation scheme changes.

The situation which relays cannot evade load impedance is corresponding to settings of zone 3 as some lines are overloading (called HF4) [12]. Figure 4 shows the quadrilateral characteristics of zone 3 and characteristics for load impedance.

The properties of zone 3 are reflected by the reliability factor K which is used to embody the ability

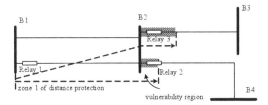

Figure 2. The HF2 mode with the setting of zone 1.

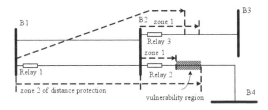

Figure 3. The HF3 mode with the setting of zone 2.

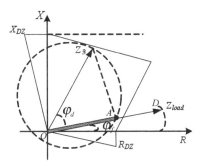

Figure 4. Quadrilateral characteristics for zone 3 of a distance protection and characteristics for load impedance.

to evade load. The factor can be calculated by ratio of load impedance and impedance of zone 3, like

$$K = \frac{OD}{OA} = \frac{Z_{\text{load}}}{Z_3 \cos(\varphi_d - \varphi_f)} \qquad (1)$$

where Z_{load} is load impedance, Z_3 is setting of zone 3. The φ_d is angle of sensitivity and φ_f is angle of load impedance.

2.3 *The breaker*

The HF of a breaker itself is influenced by system operating and weather condition, which varies with stations and bus configurations (called HF5). If a breaker fails to act, it is necessary to trip all of the breakers on its associated bus and the bus tie because of the action of circuit breaker failure relays [12]. Figure 5 shows a half station

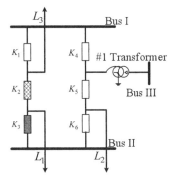

Figure 5. The HF5 mode of breaker with a half station configuration.

configuration, which has been commonly adopted in high voltage power transmission network.

When initial fault occurs on line L_1, breaker K_2 and K_3 act to clear the fault. But if breaker K_3 fails to act because of HF5, the breaker failure relay will trip all of the associated breakers, such as breaker K_6, to isolate the fault from system. However when breaker K_2 fails to act, breaker K_1 acts on trip and the station will trip line L_3. Particularly when the outgoing line connects a transformer showed in Figure 5, the station will lose a transformer if faults occur on line L_2 as breaker K_5 fails to act. Different bus configurations and the location of breakers with HF5 lead to different severity of consequences. Other station configurations such as double bus and single breaker configuration also come with similar conclusion.

Several typical HF modes have been investigated in this section. As an initial fault occurs, the protection system acts according to the time sequence. Different HF modes are corresponding to different time nodes. For example, at the time as fault occurs, HF1 and HF2 modes should be considered first, because the relay with these HF modes acts incorrectly instantaneously. The HF4 mode takes place after some lines tripped and can be considered at time t_1 ($t_1 > 0$). Based on the multiple HFs mode, an analysis of probability of N-k contingency is proposed in section III.

3 PROBABILITY ANALYSIS OF N-K CONTINGENCY

HFs often lead to an unwanted trip and worsen a condition after the initial fault. It is clear that the chains of N-k contingency are corresponding to HFs in protection system. In section II, five HF modes are given. The effect of these HF modes can be reflected through N-k contingency. Because many N-k contingencies exit with multiple HF modes, it becomes necessary to determine which N-k contingencies caused are potentially most damaging to power system. In this section, we give the probability analysis of N-k contingency under multiple HF modes. Probability of each HF mode is given. A method based on Function Group is proposed to analyze the procedure of N-k contingency [8].

3.1 Probability of different HFs

It is worthwhile to mention that occurrence of HFs come with the initial fault of a line. Initial fault probability occurring on a line according to exponential density function can be calculated as follows:

$$p(L,t) = \lambda e^{-\lambda t}, t \geq 0 \qquad (2)$$

where t is the age of line in year [14]. The fault location deems to be uniform distribution.

The probability of HF1, HF2 and HF3 mode is decided by the scope of vulnerability region after this assumption, which can be calculated by the following equation like:

$$p(HFx) = p(L) \cdot (L_{\text{region}}/L_{\text{line}}) \quad (x = 1,2,3) \qquad (3)$$

where L_{region} and L_{line} are the length of vulnerable zone and line. The system operation scheme changes all the time as the power flow or the grid topology change. The scope of vulnerability zone related with the operation scheme are uncertain and variation. In Figure 1, when line B2–B3 is tripped, the assistance current of the other lines on the line B2–B4 is changing and the scope of HF1 mode also change. It is hard to checkout each relay setting and each system operation scheme to determine the region in a grid with a large number of nodes and lines.

Thus an on-line method is a necessary way to determine the vulnerability scope. An on-line verification system has been researched and developed to verify the lines with inappropriate settings. From the EMS (Energy Management System) topology processing, which is available in most control centers, the data of the operation scheme of power system can be got. According to the on-line operation scheme, the computing values of each relay can be calculated through the verification principles, which is the inverse process of protection setting. Through the comparison between the computing values with the operation settings, which can be obtained from the FIS (Fault Information System), the system verification results can be received with lines with inappropriate relay settings. When the lines with inappropriate relay settings are selected through the verification system, the scope of each setting can be calculated by a dichotomy method through comparing measure impedance with relay settings.

The probability of HF4 can be got according to the definition of reliability factor which is related by power flow on each line. In consideration of the measure error and the system operation conditions, factor K often corresponds to a minimum value. As the factor is less than the minimum value K_{min}, the setting of the zone 3 is coming with a HF and has a probability for the distance relay to act incorrectly with no fault occurs. The probability of HF4 mode is related to the deviate degree between the reliability factor and the minimum value, which can be calculated by:

$$p(\text{HF4}) = \frac{K_{min}Z_3 - Z_m}{K_{min}Z_3 - Z_3} \quad Z_3 < Z_m < K_{min}Z_3 \qquad (4)$$

where Z_m is measure impedance of distance protection. In addition, when Z_m is less than Z_3, the probability to act incorrectly is 1.0 and when Z_m is more than $K_{min}Z_3$, the probability deems to be 0.0. The probability of a breaker with HF5 can be received by the curve of the pelvis. In this paper, the probability is replaced by a constant value received from the real operation statistical data.

3.2 Headings probability of N-k contingencies

Based on the multiple HFs model, probability of N-k contingencies are investigated through Function Group method [11] and event tree method [15]. Ref. [11] proposed a Function Group method that takes the break as the interface between two different function groups showed in Figure 6. Lines and transformers consist of different groups. The method to form function group can be found in [11]. By calculating the fault probability of each group and the breaker connecting them, the probability of N-k contingencies can be received.

However, there is a problem exiting in this method: failing to act of breaker is a reason causing N-k contingency but not an only one [16]. For example, HF1 may occur to lead to an N-2 contingency without the break's failing to act. Through analysis of multiple HFs mode, the HFs should be the virtual component combining different groups.

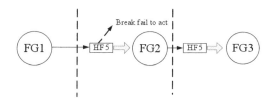

Figure 6. The HF5 mode of breaker with a half station configuration.

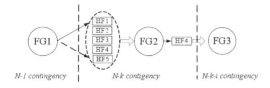

Figure 7.　The HF5 mode of breaker with a half station configuration.

Figure 7 shows the method to analysis N-k contingency and N-k-i contingency under multiple HFs mode.

Different HFs come to replace the breaker's fail to act. Different N-k contingencies are corresponding to some HFs. Based on the function group method, the event tree method can be used to calculate probability of each contingency. For example, when an initial fault occurs, the HF1 mode is considered and calculated first according to the time sequence mode. Then the HF modes coming from the relay settings are decided by the position of the fault. If HF3 mode occurs after the HF1 mode, N-2 contingency will take place. After the trips of some lines in the grid, some lines may be overloading, which lead to HF4 mode of the protection. The relays with HF4 modes are often corresponding to different probability, which the maximum one needs to be tripped. If breakers act to trip successfully, the probability of N-k contingency can be calculated by:

$$P(N-k) = p(L) \cdot p(HF1) \cdot p(HF4)$$
$$\cdot p(HF3) \cdot (1 - p(HF5)) \qquad (5)$$

Based on the above method, the procedure to identify the biggest probability of N-k contingencies is proposed as below:

a. Select a transmission line for an initial three-phase permanent fault and calculating the fault probability for this line.
b. Determine the lines which are connected adjacent to the fault line and form function groups.
c. Perform the power flow solution prior to the first tripping and Calculate the probability of HF1~HF3 mode.
d. Perform the tripping on the line in the adjacent group in turns of HF1~HF3 mode. Then perform the power flow solution and calculate the probability of HF4 of each line. If the power flow is unsolvable, then stop and exit.
e. Select the line to trip with maximum probability of HF4 and perform in step (d).
f. Identify the configuration of the station with the line. Then calculate the HF5 mode and perform in step (d).

g. Use the equation (5) to calculate the probability value. Then perform in step (a)
h. Sort the probability in a descending order.

4　CASE STUDY AND DISCUSSION

4.1　Case study

In this section, a practice power grid in China is studied to illustrate the proposed method. This network is a 500 kV system with 39 nodes and 92 branches. The topology of the grid is depicted in Figure 7. All the bus configurations in stations are a half configuration. The relay settings obtained from the FIS are set by the planners in the control center according to the real protection setting rules. The parameter λ in equation (3) is 0.001. The probability of the failure of the communication channel is 0.005. The probability for a breaker failing to act is 0.0015 according the statistic data.

The multi-mode containing HF modes and the time-sequence mode provided in section II and section III is established to analyze the probability of N-k contingency. Table 1 shows the probability of some N-2 contingencies, which consider single HF mode and multiple HF mode separately.

In some references, the probability of HF is often decided by the value of the current or power flow on the line, which seem incomplete. Through the values showed in the Table 1, compared with the traditional single HF mode, the multiple HF mode can improve the accuracy in the calculating of the probability of N-k contingency, which is the sum of probability of each HF mode and can be illustrated through the underline values. The improved accuracy of the probability can improve the accuracy of the risk assessment and the significance of some contingencies.

Considering HF4 and HF5 modes, some special N-k contingencies can be found. Table 2 shows that some N-2 contingencies with transformers tripped can be recognized. For example, when line 1–7 is in

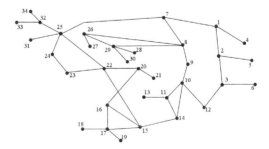

Figure 8.　Single-line diagram of the practice system under study.

13

Table 1. Probability of N-2 contingencies with single HF Mode and multiple HF mode.

Initiating fault line	Extra line tripping	Pro. of HF1	Pro. of HF5	Pro. of HF3	Pro. of N-2 contingency
10–14	10–12	4.5E-6	1.5E-6	0	6E-6
18–17	17–15	1E-6	1.5E-6	0	2.5E-6
2–3	2–5	1.5E-6	1.5E-6	0	3E-6
8–9	8–7	4.8E-5	1.5E-6	1.3E-6	5.8E-5

Table 2. Some N-2 contingencies with transformer tripped.

Line no	Contingencies	Pro. of contingencies
1–7	1–7—7#1	1.5E-6
8–7	8–7—7#2	1.5E-6
25–22	25–22—22#1	1.4E-6
22–15	22–15—22#2	1.2E-6

Table 3. Some search results for some lines with irrational zone 3.

N-k contingencies	Irrational zone 3	Pro. of contingencies
10–12—9–10	8–9	0.8E-6
9–10—8–9	10–14	1.6E-6
8–9—8–7	8–7	5.08E-5

Table 4. Some serious accidents of N-k contingencies.

Initiating fault line	Extra elements tripped	Irrational zone 3	Pro. value
8–9	8–7	8–7	5.08E-5
9–10	8–9	10–12	1.6E-6
8–7	7#2	–	1.5E-6
1–7	7#1	–	1.5E-6
25–22	22#2	–	1.4E-6
22–15	22#1	–	1.2E-6
10–12	9–10	9–10	0.8E-6
8–9	8–7—7#2	–	1.6E-7

the same ring with #1 transformer in station 7, the occurrence of HF5 mode will lead to the trip of #1 transformer and the system will lose some power associated with the transformer.

Table 3 shows some N-k contingencies with inappropriate settings of zone 3. If the characteristic of the settings of zone 3 is not analyzed in the view investigated in section II. The contingencies showed in Table 3 are easily neglected.

The contingencies showed in Tables 2 and 3 are critical ones in the grid, which are easily neglected by the planners of the system, should be paid more attention to. In the process of searching the contingencies in the grid, some serious N-k contingencies with high-risk value can be identified. Table 4 shows some serious accidents in the system.

From the Table 4, the weakness of the grid can be recognized, which needs to be under heavy surveillance. If we reduce the risk of the N-k contingencies in Table 4, the reliability of the system can be improved, which is useful to prevent the cascading failures.

4.2 Discussion

The analysis of N-k contingencies based on the multiple HFs mode can improve the accuracy of the probability of N-k contingencies and identify more critical N-k contingencies with high probability value. It is useful to apply the method on the identifying of N-k contingencies on-line. The analysis of HFs mode associated with inappropriate settings and the online verification system can be the basic tools of adjusting settings in computing settings on-line. In addition, under the economy constraint, the identifying results can maximum the operation efficiency in improving the protection system by upgrading relays on lines with high-risk values.

5 CONCLUSION

The increasing number of cascading failures caused by HFs recently has shown that there is an urgent need of the method developed for investigation of HFs in protection system. This paper has presented an analysis of some typical HF modes in protection system and established a method to calculate and identify N-k contingencies. The results show that the proposed method can improve the accuracy of the probability of N-k contingencies, identify more critical N-k contingencies with high probability value and recognize the weakness of the grid, which is important to the system planners to deal with the HFs in protection system and prevent the cascading failures. The risk assessment of N-k contingencies and control measures will be analyzed in future work.

REFERENCES

1. G. Anderson, P. Dnalek, R. Farmer, N. Hatziargyriou, I. Kamwa, P. Kundur, N. Martins, J. Paserba, P. Pourbeik, J. Sanchez-Gasca, R. Schulz, A. Stankovic, C. Taylor, and V. Vittal, "Causes of the 2003 Major Grid Blackouts in North America and Europe, and Recommended Means to Improve System Dynamic Performance," *IEEE Transaction on Power Systems*, 2005, pp: 1922–1928.
2. Phadke A.G., Thorp J.S. "Expose Hidden Failures to Prevent Cascading Outages," *IEEE Computer Application in Power*, 1996, 9(3), pp: 20–23.
3. S. Tamronglak, "Analysis of power system disturbances due to relay hidden failures," Ph.D. dissertation, Virginia Polytechnic Institute State Univ, Blacksburg, 1994.
4. Yu Xingbin, Singh C., "A Practical Approach for Integrated Power System Vulnerability Analysis with Protection Failures," *IEEE Transactions on Power Systems*, 2004, 19(4), pp: 1811–1820.
5. Nur A.S., Muhammad M.O., Mohd S.S., "Risk assessment of cascading collapse considering the effect of hidden failure," *in IEEE International Conf. Power and Energy*, Malaysia, 2012, pp. 778–783.
6. Jie Chen, James S. Thorp, Ian Dobson, "Cascading dynamics and mitigation assessment in power system disturbances via a hidden failure model," *International Journal of Electrical Power & Energy Systems*, 2005, 27(4), pp: 318–326.
7. Jaime De La Ree, YiLu Liu, Lamine Mili, Arun G. Phadke, Luiz DaSilva, "Catastrophic Failures in Power Systems: Causes, Analyses, and Countermeasures," *Proceedings of the IEEE*, 2005, 93(5), pp: 956–964.
8. C. Tamronglak, S.H. Horowitz, A.G. Phadke, and J.S. Thorp, "Anatomy of power system blackouts: preventive relaying strategies," IEEE Transaction on Power Delivery, 11(2), 1996, pp: 708–715.
9. Wu Wenchuan, Lv Ying, Zhang Boming, "On-line operating risk assessment of hidden failures in protection system," *Proceedings of the CSEE*, 2009, 29(7), pp: 78–83.
10. Fang Yang, A.P. Sakis Meliopoulos, George J. Cokkinides, Q. Binh Dam, "Effects of protection system hidden failures on bulk power system reliability," in *38th North American Power Symposium*, USA, 2006, pp: 517–523.
11. Qiming Chen, James D. McCalley, "Identifying high risk N-k contingencies for online security assessment," *IEEE Transactions on Power System*, 20(2), 2005, pp: 823–834.
12. S.H. Horowitz, A.G. Phadke, "Third Zone Revisited," *IEEE Transaction on Power Delivery*, 21(1), 2006, pp: 23–19.
13. Yongjun Xia, Gang Hu, Xianggen Yin, Zhe Zhang, Wei Chen, "The protection principle and engineering application of double-circuit transmission lines in China," in *43rd International Universities Power Engineering Conference*, Italy, 2008, pp: 1–5.
14. Mojtaba Gilvanejad, Hossein Askarian Abyaneh, Kazem Mazlumi, "Fuse cutout allocation in radial distribution system considering the effect of hidden failures," *International Journal of Electrical Power and Energy Systems*, 42(1), 2012, pp: 575–582.
15. Chen Q., Zhu Kun, McCalley J.D., "Dynamic Decision-event Trees for Rapid Response to Unfolding Events in Bulk Transmission Systems," *in IEEE Porto Power Tech Conference*, porto, Portugal, 2001.
16. I. Dobson, V.E. Lynch, D.E. Newman, "Complex systems analysis of series of blackouts: Cascading failure, critical points, and self-organization," *Int. J. American Institute of Physics*, 17, 2007.

Power and Energy – Kong (Ed.)
© 2015 Taylor & Francis Group, London, ISBN 978-1-138-02782-4

Load balancing optimization of parallel Monte-Carlo probabilistic load flow calculation based on OpenMP

F. Lu & Y.Z. Li
State Grid Jibei Electric Power Company Limited, Beijing, China

S.L. Lu, B.H. Zhang, X.K. Dai & J.L. Wu
State Key Laboratory of Advanced Electromagnetic Engineering and Technology, Wuhan, China

ABSTRACT: The algorithm of probabilistic load flow simulated by serial Monte-Carlo is more time-consuming due to the more simulation times. It's difficult to meet the demand for rapid analysis and calculation of large-scale power grid. So it needs to change the serial algorithm into a parallel algorithm in order to shorten the time. But there is the problem of load unbalance with parallel algorithm. This paper first introduces that the probabilistic load flow simulated by Monte-Carlo is more time-consuming. Then based on the OpenMP parallel programming model (Open Multi-Processing, shared memory multi-threaded compiler directive), the probabilistic load flow simulated by Monte-Carlo is coarse-grained parallel processing. To deal with the problem of load imbalance in parallel process, we need to optimize the OpenMP parallel programming model through the OpenMP instruction scheduling. With the IEEE162 standard bus system as an example, we simulated the probabilistic load flow based on serial Monte-Carlo for thirty thousand times. The results show that the Load balancing calculated by coarse-grained parallel probabilistic load flow was better, which used OpenMP dynamic scheduling strategy, and we received considerable speedup and parallel efficiency.

1 INTRODUCTION

In recent years, the maturity of wind power technology and the rapid development of the wind power grid, the intermittent and scheduling of wind power, and the geographical dispersion and volatility of electricity load make the flow distribution more uncertain in power system. The load flow of the power system essentially is uncertain, need to use probability theory to describe this uncertainty, Called probabilistic load flow[1]. Monte-Carlo simulation can be used to describe this uncertainty.

Load balancing is a hot issue of research in the field of high performance parallel computing. Parallel computing is to solve the problem of major tasks or high complexity in a cooperative manner by calculating the resource coordination of multiple processors. It can quickly solve some problem which is difficult to solve for single core CPU. Task scheduling has an important impact on load balancing and parallel overhead. Scheduling policy includes the selection of execution mode for each phase and the determination of operation parameters. Under the different scheduling policies, the execution of a program in parallel overhead and load balance degree of each processor is different.

In this paper, combined with the characteristics of OpenMP suitable for shared memory computer,

based on OpenMP parallel programming model we deal the probabilistic load flow simulated by serial Monte-Carlo with coarse-grained parallel. With the IEEE162 standard bus system as an example, we had parallel simulation about probabilistic load flow for thirty thousand times, using static and dynamic task scheduling strategy respectively. The results show that coarse grain parallel probabilistic load flow simulated by OpenMP Monte-Carlo can solve the problem of time-consuming which exists in probabilistic load flow simulated by Monte-Carlo. And dynamic scheduling strategy can solve the problem of load imbalance in parallel. We received considerable speedup and parallel efficiency by using parallel algorithm, which reduced the computing time greatly, and improved the computational efficiency of the probabilistic power flow simulated by Monte-Carlo.

2 BASE ON OPENMP PARALLEL REALIZATION OF PROBABILISTIC LOAD FLOW SIMULATED BY MONTE-CARLO

2.1 *Probabilistic load flow simulated by Monte-Carlo*

Probabilistic load flow simulated by Monte-Carlo[2] is a cumulative distribution function method that can get the state variables and branch power flow. Taking into account the access of fan and the fluctuations of load, we multiple value according to the random distribution of input variables (the reactive power and active power of load, and the active power of fan access). Then we calculate the values of state variables and branch power flow with the calculation method of deterministic trend. Finally, we got the random distribution of the state variables and the branch power flow from the times of calculation results. In this paper, considering the fan power and load fluctuation is approximate normal distribution $N(\mu, \sigma^2)$, the probability density function is:

$$f(x) = \frac{1}{\sqrt{2\pi}\sigma} \exp\left(-\frac{(x-\mu)^2}{2\sigma^2}\right) \qquad (1)$$

In the formula (1), μ and σ respectively represent the wind active power, load active power and reactive power's mean value and standard deviation. For each simulated data difference in probabilistic load flow simulated by Monte-Carlo, there are different iterations in Jacobi matrix, and the degree of convergence is different every time. Then the problem of load imbalance is obvious, so the serial program is time-consuming.

2.2 OpenMP parallel programming model and typical scheduling policy

2.2.1 OpenMP parallel programming model
OpenMP[3] is a thread level parallel programming standard, which supports C/C++ language and is suitable for shared memory structure. OpenMP is based on the fork-join parallel programming model. The basic idea of Fork-join parallel model is Fork, which breaks the task into smaller tasks, and uses this framework to perform. Join operation is defined as a task that is waiting for the end of the task it creates. Figure 1 is the Fork-Join OpenMP parallel programming schematic.

2.2.2 Load balancing and OpenMP typical scheduling pattern
OpenMP[4] oriented to circular structure for parallel. In the course of the parallel processing, scheduling

policies directly affect the utilization of system resources and parallel efficiency. The scheduling policy in OpenMP specification includes static scheduling, dynamic scheduling and dispatching guidance. And the result of dispatching guidance is between static scheduling and dynamic scheduling. Therefore, this paper focuses on the two typical kinds of static and dynamic scheduling strategy.

Static scheduling refers to the cycle times of calculation is divided into equal size blocks. When the cycle times of calculation can't be divisible by the product of threads number and blocks. We divide it into equal blocks as far as possible. By default, there are no block size specified, and static scheduling will be assigned an equal number of blocks for each thread until all computing tasks are allocated.

Dynamic scheduling is different from the static scheduling. Dynamic scheduling divides the cycle times of calculation into equal blocks dynamically, and the thread has completed a task block of data and then applies for the next data block. Dynamic scheduling is unpredictable, and the flexible task partitioning can reduce the waiting time of thread, which can alleviate the problem of load imbalance.

For example, the following Figure 2 created 12 assignments, and four threads. The computing time of each task is proportional to the width of 12 tasks. The first task spends one unit of time, and second takes two units of time, and so on. The last iteration takes twelve units of time. Serial program spends a total of 78 units of time to complete the task.

In the ideal case, the required time is 78/4 = 19.5 units of time for the dual-core four threads parallel computing. However, the situation is not so. Static scheduling makes loop iteration assigned to four threads. Thread 1 was assigned 1–3 tasks and thread 2 was assigned 4–6 tasks. Thread 3 was assigned 7–9 tasks and thread 4 was assigned 10–12 tasks. Figure 3 is a diagram of static thread scheduling assignments (the purple color block is idle waiting time for the thread).

Figure 3 shows the thread 4 requires the longest time, which requires 33 units of time. Concurrent threads 3 needs 24 hours to complete the task. Thread 2 requires 15 units of time, and thread 1

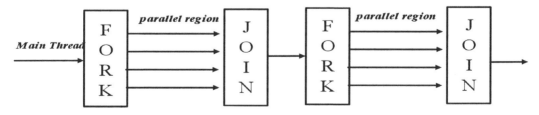

Figure 1. OpenMP fork-join model of parallel execution.

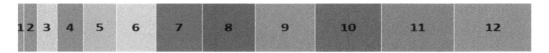

Figure 2.　Serial 12 unbalanced load assignments.

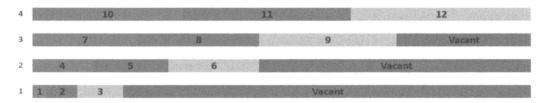

Figure 3.　Parallel static scheduling threads assignment chart.

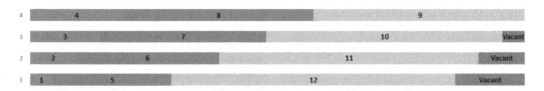

Figure 4.　Parallel dynamic scheduling threads assignment chart.

requires six units of time. All threads must be synchronized at the end of the parallel loop, then continue execution, which means that the thread 1–3 must wait until the end of thread 4. Many of the available time spent in the waiting. So load imbalance phenomenon of the static scheduling in parallel tasks mode is serious.

The dynamic scheduling can set how many assignments for each thread, and specified the task block size. Each thread completes assigned tasks first, then request a subsequent task from the scheduler. Figure 4 shows the dynamic scheduling threaded task assignments that scheduling parameters task block is 1.

In the dynamic scheduling mode in Figure 4, thread 4 takes 21 units of time. Parallel thread 3 needs 20 units of time, and parallel thread 2 spends 19 units of time. The execution time of threads 1 is 18 units of time. So the load is more balanced between four threads. Compared with the time of serial, parallel static scheduling and parallel dynamic scheduling, the execution time of serial is 78 units of time, which is more than the time of static concurrency operation that is 33 units of time, and it's also more than the time of dynamic concurrency operation that is 21 units of time. At the same time, dynamic scheduling can make load balancing better and shorten the computing time as far as possible.

2.3　Parallel probabilistic load flow simulated by OpenMP Monte-Carlo with different scheduling strategies

OpenMP parallel programming model is suitable for shared memory computers[5]. The typical features of shared memory programming is that any variables declared are automatically treated as local variables in thread of execution variables declaration statement, and there is memory effect (that is write conflict, data competition, false sharing, and so on). The main content of parallel processing includes:

1. Join the OpenMP parallel support header file in the program, and set the OpenMP support in the computer language.
2. Modify the data variables of serial probabilistic load flow calculation program, and set private and shared variables in order to avoid data conflict among each thread in parallel computing.
3. Write the main part of probabilistic load flow calculation in for circulation form, that the calculation needs repeated simulations.
4. Join the OpenMP parallel guidance statements and data handling clause before for circulation form.
5. With the OpenMP static and dynamic scheduling strategy, optimize the problem of load balancing which is among parallel threads in probabilistic load flow.

3 EXAMPLE ANALYSIS

In this paper, coarse grain parallel computing program of the probabilistic power flow based on OpenMP which uses C++ language. Parallel test environment is performance shared memory computer. The kernel number of CPU is 32, which can support up to 64 threads. Programming environment is Visual Studio 2012. By the parallel test software that are Intel Parallel Studio and Vtune Amplifier. The degree of load balancing in parallel can be analyzed, and we can also evaluate the degree of parallel algorithm. Through the analysis of the proportion of allocating task for each thread, we can judge the load balancing problem of parallel threads.

We take into account all the load fluctuation range is ± 10%, and Obey normal distribution N (P_{Li}, 0.12). P_{Li} is the rated load active power of node i. Consider the fan which is 10% of system capacity is incorporated in the 1 PV node, wind power fluctuation range is ± 15%. It obeys normal distribution N (0.1P_G, 0.15^2), and P_G is rated active capacity of node system. For the IEEE162 nodes standard test system, we had parallel probabilistic load flow simulated by OpenMP Monte-Carlo with OpenMP parallel programming model. The number of simulations is thirty thousand, and the parallel distributed thread number is 10. We explore parallel performance and load balancing problems by using static scheduling and dynamic scheduling.

3.1 The parallel performance evaluation

The following factors will influence the performance of a parallel program. We commonly used algorithm execution time, speedup and efficiency to assess the degree of load balancing algorithm performance.

a. Speedup of parallel algorithms is S

$$S = T_s/t_p \qquad (2)$$

In the formula (2), T_s is the serial computation time, T_p is the parallel computing time.

The parallel speedup is used to evaluate the improvement performance of algorithm.

b. The parallel efficiency E

$$E = S/P \times 100\% \qquad (3)$$

In the formula (3), P is the processor core number. For multi core CPU, P is the core of CPU number. P is 32 in this paper, and the parallel efficiency reflects resource utilization in parallel system.

c. Load balancing

Load balancing reflects the idle and utilization ratio of parallel processing thread. Utilization of different threads is closer shows that the load balancing is better.

3.2 Algorithm performance analysis

In ten parallel threads, for the IEEE162 nodes standard test system, we had the probabilistic load flow simulated by Monte-Carlo serial program parallelization for thirty thousand times. We contrast speedup and parallel efficiency in parallel static scheduling and parallel dynamic scheduling modes, and parallel performance (time is the average of 10 times statistics) is as shown in Table 1. Vtune Amplifier parallel testing tool[6], we got load balancing of each thread in static scheduling and dynamic scheduling modes, which is shown in Figure 5 and Figure 6, and the statistics is in Table 2.

From Table 1 and Table 2, we can know that:

1. We had the probabilistic load flow simulated by Monte-Carlo serial program parallelization. In the IEEE162 nodes standard test system, the

Table 1. The test results of speedup and parallel efficiency in IEEE162 standard system.

Program execution mode	Computing time/s	Speedup S	Parallel efficiency E
Serial execution	15160.95	–	–
Parallel execution			
Static scheduling	1427.36	10.62	33.19%
Dynamic scheduling	1041.94	14.55	45.47%

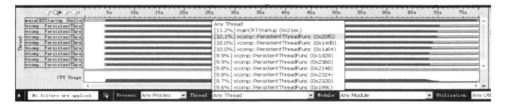

Figure 5. Assignment of 10 threads in static scheduling.

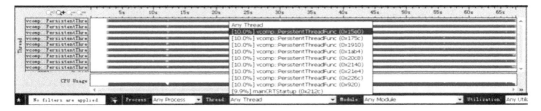

Figure 6. Assignment of 10 threads in dynamic scheduling.

Table 2. Utilization of different threads in static and dynamic scheduling modes.

Distribution type	10 parallel threads									
	Thread 1	Thread 2	Thread 3	Thread 4	Thread 5	Thread 6	Thread 7	Thread 8	Thread 9	Thread 10
Static scheduling	11.2%	10.1%	10.0%	10.0%	9.9%	9.9%	9.8%	9.8%	9.7%	9.6%
Dynamic scheduling	10.1%	10.0%	10.0%	10.0%	10.0%	10.0%	10.0%	10.0%	10.0%	9.9%

computing time that we simulated the probabilistic load flow for thirty thousand times was reduced from 15160.95 s (4.21 h) to 1427.36 s (0.40 h). It drastically reduced the serial execution time, so we can realize the rapid analysis of trend.

2. Comparing static and dynamic scheduling parallel strategy, we can know that load balancing among threads in dynamic scheduling parallel implementation is better. For the utilization of each thread is higher, the computing time is shorter than the time in static scheduling and we got a better speedup and parallel efficiency.

4 CONCLUSION

This paper solved long time-consuming the problem of the probabilistic load flow simulated by Monte-Carlo serial program, because it simulates too much times. With the IEEE162 standard bus system as an example, we had parallel simulation about probabilistic load flow based on OpenMP for thirty thousand times, which can greatly reduce the computing time. With dynamic scheduling strategy, we solved the problem of load imbalance in parallel thread task allocation, and used the performance of parallel machine adequately. It reduced the computing time of probabilistic load flow simulated by Monte-Carlo, and we got a better speedup and parallel efficiency.

ACKNOWLEDGMENT

In this paper, the work done is supported by the national high technology research and development program (863 program, project number: 2011AA05A101).

REFERENCES

1. Ding M, Wang J, Li S.H. Probabilistic Load Flow Evaluation With Extended Latin Hypercube Sampling [J]. Proceedings of the CSEE, 2013,33(4):163–170.
2. Ding M, Li S.H, Hong M. The K-means Cluster Based Load Model For Power System Probabilistic Analysis [J]. Automation of Electric Power Systems.
3. The Open MP Architecture Review Board [Z]. Http://www.openmp.org/drupal/.
4. Chapman B, Jost G, Van Der Pas R. Using OpenMP: portable shared memory parallel programming [M]. MIT press, 2008.
5. Luo Q.M, Ming Z, Liu G, et al. OpenMP compiler And implementation [M]. Beijing: Tsinghua University press, 2012.
6. Stephen Blair-Chappell Andrew Stokes. Intel Parallel Studio. Tsinghua University Press, 2013.4.

Power and Energy – Kong (Ed.)
© 2015 Taylor & Francis Group, London, ISBN 978-1-138-02782-4

A new fast screening method of multiple contingencies

J. Hu, D.H. You, C. Long, Y. Zou, Q.R. Chen & W.H. Chen
State Key of Advanced Electromagnetic Engineering and Technology of Huazhong University of Science and Technology, Wuhan, Hubei, China

L.L. Pan & P. X
China Electric Power Research Institute (Nanjing), Nanjing, Jiangsu, China

ABSTRACT: Contingency screening and ranking is one of the important components of the power system stability assessment. The object of contingency screening is to quickly and accurately select a short list of critical contingencies from a large list of potential contingencies according to the power system stability criteria and rank them according to their severity. As the N–1 check has been performed when planning the grid, this paper proposes a fast screening method of multiple faults based on the principle of concentric relaxation and the severity index of contingency. The method quickly selects a short list of critical contingencies from a large list of potential contingencies. The effectiveness is proved by the system of Ningxia grid. Calculation results show that the proposed method could provide effective information for the system security and stability analysis.

1 INTRODUCTION

The static security analysis of power system is required to analyze each contingency, and check the stability of the system after the failure to evaluate the grid's safety. However, for a complex system, only a limited number of contingency would endanger the grid's safety, for this reason, the contingency screening and ranking method is proposed to screen the critical contingency[1] and each contingency has a behavioral indictors to represent its severity to the grid on which the ranking is based. The common used indexes[2,3] can be divided into two categories: state index and margin index. The state index only takes the information of the current operating status which is easy to calculated but sometimes nonlinear, which couldn't tell the dispatchers the exact distance between the current operating status and the critical state; the margin index is the difference value of some physical quantity between normal and fault state whose information is relatively larger which is complex to calculate. In addition, active and reactive power behavioral indicators could characterize the severity of the fault by calculating the violations of lines' power flow and nodes' voltage.

The reference[4] divided the screening methods into offline and online analytical methods; the reference[5] sets the load margin as the severity index of the fault. These methods only consider single faults and don't apply to the screening of multiple faults. With the N–1 check performed when planning the grid, a greater number of multiple faults are needed to be considered which could endanger the safe operation of the grid.

The reference[6] proposed a method to calculate the probability of $N − k$ fault from the perspective of topology; the reference[7] presents a novel selection algorithm for $N − 2$ contingency analysis problem based on the iterative bounding of line outage distribution factors and successive pruning of the set of contingency pair candidates, without considering the fault of generators and the fault probability.

This paper considers the multiple faults, and set the violation of the active power flow and system load as the fault severity index. The contingency is ranked by the risk value combined the severity index and fault probability. The simulation of Ningxia power grid shows the algorithm's efficiency.

2 CONTINGENCY SCREENING

2.1 The basic idea of contingency screening

The basic idea of this contingency screening method introduced in this paper is as follows: first, the failure affecting domain is computed based on the principle of concentric relaxation[9] and the Power Transfer Distribution Factor (PTDF); then the initial failure set is formed based on the overlapping of FID; finally, the risk index of the fault is calculated combined the calculated fault

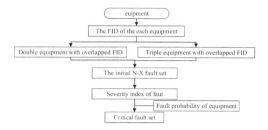

Figure 1. The flow chart of the contingency screening based on the equipment probability.

severity index and the failure probability and the critical failure set is gotten according the failure list ranked by the risk index. Transmission lines, transformers and generators are included in the considered fault equipment, and the fault of branch type transformers is merged into the fault of branches, the fault of generator-transformer unit is merged into the fault of generators.

When the multiple contingency is original fault, assuming that multiple equipment failure occurs simultaneously, the fault probability is the product of the probability of each equipment failure. Due to the low probability of double failure and extremely low probability of triple failure, only double failure and partial triple failure of important equipment are considered in this paper.

The specific flow chart is shown in Figure 1.

2.2 The calculation of equipment' FID

2.2.1 The calculation of lines' FID
The concentric relaxation principle of the power system: in a large interconnected power system, when one of the equipment failure occurs, usually the affected equipment usually lies near the faulty equipment, the impact is relatively smaller or nearly zero in the distant zone. Xiaoyan et al. have verified the validity of the concentric relaxation principle with the Northeast power system.

Based on the concentric relaxation principle, this paper takes the front four grades lines directly connected to the equipment as the equipment's initial fault affecting domain. Then based on the initial FID, the method uses PTDF to perform the second screen to get the final FID. Affected by the branch l's break the power flow change of the lines in the equipment's FID L before and after the can be calculated. Those branches that has little flow change or has a decline power flow are not cared and filtered out of l's FID, only these branches whose power flow has a significantly increase after l's break are reserved in this equipment's FID L. Whether the power flow of the branches increases significantly or not is decided

by the PTDF of the equipment and the power flow margin of the branches in the base status. Suppose the branch l's power flow is $P_\alpha^{(0)}$ before its break. Based on the PTDF, the increase of active power flow can be calculated after the break of branch l_α. The branches whose power flow increases a lot is reserved in l_α's FID. Suppose branch l_x is one branch in l_α's initial FID, whose active power flow is $P_x^{(0)}$ before branch l_α's break and $P_x^{(\alpha)}$ after branch l_α's break, then:

$$P_x^{(\alpha)} = P_x^{(0)} + \lambda_x^{(\alpha)} P_\alpha^{(0)} \tag{1}$$

where $\lambda_x^{(\alpha)}$ is the PTDF of branch l_α to branch l_x.

The screen is according to the proportion of the increase value of active power flow after the fault to its active power margin, combined with the active power flow margin. The screen threshold of the proportion is constant given in advance, the specific formula is as follows:

$$d = \frac{dP}{dP_{max}} = \frac{\left|P_x^{(\alpha)}\right| - \left|P_x^{(0)}\right|}{P_{x,max} - P_x^{(0)}} \geq d_{pre} \tag{2}$$

where, $P_{x,max}$ is the active limit of branch l_x; $dP = |P_x^{(\alpha)}| - |P_x^{(0)}|$ is the increase value of branch l_x's active power influenced by branch l_α's break; $dP_{max} = P_{x,max} - P_x^{(0)}$ is branch l_x's active power margin in the base state; d_{pre} is the screen threshold, given in advance.

The branches in the initial FID which doesn't satisfy condition (2) are deleted from this initial FID to get the final FID.

2.2.2 The calculation of generators' FID
The active incremental of all the branches in the grid is calculated based on the generators' PTDF. Then the branches whose active power increases a lot are screened into the generator's FID.

This method assumes that after a generator being out of operation, the loss of this generator output is balanced by the generators connected to the balance bus. Thus, the parameter $\lambda_x^{(\alpha)}$ means the influence on branch l_x after the node α's injected power changes. Then the power of branch l_x after the failure of the generator connected to node α can be calculated using formula (1), in which $P_\alpha^{(0)}$ is the output of the failure generator connected to node α. Then the branches that are influenced more could be screened using formula (2).

2.3 The basic idea of contingency screening

Under normal circumstance, when multiple components' failure occur simultaneously in a system, the consequences are often more much severer than the sum of the consequence of the individual component's failure. However, if the electrical distance between fault components is remote, then the influence between them is rather small, the consequence of the fault can be approximated by the overlay of the consequence of the individual component's failure. The planning of power system has guaranteed that all the $N-1$ failure would not affect the normal operation of the grid. Therefore, the initial failure set that needs attention can be formed by the overlapping of the components' FID or not.

This paper considers all the double fault and partial triple fault of important equipment. The two components of double fault in the multiple fault set satisfy that their FID is overlapped. Since the probability of multiple failure is very small, therefore, this paper only considers the triple fault of important equipment, and satisfies the condition of FID's overlapping. The condition that the double and triple faults need to satisfy to be included by the expected fault set is shown in Figure 2a and b.

2.4 *The severity index of fault*

In order to comprehensively reflect the degree of transmission lines overloaded and system overloaded, the behavioral index characterizing the severity of the failure is defined as:

$$PI = \sum_{\alpha} \omega_l \left(\frac{P_1}{P_1^{\max}} \right)^2 + \omega_p * \left(1 + \frac{P_{loss}}{0.96 * P_G} \right)^2 \quad (3)$$

In the above formula, ω_l—the active weight of branch l, which is often set to 1; P_1—the power flow of branch l; $P_1^{\max}$—the maximum power flow of branch l; α—the set of overloaded transmission lines; ω_p—the weight of system's overloaded; P_{loss}—the load that the generator cannot balance; if $P_{loss} < 0$, the system could balance the lose power, and the second part in formula (3) is 0. As the power loss of the transmission grid is controlled in 4%, therefore the maximum active power of the grid is $0.96 P_G$, according to the requirement of the grid operation, the severity of the system overloaded is higher than the branch overloaded, and therefore the weight ω_p is set to 5.

The severity index of each $N-k$ failure in the expected failure set has to be calculated. Considering both accuracy and speed, this paper uses the DC power flow algorithm or fast decoupled method with once iteration[12] to calculate the power flow after the contingency, then the severity of this contingency could be calculated with the flow result.

2.5 *The risk index of fault*

Risk theory is the theory that considers both uncertainties and severity, and combines both the severity and probability of the fault. Both the probability and the severity of contingency could determine the safety of the system, and the risk index of power system could quantitatively grasp both of the two factors, and fully reflect the impact of the contingency to the entire grid.

In this paper, the risk index[9] which indicates the severity degree of the fault is defined as:

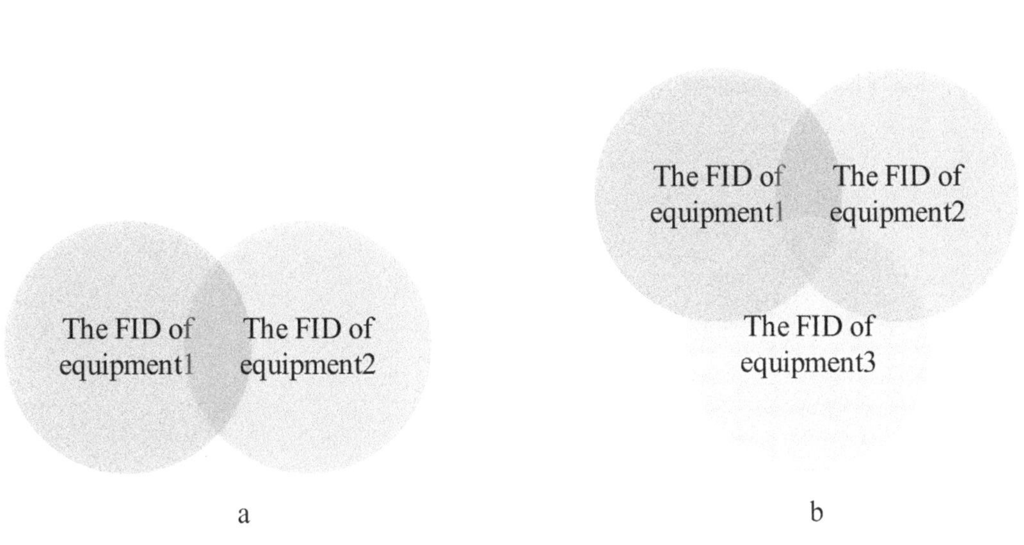

The FID of equipment1   The FID of equipment2

The FID of equipment1   The FID of equipment2

The FID of equipment3

a   b

Figure 2.   The screening of multiple fault with overlapped FID.

$$R_{\text{failure-}i} = PI_{\text{failure-}i} * P_{\text{failure-}i} \qquad (4)$$

In the above formula, $R_{\text{failure-}i}$—the risk index of the i-th multiple fault; $PI_{\text{failure-}i}$—the severity degree of this failure; $P_{\text{failure-}i}$—the probability of this failure.

The contingency screening and ranking is based on the risk index of the failure. The failure list is formed according to the risk index, and only the front n failure is needed to be considered in the following analysis, which could greatly reduce the analysis time.

3 SIMULATION ANALYSIS

The simulation system is Ningxia power grid, which includes 50 generators, 190 nodes and 299 transmission lines. The probability[10] of equipment failure is calculated under the good weather condition and the magnitude of failure probability is 10^{-3}.

Table 1 shows the number of double fault screened with the introduced method.

It is shown from Table 1 that this method could filter out a large number of slight fault, and greatly reduce the fault set, improve the efficiency of the subsequent analysis.

According to the introduced method, when screening triple fault, the transmission lines whose voltage class is 330 kV or above are considered,

and the number of triple fault is 2448 (the total number of triple fault is 25 953 389).

Figure 3 depicts the front 500 fault ranked by fault severity index, and the front 100 fault ranked by the fault risk index, the blue o is the screened fault at with severity index and the red * with risk index.

The analysis of the simulation result: the failure risk is associated with the severity index and the failure probability. As the magnitude of failure probability is 10^{-3}, therefore the screened contingency of higher risk is double fault with a relatively higher fault probability. It can be observed from Figure 3 that the fault with very small severity and low probability is screened out.

4 CONCLUSION

This paper gets the FID based on the principle of concentric relaxation and the PTDF, then the multiple fault is screened according to the overlapping of FID or not. The risk index of the fault is calculated with the calculated fault severity index and the failure probability. The fault list is ranked by their risk index. The fault severity index is selected considering both the overloaded of transmission lines and the entire system, which is more accurate to characterize the severity of the failure. The simulation of Ningxia Power Grid has shown that the algorithm is very effective.

Table 1. The screening result of double fault.

Fault type		L-L		G-L		G-G		Total number	
The number and percentage of the remaining fault	All the double failure	44 551	100%	15 249	100%	1 275	100%	61 075	100%
	The screening result with this method	1581	3.54%	455	2.98%	33	2.59%	2069	3.4%

* L—transmission line, G—generator.

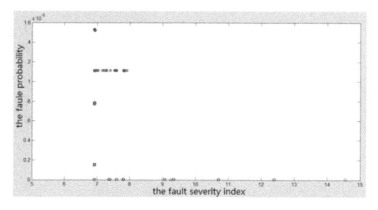

Figure 3. The screening result with risk index.

ACKNOWLEDGMENTS

This work is supported by the Technology Program of the State Grid Corporation of China (the Research of the Power Grid Operation Trajectory Characterization Based on the Situational Awareness).

REFERENCES

1. Bao Xiaohui, Hou hui. A review of power system reliability assessment [J]. Engineering Journal of Wuhan University. 2008(04): 96–101.
2. Bao Lixin, Zhang Buhan, et al. The summery of the voltage stability margin index analysis method[J]. Automation of Electric Power Systems. 1999(08): 52–55.
3. Duan Xianzhong, He Yangzan, et al. On several electric utility system voltage stability criterion and safety indicators [J]. Automation of Electric Power Systems. 1994(09):36–41.
4. Zhang Yongjun, Cai Guanglin, et al. Contingency Screening and Ranking Based on Optimal Multiplier Power Flow Evaluation [J]. Transactions of China Electrotechnical Society. 2010(01): 123–128.
5. Feng Z, Xu W. Fast computation of post-contingency system margins for voltage stability assessments of large-scale power systems [J]. Generation, Transmission and Distribution, IEE Proceedings-. 2000, 147(2): 76–80.
6. Qiming C, Mccalley J.D. Identifying high risk N-k contingencies for online security assessment [J]. Power Systems, IEEE Transactions on. 2005, 20(2): 823–834.
7. Turitsyn K.S, Kaplunovich P.A. Fast Algorithm for N-2 Contingency Problem [C]. Wailea, HI, USA: 2013.
8. Xiao Kai, Guo Yongji, et al. New algorithm for large power system reliability evaluation [J]. Journal of Tsinghua University (Science and Technology). 1999(01): 13–16.
9. Zhang Guo-hua, Zhang Jian-hua, et al. Risk Assessment Method of Power System N-K Contingencies [J]. Power System Technology. 2009(05): 17–21.
10. Tian Ke D.Y.L.L. Analysis of a Transformer Time-varying Outage Model for Operational Risk Assessment [J]. Thermal, Power and Electrical Engineering. 2013(732–733): 993–998.

Power and Energy – Kong (Ed.)
© 2015 Taylor & Francis Group, London, ISBN 978-1-138-02782-4

The comprehensive evaluation method of high wind power penetration transmission grid based on maximum entropy criterion

Haibo Li, Zongxiang Lu & Ying Qiao
Department of Electrical and Electronic Engineering, Tsinghua University, China

Han Huang & Yang Wang
State Grid Energy Research Institute, China

ABSTRACT: With the shortage of energy and the appeal to environmental protection, the environmental benefit of transmission grid is gradually of the same importance with the safety. With the rapid development of wind power, the penetration ratio of wind power in transmission grid is high. The variation and uncertainty of wind power may bring more safety risk to transmission grid. Therefore, the quantitative evaluation is important for the safety and economic operation of transmission grid. However, the number of evaluation index is too large which results the difficulty of comprehensive evaluation method of transmission grid. The key problem is how to obtain the optimal weights of each index. This paper proposes a comprehensive evaluation method for high wind power penetration transmission grid based on maximum entropy method. Finally, a case study based on a regional power grid of Southern China is researched and proves the effectiveness of the proposed method.

1 INTRODUCTION

As the core link of power delivery, the safety of transmission grid is of great significance. With the appeal to environmental protection and shortage of energy, the environmental benefit of transmission grid is gradually of the same importance with the safety. With the rapid development of wind power, the penetration ratio of wind power in transmission grid is high. The variation and uncertainty of wind power may bring more safety risk to transmission grid. Therefore, the quantitative evaluation is important for the safety and economic operation of transmission grid.

Plenty of researches have been done about the evaluation of power grid but most work is focused on part of the picture, such as safety assessment (Wan et al, 2008, Zhang et al, 2009) or environmental benefit evaluation. Thus it is significant to propose a comprehensive evaluation method considering all aspects of transmission grid. However,

the number of evaluation index is too large which results the difficulty of comprehensive evaluation method of transmission grid (Wang et al, 2011, Li et al, 2008). Therefore, it is necessary to propose a proper method to select the main indices and normalize them, then calculate the optimal weight factor of each index.

There are many researches focused on the standardization (Li et al, 2004) and weight calculation of the comprehensive indices (Lu et al, 2008). For the index standardization process, different methods such as Z-Score method, range transform method, maximization method, minimization method and so on are compared and discussed in different scenarios (Zhang et al, 2010).

In the weight calculation field, many methods are proposed and applied in the comprehensive evaluation. AHP method (Guo et al, 2008) and Delphi method (Chen et al, 2003) are convenient to be realizes and applied in different cases. But the methods are subjective and cannot give an objective evaluation. In order to cover the shortage, many objective weight calculation methods are proposed, such as evidence distance method (Lu et al, 2008), maximum entropy method (Huo et al, 2005), and the method based on rough sets (Bao et al, 2009). However, there is less application about the above methods in the comprehensive evaluation of transmission grid. This paper applies the maximum entropy criterion in the comprehensive evaluation

This work was supported by The National High Technology Research and Development Program of China (2011AA05A103) and the State Grid Shandong Company Economic Research Institute ('Research of Power System Technical and Economic Evaluation Method, Model and Application Adapting to Strong and Smart Grid Development').

of high wind power penetration transmission grid and obtains the weights of each index.

This paper proposes an index hierarchy including various aspects of transmission grid. Except for the conventional evaluation index, some new indices considering the characteristics of future transmission grid are proposed in this paper, such as wind power, environmental benefit and so on. A standardization method is used to normalization the selected indices, which makes the comparable of different indices. Then the maximum entropy criterion based on Lagrange method is applied to obtain the optimal weight of different indices. Finally, a case study is researched based on the data of a regional power grid of China. From the case study it can be observed that the proposed method can effectively screen the evaluation indices and normalize them into interval [0, 1]. The optimal weights of all evaluation indices are calculated and the comprehensive evaluation system of the high wind power penetration transmission grid is designed.

2 METHODOLOGY

2.1 Conventional evaluation index system of transmission grid

The conventional evaluation indices for transmission grid can be classified into two aspects: the indices for power system planning and operation. The indices in planning stage are mainly used for the economic and technical comparison of different projects, which is the criterion to select an optimal planning project. The indices for power system planning include technical indices and economic indices as follows:

1. Technical Indices
   a. Power Flow Distribution Index:

$$PFDI = \sum_{i=1}^{N} \frac{T_{i\max} - P_i}{N} \qquad (1)$$

where T_{imax} is the up limit of line i; P_i is the power flow of line i; N is the total number of line.

   b. Short Circuit Level Index:

$$SCLI = \sum_{i=1}^{N} \frac{I_{i\max} - I_{SCi}}{N} \qquad (2)$$

where I_{imax} is the up limit of the short circuit of line i; I_{SCi} is the actual short circuit of line i; N is the total number of line.

   c. Stable Level Index:

$$SLI = \sum n(\theta > \theta_{\max})/n_G \qquad (3)$$

where $n(\theta > \theta_{\max})$ represents the number of generation that the power angle θ is larger than $\theta_{\max}$; n_G is the total number of generation.

   d. Reliability Index:

$$RI = \sum_{i=1}^{n_L} \lambda_i \cdot \overline{L_i} \Big/ \sum_{i=1}^{n_L} \overline{L_i} \qquad (4)$$

where λ_i is the reliability index of load i; $\overline{L_i}$ is the average load demand of load i; n_L is the total number of load point.

   e. Operation Flexibility Index:

$$OFI = \frac{\sum m(N-2)}{N} \qquad (5)$$

where N is the total number of line; $m(N–2)$ represents the number of operation modes that no line violates the power limit.

   f. Network Adaptation Index:

$$NAI = \sum_{i=1}^{n_L} \Delta L_i \Big/ \sum_{i=1}^{n_L} \overline{L_i} \qquad (6)$$

where ΔL_i is the max load increment of line i; $\overline{L_i}$ is the average load demand of load i; n_L is the total number of load point.

2. Economic Indices
   a. Investment Cost:

$$IC = \left(m_0 L + \sum_{i=1}^{n} N_i m_i \right)\left(1 + \frac{a}{100} \right) \qquad (7)$$

where a is the additional expense coefficient of the project, which is 70 for 220 kV project while 80 for 500 kV project.

   b. Operating Cost:

$$OC = S + \beta L \qquad (8)$$

where S is the power loss expense for transmission grid; βL is the maintenance costs for lines.

2.2 Additional indices for wind power

With the high penetration of wind power, the risk may increase to the operation of the power system. Therefore, the evaluation indices for wind power are necessary for transmission grid. The risk is mainly caused by the variation of wind power, which increases the difficult for power balance. Thus the evaluation indices can be defined according to the variation of wind power as follows:

   a. Change Rate of Wind Power Output:

$$CR_{WPO}(t,T) = \frac{P(t+T) - P(t)}{P_{base}} \times 100\% \qquad (9)$$

where $P(t)$ and $P(t+T)$ denote the output of wind power in time t and $t+T$; P_{base} is the capacity of wind farm.

b. Smooth Effect Coefficient:

$$SEC = \frac{\hat{\sigma}_{\sin gle} - \hat{\sigma}_{cluster}}{\hat{\sigma}_{\sin gle}} \quad (10)$$

where $\hat{\sigma}_{\sin gle}$ and $\hat{\sigma}_{cluster}$ represent the standard deviations of the normalized wind power output for single wind farm and cluster wind farms respectively.

c. Theoretical Wind Power Penetration Rate:

$$WPPR_{th}(t) = \frac{P_{base}(t)}{P_{load}(t)} \times 100\% \quad (11)$$

where $P_{base}(t)$ and $P_{load}(t)$ denote the capacity of wind farms and load demand in time t.

d. Actual Wind Power Penetration Rate:

$$WPPR_{ac}(t) = \frac{P_{wind}(t)}{P_{load}(t)} \times 100\% \quad (12)$$

where $P_{wind}(t)$ and $P_{load}(t)$ represent the actual output of wind farms and load demand in time t.

e. Peak Shaving Index:

$$PSI = \frac{\left(P_{i\max} - P_{i\min}\right) - \left(P'_{i\max} - P'_{i\min}\right)}{P_{load}} \quad (13)$$

where P_{imax} and P_{imin} represent the peak and valley load.

f. Energy Output of Wind Power:

$$EOWP = \int_0^T \sum_{i=1}^{N_W} f_{wind,i}(v_i(t)) dt \quad (14)$$

where $f_{wind,i}(x)$ is the function of wind power output vs wind speed and $v_i(t)$ is the wind speed in time t; N_W is the number of wind farms.

2.3 Comprehensive evaluation system

According to the above listed indices, the comprehensive evaluation system can be constructed as shown in Figure 1. The weight factors $w_1 \sim w_{13}$ are calculated by the maximum entropy method which is introduced in the next section.

2.4 Index screening and standardization model

There are many evaluation indices for the transmission grid and can be divided into many types. Each category contains a large number of secondary indices. The different indices of the transmission grid may have a large difference in value and different dimensions, which makes it difficult to composite the indices to a comprehensive index. Thus screen the suitable indices is the foundation to construct the index system for transmission grid.

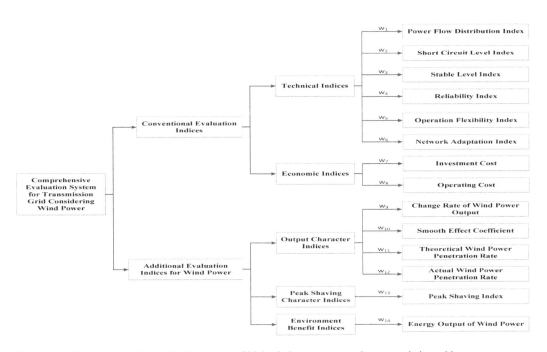

Figure 1. The comprehensive evaluation system of high wind power penetration transmission grid.

In this paper, an index screening method is applied to choose the main indices and neglect the subordinate indices. Then a standardization method is used to normalize the different indices to the interval [0,1], which makes the different indices have the comparison.

The screening method for each index should estimate the importance, independence and measurability of it. The standardization methods include range transformation method, linear scale transformation method, vector normalization method and so on. The last method above is used in this paper and the mathematical form is as follows. Assuming that the indices matrix is $X = (x_{ij})_{m \times n}$, the following $Y = (y_{ij})_{m \times n}$ is the normalized indices matrix:

$$y_{ij} = \frac{x_{ij}}{\sqrt{\sum_{i=1}^{m} x_{ij}^2}}(1 \le i \le m, 1 \le j \le n) \quad (15)$$

According to the above normalization method, the different indices can be transferred to per unit value in the interval [0,1].

2.5 The maximum entropy criterion

The model of maximum entropy criterion is shown in equation (3). The optimal weight factors can be obtained by solving the optimization problem based on Laplace theorem.

$$\begin{cases} \max B^T H B \left\{ -\sum_{i=1}^{n} P_i \log P_i \right\} \\ s.t.\ F_k = \sum_{i=1}^{n} f_k(x_i) P_i, k = 1, 2, , m(m < n) \\ \sum_{i=1}^{n} P_i = 1, P_i \ge 0, \forall i \end{cases} \quad (16)$$

where F_k is the average of the function $f_k(x_i)$ with the probability P_i. According to the Laplace theorem, the optimal weight can be calculated by the following equation:

$$w_j = \frac{\exp\left(-\left[1 + \delta \sum_{i=1}^{n} (1 - r_{ij})^2 \Big/ (1 - \delta)\right]\right)}{\sum_{j=1}^{m} \exp\left(-\left[1 + \delta \sum_{i=1}^{n} (1 - r_{ij})^2 \Big/ (1 - \delta)\right]\right)} (\forall j) \quad (17)$$

where r_{ij} is the normalized value of index j and δ is a parameter in interval [0, 1], which is selected according to the actual scenario.

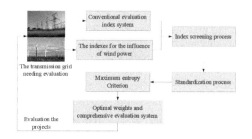

Figure 2. The flowchart of comprehensive evaluation algorithm.

2.6 The comprehensive evaluation algorithm

According to the above analysis, the comprehensive evaluation algorithm mainly includes the following steps:

Step 1: select proper indices as the evaluation index system of transmission grid considering wind power;

Step 2: run the index screening process to obtain the useful indices and normalize them by standardization methods;

Step 3: calculate the optimal weight factors based on the maximum entropy criterion;

Step 4: construct the comprehensive evaluation system according to each index's weight and apply the system to the evaluation of the transmission grid.

The flowchart of the comprehensive evaluation algorithm is shown in Figure 2.

3 CASE STUDY

3.1 Simulation system

The simulation case in this paper is based on a regional power grid of Southern China. The general topological structure of the planned grid in year 2015 is shown in Figure 3. There are 3075 bus nodes in the system, which includes 2883 transmission components (transmission lines and transformers) and 747 generation components. Based on the results of power flow in max operating plan, the generation capacity is 38296 MW and the load demand is 22979 MW. The specific data about the grid (such as the impedance of the lines and transformers, the length of the lines, and so on) can be obtained from its BPA file.

The value of the selected conventional indices during year 2010~2012 is shown in Table 1.

From Table 1 it can be observed that the value between different indices may have a large difference. Thus the normalization process is necessary for the calculation of weight factors.

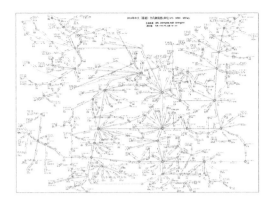

Figure 3. The general topological structure of the power grid in the simulation case.

Table 1. Selected conventional indices in simulation system during year 2010~2012.

Indices	2010	2011	2012
PFDI (MW)	1.73×10^5	1.69×10^5	1.81×10^5
SCLI (A)	5.77×10^4	4.83×10^4	6.32×10^4
SLI (%)	92.2	94.1	88.6
RI (fr/year)	0.03	0.029	0.035
OFI	329	279	380
NAI (%)	13.6	15.2	13.1
IC (yuan/year)	1.12×10^8	1.18×10^8	1.03×10^8
OC (yuan/year)	7.24×10^6	6.93×10^6	7.62×10^6
CR_{WPO} (%)	16.2	18.3	14.9
SEC	2.57	2.77	2.64
$WPPR_{th}$ (%)	28.4	30.2	32.4
$WPPR_{ac}$ (%)	13.1	12.7	14.3
PSI (%)	−5.9	−6.2	−4.9
EOWP (MWh)	4.46×10^6	4.32×10^6	4.87×10^6

Table 2. Normalized indices during 2010~2012.

Indices	2010	2011	2012
PFDI	0.573	0.559	0.599
SCLI	0.587	0.492	0.643
SLI	0.581	0.593	0.558
RI	0.551	0.532	0.643
OFI	0.572	0.485	0.661
NAI	0.561	0.627	0.540
IC	0.582	0.613	0.535
OC	0.575	0.550	0.605
CR_{WPO}	0.566	0.639	0.521
SEC	0.558	0.601	0.573
$WPPR_{th}$	0.540	0.574	0.616
$WPPR_{ac}$	0.565	0.548	0.617
PSI	−0.598	−0.629	−0.497
EOWP	0.565	0.547	0.617

Table 3. Optimal weights of each indices.

Indices	Weights	Indices	Weights
PFDI	0.121	OC	0.108
SCLI	0.018	CR_{WPO}	0.041
SLI	0.126	SEC	0.118
RI	0.042	$WPPR_{th}$	0.085
OFI	0.010	$WPPR_{ac}$	0.090
NAI	0.069	PSI	0.000
IC	0.082	EOWP	0.089

Table 4. Comprehensive index during 2010~2012.

2010	2011	2012
0.567	0.576	0.587

3.2 Normalization results

Using the normalization method proposed in section 2.4, the above indices can be normalized and shown in Table 2.

3.3 Weight factors calculation

Based on the maximum entropy method, set $\delta = 0.6$, then the optimal weights can be obtained by equation (17), as shown in Table 3.

Thus the comprehensive index can be calculated by the above weights. From the results it can be observed that the SLI index has the largest influence to the comprehensive index while the PSI index has little influence.

According to the above results, the comprehensive evaluation index of the grid in 2010~2012 can be calculated as shown in Table 4.

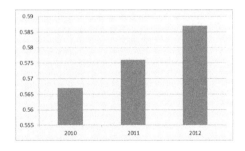

Figure 4. The comparison of comprehensive evaluation index during 2010~2012.

The comparison of the comprehensive evaluation index during 2010~2012 is shown in Figure 4. It can be observed with the development of the grid, the index is increasing, which proves the effectiveness of the proposed method.

4 CONCLUSIONS

From the research about the comprehensive evaluation method of high wind power penetration transmission grid and the case study, the following conclusions can be drawn:

1. The proposed method can effectively screen the evaluation indices and normalize them into interval [0, 1];
2. The optimal weights of all evaluation indices are calculated based on the maximum entropy method successfully;
3. The comprehensive evaluation system of the high wind power penetration transmission grid is designed, which can be applied in the department of power grid dispatching and planning.

REFERENCES

1. Wan Wei, Wang Chun, Cheng Hong, Zhao Yan, "Preliminary frame of index system for evaluating power network," Power System Protection and Control, Vol. 36, No. 24, 2008, pp. 14–18.
2. Zhang Guo-hua, Zhang Jianhua, Peng Qian, Duan Manyin, "Index System and Methods for Power Grid Security Assessment," Power System Technology, Vol. 33, No. 8, 2009, pp. 30–34.
3. Wang Wan, Liu Zongqi, Zeng Bo, Yin Hongxu, Liu Jia, Zhang Jianhua, "Comprehensive Index System and Evaluation Model of Power Grid for Metropolitan Cities," Modern Electric Power, Vol. 28, No. 4, 2011, pp. 24–28.
4. Li Xiaohui, Zhang Lai, Li Xiaoyu, Chen Zhu, "The research on the evaluation system for existing network based on analytic hierarchy process and Delphi method," Power System Protection and Control, Vol. 36, No. 14, 2008, pp. 57–61.
5. Li meijuan, Chen guohong, Chen Yantai, "Study on Target Standardization Method of Comprehensive Evaluation," Chinese Journal of Management Science, Vol. 12, 2004, pp. 45–48.
6. Lu Wenxing, Liang Changyong, Ding Yong, "A Method Determining the Objective Weights of Experts Based onidence Distance," Chinese Journal of Management Science, Vol. 16, No. 6, 2008, pp. 95–99.
7. Zhang Lijun, Yuan Nengwen, "Comparison and Selection of Index Standardization Method in Linear Comprehensive Evaluation Model," Statistics & Information Forum, Vol. 25, No. 8, 2010, pp. 10–15.
8. Guo Jinyu, Zhang Zhongbin, Sun Qingyun, "Study and Application of Analytic Hierarchy Process," China Safety Science journal, Vol. 18, No. 5, 2008, pp. 148–153.
9. Chen Wei, Fang Tingjian, Ma Yongjun, Ma Zhongmo, Jiang Xudong, "Research on Group Decision Based on Delphi and AHP," Computer Engineering, Vol. 29, No. 5, 2003, pp. 18–20.
10. Huo Yingbao, Han Zhijun, "Research on Giving Weight for Multi-indicator Based on GME Principle And GA," Statistics and Management, Vol. 24, No. 3, 2005, pp. 39–50.
11. Bao Xinzhong, Zhang Jianbin, Liu Cheng, "A New Method of Ascertaining Attribute Weight Based on Rough Sets Conditional Information Entropy," Chinese Journal of Management Science, Vol. 17, No. 3, 2009, pp. 131–135.

Power and Energy – Kong (Ed.)
© 2015 Taylor & Francis Group, London, ISBN 978-1-138-02782-4

A way to select weights for bus power injection measurements in WLS state estimation

Yubin Yao, Zhiliang Wu & Dan Wang
School of Marine Engineering, Dalian Maritime University, Dalian, China

ABSTRACT: State estimation provides data for several online system studies in energy control center. To achieve a higher degree of accuracy of the output data is the aim of state estimation. State estimation is usually solved by weighted least squares estimation, and chooses the variance of the measurement as its weight. But same error of bus power injection measurement has different effect on the accuracy of state estimation from that of branch power flow measurement. For same error, the weight for power injection measurement should be different from that for branch power flow measurement, so a method of selecting weights for bus power injection measurements is presented. The presented method divides weight of the bus power injection measurement by the number of the lines from the bus. IEEE30 system is analyzed by the proposed method, and the results prove the effectiveness of the proposed method.

1 INTRODUCTION

State estimation plays a very important role in the monitoring and control of power systems. It provides accurate data obtained by processing the less accurate and incomplete network data for several online power system studies such as dispatcher power flow, and security analysis. Several methods have been presented since state estimation was introduced in power systems. Among them, the predominant approach is the Weighted Least Squares (WLS) method (Holten et al. 1988).

The aim of state estimation is to obtain best possible data. The best way to acquire accurate data is solving the problem by the weighted least squares method and the degree of accurate of state estimation is major focus. Factors influenced the accuracy of state estimation have been investigated, and many methods to improve the accuracy have been presented. Asprou et al. (2013) analyzed effect of instrument transformer accuracy class on the WLS state estimator accuracy. Baran et al. (2009) included voltage measurements in branch current state estimation for distribution systems to improve the accuracy of state estimation. Study of Pau et al. (2013) suggested that branch-current estimators can give estimation accuracies comparable to the voltage ones if an extended version includes the slack bus voltage into the state vector.

Choice of measurement weights is an important consideration for state estimators, and the reciprocals of measurement error variances are commonly chosen as the weights for the measurements. Because measurement error varies due to variation of operating conditions of telecommunication systems and aging of the instruments, the weights need to be continuously updated. Liu et al. (1998) presented a novel algorithm to estimate and adaptively update measurement variances via sensitivity matrix. Zhong & Abur (2004) presented a simple method based on the sample variances of the measurement residuals calculated in historical records, which avoids time-consuming calculation of the sensitivity matrix.

The measurement weights are usually chosen to be the reciprocals of measurement error variances without considering the type of measurements. For different types of the measurement, same measurement error has different influence on the accuracy of state estimation. The degree of the effect of bus power injection measurement error is related to the number of the lines from the bus. So this paper compares the effect of power injection measurement error on the accuracy of state estimation with that of branch power flow measurement, and introduces reciprocal of the number of the lines from the bus as coordination factor to adjust the weights for power injection measurements. IEEE30 network is analyzed by the proposed method in this paper, and the results prove the effectiveness of the proposed method.

2 FUNDAMENTAL EQUATIONS OF STATE ESTIMATION

The purpose of state estimation is to acquire state variables by measurements. There is error between the measurement and its true value as follows

$$\mathbf{z} = \mathbf{z}_0 + \mathbf{e} \qquad (1)$$

where $\mathbf{z}$ = measurement vector of dimension m; $\mathbf{z}_0$ = true value vector of dimension m; $\mathbf{e}$ = m-dimensional measurement error vector with zero mean.

In fact true value is unknown in a real system, so it is substituted by its estimate value and gives the relation of state variables and measurements as follows:

$$\mathbf{z} = \mathbf{h}(\mathbf{x}) + \mathbf{r} \qquad (2)$$

where $\mathbf{x}$ = state vector of dimension n consisted of voltage magnitudes and phase angles; $\mathbf{h}(\mathbf{x})$ = m-dimensional function vector of state vector represents the desired estimation of $\mathbf{z}$; $\mathbf{r}$ = m-dimensional measurement residual vector with zero mean.

The measurement errors are commonly assumed to have a Gaussian distribution with unknown parameters. The estimate vector $\hat{\mathbf{x}}$ can be solved by Weighted Least Squares (WLS) method.

The objective function of the WLS method is as follow

$$J = [\mathbf{z} - \mathbf{h}(\hat{\mathbf{x}})]^{T} \mathbf{W} [\mathbf{z} - \mathbf{h}(\hat{\mathbf{x}})] \qquad (3)$$

where $\mathbf{W}$ = the weighting matrix of dimension $m \times m$; the superscript (T) denotes transposition of the matrix.

The weighting matrix is diagonal matrix the element w_{ii} of which is as follows

$$w_{ii} = 1/\sigma_i^2 \qquad (4)$$

where σ_i^2 = measurement error variance of the ith measurement.

The function $\mathbf{h}(\hat{\mathbf{x}})$ in (3) is the nonlinear function about $\hat{\mathbf{x}}$ which leads nonlinear objective function. $\mathbf{h}(\hat{\mathbf{x}})$ is linearized as

$$\mathbf{h}(\hat{\mathbf{x}}) = \mathbf{H}\hat{\mathbf{x}} \qquad (5)$$

where $\mathbf{H} = \partial \mathbf{h}(\hat{\mathbf{x}})/\partial \hat{\mathbf{x}}$ is the Jacobian matrix of dimension $m \times n$ at a given value of the vector $\hat{\mathbf{x}}$.

To minimize (3) and take into account (5) give the normal equation as follows:

$$\mathbf{H}^{T} \mathbf{W} \mathbf{H} \Delta \hat{\mathbf{x}} = \mathbf{H}^{T} \mathbf{W} [\mathbf{z} - \mathbf{h}(\hat{\mathbf{x}})] \qquad (6)$$

To solve (6) for $\Delta \hat{\mathbf{x}}$ and update the estimate $\hat{\mathbf{x}}$. Repeat this procedure until $\Delta \hat{\mathbf{x}}$ satisfies the given tolerance.

3 EFFECT OF MEASUREMENT ERROR ON ACCURACY OF STATE ESTIMATION

Part of a simple network is shown in Figure 1. There are three lines from bus 1 to bus 2, bus 3

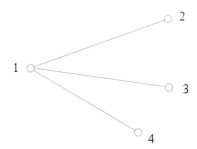

Figure 1. Part of a simple network.

and bus 4 respectively. All these three lines are furnished with branch power flow measurements, and bus 1 with power injection measurement.

To make analysis simple, set weights for measurements in (6) are 1.0, we have

$$\mathbf{H}^{T} \mathbf{H} \Delta \hat{\mathbf{x}} = \mathbf{H}^{T} \mathbf{r} \qquad (7)$$

While the left side of (7) is unchanged for a given condition with different measurement errors, the right side vector changes with the variation of measurement errors. The effect of measurement error on accuracy of state estimation can be known by investigating the right side vector of (7). We study the effect of measurement error in following three cases.

3.1 Errors exist in power injection measurement

If the active power injection measurement at bus 1 has a small error of r_0, others are error-free measurements; the right side vector of (7) is as follows

$$
\begin{bmatrix}
\partial P_1/\partial V_1 & \partial P_{12}/\partial V_1 & \partial P_{13}/\partial V_1 & \partial P_{14}/\partial V_1 & \cdots \\
\partial P_1/\partial V_2 & \partial P_{12}/\partial V_2 & 0 & 0 & \cdots \\
\partial P_1/\partial V_3 & 0 & \partial P_{13}/\partial V_3 & 0 & \cdots \\
\partial P_1/\partial V_4 & 0 & 0 & \partial P_{14}/\partial V_4 & \cdots \\
\partial P_1/\partial \theta_1 & \partial P_{12}/\partial \theta_1 & \partial P_{13}/\partial \theta_1 & \partial P_{14}/\partial \theta_1 & \cdots \\
\partial P_1/\partial \theta_2 & \partial P_{12}/\partial \theta_2 & 0 & 0 & \cdots \\
\partial P_1/\partial \theta_3 & 0 & \partial P_{13}/\partial \theta_3 & 0 & \cdots \\
\partial P_1/\partial \theta_4 & 0 & 0 & \partial P_{14}/\partial \theta_4 & \cdots \\
\vdots & 0 & 0 & 0 & \cdots
\end{bmatrix}
\begin{bmatrix}
r_0 \\ 0 \\ 0 \\ 0 \\ 0 \\ 0 \\ 0 \\ 0 \\ 0
\end{bmatrix}
$$

$$
= r_0
\begin{bmatrix}
\partial P_1/\partial V_1 \\
\partial P_1/\partial V_2 \\
\partial P_1/\partial V_3 \\
\partial P_1/\partial V_4 \\
\partial P_1/\partial \theta_1 \\
\partial P_1/\partial \theta_2 \\
\partial P_1/\partial \theta_3 \\
\partial P_1/\partial \theta_4 \\
0
\end{bmatrix}
\qquad (8)
$$

where P_1 = active power injection at bus 1; P_{12} = active power flow from bus 1 to bus 2; V_1 = voltage magnitude at bus 1; θ_1 = phase angle at bus 1.

3.2 *Errors exist in power flow measurement*

If the branch active power flow measurement through the line from bus 1 to bus 2 has error of r_0, others are error-free measurements; the right side vector of (7) is as follows

$$
\begin{bmatrix}
\partial P_1/\partial V_1 & \partial P_{12}/\partial V_1 & \partial P_{13}/\partial V_1 & \partial P_{14}/\partial V_1 & \cdots \\
\partial P_1/\partial V_2 & \partial P_{12}/\partial V_2 & 0 & 0 & \cdots \\
\partial P_1/\partial V_3 & 0 & \partial P_{13}/\partial V_3 & 0 & \cdots \\
\partial P_1/\partial V_4 & 0 & 0 & \partial P_{14}/\partial V_4 & \cdots \\
\partial P_1/\partial \theta_1 & \partial P_{12}/\partial \theta_1 & \partial P_{13}/\partial \theta_1 & \partial P_{14}/\partial \theta_1 & \cdots \\
\partial P_1/\partial \theta_2 & \partial P_{12}/\partial \theta_2 & 0 & 0 & \cdots \\
\partial P_1/\partial \theta_3 & 0 & \partial P_{13}/\partial \theta_3 & 0 & \cdots \\
\partial P_1/\partial \theta_4 & 0 & 0 & \partial P_{14}/\partial \theta_4 & \cdots \\
\vdots & 0 & 0 & 0 & \cdots
\end{bmatrix}
\begin{bmatrix} 0 \\ r_0 \\ 0 \\ 0 \\ 0 \\ 0 \\ 0 \\ 0 \\ 0 \end{bmatrix}
$$

$$
= r_0
\begin{bmatrix}
\partial P_{12}/\partial V_1 \\
\partial P_{12}/\partial V_2 \\
0 \\
0 \\
\partial P_{12}/\partial \theta_1 \\
\partial P_{12}/\partial \theta_2 \\
0 \\
0 \\
0
\end{bmatrix}
\tag{9}
$$

If the branch active power flow measurements through all lines from bus 1 have error of r_0, others are error-free measurements; the right side vector of (7) is as follows

$$
\begin{bmatrix}
\partial P_1/\partial V_1 & \partial P_{12}/\partial V_1 & \partial P_{13}/\partial V_1 & \partial P_{14}/\partial V_1 & \cdots \\
\partial P_1/\partial V_2 & \partial P_{12}/\partial V_2 & 0 & 0 & \cdots \\
\partial P_1/\partial V_3 & 0 & \partial P_{13}/\partial V_3 & 0 & \cdots \\
\partial P_1/\partial V_4 & 0 & 0 & \partial P_{14}/\partial V_4 & \cdots \\
\partial P_1/\partial \theta_1 & \partial P_{12}/\partial \theta_1 & \partial P_{13}/\partial \theta_1 & \partial P_{14}/\partial \theta_1 & \cdots \\
\partial P_1/\partial \theta_2 & \partial P_{12}/\partial \theta_2 & 0 & 0 & \cdots \\
\partial P_1/\partial \theta_3 & 0 & \partial P_{13}/\partial \theta_3 & 0 & \cdots \\
\partial P_1/\partial \theta_4 & 0 & 0 & \partial P_{14}/\partial \theta_4 & \cdots \\
\vdots & 0 & 0 & 0 & \cdots
\end{bmatrix}
\begin{bmatrix} 0 \\ r_0 \\ r_0 \\ r_0 \\ 0 \\ 0 \\ 0 \\ 0 \\ 0 \end{bmatrix}
$$

$$
= r_0
\begin{bmatrix}
\partial P_{12}/\partial V_1 + \partial P_{13}/\partial V_1 + \partial P_{14}/\partial V_1 \\
\partial P_{12}/\partial V_2 \\
\partial P_{13}/\partial V_3 \\
\partial P_{14}/\partial V_4 \\
\partial P_{12}/\partial \theta_1 + \partial P_{13}/\partial \theta_1 + \partial P_{14}/\partial \theta_1 \\
\partial P_{12}/\partial \theta_2 \\
\partial P_{13}/\partial \theta_3 \\
\partial P_{14}/\partial \theta_4 \\
0
\end{bmatrix}
\tag{10}
$$

According to relation of buses, the active power injection at bus 1 and active power flow through the lines from bus 1 have the following relation

$$
P_1 = P_{12} + P_{13} + P_{14} \tag{11}
$$

To calculate partial derivative to state variables respectively for both sides of (11), we have

$$
\begin{cases}
\partial P_1/\partial V_1 = \partial P_{12}/\partial V_1 + \partial P_{13}/\partial V_1 + \partial P_{14}/\partial V_1 \\
\partial P_1/\partial V_2 = \partial P_{12}/\partial V_2 \\
\partial P_1/\partial V_3 = \partial P_{13}/\partial V_3 \\
\partial P_1/\partial V_4 = \partial P_{14}/\partial V_4 \\
\partial P_1/\partial \theta_1 = \partial P_{12}/\partial \theta_1 + \partial P_{13}/\partial \theta_1 + \partial P_{14}/\partial \theta_1 \\
\partial P_1/\partial \theta_2 = \partial P_{12}/\partial \theta_2 \\
\partial P_1/\partial \theta_3 = \partial P_{13}/\partial \theta_3 \\
\partial P_1/\partial \theta_4 = \partial P_{14}/\partial \theta_4
\end{cases}
\tag{12}
$$

Substituting (12) into (8) gives (10), i. e. (8) and (10) are same. The effect of bus power injection measurement error on accuracy of state estimation is greater than that of branch power flow measurement error and the degree of the effect of bus power injection measurement error is related to the number of the branches from the bus.

4 SELECT WEIGHTS FOR BUS POWER INJECTION MEASUREMENTS

4.1 *Procedure of the presented weights setting method*

Based on analysis in section 3, a new way to set weights for power injection measurement is presented.

To coordinate power injections measurement with branch power flow measurement, the presented method introduce a coordination factor for weight setting of power injection measurement, the coordination factor for a bus power injection is the reciprocal of the number of the branches from the bus.

With the coordination factor we set weights for power injection measurement as follows:

$$
w_{ii} = 1/(\alpha_i \sigma_i^2) \tag{13}
$$

where α_i = number of the branches from the bus i.

The flow chart of the presented weights setting method is shown in Figure 2, the number of measurements is m.

4.2 *Accuracy index of state estimation*

To evaluate accuracy of state estimation, we define a accuracy index J_a as follows:

$$
J_a = [\mathbf{z}_0 - \mathbf{h}(\hat{\mathbf{x}})]^{\mathrm{T}}[\mathbf{z}_0 - \mathbf{h}(\hat{\mathbf{x}})] \tag{14}
$$

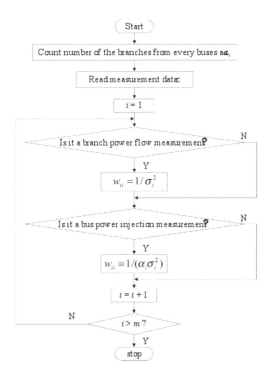

Figure 2. Flow chat of setting weights for measurements.

Table 1. Comparison of accuracy index by different methods.

| Cases | Accuracy index | | Accuracy increased |
	Traditional method	Presented method	
Case 1	0.0000239	0.0000232	2.93%
Case 2	0.0000970	0.0000946	2.47%
Case 3	0.0002217	0.0002137	3.61%
Case 4	0.0004008	0.0003875	3.32%
Case 5	0.0006371	0.0006178	3.03%

Case 4: The error of each measurement is 0.004.
Case 5: The error of each measurement is 0.005.

The accuracy index by different methods is listed in Table 1. For the traditional method, weight of every measurement is 1.0; for the presented method, weight of branch power flow measurement is 1.0, weight of bus injection measurements is reciprocal of the number of the lines from the bus.

As shown in Table 1, the presented method is more accurate than traditional method.

where $\mathbf{z}_0$ = true value vector of dimension m; $\mathbf{h}(\hat{\mathbf{x}})$ = the desired estimation of $\mathbf{z}$ calculated by estimate state variable vector $\hat{\mathbf{x}}$.

The true value vector $\mathbf{z}_0$ in (14) is unknown for a real power system, so J_a can not evaluate accuracy of state estimation for a real power system. Definition of accuracy index J_a in this paper is just to evaluate accuracy of state estimation with different weights for test systems where the output of power flow can be used as true values and testify the effectiveness of the proposed method.

5 CASE STUDY

IEEE30 system is used to testify our conclusions. There are 30 buses, 41 branches (include transmission lines and transformers), 6 power sources, and 2 shunt capacitors. There are 164 branch power flow measurements and 60 bus injection measurements.

The test system is carried out by traditional method and the proposed method respectively for five cases (all electrical quantities are in per unit):

Case 1: The error of each measurement is 0.001.
Case 2: The error of each measurement is 0.002.
Case 3: The error of each measurement is 0.003.

6 CONCLUSION

Measurements of different type have different effect on the accuracy of state estimation even if they have same errors. The effect of bus power injection measurement error on accuracy of state estimation is greater than that of branch power flow measurement error and the degree of the effect of bus power injection measurement error is related to the number of the branches from the bus. To divide weight of the bus power injection measurement by the number of the lines from the bus can improve the accuracy of state estimation. The results of a test system prove the effectiveness of the proposed method.

ACKNOWLEDGMENT

This work was supported in part by NSFC under Grant 61273137 to Dan Wang.

REFERENCES

1. Markos Asprou, Elias Kyriakides & Mihaela Albu. 2013. The effect of instrument transformer accuracy class on the WLS state estimator accuracy. *proc IEEE Power and Energy Society General Meeting*, July 2013: 1–5.

2. M. Baran, J. Jung & T. McDermott. 2009. Including voltage measurements in branch current state estimation for distribution systems. *proc IEEE Power Energy Society General Meeting*, July 2009: 1–5.
3. L. Holten, A. Gjelsvik, S. Aam, F.F. Wu & W.H.E. Liu. 1988. Comparison of different methods for state estimation. *IEEE Trans. on Power Systems* 3(4): 1798–1806.
4. Guangyi Liu, Erking Yu & Y.H. song. 1998. Novel algorithms to estimate and adaptively update measurement error variance using power system state estimation results. *Electric Power System Research* 47(1): 57–64.
5. M. Pau, P.A. Pegoraro & S. Sulis. 2013. WLS distribution system state estimator based on voltages or branch-currents: accuracy and performance comparison. *proc IEEE Instrumentation and Measurement Technology Conference*, May 2013: 1–6.
6. Shan Zhong & Ali Abur. 2004. Auto tuning of measurement weights in WLS state estimation. *IEEE Trans. on Power Systems* 19(4): 2006–2013.

Power and Energy – Kong (Ed.)
© 2015 Taylor & Francis Group, London, ISBN 978-1-138-02782-4

Research on the delay calculation method of partial discharge source localization

Min Chen & Jun Chen
Electric Power Research Institute of Hubei Electric Power Company, Wuhan, Hubei, China

Jun Lu
Hubei Electric Power Company, Wuhan, Hubei, China

ABSTRACT: Delay calculation is extremely important for localization based on the time difference of UHF signals arriving, when the time difference of multiple sensors receiving signals is used to determine the discharge source location. The delay accuracy directly determines the positioning accuracy. By the artificial analog defects ways of discharge model placed in GIS device, the time delay calculation method of partial discharge source location technology based on the time difference in GIS are researched. The experimental results show that, the accuracy of the waveform characteristic point method is not high, the stability of the accumulated energy curve inflection point method is not high, the generalized correlation method must reasonably choose weighting function to considering both high resolution and stability problems, also need stationary signals and statistics prior knowledge of signal and noise, However, the adaptive time delay calculation method based on LMS which can provide a higher accuracy is more ideal.

1 INTRODUCTION

Gas Insulated Switchgear (GIS) is used widely more and more in electric power system because of its high reliability, compact structure, easy maintenance, and so on. But sometimes some small defects such as metal particles, insulating air gap and so on, is left in the manufacturing and assembly process of GIS because of the technology problem or the other problems. These tiny defects may progress to dangerous discharge channel during GIS operation, and eventually cause insulation breakdown accident[1–6]. Therefore, partial discharge monitoring during GIS operation is becoming more and more important in order to prevent the insulation fault in GIS. And in the process of the GIS partial discharge monitoring, partial discharge quickly and accurately locating has the vital significance to quickly eliminate breakdown hidden danger and ensure the system safety after the existence of partial discharge is found in the equipment.

The method of detecting the electromagnetic wave signal [Q1] excited from GIS inner partial discharge by the internal or external sensor is more practical than conventional methods (such as gas analysis and photoelectric detection method)[7–9]. For example, Britain's Hampton et al have made independent partial discharge sources, and calibrated partial discharge amount of each defect model, then tested the discharge signal using the UHF method, depicted the discharge characteristics of different discharge source, and researched on the signal source locating by using the method of time difference on signal arrival[10]. Ultra high frequency method, which has strong anti-interference ability, high sensitivity, can realize the PD source locating and fault type identification. It is suitable for on-line monitoring and has been widely concerned[11–19].

Multiple time delay calculation methods of partial discharge source UHF location technology were studied. The results show that the adaptive time delay algorithm based on LMS is an ideal method of delay.

2 PARTIAL DISCHARGE SOURCE LOCALIZATION METHOD BASED ON THE TIME DIFFERENCE IN GIS

When positioning of the PD source by the UHF method, first, the PD sources is rough located through the measurement of the insulator. After that, the two sensors are placed in the two test points to accurate positioning of the PD source, the basic method is TDOA location method[20], which determines the PD source position by using the time difference of sensor accepting the UHF signal emitted from the PD source.

The calculation for postponing is very important to determine the PD source position by using the time difference of multiple sensors accepting the signal, and the delay calculation accuracy directly determines the positioning accuracy. There are two basic methods of time delay calculation: one is the signal feature extraction method, namely measuring a certain time that the signal has a characteristic, then directly subtracting the corresponding feature point time of the different signal, another is the method to calculate the delay based on signal correlation analysis.

3 THE SIGNAL FEATURE EXTRACTION METHOD

3.1 *The waveform characteristic point method*

The signal characteristic points include the starting point and peak point of the waveform rising edge. According to the fee code shortest optical path principle, the signal starting point is wavefront reflection of the wavelet that arrives along the shortest path in the minimum propagation time. Therefore, the starting point of the rising edge should be the reference point of the signal starting position. The time difference between two signals reference point is the arrival time delay. But, the starting point of the rising edge is very difficult to determine because of some influencing factors in practice, such as the background noise, the refraction and reflection in the signal propagation process, and so on. In order to reduce the error, the first high peak is selected as a basis for calculating time delay. The measured discharge waveform in the laboratory is shown in Figure 1, and the delay is calculated in the signal waveform characteristic point method.

The waveform measured from field location of the PD source in GIS by the oscilloscope is shown in Figure 2. The UHF signal emitted from the PD source is received by the sensor and converted to a voltage signal, which is measured by the oscilloscope. The ordinate is the voltage value in units of mv, and the abscissa is the time in units of ns. If the oscilloscope interface center line is the time coordinate origin, the time coordinate

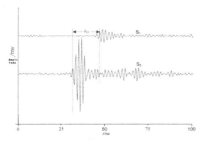

Figure 1. The signal waveform characteristic point method for delay calculation (in laboratory).

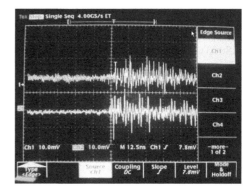

Figure 2. The signal waveform characteristic point method for delay calculation (field measurement).

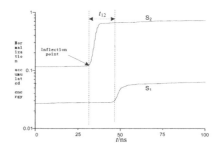

Figure 3. The accumulated energy curve inflection point method for delay calculation.

corresponds to the first peak value of the oscilloscope channel 1 is about 3ns, and channel 2 is about 0.5ns, Therefore, the delay is 2.5ns.

From Figures 1 and 2 it can be observed that, as the violent oscillation of the UHF signal, the starting part is not apparent, and the starting point can't be defined by a standard pulse, which causes the difficulties of delay computation by using the signal waveform characteristic.

3.2 *The accumulated energy curve inflection point method*

The voltage waveform of UHF signal can be transformed into the accumulated energy curve to determine the time difference between the signal. The cumulative energy X can be carried out in accordance with the following formula: $X_j = \sum_{k=0}^{j} V_k^2$, in which V_k is the k point voltage of the signal waveform.

After that the accumulated energy X is calculated, the cumulative energy curve to the signal is painted and shown in Figures 6–7 with time as abscissa and accumulated energy as ordinate. Signal energy always attenuates stably, so the accumulated energy curve finally tends to level, in which the ultimate longitudinal coordinate values represent the all energy received by the receiving antenna. The curve trend is rising slowly at first, then rising sharply at one point, and eventually become level. The curve sharply rising point is known as the inflection point, in which the discharge signal becomes stronger than the background noise. The signal time delay can be obtained by comparing the time of each curve inflection point.

Compared with the waveform characteristics method, in the cumulative energy inflexion method, the energy curve frontier doesn't have oscillation, but its rise is more slowly, which is not conducive to read the reference time.

4 THE DELAY CALCULATION METHOD BASED ON SIGNAL CORRELATION ANALYSIS

4.1 The generalized correlation method

The traditional methods used to solve the calculation of the time difference TDOA that signal arrives at two sensors is the generalized correlation method (GCC). Before the correlation algorithm of signal, First, the signal is prefiltered on, and then the time difference TDOA Δt that signal received by two sensor arrives is calculated, Δt is the time that the cross-correlation function has the maximum value. Prefilter is designed to improve the signal-to-noise ratio and reduce the noise power before the signal is send into the correlator.

The principle of delay calculation by the method is shown in Figure 4. Signal x1(t) and x2(t) respectively go through the filter of H1(f) and H2(f), and then the correlation operation, integral and square is performed. The operation process goes on in a certain time offset range τ until obtaining related peak. The value of τ that obtaining the peakvalue is calculated value of TDOA Δt. The process of delay calculation by the method is shown in Figure 5.

4.2 The adaptive time delay calculation method based on LMS

Adaptive LMS filter can adjust adaptively the filter coefficient according to the current input signal sampling, so that the output error signal reaches the minimum, without the need for the priori

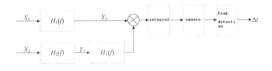

Figure 4. The schematic diagram of delay calculation by the generalized correlation method.

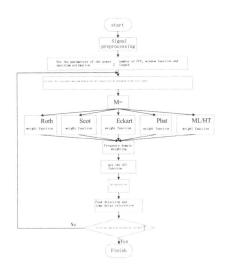

Figure 5. The process diagram of delay calculation by the generalized correlation method.

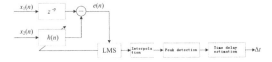

Figure 6. The schematics of adaptive time delay calculation method based on LMS.

knowledge of the input signal spectrum. Therefore, the LMS algorithm is widely used in the situation that the input signal statistic characteristics is unknown. And LMS algorithm also has been successfully introduced into the delay calculation based on the same considerations. The principle of the method is shown in Figure 6, and the process of the method is shown in Figure 7.

In order to ensure the causality of the system, z-p is introduced to calculate negative delay in the graph. It can be observed from the graph, LMS delay calculator automatically adjust h(n), so that the output approximate x1(n), which is essentially equal to inserting a delay in the signal x2(n) to make sure two channel signals align. In ideal conditions, the corresponding weighting coefficient to the actual delay will converge to 1 in H(n), but the other part will converge to 0. Finally, in order to obtain the fractional multiple sampling delay, the interpolation to h(n) can be operated. In the case of a sufficient number of observation data, LMS may achieve the optimal filter in the statistical meaning, namely Wiener filter, and the frequency domain expression is: $H(\omega) = G_{x_2 x_2}^{-1}(\omega) G_{x_1 x_2}(\omega)$.

In the statistical sense, the LMS method and Roth weighted GCC method are similar, but their

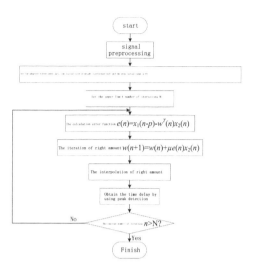

Figure 7.   Flow chart of adaptive time delay calculation method based on LMS.

Figure 8.   GIS cavity for testing.

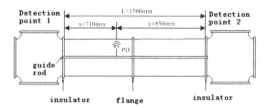

Figure 9.   Position and size schematic diagram of partial discharge source.

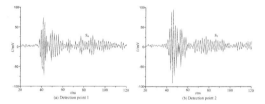

Figure 10.   Time-domain waveforms.

starting points and conditions are different. Time delay is calculated from the perspective of the signal cross-correlation by GCC, which is based on the prior knowledge of signal and noise, and need a large number of data using statistical method to draw. But in the actual operation, signal power spectrum and cross power spectrum calculation are often received by only one frame data in the GCC method, so the calculation accuracy is not high. However, the LMS adaptive filtering make a signal channel approach another through certain error criteria, and time delay is calculated in the convergence conditions. It does not need any prior knowledge of the signal spectrum, so the LMS time delay calculation method can be seen as adaptive realization of the Roth processor.

5   COMPARISON BY TEST IN LABORATORY

Test is carried out by using GIS entity model in the laboratory. The GIS model for testing is shown in Figure 8.

Position and size schematic diagram of partial discharge source is shown in Figure 9.

Time-domain waveforms test at detection point 1 and 2 is shown in Figure 10.

The delay calculation results are shown in Table 1.

It can be observed that the calculation precision by the adaptive time delay estimation method with high accuracy based on LMS can reach about 1 cm.

Analysis results show that, although the waveform characteristic point as a basis for calculating delay method is simple, but its accuracy affected by the noise and the signal catadioptric is not high,

and it isn't up to the requirements in high precision positioning. However, in the inflection point of the accumulated energy curve as a benchmark method, there is no uniform standard for the inflection point selection, so its stability is not high. The generalized correlation method and adaptive time delay calculation method based on LMS can improve the time delay calculative precision, especially the latter. But there are some defects to the generalized correlation method, which must reasonably choose weighting function to considering both high resolution and stability problems, also need stationary signals and statistics prior knowledge of signal and noise. Therefore, the adaptive time delay calculation method based on LMS is more ideal.

What need to be pointed out is that, the experiments were test in the laboratory, signal to noise ratio is relatively high, and it didn't have influence of external noise. Therefore, positioning accuracy is very high. The positioning accuracy has a great relationship with signal to noise ratio in practical application.

Table 1. Delay calculation result (ns).

The academic time delay	The waveform characteristic point method	The accumulated energy curve inflection point method	The generalized correlation method (Roth weighted)	The adaptive time delay calculation method based on LMS
0.47	0.82	−0.21	0.74	0.41

6 CONCLUSION

Delay calculation is extremely important for localization based on the time difference of UHF signals arriving, when the time difference of multiple sensors receiving signals is used to determine the discharge source location. The delay accuracy directly determines the positioning accuracy. By the artificial analog defects ways of discharge model placed in GIS device, the time delay calculation method of partial discharge source location technology based on the time difference in GIS are researched, including signal waveform characteristic point method and adaptive time delay calculation method based on LMS. The experimental results show that adaptive time delay calculation method based on LMS which can provide a higher accuracy is more ideal.

REFERENCES

1. Pryor B.M. Review of partial discharge monitoring in gas insulated substations [C] // Science, Education and Technology Division Colloquium on Partial Discharges in Gas Insulated Substations. London, U K: [s. n.], 1994: 122.
2. Anon. Partial discharge testing of gas insulated substations [J]. IEEE Transactions on Power Delivery, 1992, 7 (2): 499–506.
3. Pearson J, Hampton B.F, Judd M.D, et al. Experience with advanced in-service condition monitoring techniques for GIS and transformers [C] // IEE Colloquium on HV Measurements, Condition Monitoring and Associated Database Handling Strategies. London, U K: IEE, 1998:8/128/10.
4. Joint Working Group 33/23.12. Insulation coordination of GIS:return of experience, on site tests and diagnostic techniques [J]. Electra, 1998, 176 (2): 67–97.
5. Liu Jun-hual, Yao Ming, Huang Cheng-jun, et al. Experimental Research on Partial Discharge Localization in GIS Using Ultrasonic Associated with Electromagnetic Wave Method. High Voltage Engineering, 2009, 35 (10):2458–2463.
6. Liu Jun-hua, Wang Jiang, Qian Yong, et al. Simulation Analysis on the Propagation Characteristics of Electromagnetic Wave in GIS [J]. High Voltage Engineering, 2007, 33 (8):139–142.
7. Xiao Yan, Yu Wei-yong. Present status and prospect of research of on-line partial discharge monitoring system in GIS [J]. High Voltage Engineering, 2005, 31 (1):47–49.
8. Qian Yong, Huang Cheng-jun, Jiang Xiu-chen, et al. Current status and development of PD monitoring technology in GIS [J]. High Voltage Apparatus, 2004, 40(6): 453–456.
9. Tang Ju, Wei Gang, Sun Cai-xin, et al. Research on the dipole antenna sensor with broadband for partial discharge detection in GIS [J]. High Voltage Engineering, 2004, 30(3):29–32.
10. Hampton B. UHF diagnostics for gas insulated substations [J]. Eleventh International Symposium on High Voltage Engineering (Conf Publ No 467), vol. 15, 1999: 6–16.
11. Yao Yong, Yue Yan-feng, Huang Xing-quan. Field application of UHF and Ultrasonic methods for detecting Partial Discharge in GIS [J]. High Voltage Engineering, 2008, 34 (2): 422–424.
12. Qian Yong, Huang Cheng-jun, Jiang Xiu-chen, et al. Present situation and prospect of ultrahigh frequency method based research of on-line monitoring of partial discharge in gas insulated switchgear [J]. Power System Technology, 2005, 29(1):40–43, 55.
13. Tang Ju, Zhu Wei, Sun Cai-xin, et al. Analysis of UHF method used in partial discharge detection in GIS [J]. High Voltage Engineering, 2003, 29 (12):22–23, 55.
14. Hoshino T, Kato K, Hayakawa N, et al. A novel technique for detecting electromagnetic wave caused by partial discharge in GIS [J]. IEEE Transactions on Power Delivery, 2001, 16(4): 545–551.
15. Tang Ju, Wei Gang, Dai Hai-jun, et al. Analysis of UHF method used in partial discharge detection in gas insulated switchgear [J]. Journal of Chongqing University, 2004, 27 (4):125.
16. Li Xin, Li Cheng2rong, Ding Li2 jian, et al. Identification of PD patterns in gas insulated switchgear (GIS) based on UHF signals [J]. High Voltage Engineering, 2003, 29 (11):26–30.
17. Xu Min-ye, Wu Xiao-chun, Lu Zhen-hua. Field application of partial discharge testing and positioning technology for GIS [J]. East China Electric Power, 2009, 37 (7):1086–1089.
18. Li Zhi-min, Feng Yun-ping. Analysis and experiment of UHF method used in partial discharge detection in GIS [J]. High Voltage Engineering, 1998, 24 (4):29–30, 38.
19. Pearson J.S, Farish O, Hampton B.F, et al. Partial discharge diagnostics for gas insulated substations [J]. IEEE Transactions on Dielectrics and Electrical Insulation, 1995, 2 (5): 893–905.
20. Liu Wei-dong, Huang Yu-long, Wang Jan-feng, et al. On-line monitoring and Localization of Partial Discharge based on UHF signals in GIS [J]. High Voltage Apparatus, 1999, 35 (1):11–15.

Power and Energy – Kong (Ed.)
© 2015 Taylor & Francis Group, London, ISBN 978-1-138-02782-4

Research on hydropower generator group equivalence and parameter identification based on PSASP calling and optimization algorithm

H.B. Shi
State Grid Sichuan Electric Power Research Institute, Chengdu, China

B.W. Hu & J.T. Sun
School of Electrical Engineering, Southwest Jiaotong University, Chengdu, China

ABSTRACT: It's difficult to establish the detailed model that contains each generating unit when power system with centralized small and medium hydropower generator group is analyzed. At present, the common practice is to analyze the equivalent system after dynamic equivalence. There are significant limitations when using traditional homology method or modal method to build the equivalent model because the structure and parameters of the hydropower generator group must be known. This paper presents a method to build a dynamic equivalent model of hydropower generating units based on estimate equivalent method for the structure and parameters are unknown. This method makes the online hydropower generator group be equivalent as a unit. Using this method, all the relevant parameters can be identified accurately. Some procedures, such as taking the dynamic response signal which exist in interconnection line after a major disturbance as the objective function value, and using particle swarm optimization algorithm to update each parameter, and calling PSASP to simulate the transient stability of the equivalent model under different parameters based on Visual Studio platform, are used in this method. For validation purposes, the proposed method is applied to the hydropower generating units in Sichuan province. Results show that the equivalent method is practical and feasible.

1 INTRODUCTION

In order to take advantage of water resources, many small and medium hydropower stations have established in abundant water resources regions recently. If the region contains a large number of small and medium hydropower stations, the dynamic performance of the power system must be affected by the hydropower generator group. It's difficult to establish the detailed model that contains each generating unit when power system with centralized small and medium hydropower generator group is analyzed. At present, the common practice is to analyze the equivalent system after dynamic equivalence. There are significant limitations when using traditional homology method or modal method to build the equivalent model because the structure and parameters of the hydropower generator group must be known. Therefore, it's significant to research the dynamic equivalent method based on estimate equivalent method while the structure and parameters of the hydropower generator group are unknown.

This paper presents a method to build a dynamic equivalent model of hydropower generating units based on estimate equivalent method for the structure and parameters are unknown. This method makes the online hydropower generator group be equivalent as a unit. Using this method, all the relevant parameters can be identified accurately. Some procedures, such as taking the dynamic response signal which exist in interconnection line after a major disturbance as the objective function value, and using particle swarm optimization algorithm to update each parameter, and calling PSASP to simulate the transient stability of the equivalent model under different parameters based on Visual Studio platform, are used in this method. The proposed method is applied to an actual power system with hydropower generator group. Results show that the equivalent model could meet the system transient stability analysis.

2 EQUIVALENT MODEL OF HYDROPOWER GENERATOR GROUP

2.1 Regional dynamic equivalent model

Regional dynamic equivalent model is shown in Figure 1, that is, a single generator. Since the external system is connected through a long interconnection line with the internal system, so the external system can be considered homology when

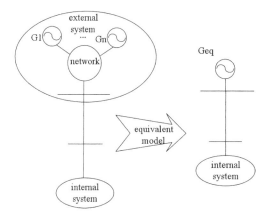

Figure 1. Regional dynamic equivalent model.

disturbances occur within the internal system. It's a premise that the external system can be equal to a generator.

2.2 Equivalent generator model

It can only reflect the steady trend and quite be different from dynamic characteristics of the prototype system when adopt the second-order generator model which takes no account of the process of the field winding and damping winding. The accuracy of reactive power is not high because the third-order generator model only considers the d-axis field winding. In this paper, fifth-order generator model which is suitable for Turbine generator is taken as the equivalent generator model. The equations of the generator model are shown below.

$$
\begin{cases}
T'_{d0}\dfrac{dE'_q}{dt} = E_f - E'_q - (x_d - x'_d)i_d \\[2mm]
T''_{d0}\dfrac{dE''_q}{dt} = -E''_q + E'_q - (x'_d - x''_d)i_d + \dfrac{dE'_q}{dt} \\[2mm]
T''_{q0}\dfrac{dE''_d}{dt} = -E''_d + (x_q - x''_q)i_q \\[2mm]
\dfrac{d\delta}{dt} = \omega - 1 \\[2mm]
T_J\dfrac{d\omega}{dt} = T_m - E''_q i_q - E''_d i_d + (x''_d - x''_q)i_d i_q \\[2mm]
\qquad\qquad - D(\omega - 1)
\end{cases}
\tag{1}
$$

where E''_q = q-axis sub-transient electromotive force; E'_q = q-axis transient electromotive force; E''_d = d-axis sub-transient electromotive force; x''_d = d-axis sub-transient reactance; x'_d = d-axis transient reactance; x_d = d-axis synchronous

reactance; x''_q = q-axis sub-transient reactance; x_q = q-axis synchronous reactance; T''_{d0} = d-axis sub-transient open circuit time constant; T'_{d0} = d-axis transient open circuit time constant; T''_{q0} = q-axis sub-transient open circuit time constant; T_J = inertia time constant; δ = power angle; ω = rotor speed; T_m = mechanical torque.

3 EQUIVALENCE METHOD OF HYDROPOWER GENERATOR

3.1 Solving the equivalent system

Prototype system can be described as the following state equations.

$$
\begin{cases}
\dot{X} = AX + Bu \\
Y = CX
\end{cases}
\tag{2}
$$

where X = system state variables; Y = measurements, that is, active and reactive power on interconnection line; u = excitation signal, that is, a major disturbance in internal system; A = system matrix; B = excitation signal matrix; C = measurement matrix.

The equivalent system can be expressed as the following equations.

$$
\begin{cases}
\dot{X}'(\alpha) = A'(\alpha)X'(\alpha) + Bu \\
Y'(\alpha) = C'(\alpha)X'(\alpha)
\end{cases}
\tag{3}
$$

where $X'(\alpha)$, $Y'(\alpha)$, $A'(\alpha)$ and $C'(\alpha)$ are the function of the parameter vector α. The outputs are the active and reactive power on interconnection line $[P_l, Q_l]$. Parameters to be identified are $\alpha = [x_d, x'_d, x''_d, x_q, x''_q, T'_{d0}, T''_{d0}, T''_{q0}]$.

The dynamic response signal after a major disturbance in prototype system is Y while another dynamic response signal after the same disturbance in equivalent system is $Y'(\alpha)$. And the deviation of them is e. The objective function J is a function of the deviation e. Update the model parameter vector α by identification method iteratively until the J satisfy the threshold value. The equation of the objective function is as follows:

$$
J = f(e) = \sum R|Y - Y'(\alpha)|^2
\tag{4}
$$

The optimum vector α is identified when α makes J reach the threshold value. The calculation process is shown in Figure 2.

Specific steps: (1) calculate power flow in prototype system to obtain the initial steady-state value; (2) simulate transient stability with a disturbance in prototype system with PSASP; (3) get dynamic

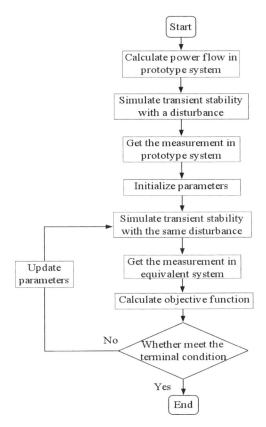

Figure 2. Process of equivalent model parameters calculation.

The optimal solution is found by initializing a group of random particles and then iterate them again and again with the method of PSO. In each iteration process, the particles are updated by tracking two "extreme": one is the optimal solution found by themselves (personal best, that is *pbest*), and the other is the optimal solution found by the entire groups (groups best, that is *gbest*). Each particle updates its velocity and position respectively according to the following equations.

$$\begin{cases} v_{k+1} = wv_k + c_1 \times rand \times (pbest_k - x_k) \\ \qquad + c_2 \times rand \times (gbest_k - x_k) \\ x_{k+1} = x_k + v_{k+1} \end{cases} \tag{5}$$

where v_k = velocity vector of the particle; w = inertia weight factor; $c_{1,2}$ = learning factor; *rand* = a random number between 0 and 1.

The number of iterations will become a substantial increasing if the population size is too small while the search efficiency will be declined if the population size is too large. So population size is taken between 30 and 50 generally. The process of solving the dynamic equivalent model parameters based on the above-mentioned method with PSO is shown in Figure 3.

Specific steps: (1) get the active power P_l and reactive power Q_l on interconnection line; (2) initialize the parameters of each particle which to be identified; (3) calculate the objective function and determine the initial value of *pbest* and *gbest*; (4) initialize iterations; (5) update iterations and velocity and position of the particle; (6) work out the output vector Y' with PSASP; (7) calculate the objective function and then update *pbest* and *gbest*; (8) determine whether meet the terminal condition or not. If the answer is Yes, stop calculating and output *gbest*, otherwise return (5).

3.3 *Calling PSASP in the application of the parameter identification of equivalent system*

PSASP is widely used in power gird planning and security and stability analysis. In this paper, PSASP is called to calculate the transient stability with the same disturbance when the parameters of the equivalent system are updated. It is not only reduces the workload, but also improves the accuracy.

For the purpose of calling the appropriate module of PSASP, the database structures, software architectures and running processes must be known. There are a lot of functional modules and data tables in the default folder WPSASP. These functional modules correspond to data entry modules, calculation modules and results modules.

response Y on interconnection line; (4) select the equivalent system model and initialize parameters to be identified; (5) simulate transient stability with the same disturbance in equivalent system with PSASP; (6) get dynamic response Y' on interconnection line; (7) calculate the objective function; (8) determine whether meet the terminal condition or not. If the answer is Yes, stop calculating, otherwise update estimated parameters with optimization algorithms and return (5).

3.2 *Particle swarm optimization algorithm in the application of the parameter identification of equivalent system*

Estimate equivalent method based on parameter identification is used to obtain the equivalent model parameters by making use of dynamic response signal on interconnection line after a major disturbance. In this paper, Particle Swarm Optimization algorithm (PSO) is used to identify the parameters of the equivalent model due to PSO has many excellent characteristics.

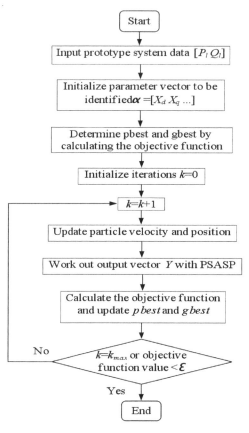

Figure 3. Process of parameter identification based on PSO.

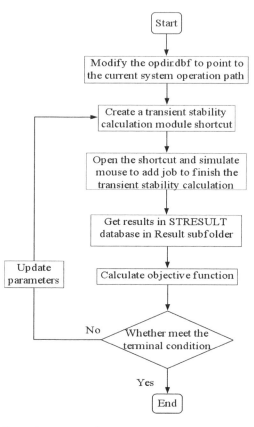

Figure 4. Process of calling transient stability calcula-tion module on PSASP.

Some intermediate variables are stored in data table files during PSASP operation. The interme-diate variable used in this paper is the path of jobs database of calculation module named opdir.dbf. The calling module in PSASP is the transient sta-bility calculation module named hstbatcal_ex. The transient stability calculation module calling can be achieved by writing the path of jobs database and calling the hstbatcal_ex. The process of calling transient stability calculation module on PSASP is shown in Figure 4.

Specific steps: (1) modify the opdir.dbf to point to the current system operation path; (2) create a transient stability calculation module shortcut in this path, namely hstbatcal_ex; (3) open hstbatcal_ex and simulate mouse to add job to finish the transient stability calculation; (4) get results in STRESULT database in Result sub-folder. (5) calculate objective function; (6) deter-mine whether meet the terminal condition or not. If the answer is Yes, stop calculating, otherwise

update the parameters with PSO method and return (2).

4 SIMULATION VERIFICATION

Take eight small and medium hydropower stations as a system to be equivalent in Sichuan province which access to the system by 220 kV substation.

Build the dynamic equivalent model with the above method. The parameters of the equivalent model are shown in Table 1. The power and volt-age curves on the interconnection line in the pro-totype system and equivalent system are shown in Figures 5, 6 and 7 with the same N-1 fault.

It can be observed from Figures 5–7, the dynamic responses are same between prototype system and equivalent system, and the results meet the equivalent requirements. The method is simple and accurate, and plays an important role in dynamic equivalent method according to the simulation results.

Table 1. Identification parameters results of dynamic equivalence.

Parameters	x_d	x''_d	x_q	x''_q
Value	0.85	0.29	0.74	0.34
Parameters	T'_{d0}	T''_{d0}	T''_{q0}	T_J
Value	5.0	0.088	0.18	8.4

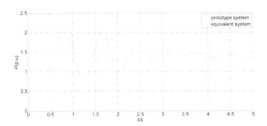

Figure 5. Active power curve on interconnection line compared prototype system with equivalent system.

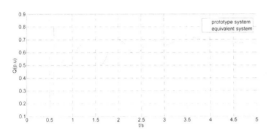

Figure 6. Reactive power curve on interconnection line compared prototype system with equivalent system.

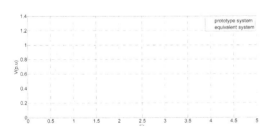

Figure 7. Voltage curve on interconnection node compared prototype system with equivalent system.

5 CONCLUSIONS

The dynamic equivalent model should be used when dynamic characteristic of power grid which contains small and medium hydropower generator group is analyzed. There are significant limitations when using traditional homology method or modal method to build the equivalent model. In response to the practical problem, a dynamic equivalent model and parameter identification method are proposed in the paper based on estimate equivalent method. The model parameters are updated with PSO and the transient stability is simulated with PSASP based on Visual Studio platform. Using this method, all the relevant parameters can be identified accurately. The proposed method is applied to the hydropower generating units in Sichuan province. Results show that the equivalent model could meet the transient stability analysis and the equivalent method is practical, simple and feasible.

REFERENCES

1. Chaoxian Han, "The Identification of Synchronous Generator Parameters Based on PMU," Qingdao University, 2012.
2. Chenxi Yue, "Research on The Identification of Synchronous Generator Parameters Based on GPS-PMU," Hohai University, 2005.
3. China Electric Power Research Institute, "PSASP 6.2 User Manual," 2004, Beijing: China Electric Power Research Institute.
4. Jingping Yang, "Study on Dynamic Equivalent Methods for Large-scale Power Systems," Zhejiang University, 2007.
5. Lixia Liu, Min Luo, Xiaohui Li, "Comparison and Improvement of Common Methods of Dynamic Equivalence in Power System," Proceedings of the CSU-EPSA, Vol. 23 No. 1, 2011, pp. 149–154.
6. Min Wang, Jinyu Wen, Wenbo Hu, "A Dynamic Equivalent Modeling for Regional Small Hydropower Generator Group," Power System Protection and Control, Vol. 41, No. 17, 2013, pp. 1–9.
7. Ning Zhang, "Research on The Identification of Synchronous Generator Parameters Based on Phasor Measurement," North China Electric Power University, 2007.
8. Pin Ju, Weihua Wang, Hongjie Xie, "Identification Approach to Dynamic Equivalents of The Power System Interconnected with Three Areas," Proceedings of the CSEE, Vol. 27, No. 13, 2007, pp. 29–34.
9. Pin Ju, "Modeling Theory and Method of Power System," 2010, Beijing: Science Press.
10. Qixin Huang, Lixia Sun, Wei Zhen, "Ant Colony Algorithm and Disturbance Analysis of Synchronous Generator Parameter Identification," Electric power automation equipment, Vol. 29, No. 11, 2009, pp. 50–53.
11. Wei Wei, Lijie Ding, Huabo Shi, "Research on Equivalent Method and Parameter Identification of Hydropower Generator Groups Based on Circuit Export Information of Substation," Sichuan Electric Power Technology, Vol. 36, No. 5, 2013, pp. 20–24.

12. Yalou Li, Zhongxi Wu, "An Approach to Interface Matlab Model with PSASP Transient Stability Module," Power system technology, Vol. 32, No. 19, 2008, pp. 31–36.
13. Yang Zhou, "A Thesis Submitted in Partial of Fulfillment of The Requirements for The Degree of Master of Engineering," Huazhong University of Science and Technology, 2012.
14. Yixin Ni, Shousun Chen, Bolin Zhang, "Dynamic Theory and Analysis of Power System," 2002, Beijing: Tsinghua University Press.
15. Yuanfan Guo, Feng Yang, Zhaoxia Dong, "Damping Competition and Damping Coordination of Additional Excitation Control in Multi-machine Power Systems," Advanced technology of electrical engineering and energy, Vol. 25, No. 4, 2006, pp. 52–57.
16. Yuansheng Zhang, "The Effects on Stable Running of Power Grid of The Distributed Network with Small Hydropower," Hunan University, 2012.
17. Yunhai Zhou, Xianshan Li, Xiangyong Hu, "Dynamic Equivalents Based on The Transient Power Flow of The Connecting Lines," Proceedings of the EPSA, Vol. 11, No. 5–6, 1999, pp. 29–33.

Power and Energy – Kong (Ed.)
© 2015 Taylor & Francis Group, London, ISBN 978-1-138-02782-4

Fast arc fault detection based on the arc resistance model

F. Du, W.G. Chen, Z. Liu & Y. Zhuo
Siemens Ltd., Shanghai, China

M. Anheuser
Siemens AG, Amberg, Germany

ABSTRACT: The arc fault in the switchgear can result in disastrous damage on the personnel, devices and system. Over the past years, researchers have developed a lot of detection methods, e.g., light plus current, sound plus current. In this paper, an arc flash detection method based on the arc resistance recognition is presented. The arc resistance is deemed as a temporal variant with specific empirical characteristics, which determines the practical current amplitude flowing through the arc fault. Taking advantage of the current sensor, and the equivalent circuit model, the prospective current flowing through the arc at each time point can be forecasted via early short circuit detection method. Then, the arc resistance dependent prospective current value in the forecasted results can be filtered by the moving average method, which is applied to calculate the empirical arc resistance. In another side, the instantaneous arc resistance can be calculated with the measured current and voltage. Once these two calculated values match each other, an arc fault can be asserted. Per to the verification with practical arc test data, the proposed solution can detect the arc fault in its initial stage.

1 INTRODUCTION

The damage caused by arc flash fault is remarkably noticed in recent years (P.E.J. Phillips 2012, H.L. Floyd 2013). Thus, it is very import to detect the arc fault as soon as possible when it occurs. Technically, an arc fault is an unintended self-sustaining discharge of electricity in a highly conductive ionized gas, which allows current flow between the conductors, limited only by circuit parameters that are predominantly resistive. In case of the arc flash fault, a lot physical phenomena including intense light, sound waves, pressure waves, and extreme high temperature are generated by the explosion of the arc energy. It has been reported that the arc fault can be detected by applying light sensor, sound sensor, or pressure sensor (J. Vico 2012, R.M. Harris 2012). Besides the complex mounting process of these kinds of sensor, the current information is often necessary to be taken as a backup measure to ensure the reliability of the detection.

Meanwhile, researchers have spent a lot of efforts on the study of the mechanism of the arc nature by creating some empirical models. In the paper (R.F. Ammerman 2008), the arc models have been classified into three categories: 1) theoretical models developed from arc physics, 2) statistical models developed from statistical analysis, and 3) semi-empirical models developed from known observations and numerical analysis. A great plenty of tests have been done to evaluate these models. However, because these models are used to calculate the incident energy aroused from the arc fault, the arc current RMS value is required. Obviously, it is impossible to calculate the current RMS value in less than 5 ms with the traditional method for a 50 Hz power signal.

In this paper, the arc fault detection method based on the arc resistance model is proposed. In the second section, the arc model based on arc resistance is introduced and the Early Short Circuit Detection (ESCD) method (F. Du 2014) is applied to detect the arc fault as soon as possible. In the third part, the moving average method is taken advantage to process the calculated arc resistance results. In the fourth part, the evaluation results based on the practical test are introduced. In the end, the summary and the further work are concluded.

2 ARC RESISTANCE CALCULATION

2.1 *Working principle*

The basic working process of the proposed solution is like the below,

Step 1: calculate the arc resistance based on the arc resistance model and obtain R_{model} in the time domain.

Step 2: calculate the impedance based on sampled voltage $V(t)$ and current $i(t)$ and obtain R_{test} in time domain.

$$R_{test}(t) = \frac{V(t)}{i(t)} \qquad (1)$$

Step 3: Check the similarity of R_{model} and R_{test}.

Exemplary criteria as shown in formula (2): within a period of time T, e.g., 1 ms, if the similarity is always within a defined percentage, e.g., 20%, the arc fault is detected, i.e.,

$$(R_{model} - R_{test})/R_{test} \le 20\% \qquad (2)$$

2.2 Arc model

By comparing a lot of models with available test data, as mentioned in the conclusion of the paper (R.F. Ammerman 2008), the Wilkins method gives the most consistent results, which had been verified by a great deal of test data. It is also pointed out that the resulting semi-empirically derived equations are relatively easy to use and are more suitable for a wide audience.

Wilkins' simple improved method is about an incident energy model (R. Wilkins 2004) which is a simplified version of an approach Wilkins, etc developed (R. Wilkins 2005). The methods correct the anomalies observed in the IEEE 1584 arcing current and incident energy equations. And no complex time-domain model is used in the simplified approach. As a result, it is suitable to more general applications. Figure 1 shows the equivalent circuit used to model a three-phase arcing fault.

The arc voltage is determined using the equation (3) below.

$$V_{arc} = 1.757 \cdot I_{arc}^{0.1457} \cdot g^{0.2476} \cdot V_{LL}^{0.4166} \qquad (3)$$

where

V_{arc} RMS arc voltage (V)
I_{arc} RMS arc current (A)
g Electrode gap (mm)
V_{LL} Line to line voltage (V)

Consequently, the arc resistance in the simplified model (4) is got from (3) by dividing the arc current,

$$R_{arc} = 1.757 \cdot I_{arc}^{-0.8543} \cdot g^{0.2476} \cdot V_{LL}^{0.4166} \qquad (4)$$

Equations (3) and (4) are applicable for arcs in open air. For arcs initiated within a box, the formulas must be multiplied by a coefficient of 0.821.

Obviously, this semi-empirical equation is fitted via the RMS arc current. In other words, the RMS value should be known to apply this equation to calculate the arc resistance. However, it is impossible to calculate the RMS value of 50 Hz signal less than 5 ms with the traditional integral method.

2.3 Early Short Circuit Detection (ESCD) method

As known that the arc is almost resistive, it is possible to convert the model in Figure 1 to the model in Figure 2. The total circuit is equivalent to a variant resistor R and inductance L under a power source $V(t)$, which is complied with the formula (5),

$$V_m \cos(wt + \varphi) = Ri + Li'$$
$$\tan\theta = \frac{R}{wL},$$
$$\cos\theta = \frac{R}{\sqrt{R^2 + (wL)^2}} \qquad (5)$$
$$I_{RMS} = \frac{V_m}{\sqrt{2R^2 + 2(wL)^2}}$$

where I_{RMS} is called as the prospective short circuit current RMS value.

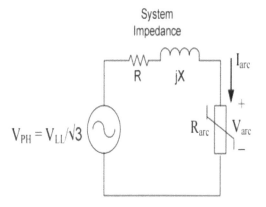

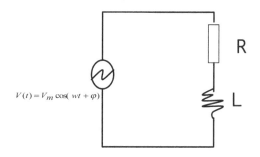

Figure 1. Per-phase equivalent circuit mode.

Figure 2. Equivalent circuit model.

Then, the method called Early Short Circuit Detection (ESCD) (F. Du 2014) can be applied to calculate the prospective RMS value according to the variant arc resistance at each time point.

Based on the formula (5), the formula below can be conducted.

$$I(t) - I(0) = -tg\varphi/w \cdot (i(t) - i(0))$$
$$+ \frac{I_{peak}}{w} \frac{\sin(wt)\cos\theta}{\cos\varphi} + \frac{I_{peak}}{w} \frac{(\cos(wt)-1)\sin\theta}{\cos\varphi} \quad (6)$$

where $\exists\ I(t) = \int i(t)\,dt$

By defining the variables (7) below, the formula (8) can be

$$R(t) = I(t) - I(0)$$
$$A(t) = \frac{i(t) - i(0)}{w}$$
$$B(t) = \frac{\sin(wt)}{w}$$
$$C(t) = \frac{\cos(wt) - 1}{w} \quad (7)$$
$$\gamma = tg\varphi$$
$$P = \frac{I_{peak} \cdot \cos\theta}{\cos\varphi}$$
$$Q = \frac{I_{peak} \cdot \sin\theta}{\cos\varphi}$$
$$R(t) = -\gamma \cdot A(t) + P \cdot B(t) + Q \cdot C(t) \quad (8)$$

used as the regression equation to calculate the γ, P, Q, which can be used to forecast the I_{RMS} in the time domain according to formula (9).

$$I_{RMS} = \sqrt{\frac{P^2 + Q^2}{2 + 2\gamma^2}} \quad (9)$$

By considering the formula (4) and (9), the R_{arc} at each time point can be calculated.

2.4 Data processing

Both the original calculated R_{arc} and R_{test} have a lot of noise due to the uncertainty of the original signal. To facilitate the matching process of these two calculated variables, the moving average method is applied to the calculated results from (1) and (4) respectively.

Supposing a data array X: X(1), X(2), X(3), …, X(N), and the processed data array Y: Y(1), Y(2), Y(3), …, Y(N), the moving average method complies to the formula below,

$$Y(i) = \sum_{j=i-m}^{i-1} \alpha_j Y(j) + \sum_{j=i-k}^{i} \beta_j X(j) \quad (10)$$

where α_j, β_j are the coefficients satisfying

$$\sum_{j=i-m}^{i-1} \alpha_j + \sum_{j=i-k}^{i} \beta_j = 1 \quad (11)$$

and m, k are the average numbers.

3 VERIFICATION

3.1 Test condition

The experiment was conducted per to IEC/TR 61641:2008-1 (Enclosed low-voltage switchgear and control gear assemblies-Guide for testing under conditions of arcing due to internal fault). The test parameters and the test setup are shown in Table 1 and Figure 3. There are 11 test cases with original current and voltage data in phase A, B and C.

The current sensor and voltage sensor are applied to capture the current and voltage information on each phase. One example of the arc voltage and current curves is shown in Figure 4.

Table 1. Test parameters.

Test voltage	462 V	725 V
Prospective short circuit current	50 KA	50 KA
Test duration	100 ms	100 ms
Ignition cross area	0.5~1.5 mm	1.5 mm

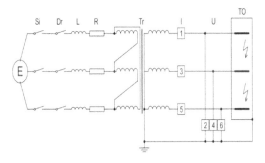

Figure 3. Test setup.

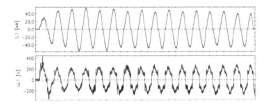

Figure 4. Test current and voltage data.

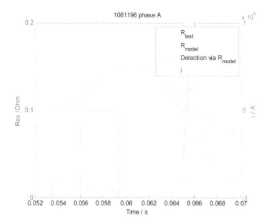

Figure 5. Verification curve.

Table 2. Summary results.

Test case	Phase	Individual phase detection time (ms)	Arc detection time (ms)
1081196	A	2.76	1.64
1081196	B	3.64	
1081196	C	1.64	
1081199	A	4.04	3.38
1081199	B	3.38	
1081199	C	9.52	
1081200	A	3.02	1.34
1081200	B	3.62	
1081200	C	1.34	
1081201	A	/	1.42
1081201	B	9.9	
1081201	C	1.42	
1081202	A	8.1	1.34
1081202	B	10.86	
1081202	C	1.34	
1081203	A	2.88	2.88
1081203	B	3.22	
1081203	C	/	
1081204	A	/	4.2
1081204	B	4.2	
1081204	C	9.02	
1081213	A	7.12	1.36
1081213	B	1.36	
1081213	C	11.98	
1081214	A	7.32	1.32
1081214	B	1.32	
1081214	C	11.7	
1081215	A	8.48	1.28
1081215	B	1.28	
1081215	C	2.48	
1081216	A	9.92	2.14
1081216	B	5.08	
1081216	C	2.14	

3.2 *Verification results*

With proposed arc resistance detection solution and data processing method, the exemplary result is shown in Figure 5. The black line is the sensed current curve, and the green line shows calculated instantaneous resistance via sensed current and voltage. And the red line is the forecasted arc resistance based on arc model. The blue line shows the detection results by comparing the matching level with a tolerance of 20%, where the high level means that the two calculated resistance values have a difference within 20% at least in a continuous time of 0.1 ms.

The Table 2 shows the summary results of the 11 test cases.

The number in individual phase detection time shows the detection time after the arc occurs in the said phase. The case with '\' mark means that the arc fault case can't be detected by the solution. Per to the above results considering 3 phases of A, B and C, all the tested cases can be detected within 5 ms.

4 CONCLUSION

In this paper, a fast arc detection method based on the verified empirical arc model has been proposed. The early short circuit detection method is applied to forecast the RMS value which is required in the empirical arc model. By comparing the forecasted arc resistance and the calculated instantaneous arc resistance, the occurrence of the arc can be checked. Per to the verification with practical arc test data, the proposed solution can detect the arc fault in its initial stage, e.g., within 5 ms.

The future research work will focus on more test and verification under different test condition, e.g., different arc fault current, different voltage class, and different gap distance.

REFERENCES

1. F. Du, W.G. Chen, Y. Zhuo and M. Anheuser. "A new method of early short circuit detection", IEEE Power & Energy Society (APPEEC2014), Shanghai, China, 2014.
2. H.L. Floyd and B.C. Johnson. "A global approach to managing risks of arc flash hazards," in PCIC Europe (PCIC EUROPE), 2013 Conference Record, 2013.
3. J. Vico, P. Parikh, D. Allcock, and R. Luna. "A novel approach for Arc-Flash detection and mitigation: At the speed of light and sound," in Petroleum and Chemical Industry Technical Conference (PCIC), 2012 Record of Conference Papers Industry Applications Society 59th Annual IEEE, 2012.

4. P.E.J. Phillips and M. Frain. "A European view of arc flash hazards and electrical safety," in Electrical Safety Workshop (ESW), 2012 IEEE IAS, 2012.

5. R.F. Ammerman, T. Gammon, P.K. Sen, and J.P. Nelson. "Comparative study of arc modeling and arc flash incident energy exposures," in Petroleum and Chemical Industry Technical Conference, 2008. PCIC 2008. 55th IEEE, 2008.

6. R.M. Harris, X. Hu, M.D. Judd, and P.J. Moore. "Detection and location of arcing faults in distribution networks using a non-contact approach," in Power Modulator and High Voltage Conference (IPMHVC), 2012 IEEE International, 2012.

7. R. Wilkins. "Simple Improved Equations for Arc Flash Hazard Analysis," posted by Mike Lang on August 30, 2004 to the IEEE Electrical Safety Forum, https://www.ieeecommunities.prg/ieee.esafety.

8. R. Wilkins, M. Allison, and M. Lang. "Calculating Hazards," IEEE Industry Applications Magazine, May/June 2005.

Power and Energy – Kong (Ed.)
© 2015 Taylor & Francis Group, London, ISBN 978-1-138-02782-4

Characteristic spectrum analysis on joint detection of PD in switchgear

Z.J. Jia, F. Liu & D.G. Gan
State Grid Sichuan Electric Power Research Institute, Chengdu, P.R. China

Y. Liu
State Grid Electric Power Research Institute, Wuhan, P.R. China

B. Zhang
Zhejiang Ningbo Electric Power Company, Ningbo, P.R. China

ABSTRACT: With the deficiency of discharge spectrum analysis and combined use of means of PD detection for switchgear, this paper has set up four kinds of PD model of switchgear according to four kinds of typical PD (corona discharge, surface discharge, internal discharge and floating potential discharge). Moreover, based on homemade PD detection system of switchgear, methods including TEV, AE and UHF are used for feature extraction of typical PD model and comparative analysis of discharge spectrum. Furthermore, multi-dimension maintenance strategy based on PD detection of switchgear is proposed.

Keywords: Switchgear cabinet; partial discharge; characteristic spectrum; phase; live detection

1 INTRODUCTION

HV switchgear cabinet is one of the important components of grid, and its operation stability directly influences grid's safe operation. Insulation aging and insulation fault of switchgear are mainly induced by Partial Discharge (PD). Moreover, test cycle of condition-based maintenance of switchgear cabinet is four years, so it has great significance to adopt proper methods of PD live detection and online monitoring during practical operation of switchgear.

The detection methods of partial discharge in switchgear cabinet mainly include ultrasonic (AE), TEV and UHF, and so on. At present, portable devices with AE sensor can detect partial discharge of switchgear at abroad, but it's lack of study on transmission characteristics of ultrasonic wave induced by PD in switchgear. Therefore, AE detection still depends on field experience and no theory now. PD detection device based on principle of TEV mainly includes production by EA and HVPD company. Some electric power companies at home have applied these products to form the operation mode of regular live detection. However, discharge characteristic spectrum of switchgear can't be displayed by humane interface. Phase characteristic and discharge time of discharge signal can't be obtained by measurement. As technology advances, the method of VHF/UHF antenna with high-speed acquisition device

is frequently used to measure electromagnetic wave excited by partial discharge at home and abroad, and different kinds of signal processing methods are also researched. Research materials on UHF method used in switchgear are rare at abroad, and only a small amount of companies have UHF live detection/online monitoring system.

With the deficiency of discharge spectrum analysis and combined use of means of PD detection for switchgear, this paper has set up four kinds of PD model of switchgear according to four kinds of typical PD (corona discharge, surface discharge, internal discharge and floating potential discharge). Moreover, based on homemade PD detection system of switchgear, methods including TEV, AE and UHF are used for feature extraction of typical PD model and comparative analysis of discharge spectrum. Furthermore, multi-dimension maintenance strategy based on PD detection of switchgear is proposed.

2 ESTABLISHMENT OF PD MODEL

Partial discharge in switchgear usually occurs in the dielectric with low breakdown strength. The occurrence of partial discharge depends on electric field distribution among internal insulation dielectric and electrical performance of dielectric. According to the position and mechanism of partial discharge in switchgear cabinet, partial discharge can be divided into four types including corona discharge, internal discharge, floating electrode discharge and surface discharge. According to these four typical PD types, four kinds of PD models in switchgear are set up.

2.1 *Model of needle-plate electrode*

According to the characteristic of corona discharge, the model simulates corona discharge induced by spring leaf loosening or manufacturing technology in switchgear, as is shown in Figure 1. The curvature radius of the tip of needle electrode is 10 μm, and distance between needle and plate electrode is 5 mm. The amplitude of applied voltage is 5–12 kV, and the frequency of it is 50 Hz.

2.2 *Model of internal defect*

The model simulates internal defect induced by dielectric defects in switchgear, as is shown in Figure 2. The round pass defect with the diameter of 5 mm and the height of 2 mm is surrounded by dielectric made of XLPE. The dielectric with the thickness of 6 mm and the radius of 40 mm is fixed between two plate electrodes. The amplitude of applied voltage is 5–10 kV, and the frequency of it is 50 Hz.

2.3 *Model of floating electrode*

The model of floating electrode is shown in Figure 3. Epoxy resin wafer with the thickness of 2 mm is placed between high voltage electrode and

Figure 1.   Model of needle-plate electrode.

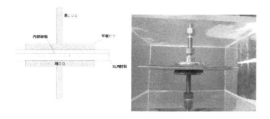

Figure 2.   Model of internal defect.

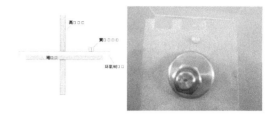

Figure 3.   Model of floating electrode.

Figure 4.   Model of surface discharge.

plate ground electrode, and brass cylinder is used as floating electrode. The amplitude of applied voltage is 5–12 kV, and the frequency of it is 50 Hz.

2.4 *Model of surface discharge*

The model simulates surface discharge along insulator or bushing in internal switchgear, which is one of the common discharge types in switchgear. As is shown in Figure 4, epoxy resin with the thickness of 2 mm is selected for surface discharge dielectric. It is made into a piece of wafer with the radius of 40 mm. The cylindrical copper electrode with the radius of 3 mm is placed at the center of epoxy resin dielectric as high voltage electrode. Epoxy resin wafer is fixed on the plate aluminum electrode and has good grounding. The amplitude of applied voltage is 5–10 kV, and the frequency of it is 50 Hz.

3   PD DETECTION OF TYPICAL DEFECTS OF SWITCHGEAR CABINET

Test and analysis are conducted for defect model of needle-plate electrode, internal defect, floating electrode and surface discharge.

3.1   *PD measurement of needle-plate electrode defect*

Fix position of TEV probe, gradually increases the test voltage U (6.0~12.0 kV), and measures discharge amplitude Um of needle-plate electrode defect model. The partial discharge severity can be reflected by discharge amplitude.

The partial discharge detection system can set up trigger threshold Y ($Y = 0$, 1%, 2%, 3%,..) according to practical situation, and for every detected signal, it indicates the existence of partial discharge by digital signal with a fixed amplitude.

First place TEV sensor, second ultrasonic sensor, and third UHF sensor. Adjust trigger threshold of PD detection device, the phase statistic of discharge signal in switchgear can be obtained and is shown in Figure 5. For TEV detection, partial discharge mainly occurs near peak of positive and negative half-cycle. In two discharge area, PD amplitude of positive half-cycle is roughly the same as that of negative half-cycle, and discharge frequency of positive half-cycle is much lower than that of negative half-cycle. For AE detection, partial discharge mainly occurs near the peak of positive half-cycle, and partial discharge does not exist in negative half-cycle. For UHF detection, partial discharge mainly occurs near peak of positive and negative half-cycle. In two discharge area, PD amplitude of positive half-cycle is slightly higher than that of negative half-cycle, and discharge frequency of positive half-cycle is equal to that of negative half-cycle.

3.2 PD measurement of internal defect

With regard to defect model of internal defect, test is conducted according to PD experiment process

of needle-plate electrodes defect. The phase statistic of discharge signal in switchgear is shown in Figure 6. For TEV detection, partial discharge mainly occurs near peak of positive and negative half-cycle. In two discharge area, PD amplitude of positive half-cycle is slightly higher than that of negative half-cycle, and discharge frequency of positive half-cycle is much higher than that of negative half-cycle. For AE detection, discharge of positive half-cycle centers between 30° and 120°, and discharge of negative half-cycle centers between 220° and 300°. In two discharge area, PD amplitude of positive half-cycle is equal to that of negative half-cycle, and discharge frequency of positive half-cycle is slightly lower than that of negative half-cycle. For UHF detection, discharge of positive half-cycle centers between 45° and 135°, and discharge of negative half-cycle centers between 225° and 315°. In two discharge area, PD amplitude of positive half-cycle is slightly higher than that of negative half-cycle, and discharge frequency of positive half-cycle is slightly higher than that of negative half-cycle.

3.3 PD measurement of floating electrode defect

With regard to defect model of floating electrode, test is conducted according to PD experiment process of needle-plate electrodes defect.

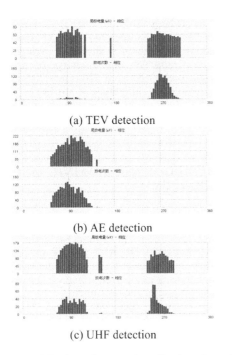

(a) TEV detection

(b) AE detection

(c) UHF detection

Figure 5. PD phase diagram of needle-plate electrode defect.

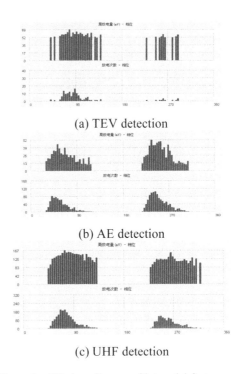

(a) TEV detection

(b) AE detection

(c) UHF detection

Figure 6. PD phase diagram of internal defect.

The phase statistic of discharge signal in switchgear is shown in Figure 7. For TEV detection, partial discharge mainly occurs near peak of positive and negative half-cycle. In two discharge area, PD amplitude of positive half-cycle is roughly the same as that of negative half-cycle, and discharge frequency of positive half-cycle is equal to that of negative half-cycle. Discharge of positive half-cycle centers between 40° and 90°, and discharge of negative half-cycle centers between 220° and 270°. For AE detection, partial discharge mainly occurs near peak of positive and negative half-cycle. In two discharge area, PD amplitude of positive half-cycle is slightly higher than that of negative half-cycle, and discharge frequency of positive half-cycle is higher than that of negative half-cycle. Discharge of positive half-cycle centers between 60° and 140°, and discharge of negative half-cycle centers between 235° and 315°. For UHF detection, partial discharge mainly occurs between 90° and 180°, and partial discharge does not exist in negative half-cycle.

3.4 PD measurement of surface discharge defect

With regard to defect model of surface discharge, test is conducted according to PD experiment

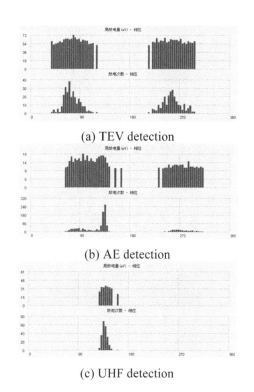

(a) TEV detection

(b) AE detection

(c) UHF detection

Figure 7.　PD phase diagram of floating electrode defect.

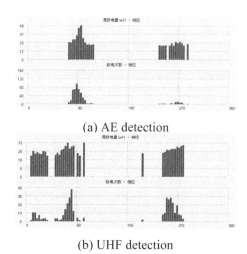

(a) AE detection

(b) UHF detection

Figure 8.　PD phase diagram of surface discharge defect.

process of needle-plate electrodes defect. The phase statistic of discharge signal in switchgear is shown in Figure 8. For AE detection, partial discharge mainly occurs near peak of positive and negative half-cycle. In two discharge area, PD amplitude of positive half-cycle is slightly higher than that of negative half-cycle, and discharge frequency of positive half-cycle is much higher than that of negative half-cycle. Discharge of positive half-cycle centers between 60° and 120°, and discharge of negative half-cycle centers between 240° and 300°. For UHF detection, discharge of positive half-cycle centers between 0° and 90°, and discharge of negative half-cycle centers between 220° and 270°. In two discharge area, PD amplitude of positive half-cycle is roughly the same as that of negative half-cycle, and discharge frequency of positive half-cycle is slightly higher than that of negative half-cycle.

4　MULTI-DIMENSION MAINTENANCE STRATEGY

According to PD generation principle and insulation defects which may exist in switchgear during practical operation, defect model is designed for test. Methods of UHF, AE and TEV are comprehensively applied in test, and characteristic spectrum analysis is carried out which can be used as application gist for the instrument.

It's effective to apply methods of AE, TEV and UHF to joint detection of PD in switchgear cabinet. Test efficiency can be promoted by maintenance strategy of general test first and diagnostic test second. First, methods of AE and TEV are applied to general detection. Second, the existence of partial discharge can be judged by the following criteria. ① If amplitude of TEV is relatively small and the ultrasonic signal is normal, PD signal does not exist in switchgear cabinet. ② If amplitude of TEV is relatively large and ultrasonic signal is normal, test cycle should be shortened. ③ If amplitude of TEV is normal but ultrasonic signal is abnormal, analysis should be carried out whether components such as plates vibrate. If judgment cannot be obtained, then test cycle should be shortened. ④ If both amplitude of TEV and ultrasonic signal are abnormal, partial discharge may exist in switchgear cabinet. Retest should be carried out by joint detection of methods of AE, TEV and UHF. If retest results are unanimous, online monitoring device should be fixed. Accurate judgment is carried out according to measuring results. In addition, according to the common problems, communication with switchgear cabinet producer and discussion about improvement scheme is necessary.

5 CONCLUSION

According to four kinds of typical insulation defects of switchgear, phase statistical spectrum of PD signal amplitude and discharge time detected by methods of TEV, AE and UHF has obvious feature, which can be used as qualitative distinguish basis of PD in switchgear cabinet.

It's effective to apply methods of AE, TEV and UHF to joint detection of PD in switchgear cabinet.

First, methods of AE and TEV are applied to detection. If amplitude of TEV is relatively large and amplitude of AE is normal, or amplitudes of TEV and AE are both abnormal, multi-channel synchronous acquisition system should be adopted immediately. Three kinds of methods should be applied to live detection and online monitoring, and accurate judgment is carried out according to measuring results.

REFERENCES

1. Ren, M. & Peng, H. & Chen, X. 2010. Comprehensive detection of partial discharge in switchgear using TEV. *High Voltage Engineering* 36(10): 2460–2466.
2. Chen, G. 2010. Application of sound electric wave detector on PD detection of 10 kV switchgear. *Electric Engineering* 7: 67–68.
3. Wang, J. & Zhang, T. & Li, G. 2011. On-line detection of 10 kV switchgear by partial discharge technology. *Insulating Materials* 44(6): 60–64.
4. Wang, Z. & Liu, E. & Zhang, R. 2011. Artificial neural network-based ultrasonic detection of partial discharge of switchboard. *East China Electric Power* 39(3): 498–500.
5. Wang, K. 2013. The performance evaluation of TEV PD testing equipments for high voltage switchgear with analog signal injection method. *Southern Power System Technology* 7(4): 43–46.
6. Zeng, X. & Jiang, J. & Hou, J. 2012. Application of TEV and UHF in PD detection for the live 10 kV switchgear. *High Voltage Apparatus* 48(1): 41–46.

Power and Energy – Kong (Ed.)
© *2015 Taylor & Francis Group, London, ISBN 978-1-138-02782-4*

Implementing network topology by multiplication of vector and matrix

Yubin Yao, Zhiliang Wu & Dan Wang
School of Marine Engineering, Dalian Maritime University, Dalian, China

ABSTRACT: The traditional matrix method for network topology is very slow since it needs to achieve full connectivity matrix by matrix multiplication repeatedly. Nodes in a connective set have same rows in the full connectivity matrix, the connective set can be determined by the first row of these same rows, and the first row is obtained by itself multiplies adjacency matrix repeatedly. If a new connective set is determined after all the nodes in the preceding connective set are found rather than all connective sets are computing in parallel, the computation of other rows except the first row in each connective set can be avoidable. Based on the above consideration, a network topology method by multiplication of vector and matrix is presented. A practical network is analyzed by the proposed method, and the results prove the effectiveness of the proposed method.

1 INTRODUCTION

To provide the bus/branch model for many power system studies such as state estimation, power flow and contingency analysis, the power system network topology algorithm plays a very important role in the monitoring and control of power systems. The traditional network topology algorithms are search method and matrix method both of which are popular methods to find connective graph in graph theory. Besides these two basic methods, Yao et al. (2009) presented a new network topology method by solving logic equations.

The matrix method implements the network topology by multiplication operation of the adjacency matrix which represents the relationship of the nodes. The matrix method requires large computer storage and takes considerable running time which prevents it from practical usage.

The matrix method computes the $n-1$ power to adjacency matrix to get full connectivity matrix which contains complete connectivity information of a network. The full connectivity matrix can be obtained by multiplying the adjacency matrix $n-2$ times for systems with n nodes at the worst, which is very time-consuming. The number of matrix multiplication decreases to maximal $\log_2(n-1)$ times by squaring the connectivity matrix successively. In fact the number of matrix multiplication can be reduced if the computation is ceased as soon as full connectivity matrix is achieved (Goderya et al. 1980). If the element of the connectivity matrix is updated and used immediately in calculation for next element, the full connectivity matrix is achieved by only two times of matrix multiplication (Yao et al. 2011), or with only one time of matrix multiplication a connectivity matrix with sufficient connectivity information is obtained, network topology is then acquired by reverse row sweep (Yao 2012). Both the above two methods increase the speed of network topology greatly. The matrix method implements the connectivity network by multiplication of matrices, the adjacency matrix as one multiplier is sparse, and the sparse matrix techniques can apply to this multiplier, the sparse matrix techniques speed up the multiplication greatly (Yao et al. 2011).

The nodes in the same connective set have identical rows in the full connectivity matrix, so the first row of these same rows is enough to find the connective set and computation of other rows whose connective set is ascertained by the preceding rows is unnecessary. (Yao et al. 2014) suggested that immediately ceasing the calculation of the row of the connectivity matrix whose related node is determined in an existing connective set, and presented a new network topology method by matrix partial multiplication.

The matrix partial multiplication method computes all connective sets in parallel, which is to search all connected graph simultaneously. In the procedure when a row of the connectivity matrix is uncertain whether it belongs to an existing connective set or is the first row of a new connective set, this row must be computed, such many rows of connectivity matrix in a connective set must be computed. If we compute new connective set after all the nodes in the preceding connective set are found rather than all connective sets are computing in parallel, the computation of other rows except the first row in each connective set can be avoidable. Accordingly a new network topology method by multiplication of vector and matrix is presented. The presented method is very fast since it computes only one row of the connectivity matrix for each connective set, and is memory-saving since storage of the entire connectivity matrix is unnecessary. The row of the connectivity matrix being computed is initialized by sparse-stored adjacency matrix. A practical network with 7097 physical nodes is analyzed by the proposed method, and the results prove the effectiveness of the proposed method. For this network, the running time by the proposed method is 0.00486 s.

2 NETWORK TOPOLOGY BY MATRIX METHOD

2.1 Network topology and connectivity matrix

The purpose of network topology is to establish the bus/branch model for other network analysis such as power flow and state estimation. It includes two main sections: substation configuration which forms buses from nodes according to switches' state; network configuration which forms islands from buses according to connection of the buses via live branches.

Although the purpose and the object of substation configuration and network configuration are different, they share same method. Because the procedure of substation configuration is only carried out at the substation voltage levels and the number of the nodes in each substation voltage level is small, tracing in a substation voltage level is very fast. But searching scope for network configuration is entire network. Searching in sorting data can greatly reduce searching scope and considerably save computing time (Yao et al. 2005).

The connection of nodes in a network can be represented by adjacency matrix. For a network with n nodes, the adjacency matrix $\mathbf{A}$ is an $n \times n$ square matrix whose element a_{ij} is 1 if node i and node j connect directly and it is 0 otherwise. The adjacency matrix represents the first level of the network connectivity.

If the adjacency matrix is multiplied by itself, the resulting matrix $\mathbf{T}$ is connectivity matrix which represents the second level of the network connectivity. $\mathbf{T}$ indicates connection of the nodes directly or indirectly. We can obtain at best the $(n-1)$th level of the network connectivity for a network with n nodes, which is called as full connectivity matrix.

2.2 The matrix method

The matrix method is described as follows:

$$\mathbf{T}^{(k+1)} = \mathbf{T}^{(k)} \cdot \mathbf{A} \tag{1}$$

where $\mathbf{A}$ = adjacency matrix; $\mathbf{T}$ = connectivity matrix; the superscript (k) denotes the kth level connectivity matrix.

The adjacency matrix represents the first level of the network connectivity, so $\mathbf{T}^{(1)} = \mathbf{A}$.

Equation (1) is repeated until the adjacency matrix to the power $(n-1)$. By squaring the connectivity matrix, we can reach the $(n-1)$th power of the adjacency matrix rapidly, so the following equation is also used to obtain the full connectivity matrix:

$$\mathbf{T}^{(2k)} = \mathbf{T}^{(k)} \cdot \mathbf{T}^{(k)} \tag{2}$$

At the worst we need multiply $n-2$ times to get the full connectivity matrix with (1) or $\log_2(n-1)$ times with (2). In fact it does not need so much times of matrix multiplication, since when the successive connectivity matrices are same, i.e. the elements of the connectivity matrix do not change any more, the full connectivity matrix is reached.

The connectivity matrix resulting from multiplication of the matrices is dense one, and the density increases rapidly along with the number of matrix multiplication. New connectivity matrix is obtained by (1) or (2). The multipliers are both dense matrices in (2), and one of the two multipliers is dense matrix in (1), such in general the sparse matrix techniques are not applied to the matrix method.

2.3 The sparse matrix method

Multiplication of two dense matrices is very time-consuming; therefore the running time of the matrix method is very long and unbearable. In (1) the connectivity matrix as one multiplier is dense, but the adjacency matrix as the other multiplier is sparse and unchanged, the sparse matrix techniques can apply to this multiplier.

In (1), when the adjacency matrix is stored in sparse form, we can apply the sparse matrix techniques to matrix method.

For sparse storage of the adjacency matrix, we can use the following two arrays:

AC is the column number of each non-zero element.
AR is the location in array AC for the first non-zero element in each row of the adjacency matrix.

When the connectivity matrix is calculated by (1), and considering symmetry of the adjacency matrix, the element of the connectivity matrix is as follows:

$$t_{ij} = \sum_{m=1}^{n} t_{im} a_{jm} \tag{3}$$

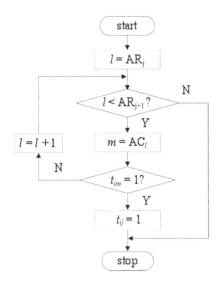

Figure 1. Flow chart of forming elements of the connectivity matrix.

where t_{ij} = element of connectivity matrix; a_{jm} = element of adjacency matrix.

The elements of the connectivity matrix are stored with an $n \times n$ array, and the adjacency matrix is stored in sparse storage.

Since the adjacency matrix is sparse one, lookup all elements in a given row of the connectivity matrix is unnecessary, we only need pay attention to a few t_{im} whose column number is same as that of a_{jm}. If any of these t_{im} is 1, and then new t_{ij} will be 1 and further computation is needless. The flow chart of the computation of the element t_{ij} of the connectivity matrix is shown in Figure 1.

3 MATRIX METHOD BY MULTIPLICATION OF VECTOR AND MATRIX

3.1 *Principle of the presented method*

The matrix method which gets full connectivity matrix by multiplication of the matrices is very time-consuming.

From (3), it needs only the old elements in the ith row of the connectivity matrix to calculate new elements in that row, the other rows have nothing to do with formation of the elements in the ith row. So the elements of every row of the connectivity matrix can be calculated independently.

The rows of the full connectivity matrix belong to the same connective set are identical, so we can find a connective set by just sweeping its first row rather than all the rows. Computation of the rows other than the first row in a connective set is unnecessary. If all connective sets are computed

in parallel it is difficult to tell whether some row belongs to an existing connective set or is the first row of a new connective set.

But if we compute the connective set one by one, begin computation of a new connective set after all the nodes in the preceding connective set are found, the first row in current connective set is certain, and the computation of other rows can be avoidable.

3.2 *Procedure of the presented method*

The flow chart of the presented method is showed as Figure 2.

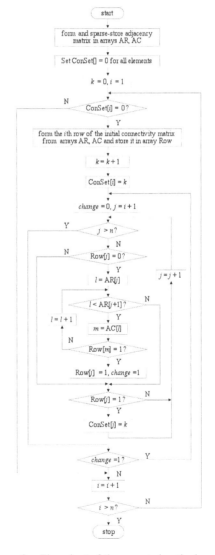

Figure 2. Flow chart of the presented method.

67

To get a new element of the connectivity matrix, we want know if the element can change to 1, so we just calculate the element whose old value is 0. The element whose old value is 1 already need not be calculated.

When the ith row of the connectivity matrix need be computed, the ith row of the initial connectivity matrix is formed from arrays AR and AC, and stored in array row.

During elements in one row of the connectivity matrix are being formed, only the elements right to the principal diagonal need be calculated, that is we just need determinate whether one node connects to the nodes whose numbers are big, the connection with the nodes whose numbers are small have be justified when the preceding rows is calculated. So only the elements right to the principal diagonal need be calculated and updated immediately.

In Figure 2, *change* recodes the variation in array row. If the elements of row do not change, set *change* = 0, which means the computation of current connective set is finish and computation of next connective set begins.

Array ConSet records the number of the connective set each node belongs to. ConSet[i] = k ($k \neq 0$) means that node i belongs to the kth connective set, the row i does not need calculating; ConSet[i] = 0 means that node i does not belong to any connective set yet, a new connective set which includes node i will be created.

4 CASE STUDY

A real network named Hangzhou network in China is used to testify our conclusions. There are 187 stations, 715 busbars, 7329 switches, 318 transmission lines, 127 two-winding transformers, 123 three-winding transformers, 11 series reactors, 232 shunt capacitors, and 27 shunt reactors in the network. There are total 7097 nodes, and 825 branches including transmission lines, series reactors and transformers (3 branches for a three-winding transformer).

The study is fulfilled on a personal computer with an Intel Pentium processor of 2.61 GHz. The procedure of substation configuration is limited at the substation voltage levels, while searching scope for network configuration is entire network. 957 buses and 49 islands are found. Only one island of these islands is live. There are 704 buses in the live island, and 122 buses in dead islands, the other buses are isolated ones.

4.1 *Comparison of running time for different matrix methods*

The network topology on this network was carried out by five different matrix methods:

Method 1: The full connectivity matrix is determined by multiplying the adjacency matrix repeatedly.

Method 2: The full connectivity matrix is determined by squaring the connectivity matrix repeatedly.

Method 3: The full connectivity matrix is determined by multiplying the adjacency matrix repeatedly, and sparse matrix techniques are applied.

Method 4: Method by matrix partial multiplication.

Method 5: The presented method.

The running time required by different matrix methods is listed in Table 1.

As showed in Table 1, the presented method is greatly fast than the method implements the full connectivity matrix by multiplying the adjacency matrix repeatedly or the method implements the full connectivity matrix by squaring the connectivity matrix repeatedly. It is also fast than the method with sparse matrix techniques and method by matrix partial multiplication.

For all methods except the presented one, most of the running time is consumed in network configuration because the matrices in network configuration are large while those in substation configuration are small.

4.2 *Comparison of running time for methods of different principle*

Two methods of different principle were carried out for this network to compare with the presented method. These methods differ one another by their principles, one is search method, another is method by solving equations, and the other is matrix method. The running time required by different methods is listed in Table 2.

Table 1. Running time required by different matrix methods.

| Methods | Running time (ms) | | |
	Substation configuration	Network configuration	Total running time
Method 1	191.32	23804.16	23995.48
Method 2	90.43	4140.90	4231.33
Method 3	9.69	34.44	44.13
Method 4	6.10	8.96	15.06
Method 5	4.32	0.54	4.86

Table 2. Running time required by different methods.

Methods	Running time (ms)
Search method	4.34
Method by solving equations	7.61
The presented method	4.86

As showed in Table 2, the presented method is much fast than the method by solving equations, and as fast as the search method.

5 CONCLUSION

The nodes belong to same connective set have identical rows in full connectivity matrix; the connective set can be found by sweep its first row in the full connectivity matrix, calculation of other rows belong to the same connective set is unnecessary. But when all connective sets are computing in parallel, it is difficult to judge whether a given row need be computed or not, many rows in a connective set are computed.

If a new connective set is determined after all the nodes in the preceding connective set are found rather than all connective sets are computing in parallel, the computation of other rows except the first row in each connective set can be avoidable.

Compute only one row of the connectivity matrix for one connective set. In this way the presented matrix method turns multiplication of matrices into multiplication of vector and matrix, and the computer time reduces greatly.

ACKNOWLEDGMENT

This work was supported in part by NSFC under Grant 61273137 to Dan Wang.

REFERENCES

1. F. Goderya, A. A. Metwally & O. Mansour. 1980. Fast detection and identification of islands in power networks. *IEEE Trans. on Power Apparatus and Systems* 99(1): 217–221.
2. Yubin Yao, Wenzhuan Jin & Li Jin. 2005. Fast network topology method for a distribution network. (in Chinese). *Relay* 33(19): 31–35.
3. Yubin Yao, Guoshun Zhou, Hong Li & Dan Wang. 2009. A network topology method by solving logic equations. *Proc. SUPERGEN 2009*, April, 2009.
4. Yubin Yao, Jian Xuan, Na Yu, Dan Wang & Zhiliang Wu. 2011. Determination of network topology by quasi-square of the connectivity matrix. (in Chinese). *Power System Protection and Control* 39(5): 31–34, 40.
5. Yubin Yao, Shuangli Ye, Zhiliang Wu & Dan Wang. 2011. Determination of network topology by the matrix method with sparse matrix techniques. (in Chinese). *Power System Protection and Control* 39(23): 1–5, 10.
6. Yubin Yao. 2012. Determination of network topology by fast quasi-square of the adjacency matrix. (in Chinese). *Power System Protection and Control* 40(6): 17–21, 29.
7. Yubin Yao, Dan Wang & Zhiliang Wu. 2014. Determination of network topology by matrix partial multiplication. *Proc. MEIC2014*, November, 2014.

Power and Energy – Kong (Ed.)
© 2015 Taylor & Francis Group, London, ISBN 978-1-138-02782-4

Wavelet analysis based PEMFC fault diagnosis

Xiao Jia & Nan Wang
College of Automotive Engineering, Tongji University, Shanghai, China

Su Zhou
College of Automotive Engineering and Sino-German Postgraduate School, Tongji University, Shanghai, China

ABSTRACT: A 3D numerical Proton Exchange Membrane Fuel Cell (PEMFC) stack model, which is able to reflect the performance otherness between the adjacent cells, is established to analyze the fuel cell characteristics. Through embedding faults into the model, anode and cathode starvation, taken as representative fault cases, are studied using User-Defined Function (UDF) in program FLUENT. A fault diagnosis method based on wavelet analysis, which only use the stack voltages to diagnose the stack states, is proposed to find some rules to judge the different fault types. This method, compared to the Total Harmonic Distortion Analysis (THDA) technique, is more flexible and less demanding in data and has considerable advantages of application to PEMFC stack fault diagnosis.

1 INTRODUCTION

Fuel cell electric vehicle is considered as the green car that settles the energy shortage and emission problem at the same time and will become the main trend of vehicle development [1–2]. Fuel cell is an electrochemical power generation device, which converts the chemical energy stored in the fuel and oxide into the electrical energy [8]. PEMFC takes the hydrogen and oxygen as reaction and produces water only which has some advantages such as high efficiency, no pollution, long lifespan and high power density [3–5].

Due to the complex structure of the fuel cell stack and the deterioration of the control process and operation environment, it is inevitable to malfunction, leading to negative results ranging from stack performance degradation to stack damage or even major security incidents [7]. Therefore, the fault diagnosis technology of the PEMFC is the key technical limitation restricting the development. The coming of the fault-diagnosis and fault-tolerant control technology provides a new method to improve the reliability of modern complex systems. To change and breach the traditional diagnosis method which is mainly based on the experience and skill, it is necessary to establish a practical, rational and complete theory system of fuel cell diagnosis. Ramschak E. et al. [6] firstly proposed a solution called the Total Harmonic Distortion Analysis (THDA) technique, providing a new way of thinking for the fuel cell stack fault diagnosis technology.

The fuel cell stack fault diagnosis method mentioned herein tries to take use of wavelet analysis theory, by detecting the voltage of the fuel cell stack to determine the fault type. This method reduces the peripheral detection circuit, and is easy and low-cost to operate as well as facilitate troubleshooting and improve the efficiency of fault diagnosis.

2 PEM-STACK MODELING AND SIMULATION

2.1 Introduction of three-dimensional model

In the pre-processing software of FLUENT, GAMBIT, a fuel cell stack model consisting of four cells is established, as shown in Figure 1.

The fuel cell stack is composed of the proton exchange membrane, the diffusion layer, the catalysis layer, the flow channel and the current collector. In this geometry model, there are two channels, including the anode channel and the cathode channel.

The specific dimension of each part is shown in Table 1.

2.2 The simulation condition settings

2.2.1 Inlet boundary conditions

Boundary conditions at the inlet include the mass flow rate of anode reaction and cathode reaction and the respective mass fraction of hydrogen, oxygen and water.

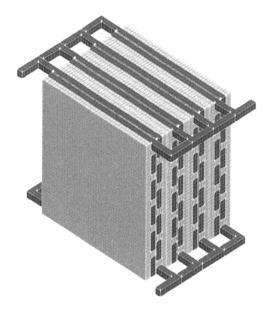

Figure 1.   3D PEM-stack model.

Table 1.   Model dimensions.

Components	Width (mm)	Length (mm)	Height (mm)
Cell	8.48	40	42
Membrane	0.05	40	42
Catalyst layer	0.015	40	42
GDL	0.2	40	42
Channel	2	40	2
Current collector	4	40	42
Stack	33.92	56	42

The inlet mass flow rate is calculated by

$$\dot{m} = \frac{\zeta \times M \times I_{ref} \times A_{mem}}{n \times F \times y^{in}} \tag{1}$$

where, ζ is the stoichiometric ratio, which is the ratio of the mass flow rate and the actual consumption, for example with air,

$$\zeta_{O_2} = \frac{\dot{m}_{O_2}^{in}}{\dot{m}_{O_2}^{op}} \tag{2}$$

where,

$$\dot{m}_{O_2}^{op} = \frac{M_{O_2}}{2F} \times I_{ref} \times A_{mem} \tag{3}$$

where, I_{ref} is the reference current density, F is the Faraday constant, A_{mem} is the area of the proton exchange membrane,

$$\dot{m}_{O_2}^{in} = \dot{m}_{ca}^{in} \times y_{O_2}^{in} \tag{4}$$

where, $y_{O_2}^{in}$ is the mass fraction of air.
Therefore, the mass flow rate of the anode and the cathode is separately as follows,
Cathode

$$\dot{m}_{ca}^{in} = \frac{\zeta_{O_2} \times M_{O_2} \times I_{ref} \times A_{mem}}{4F \times y_{O_2}^{in}} \tag{5}$$

Anode

$$\dot{m}_{an}^{in} = \frac{\zeta_{H_2} \times M_{H_2} \times I_{ref} \times A_{mem}}{2F \times y_{H_2}^{in}} \tag{6}$$

The mole fraction at inlet shall be converted from the mass fraction as follow,

$$x_i = \frac{M}{M_i} y_i \tag{7}$$

where, M is the average mole mass of all the species,

$$M = \sum x_i M_i \tag{8}$$

2.2.2   Outlet boundary conditions

For the parameters of P, T and Y, the second boundary condition are applied, that is $\partial P / \partial \mathbf{n} = 0$, $\partial T / \partial \mathbf{n} = 0$, where $\mathbf{n} = (n_1, n_2, n_3)$ is the unit vector of the external normal line, Y is the gas mass fraction.

2.2.3   Operating boundary conditions

In FLUENT simulation, there are a series of boundary conditions remained to be set. Some of

Table 2.   Operation conditions.

Reference current density (mA · cm⁻²)	500	Excess coefficient λ	3
Pressure bar	1.5	Anode humidity	15%
Temperature K	353	Cathode humidity	80%

Table 3.   Parameters in anode and cathode inlet.

	$\dot{m}_{in}$ (kg · s⁻¹)	x_{H_2}	x_{O_2}	x_{H_2O}
Cathode	3.59×10^{-4}	0	0.8488	0.1512
Anode	6.12×10^{-6}	0.7181	0	0.2819

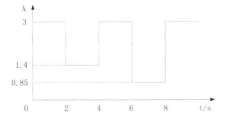

Figure 2. λ variation.

Table 4. Anode and cathode inlet flow rate.

λ	3	1.4	0.85
Cathode $\dot{m}_{in}$ (kg·s⁻¹)	3.59×10^{-4}	1.68×10^{-4}	1.02×10^{-4}
Anode $\dot{m}_{in}$ (kg·s⁻¹)	6.12×10^{-6}	2.86×10^{-6}	1.74×10^{-6}

Table 5. Simulation parameters.

Open-circuitvoltage (V)	0.95	Molar equivalent of PEM (kg·kmol⁻¹)	1100
Exchange coefficient	2	Conductivity of PEM (omh⁻¹·m⁻¹)	1×10^{-16}
Diffusivity of hydrogen (m²·s⁻¹)	1.10×10^{-4}	Porosity of catalyst layer	0.5
Diffusivity of air (m²·s⁻¹)	7.35×10^{-5}	Conductivity of catalyst layer (omh⁻¹·m⁻¹)	5000
Diffusivity of oxygen (m²·s⁻¹)	3.23×10^{-5}	Viscous resistance of catalyst layer (m⁻²)	2×10^{5}
Conductivity of collector (omh⁻¹·m⁻¹)	3.54×10^{7}	Anode concentration exponent	1
Porosity of diffusion layer	0.5	Cathode concentration exponent	1
Viscous resistance of diffusion layer (m⁻²)	5.68×10^{10}	Anode reference current density (A·m⁻²)	7500
Conductivity of diffusion layer (omh⁻¹·m⁻¹)	5000	Cathode reference current density (A·m⁻²)	20

the conditions shall be set according to the practical operation conditions and the others can be calculated from the setting conditions.

Referring to the actual operational conditions of the fuel cell stack, the external operating conditions of the simulation on the fuel cell stack model are set as shown in Table 2.

In addition, in the FLUENT simulation, a sinusoidal signal with small amplitude need to be superimposed on the current boundary conditions, here a sinusoidal signal with an amplitude of 20 mA·cm⁻² and a frequency of 1 Hz is superimposed on the current boundary condition, then the actual current boundary condition is set as [500+20 sin(2πt)] mA·cm⁻².

According to the above Equations (1)–(8) and external operating conditions, the inlet mass flow rate of anode and cathode can be calculated, as well as the mass fraction of each species, as shown in Table 3.

Here the $\dot{m}_{in}$ means the mass flow rate at inlet, the x_i means the mole fraction of the related species i.

For the anode and the cathode inlet fault conditions, set the starvation fault by adjusting the excess coefficient of anode and cathode, separately, as is shown in Figure 2.

According to the excess coefficient the relevant inlet flow rate can be calculated as shown in Table 4.

2.2.4 *Operating boundary conditions*
In FLUENT, the solver is pressure and transient, loading the add-on module Fuel Cells and Electrolysis-PEMFC, and the discrete scheme is SIMPLE algorithm using the flux calculation based on the first-order upwind method. Simulation parameters set in FLUENT are shown in Table 5.

However, in the process of the simulation, since the transient model need not be initialized, in order to make the simulation run from better initial values it is necessary to run the simulation in steady model to convergence and then based on the results the simulation is continued in transient model.

Taking use of the User-Defined Function in FLUENT, the inlet flow rate conditions are programed, compiled and loaded into FLUENT. Since the anode starvation fault and the cathode starvation fault are investigated in this paper, three fault conditions are set which include cathode starvation only, anode starvation only and both anode and cathode starvation.

3 WAVELET ANALYSIS AND FAULT DIAGNOSIS

There are different kinds of wavelets which can be used for wavelet analysis, so it is significant to select a right wavelet for the analysis. However, there is not a uniform standard for how to select a wavelet function. In general, the scale shall be selected according to the purpose of the signal processing, if the approximate overall signal characteristics are needed, a larger scale shall be chosen, while a smaller scale will reflect the detail of the signal.

Here three signals are decomposed into three orders. The processing results are shown in Figure 3.

Then the db wavelet is applied, which contains a series of wavelets, so each simulation signal is allowed to be decomposed into a number of sub-signals corresponding to the db component. Moreover, the db wavelet transform allows the signal to be processed to be decomposed into any order according to the actual situation. Generally, db3 wavelet transforms the signal into approximate triangle wave, and the signal transformed by the db4, db5 and db8 wavelet is more smooth and approximate to a sine wave.

From the results of the db decomposition above, there exist some differences between the faults in the components of ca3 and d3 during the fault period. And then the statistical information can be used to identify the fault type, as shown in Figure 4.

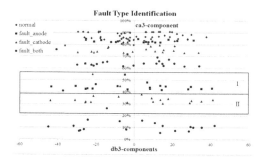

Figure 4. The statistical information of db wavelet transform.

It is obvious that the different fault types correspond to different areas and we can reach the preliminary conclusion of the fault diagnosis based on the statistical figure. The anode and cathode starvation can be distinguished according to the statistical figure (Figure 4), that is, by make a comparison between the wavelet transform results of the stack output signal with db wavelet functions and Figure 4, if most of the component indexes are located in zone I/II, it is possible to result from the anode/cathode starvation or partly at least.

As can be seen from the above conclusion, wavelet theory can be used in the field of fuel cell stack fault diagnosis. This paper analyzes and validates the feasibility of the application of wavelet theory to the simple faults of anode and cathode starvation. While a large amount of simulation, analysis and statistical work, including the starvation fault under different operating conditions and different coefficients has been done, other fault types like membrane drought and flooding etc. remain to be done to form a comprehensive and reliable diagnosis method.

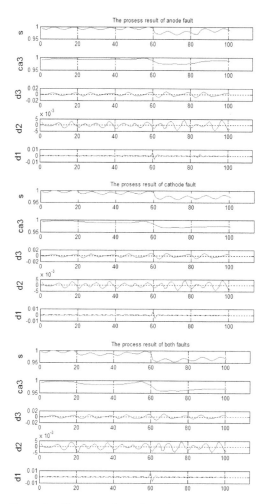

Figure 3. The comparison of wavelet decompositon.

4 COMPARATIVE ANALYSIS WITH THDA

In the THDA (the Total Harmonic Distortion Analysis) method, as shown in Figure 5, a small amplitude sinusoidal current signal is superposed continuously to the stack current during fuel cell operation. According to the U–I curve of the fuel cell stack, the fuel cell stack can be regarded as a linear system within the normal operation range, the responding voltage signal consists of the same frequency spectrum as the original superposed one, while in case of any critical cell operation, the distortion forms.

Derived from the power system, the concept of the Total Harmonic Distortion is usually expressed by the THD value, which is defined as the harmonic

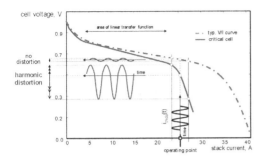

Figure 5. Distortion of a sinusoidal signal in a fuel cell [6].

root mean square, a percentage of the fundamental component, that is,

$$THD = \sqrt{\sum_{n=2}^{N} V_n^2} / V_1 \qquad (9)$$

where, V_1 is the effective value of the phrase voltage of the fundamental wave, V_n is the effective value of the phrase voltage of the N_{th}-degree harmonic, N is the highest harmonic considered.

4.1 *Fault identified*

In the THDA method, since response to a single frequency is not sufficient to determine the type of fault, it is appropriate to obtain comprehensive judgment under several other frequencies.

The selection of frequencies for the superimposed signal is significant according to the EIS (Electrochemical Impedance Spectroscopy) curves. Different reasons of critical fuel cell operations have obvious influences at different frequency ranges. According to the experiment results, the mass transfer relevant faults for both anode and cathode are related to the low frequency range, which are less than 40 Hz.

Based on the experiment results, two major frequencies are chosen, that is the 4 Hz and 10 Hz, for other simulations under the same fault conditions. The study adopts the input AC (Alternating Current) sinusoidal disturbances of 4 Hz (1st Harmonic: 8 Hz, 2nd Harmonic: 12 Hz) and 10 Hz (1st Harmonic: 20 Hz, 2nd Harmonic: 30 Hz) which would cover the low frequency range (less than 40 Hz) corresponding to the mass transfer faults, according to the frequencies range from the EIS curves.

Based on the existing simulation data and the responding voltage signal of the stack under the small amplitude sinusoidal current signal of 1 Hz, the THD values and the frequencies corresponding

to the harmonics can be obtained via the Fourier transform and the Equation (9). Then, the sinusoidal current signal of the frequencies corresponding to the harmonics is superposed on the stack current and the simulation is run under the same fault conditions. Finally, based on the simulation data of the stack responding voltage, all the THD values can be worked out which can be used as the diagnostic index to identified the fault types.

Next are the responding voltage signals of the fuel cell stack obtained with the small amplitude sinusoidal signals of 4 Hz and 10 Hz superposed separately on the stack current.

Based on all the three groups of the simulation data, the THD values can be calculated as shown in Table 6. For all the three frequencies, the THD of the cathode starvation is greater than that of the anode starvation, so the cathode starvation has a severe distortion than the anode starvation.

From Figure 6, it is easy to identify the fault type based on the THD values.

4.2 *Comparative analysis*

Generally speaking, both of the diagnostic methods are able to achieve the identification of the fuel cell faults, promising for the fuel cell diagnosis because of the feature of easy operation and low cost.

The THDA method is more mature and reliable. However, this method has a higher demand for the data. Instead, the wavelet analysis method is more flexible which has the potential of detecting the

Table 6. The THD values.

Frequency (Hz)	Cathode	Anode	Both
1	1.11%	0.88%	1.44%
4	1%	0.74%	1.25%
10	0.92%	0.79%	1.2%

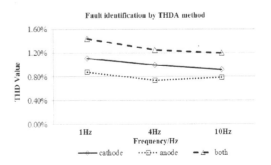

Figure 6. The THD values under all three frequencies.

status of the fuel cell stack with the stack responding voltage signal under general working conditions. It remains to be verified under other critical operation conditions, moreover, a systematic procedure for the fault diagnosis of fuel cell stack need to be developed based on both simulation and experiment under all the possible fault conditions.

5 CONCLUSION

This paper focuses on two aspects mainly, including the three-dimensional dynamic model of the fuel cell stack and fuel cell stack fault diagnosis method based on the wavelet analysis.

According to the dynamic model of the fuel cell stack and simulation results, the wavelet theory is used to explore a new method of the fuel cell stack fault diagnosis.

According to the wavelet theory, the fault signal is handled to extract characteristic quantities like the average value, standard deviation and the level three detail waveform etc. based on which some rules to judge the different fault types can be found.

The method of fuel cell stack fault diagnosis in this paper is based on wavelet analysis theory, which is more flexible and less demanding in data than the THDA method. At present, there has not appeared effective diagnostic method based on wavelet theory in the field of fuel cell stack fault diagnosis. Based on the method developed in this paper, further exploration for the application of wavelet theory on fuel cell stack diagnosis will be carried out.

REFERENCES

1. Chang P., Kim G.S., Promislow K., Wetton B. 2007. Reduced dimensional computational models of polymer electrolyte membrane fuel cell stacks. Journal of Computational Physics 223:797–821.
2. Kolodziej Jason R. 2007. Thermal Dynamic Modeling and Nonlinear Control of a Proton Exchange Membrane Fuel Cell Stack. Journal of Fuel Cell Science and Technology 4:255–260.
3. Le A.D., Zhou B. 2009. A 3D single-phase numerical model for a PEMFC stack. Proceedings of ASME 2009 Seventh International Fuel Cell Science, Engineering and Technology Conference Fuel Cell 2009, June 8–10, Newport Beach, California, USA.
4. Ly H., Birgersson E., Vynnycky M. 2010. Asymptotically Reduced Model for a Proton Exchange Membrane Fuel Cell Stack: Automated Model Generation and Verification. Journal of the Electrochemical Society 157(7): B982–B992.
5. Muller Eric A., Stefanopoulou Anna G. 2006. Analysis, Modeling and Validation for the Thermal Dynamics of a Polymer Electrolyte Membrane Fuel Cell System. Journal of Fuel Cell Science and Technology 3:99–110.
6. Ramschak E., Peinecke V., Prenninger P., Schaffer T., Hacker V. 2006. Detection of fuel cell critical status by stack voltage analysis. Journal of Power System 157:37–840.
7. Wu K., Ma T.C., Zhou Y., Shen X.Y., Zhou S. 2008. Cell Monitoring and Its application on a Fuel Cell Engine. Journal of Qingdao University (Engineering & Technology Edition) 23(4):20–25.
8. Yalcinoz T., Alam M.S. 2008. Dynamic modeling and simulation of air-breathing proton exchange membrane fuel cell. Journal of Power Sources 182: 168–174.

Power and Energy – Kong (Ed.)
© 2015 Taylor & Francis Group, London, ISBN 978-1-138-02782-4

Research on coarse-grained parallel algorithm of the Monte-Carlo simulation for probabilistic load flow calculation

Z.K. Wei & J.F. Xu
State Grid Jibei Electric Power Company Limited, Beijing, China

X.K. Dai, B.H. Zhang, S.L. Lu & W.S. Deng
State Key Laboratory of Advanced Electromagnetic Engineering and Technology, Wuhan, China

ABSTRACT: In order to improve the probabilistic load flow calculation efficiency of Monte-Carlo simulation to meet the large-scale power system analysis and calculation time requirements, this paper introduced the coarse-grained parallel algorithm of probabilistic load flow calculation. This paper first introduces the probabilistic load flow calculation simulated by the Monte-Carlo method, then based on the OpenMP (Open Multi-Processing, shared memory multi-threaded compiler directive) parallel programming model, focuses on the main problems existed when to parallel processing of the serial probabilistic load flow program and its parallelizing. Taking IEEE14, 30, 57, 118, 162, 300, 600 bus system for example, carried out 30000 times Monte-Carlo probabilistic load flow parallel simulation, results show that the coarse-grained parallel Monte-Carlo simulation of probabilistic load flow calculation gained considerable speedup and parallel efficiency.

Keywords: Monte-Carlo simulation; probabilistic load flow; parallel computing; OpenMP

1   INTRODUCTION

In recent years, with the rapid development of wind power generation technology, the percentage of wind power in total grid increased, the network flow distribution uncertainty greatly enhanced, which will have an important impact on the planning and operation of power system. Probabilistic Load Flow (PLF) commonly used in power system for steady state analysis of uncertainty, the basic principle is considering load, generator and the network structure as uncertainties to obtain the probability distribution of node voltage and line power flow results. Thus, obtains the probability index such as voltage violations and circuit overload. PLF calculation is divided into two main method, simulation and analysis. The simulation method can obtain the expectation, variance and the distribution function of the output variables with high precision [1], and easy to process the input variables correlation (e.g. between load [2] and renewable energy's output [3][4]), is the standardized approach of PLF calculation. However, the simulation method requires a large amount of simulation computation, it will consume more time than analytical method, so this paper introduced the coarse-grained parallel computing to improve the computational efficiency.

Parallel computing is calculated using multiple processors concurrent execution, according to different computer memory structures is mainly divided into two kinds. One kind is the cluster distributed memory computing, structure characteristics of distributed memory determines the data independence is very good, the data sharing in network messages form by MPI (Message Passing Interface). Another kind is the shared memory parallel computing, shared memory structure characteristics determines that it has very good data sharing, data independence between threads is realized in private variables form by OpenMP.

This paper considers that OpenMP is very suitable for shared memory computer structure, based on the OpenMP parallel programming model, mainly focuses on the main problems existed when to parallel processing of the serial PLF program and its parallelizing. The testing results of IEEE14, 30, 57, 118, 162, 300, 600 bus system obtains good speedup and parallel efficiency.

2   THE REALIZATION OF PLF SIMULATED BY MONTE-CARLO METHOD

The large-scale wind power integration and load fluctuation effect of the power system will greatly enhance the network flow distribution in uncertainty. In this paper, the PLF calculation of wind turbine's output and the uncertainty of load effects is used to analyze the effect performance of grid indicators. Firstly, the operation modes of power system are generated by the Monte-Carlo simulation, and the power flow calculation of each group is realized by using the Newton-Laphson method.

Monte-Carlo simulation is a common method for PLF calculation. The method is based on the estimation of probability distribution of random variables, using random sampling method to obtain a group of pseudo random number matching particular distribution. Then entering the pseudo random number to calculate the evaluation indexes, getting the probability distribution, expected statistical characteristic value, the standard deviation by multiple simulations. This paper considering that the wind active power and load fluctuation approximately obey the normal distribution N (μ, σ^2). The probability density function is

$$f(x) = \frac{1}{\sqrt{2\pi}\sigma} \exp\left(-\frac{(x-\mu)^2}{2\sigma^2}\right) \tag{1}$$

In the formula (1), μ and σ respectively represent wind active power, load active power and reactive power's mean value and standard deviation.

3 OPENMP BASED PARALLEL PROCESSING METHOD

The PLF computation needs thousands of Monte-Carlo simulation, and the only difference among every power flow calculation is the input data, the calculation method is the same. The PLF computation is very suitable for coarse-grained parallel computing, so this paper introduced the OpenMP coarse-grained parallel computing method.

3.1 OpenMP parallel programming model

OpenMP is a set of compiler directives and library function, the compiler guidance statement explicit instruction the parallel. As a shared parallel program standard, OpenMP stepped into the era when multi-core processors coming out, before it the parallel computing is mainly provided by windows API function [5]. OpenMP implementation of shared memory parallel program through the fork/join mode, as shown in Figure 1. The program began from a main thread, when running to the parallel computing part of the code, the program automatically derived (Fork) multiple threads to execute the parallel tasks, parallel code finishes execution, the thread exits or hung up when parallel

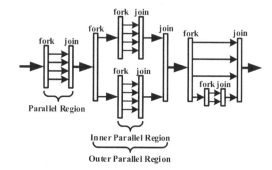

Figure 1.   OpenMP Fork-Join parallel execution model.

code finishes, and then back to the main thread in individual (Join) [6].

3.2 Parallel processing method

The OpenMP parallel programming model is suitable for shared memory computer, Shared memory programming have the memory effect (i.e. write conflict, data competition, false sharing data etc.). Combining PLF calculation program with Monte-Carlo simulation, parallel processing need to consider and solve the following key problems.

3.2.1 Parallel random number generation

The PLF computation needs to obey the normal distribution of pseudo-random array to simulate the network with uncertainty. The random number generating function is the clock function. From the statistical sense that multiple random numbers generated one moment may have the same situation. Then in the multi thread parallel calculation, if different threads to access a random number generating function at the same time, it is possible to obtain the same random number. In order to ensure that all random numbers are independent, the program design through the serial mode to generate random numbers before the parallel execution of random number generation and stored all random numbers in an array in memory. When OpenMP has created a multithreaded concurrent execution, the pseudo random array generated in serial mode will copy to each thread and backup in private by OpenMP data handling clause firstprivate (it can parallel foreign data to each parallel thread and does not need to initialize variables), thus to ensure the independence of different random numbers.

3.2.2 Avoid data conflict

Parallel programming needs to pay special attention to reduce the data coupling relationship to avoid data conflict, namely minimize the scope of each variable. Data structure of the Newton-Laphson method of power flow calculation is mainly divided into the following 3 categories: ① Bus data: Including the node number, node type, load, generator power and node voltage, etc; ② Branch data: Including branch start number and stop number, branch parameter, transformer, power and ground admittance, etc; ③ The grounding branch data: including the grounding branch parameter and grounding branch power etc. These 3 types of data have collection characteristics, should be defined as the class formal and all kinds of data contained in a class is defined as a member. Because of the existence of the weak coupling relationship between different classes in PLF calculation, we should try to reduce its scope. Therefore, the class should be defined as private variables for the PLF computation directly call each function in the program. The coupling relationship between different kinds of data is very strong, so the data in each class should be defined as shared variables.

3.2.3 The best parallel threads

The OpenMP parallel programming model to create threads need to consume a certain processor time, task allocation among threads also need some time to coordinate, the sum is the thread overhead time t_{TO}. The more the number of threads are created (T), the more the load imbalance is, then t_{TO} is more large. Considering each PLF calculation process is the same only the data vary with load, so t_{TO} is positively correlated with T.

At the same time, the more the number of threads created (not to exceed the maximum number of threads the computer can support), then parallel calculation time t_{TC} is smaller. Assuming the total number of the Monte-Carlo PLF calculation is M. Due to the load is balanced, each time the PLF calculation time is approximately equal, assume as t_0. The theory computation time for parallel computing is approximated by

$$t_{TC} = t_0 \cdot \mathrm{ceil}\left(\frac{M}{T}\right) \tag{2}$$

In the formula (2), ceil is the function of rounding up. Then get the total time of parallel computing t_{PC} as

$$t_{PC} = t_{TO} + t_{TC} \tag{3}$$

In order to improve the speedup and parallel efficiency of parallel computing, we need to reduce the total time of parallel computing as much as

possible. So we need to coordinate calculation time and thread overhead time, namely, to seek the best parallel threads number T_{best} of different single computing time t_0. Based on a large amount of test data of "for cycle", we can get that t_0 and T_{best} approximately satisfy the power function relationship as

$$T_{best} = \begin{cases} \mathrm{round}(27.2 \cdot t_0^{0.3102} + 0.8439) & (t_0 < 15.1143s) \\ 64 & (t_0 \geq 15.1143s) \end{cases}$$

$$\tag{4}$$

In the formula (4), round is the four to five rounding function, 64 to support the test computer maximum concurrent threads. According to that, equation (4) can be the best thread computing different computational tasks for creating number.

Based on the parallel processing of pseudo random numbers, processing of data structure and the optimal number of threads, the OpenMP parallel programming model can be used to parallel processing the serial program. The main content of parallel processing is:

1. Join the OpenMP parallel support header file in the program, and set OpenMP supported in the computer language;
2. Modify the data structure of serial PLF calculation program, optimize the data call relationship of functions to avoid a conflict when the thread of the data is in the parallel region;
3. The part that need to repeat calculation of PLF should be written in "for cycle" model;
4. Join the OpenMP parallel guidance statements and data handling clause before the "for loop";
5. Calculation of optimal parallel threads according to a single serial calculation time.

4 EXAMPLE ANALYSIS

In this paper, the program to calculate the coarse grain parallelism of PLF is based on the use of C++ and OpenMP, the test computer configured for shared memory computer Intel Xeon E7-8837 2.67 GHz CPU, 128G memory, Windows 2008 x64 operating system, CPU core number is 32, the maximum number of concurrent threads can support 64, programming environment is Visual Studio 2010. In this paper, considering all the load fluctuations of 15% obey normal distribution of $N(P_{Li}, 0.15^2)$, P_{Li} is the rated load of bus i, the bus that without load is not considering the load fluctuation. At the same time to consider the access of wind power in a PV node with the capacity of 10% of system, the fluctuation range of wind power is 20%, obey normal distribution of N (0.1 P_S, 0.2²), P_S is the rated capacity of the system. The IEEE14, 30, 57, 118, 162, 300 node test system and 600 node test system, carried out 30000 times Monte-Carlo simulation of concurrent PLF calculation. 600 node test system consists of IEEE300 node system replicated 1 times with 2 lines of communication: Set No. of each region is n (as shown in Fig. 2), the node number of IEEE300 node system corresponding to each region j is $j+(n-1) *300$. The parameters R and X of Liaison line are 0.048 p.u. and 0.196 p.u.[7] Balance node is in the area 1 and its node number is 257. The original balance node 557 of sub region 2 is changed into PV node and its injection power is 0. In order to reduce the data affected by other procedures, we let the mean of ten tests as comparative objects.

4.1 *Performance evaluation index*

The performance of a program affected by the following factors: the algorithm structure, the thread overhead, scale and classification of the problem and parallel computing performance. For a given problem, usually use speedup and parallel efficiency to evaluate the performance of the algorithm.

1. Speedup of the parallel algorithm S_n

$$S_n = \frac{t_s}{t_p} \tag{5}$$

In the formula (5), t_s is the serial computation time, t_p is the parallel computing time. The parallel speedup is used to evaluate the improvement performance of algorithm.

2. Parallel efficiency E_n

$$E_n = \frac{S_n}{n} \times 100\% \tag{6}$$

In the formula (6), n is the core number, n refers to the core number of CPU for the nuclear CPU, in this paper $n = 32$. The parallel efficiency is used to evaluate the utilization rate of the system resources.

4.2 *Algorithm performance analysis*

Because of the power flow calculation program doesn't make full use of sparse processing technology, which causes the single computation time in Table 1 is a little long, leading to serial execution time and the parallel execution time is also long. The purpose of this paper is to use the serial, parallel execution time ratio to analysis of the speedup and parallel efficiency of each node system, when using sparse processing technology can significantly improve the computation time of the program.

By the results shown in Table 1, with the increase of the number of nodes, a single serial computation time is approximately exponential growth, the speedup of the algorithm increases gradually, the parallel processing efficiency increases. Analysis of operation mode of OpenMP we can get:

The OpenMP uses a multi thread parallel method to improve operating efficiency, to create and terminate threads, and the threads hang will occupy part of system resource. Because the system of IEEE14 30, 57 node system is small in scale, thread overhead relative to the total time of parallel computing time proportion is larger, so that the utilization rate of resources is very low, which lead to parallel speedup ratio is not high. The network size has become a certain extent of IEEE118 162, 300 node system, the parallel thread overhead time relative to the total parallel duration time decreases, so a considerable speedup and parallel efficiency is obtained.

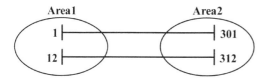

Figure 2. 600 node test system.

Table 1. The test result of speedup and parallel efficiency of IEEE standard system.

IEEE system	14	30	57	118	162	300	600
t_0 (s)	0.0054	0.0122	0.0385	0.237	0.505	2.164	23.004
T_{best}	6	8	11	18	23	35	64
t_s (s)	162.04	364.63	1155.23	7100.18	15160.95	64915.80	689806.5
t_p (s)	64.43	115.61	173.44	660.68	1442.06	2803.68	27234.5
S_n	2.515	3.154	6.661	10.741	13.259	23.154	25.328
E_n	7.853	9.856	20.816	33.566	41.434	72.358	79.151

5 CONCLUSION

Based on OpenMP parallel programming model, this paper focuses on the PLF of Monte-Carlo simulation of coarse-grained parallel computational method and the parallel processing of serial program. On IEEE14, 30, 57, 118, 162, 300 nodes test system and 600 nodes test system, 30000 series parallel Monte Carlo probabilistic load flow calculation, by comparing the speedup and parallel efficiency, draws the following conclusion:

1. For a given task, since the weight to create different thread parallel overhead time and the weight of theoretical calculation is different, we should set appropriate parallel threads to improve the parallel speedup and parallel efficiency as far as possible.
2. With the increase of the number of nodes, a single computational time PLF calculation program based on Monte-Carlo simulation is bigger, weight of thread overhead time decreases, the speedup and parallel efficiency becomes larger, so the system is suitable for coarse-grained parallel computing. And if the computer memory and computing performance allows, the large system is more suitable for coarse-grained parallel computing.

The Monte-Carlo simulation of the coarse-grained parallel algorithm is not only suitable for PLF, and is also suitable for other need parallel random number generation, or have the repeated numerical case, such as the economic allocation, optimal power flow calculation. In addition, because the single computing time of PLF of Newton-Laphson method is long, there is much space for improvement.

ACKNOWLEDGMENT

In this paper, the work done is supported by the national high technology research and development program (863 program, project number: 2011 AA05 A101).

REFERENCES

1. Ding M., Wang J.J, Li S.H. Probabilistic Load Flow Evaluation With Extended Latin Hypercube Sampling [J]. Proceedings of the CSEE, 2013, 33(4):163–170.
2. Ding M., Li S.H., Hong M. The K-means Cluster Based Load Model For Power System Probabilistic Analysis [J]. Automation of Electric Power Systems, 1999, 23(19):51–54.
3. Chen Y., Wen J.Y, Cheng S.J. Probabilistic Load Flow Analysis Considering Dependencies Among Input Random Variables [J]. Proceedings of the CSEE, 2011, 31(22):80–87.
4. Ai X.M, Wen J.Y, Wu T, et al. A Practical Algorithm Based on Point Estimate Method and Gram-Charlier Expansion for Probabilistic Load Flow Calculation of Power Systems Incorporating Wind Power [J]. Proceedings of the CSEE, 2013, 33(16):16–22.
5. Li Q. The API function based windows multi thread serial communication [J]. Science & Technology Information, 2008, 17:402–404.
6. Luo Q.M, Ming Z, Liu G, et al. OpenMP compiler And implementation [M]. Beijing: Tsinghua University press, 2012.
7. Liu M.B, Xie M, Zhao W.X. Large power system optimal power flow calculation [M]. Beijing: Science Press, 2010.

Power and Energy – Kong (Ed.)

Fuzzy modeling and solution of load flow considering uncertainties of wind power

D.L. Wu, B.H. Zhang, J.L. Wu, K.M. Zhang, W.S. Deng & B.J. Jin
*State Key Laboratory of Advanced Electromagnetic Engineering and Technology,
Huazhong University of Science and Technology, Wuhan, Hubei, China*

K. Wang & D. Zeng
China Electric Power Research Institute (Nanjing), Nanjing, Jiangsu, China

ABSTRACT: Power system load flow calculation is one of the most important types of electrical operations. However, due to the uncertainty presence of load and the growing attention to wind resource, traditional deterministic load flow analysis cannot meet the requirements with severe uncertainty. In the uncertainty load flow calculation methods, fuzzy load flow calculation method can give the possibility of load flow distribution in a very short time. So it plays a big role in the online grid computing. This paper has mentioned about the details of the fuzzy load flow calculation using the incremental method. And then it has applied to practical system—Ningxia Power Grid 151-bus system. By comparing with the results of Monte Carlo simulation method, numerical results have demonstrated the validity and effectiveness of the proposed model and method.

Keywords: Fuzzy load flow; wind power; fuzzy-set theory; incremental method

1 INTRODUCTION

Due to lack of valid statistics, some of uncertain variables are difficult to describe by the random variable. In this case the fuzzy load flow analysis developed by fuzzy set theory may be more reasonable[1,2].

At present, the fuzzy load flow calculation's main mathematical theory is based on the Zadeh's fuzzy set theory[3]. There are many fuzzy load flow calculations' methods built on it, and the most common is the incremental method[4-6]. The earliest use of fuzzy theory to solve the problem is in the literature[1]. The active power flow is represented by DC power flow while the reactive power flow is represented by AC power flow using the PQ decomposition method. It just uses the fuzzy arithmetic rules for solving equations for only one time but it has some limitations. In the literature[7], generator output and load forecast uncertainty was described by trapezoidal fuzzy number. It is a good way to solve the AC power flow with a central value plus fuzzy incremental number.

2 FUZZY-SET THEORY

2.1 *Definition of fuzzy numbers*

The most basic concept of fuzzy mathematics theory is the membership function $\mu_{\tilde{L}}(x)$. It indicates the degree of membership of an arbitrary variable x to the fuzzy sets L on the domain U. The membership function $\mu_{\tilde{L}}(x)$ of the fuzzy set which domains in real numbers R is called fuzzy distribution.

Definition 1. Note $L = (L_1, L_2, L_3, L_4)$, $-\infty < L_1 \leq L_2 \leq L_3 \leq L_4 < \infty$, L is called a trapezoidal fuzzy number. If $L_2 > 0$, it is a positive trapezoidal fuzzy number. The membership function $\mu_{\tilde{L}}(x): R \rightarrow [0,1]$ of the trapezoidal fuzzy number can be defined as equation 1.

$$\mu_{\tilde{L}}(x) = \begin{cases} 0 & x < L_1 \\ \dfrac{x - L_1}{L_2 - L_1} & L_1 \leq x < L_2 \\ 1 & L_2 \leq x < L_3 \\ \dfrac{L_4 - x}{L_4 - L_3} & L_3 \leq x < L_4 \\ 0 & x \geq L_4 \end{cases} \quad (1)$$

Its membership function as shown in Figure 1.

Trapezoidal fuzzy number indicates the set ranges between $L_1 \sim L_4$, and it has a maximum possibility value between $L_2 \sim L_3$. Its central value is the average $(L_2 + L_3)/2$ of the cut $\mu_{\tilde{L}}(x) = 1$.

2.2 *Fuzzy number arithmetic*

Calculation of the expected value of trapezoidal fuzzy numbers is defined as follows. In order

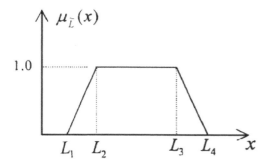

Figure 1. The membership function of a trapezoidal fuzzy number.

to facilitate the representation, in the trapezoidal fuzzy number $A = (a,b,c,d)$, note $b - a$ as α and $d - c$ as β, then the expected value is

$$E(A) = \frac{1}{2}(b+c) + \frac{1}{6}(\beta - \alpha) \qquad (2)$$

The possibility of variance expressed as

$$Var(A) = \frac{1}{4}(c-b)^2 + \frac{1}{24}(\alpha + \beta)^2 + \frac{1}{6}(c-b)(\alpha + \beta) \qquad (3)$$

2.3 Fuzzy number algorithms

According to the principle of credibility theory expansion, for any two trapezoidal fuzzy numbers $A_1 = (a_1, b_1, c_1, d_1)$ and $A_2 = (a_2, b_2, c_2, d_2)$ related algorithms as follows.

$$A_1 \oplus A_2 = (a_1 + a_2, b_1 + b_2, c_1 + c_2, d_1 + d_2) \qquad (4)$$

$$A_1 - A_2 = (a_1 - d_2, b_1 - c_2, c_1 - b_2, d_1 - a_2) \qquad (5)$$

$$\begin{aligned} A_1 \times A_2 = [&\min(a_1 \times a_2, a_1 \times d_2, d_1 \times a_2, d_1 \times d_2), \\ &\min(b_1 \times b_2, b_1 \times c_2, c_1 \times b_2, c_1 \times c_2), \\ &\max(b_1 \times b_2, b_1 \times c_2, c_1 \times b_2, c_1 \times c_2), \\ &\max(a_1 \times a_2, a_1 \times d_2, d_1 \times a_2, d_1 \times d_2)] \end{aligned} \qquad (6)$$

$$\begin{aligned} A_1 / A_2 = [&\min(a_1/a_2, a_1/d_2, d_1/a_2, d_1/d_2), \\ &\min(b_1/b_2, b_1/c_2, c_1/b_2, c_1/c_2), \\ &\max(b_1/b_2, b_1/c_2, c_1/b_2, c_1/c_2), \\ &\max(a_1/a_2, a_1/d_2, d_1/a_2, d_1/d_2)] \end{aligned} \qquad (7)$$

$$\sqrt{A} = (\sqrt{a}, \sqrt{b}, \sqrt{c}, \sqrt{d}) \qquad (8)$$

In the above formula, '$\oplus$' and '$\times$' denote the addition and multiplication of the trapezoidal fuzzy numbers.

3 FUZZY LOAD FLOW CALCULATION METHOD

First, we introduce fuzzy modeling process of loads and wind power, and then use the incremental method for solving the fuzzy load flow model.

3.1 Fuzzy modeling of load and wind power

The given loads and wind power prediction in a power system are probabilistic datum meet certain probability distribution. Assuming that loads and wind power prediction errors generally follow a normal distribution, we can get their fuzzy model based on normal probability density function.

Normal probability density function is

$$f(x) = \frac{1}{\sqrt{2\pi}\sigma} e^{-\frac{(x-\mu)^2}{2\sigma^2}}, -\infty < x < +\infty. \qquad (9)$$

Its second derivative is

$$\begin{aligned} f''(x) = &-\frac{1}{\sqrt{2\pi}\sigma^3} e^{-\frac{(x-\mu)^2}{2\sigma^2}} \\ &\times \left(1 + \frac{x-\mu}{\sigma}\right)\left(1 - \frac{x-\mu}{\sigma}\right) \end{aligned} \qquad (10)$$

Let the $f''(x) = 0$, it has two roots $x_1 = \mu - \sigma$, $x_2 = \mu + \sigma$ located at both sides of expectations. In this interval, $f''(x) < 0$, so this is obviously a interval with the maximum possibility. And according to the integral of the normal probability density function, in the interval $[\mu - 2\sigma, \mu + 2\sigma]$, probability value is 95.4% while in the interval $[\mu - 3\sigma, \mu + 3\sigma]$, probability value is 99.7%. Therefore, in this paper, if the prediction value is $\hat{y}$ with a standard deviation value $\hat{\delta}$, we can use $\tilde{y} = [\hat{y} - 3\hat{\delta}, \hat{y} - \hat{\delta}, \hat{y} + \hat{\delta}, \hat{y} + 3\hat{\delta}]$ to build the fuzzy model process of loads and wind power.

3.2 Fuzzy model solution

Using fuzzy incremental method to solve this model, the solution process is as follows.

1. Find the center value of the fuzzy numbers of state variables (voltage magnitude and voltage phase) and output variables (active power and reactive power). Solve the deterministic power flow equations by the central value $[P_d]$, $[Q_d]$ of the fuzzy injection $[P]$, $[Q]$, in order to obtain the corresponding determine the value of state and output variables $[X_d]$, $[Z_d]$.

2. Solve fuzzy increment of the state variables. The load flow equations of power system is

$$\begin{cases} P_{is} = V_i \sum_{j \in i} V_j (G_{ij} \cos \theta_{ij} + B_{ij} \sin \theta_{ij}) \\ Q_{is} = V_i \sum_{j \in i} V_j (G_{ij} \sin \theta_{ij} - B_{ij} \cos \theta_{ij}) \end{cases}$$ (11)

It can be summarized as $W = f(X)$. W is the bus power injection and X is the bus state variable. We consider W as $W = W_d + \Delta W$ and X as $X = X_d + \Delta X$. Obtained by the Taylor series expansion and ignore higher order terms, we have

$$W_d + \Delta W = f(X_d + \Delta X) = f(X_d) + J_0 \Delta X + \cdots$$ (12)

From $\Delta W \cong J_0 \Delta X$, we can calculate $\Delta X = J_0^{-1} \Delta W = S_0 \Delta W$.

3. Solve fuzzy increment of the output variables. We can also carry out a linearization on branch flow equations as above. When we know the state variables, branch flow equations is

$$\begin{cases} P_{ij} = -V_i V_j (G_{ij} \cos \theta_{ij} + B_{ij} \sin \theta_{ij}) + V_i^2 G_{ij} \\ Q_{ij} = -V_i V_j (G_{ij} \sin \theta_{ij} - B_{ij} \cos \theta_{ij}) - V_i^2 (B_{ij} + b_{i0}) \end{cases}$$ (13)

It can be summarized as $Z = g(X)$. As above, we get $\Delta Z = G_0 \Delta X$. This G_0 is the partial derivative matrix. So the fuzzy increment of the output variables can be calculate by

$$\Delta Z = G_0 S_0 \Delta W = T_0 \Delta W$$ (14)

4. Solve the fuzzy numbers of the state and output variables. The fuzzy numbers of the state and output variables are calculated by the fuzzy incremental value obtained from step 2) or 3) superimposed with the center value obtained from step 1).

$$\begin{cases} [X] = [X_d] + [\Delta X] \\ [Z] = [Z_d] + [\Delta Z] \end{cases}$$ (15)

4 CASE STUDY

The fuzzy flow calculation applied to practical system—Ningxia Power Grid 151-bus system. Then compare with the results of Monte Carlo simulation method simulated 10000 times from the curve and the numerical interval probability.

4.1 Compare the curve

At a certain moment, the total active load of Ningxia Power Grid 151-bus system is about 12000 MW, and the total wind power is about 1300 MW. Assuming that load-forecasting error is 5%, wind power forecasting error is 30% and wind generations are run in a constant power factor. Results of fuzzy load flow calculation and the results of Monte Carlo simulations plotted in the same coordinate axes and use a uniform height in order to be able to clearly contrast. Parts of the results are shown in Figure 2 to Figure 5. Figure 2 and Figure 3 represented the active power of most branches, Figure 4 represented the voltage magnitude of most buses and Figure 5 represented the voltage phase of most buses. The red line indicates the results of fuzzy load flow calculation and the blue line indicates the results of Monte Carlo simulation method.

4.2 Compare the numerical interval probability

The fuzzy numbers can show a maximum possibility interval and an all possibility interval. When we compare the numerical interval probability, we can sum up all the probability value in these two intervals. Ideally, the narrower interval contains about

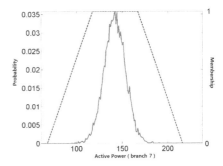

Figure 2. Distribution of active power flow (branch 7).

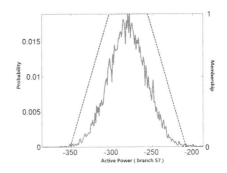

Figure 3. Distribution of active power flow (branch 57).

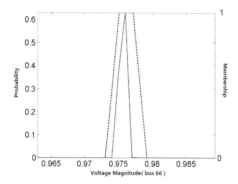

Figure 4.   Distribution of voltage magnitude (bus 66).

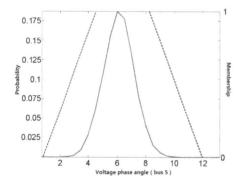

Figure 5.   Distribution of voltage phase (bus 5).

Table 1.   Interval probability of active power.

Branch	Narrow interval	Wide interval	Branch	Narrow interval	Wide interval
60	0.8761	1	191	1	1
122	0.8769	1	80	0.9912	1
1	1	1	81	0.9909	1
2	1	1	58	0.9748	1
84	1	1	57	0.7002	0.9974
74	0.8439	0.998	79	0.6989	0.9974
171	0.846	0.9976	117	0.7052	0.9974
178	0.9845	1	127	0.9924	1
204	0.9759	1	128	0.9918	1
207	0.9747	1	129	0.9913	1
214	0.9955	1	184	0.7161	0.999
217	0.9946	1	187	0.7158	0.999
46	0.9995	1	131	0.6736	0.996
47	0.9995	1	111	0.7377	0.9993
19	0.9989	1	118	0.7364	0.9993
93	0.9989	1	131	0.6736	0.996
145	0.9967	1	131	0.6736	0.996
146	0.9971	1	130	0.8029	0.9998
5	0.9655	1	190	1	1

70% probability and the wider interval contains all probability. Here we show the results of the branches which are concerned as below. For example, we see the branch 60, there are about 87.61% of the results of Monte Carlo simulation are located in the narrow interval, and all the results of Monte Carlo simulation are located in the wide interval.

5   CONCLUSIONS

It can be seen from the analysis above; the fuzzy load flow calculation can accurately delineate the scope of the possibility distribution of the results of load flow with uncertainties. Illustrated the validity of this method. But the interval has a little wider than reality so it is a conservative estimate. Because the wide interval is too conservative, policymakers can directly use narrow interval to judge. Besides, fuzzy load flow calculation only needs about 4 seconds while Monte Carlo simulations takes about 8 minutes on the same computer. All in all, the fuzzy load flow with its correctness and efficiency will play an important role in grid online computing.

ACKNOWLEDGMENTS

This work was supported by State Grid Corporation of China (the Study of Steady-state Analysis Method in the 'Source-Grid-Load' Interactive Environment) (dz71-13-036).

REFERENCES

1. Miranda V., Matos M.A. Distribution system planning with Fuzzy models and techniques. CIRED 1989, 10th International conference on electricity distribution, 7, 472476 6th Mediterranean electro technical conference proceedings, 2, 13391342.
2. Saraive J.T., Miranda V., Matos M.A. Generation and load uncertainties incorporated in load flow studies[J].
3. Zadeh L.A. Fuzzy sets as a basis for a theory of possibility[J]. Fuzzy Sets and Systems, 1978, 1(1):3–28.
4. Pajan P.A., Paucar V.L. Fuzzy power flow: considerations and application to the planning and operation of a real power system[J]// Proceedings of International Conference on Power System Technology, Oct 13–17, 2002, Kunming, China: 433–437.
5. Saraiva J.T., Fonseca N., Matos M.A. Fuzzy power flow: an AC model addressing correlated data[J]// Proceedings of 8th International Conference on Probabilistic Methods Applied to Power Systems, Sep 12–16, 2004, Ames, IA, USA: 519–524.
6. Zhang Y., Chen Z.C. Fuzzy load flow calculation in grid planning[J]. Automation of Electric Power Systems,1998, 22(3):20–22.
7. Zhang Y., Chen Z.C. Algorithm of AC fuzzy load flow incorporating uncertainties. Power System Technology. 1998, 22(2):20–25.

Power and Energy – Kong (Ed.)
© 2015 Taylor & Francis Group, London, ISBN 978-1-138-02782-4

Dynamic modeling and simulation of parallel operation of ship diesel generators

Chunhua Li
China Satellite Maritime Tracking and Controlling Department, Jiangyin, Jiangsu, China

Yuewei Dai
Jiangsu University of Science and Technology, Zhenjiang, Jiangsu, China

Jiahong Chen & Zhifeng Zhou
China Satellite Maritime Tracking and Controlling Department, Jiangyin, Jiangsu, China

ABSTRACT: To analyze and study the process of the parallel operation of the ship diesel generators, the dynamatic model of the ship power system including two parallel diesel generators is established and simulated. Virtual difference method based on the PID strategy is used to realize the average distribution of active and reactive power among synchronous generators and ensure the stabilization of voltage and frequency of the ship power system. In the Matlab/Simulink environment, the ship power system including two parallel diesel generators is developed. The dynamic simulation results cerificate the high accuration of the ship power system model, and show that the proposed controller can efficiently allocate the active and reactive power among synchronous generators and maintain the stability of voltage and frequency of the ship grid.

1 INTRODUCTION

Ship power system is a independent power system with limited capacity, and composed of ship power station, ship power grid, ship load and so on. Most of the ship loads are asynchronous motor loads (Shi et al. 2003; Li et al. 2008). The rated power of some ship load is comparable to the rated power of the single generator (Li et al. 2003). The startup of this kind of load will produce a big disturbance to the ship power system and let the ship voltage and frequency undergo a big fluctuation process (Yong & Hu 2010).

Generally, several synchronous power generators in ship power station parallelly operate to provide the ship load with the power (Li 2009). The number of the parallel operating generators will change according to the required electricity of ship loads. When the startup or stop of the ship generators, the load power needs to be proportionally

allocated between generators as the same time that the voltage and frequency of ship power system must keep steady.

However, due to the nonlinear characteristics of ship power system and the power transfer between parallel operating generators, it is easy for the ship power system to run into the repeated power regulating state (Huang 2013). The above phenómena will lead to the big power fluctuation, deteriorate the supplied power quality, and have a bad effect on the electrical equipment in the ship grid. Therefore, the automatic frequency and volate modulation controllers are indispensable to effectively realize the power allocation and keep the system frequency and voltage steady.

In this work, based on the simulation platform of the ship power system including two parallel synchronous generators, two virtual difference PID controllers are designed as the parallel operating control strategies of parallel synchronous generators. Several typical operating conditions of the ship power system is simulated. The dynamic performance of the rotation speed of the diesel prime mover, and the generator input torque, excitation voltage and stator output voltage are analyzed to validate the proposed controller under the contions of the parallelly operating generators.

This work is supported by the National Natural Science Foundation (NNSF) of China (Grant 51307074), the natural science foundation of Jiangsu Province of China (Grant BK20130466) and the China postdoctoral science foundation (No. 2014m562615).

2 CONTROL STRATEGY OF PARALLEL DIESEL GENERATORS

There are two main steps for the generators operating in parallel: the synchronous operation and the subsequent power distribution between parallel generators. The synchronous operation is to realize the consistency of the phase sequence, frequency, voltage, and phase between the backup and work generatores (Li 2009; Sun et al. 2010). The connection of the backup generator to the ship grid without satisfying the synchronous standard of the parallel operation will produce the big impact current which will damage the generators and the grid devices. Under the premise of the phase sequence consistency, the synchronous operation conditions proved by the practical experience are: voltage RMS deviation within ±10%, frequency deviation within ±1%, and phase deviation within ±15%.

So as to realize the power allocation between the parallel synchronous generators and keep the system frequency and voltage steady, the virtual difference PID controllers are designed to control the output frequency and voltage of the diesel synchronous generators.

The input signal of the PID controller for the speed governor of the diesel:

$$E_{\omega p} = K_\omega \Delta \omega + K_p \Delta P$$
$$= K_\omega (\omega_i - \omega_e) + K_p (P_i - P_e) \qquad (1)$$

$$P_e = \frac{1}{n} \sum_{i=1}^{n} P_i \qquad (2)$$

where: ω_i is the actual frequency of the ith generator; ω_e is the rated frequency of ship grid; P_i is the output power of the ith generator; K_ω and K_p are the proportion coefficients of the frequency and active power deviations, respectively; P_e is the average active power calculated based on the total active power of the ship grid; n is the number of generators operating in parallel. When E_{fp} is less than zero, it is necessary to increase the throttle opening degree of the generator and increase the diesel output torque; when E_{fp} is greater than zero, the throttle opening degree will be reduced to decrease the diesel output torque.

Figure 1 shows the model structure of the diesel and its speed governor. The PID controller is used to regulate the rotation speed of diesel by adjusting engine throttle. The diesel and its throttle actuator are modelled by using a second order transfer function. Output torque of the diesel is obtained by integrating and delaying the output of the second order system. The mechanical power is obtained by multiplying the output torque and

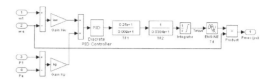

Figure 1. Model structure of diesel and its speed governor.

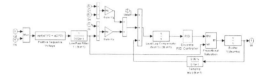

Figure 2. Excitation system model of the diesel generator.

rotation speed measurement value, and then output to drive generator.

The excitation current of the excitation device can be changed to adjust the reactive power output and the terminal voltage of the synchronous generator, which process can realize the reactive power allocation among the parallel generators. The input signal of the PID controller for the excitation device of the generator:

$$E_{uq} = K_u \Delta U + K_q \Delta Q$$
$$= K_u (U_i - U_e) + K_q (Q_i - Q_e) \qquad (3)$$

$$Q_e = \frac{1}{n} \sum_{i=1}^{n} Q_i \qquad (4)$$

where: U_i is the actual value of the terminal voltage of the generator; U_e is the rated value of the ship grid voltage; Q_i is the set output value of the reactive power for the ith-generator; K_u and K_q are the proportion coefficients of the voltage and reactive power deviations, respectively; Q_e is the average reactive power calculated based on the total reactive power of the ship grid.

Figure 2 shows the excitation system model of the generator. U_d is the d-axis voltage of the generator; U_q is the q-axis voltage of the generator; U_{stab} is ground zero voltage; U_f is the excitation voltage. The voltage deviation value and the reactive power deviation value are input into the PID controller so as to output the compensating signal. The excitation voltage is attained after the compensating signal going through the saturation limit and exciting parts. The differential part in the feedback loop plays a dampling and stabilizing role on the system.

Using the voltage and reactive power deviation to change the exciting current and achieve the automatic allocation of the reactive power, not only can improve the control precision of the terminal voltage of the generator sets, but also can improve the stability of the generators operating in parallel.

3 SYSTEM SIMULATION

Figure 3 shows the simulation structure of the ship power system. The ship power grid which is a AC three-phase AC400 V, 50 Hz power system, is equipped with two diesel synchronous generators (AC400 V, 770 kW, 1389 A, 750 r/min). Two 640 KW, 160 kvar loads are connected to the ship power grid through two isolation transformer, respectively.

Figure 4 shows the dynamic response of the first and second diesel generator sets. At the beginning, the rotation speed ω of the first generator starts to rise with the output mechanical power P_{mec} increasing, the excitation voltage U_f decreases from a big value to a steady value, and the terminal voltage U_t of the generator needs 2s to reach a stable state. At the 3th second, the second generator begins to run in parallel with the first generator. The frequency, voltage and phase bias between two generators make it necessary to adjust the system parameters by the proposed controllers. The system parameters reach the set steady state by using 1.5 s. The rotation speed ω has 3% instantaneous overshoot, and the terminal voltage U_t has 1.5% instantaneous overshoot.

As shown in Figure 5, the active and reactive powers have been averagely allocated between the two generators. At the 6th second, the ship load is increased by 100%. The active and reactive powers are averagely allocated between two diesel generators and the ω and U_t return to steady state by using 1.5 s. The rotation speed ω has 5% instantaneous overshoot, and the terminal voltage U_t has 1% instantaneous overshoot. At the 9th second, the ship load are decreased by 50%. The rotation

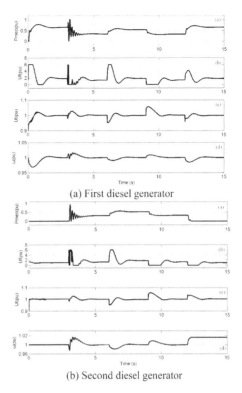

(a) First diesel generator

(b) Second diesel generator

Figure 4. Dynamic responses of the parallel diesel generators.

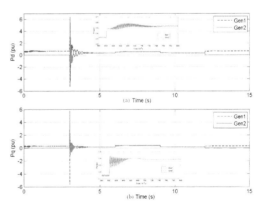

Figure 5. Active and reactive power allocation between two diesel generators.

speed ω have 6% instantaneous overshoot, and the terminal voltage U_t have 1% instantaneous overshoot. At the 12th second, the second diesel generator is disconnected to the ship grid, and all of the ship loades are satisfyed by the first diesel generator.

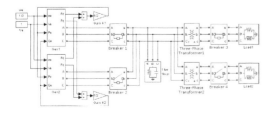

Figure 3. Simulation structure of ship power system.

The above simulation results show: the ship power controller can quickly and stably realize the power allocation or transfer, and ensure that the operating frequency and terminal voltage of the generator keep steady under the big disturbance from the ship loades.

4 CONCLUSIONS

In this work, the virtual difference PID controllers are proposed to realize the automatic allocation of the active and reactive powers among the parallel diesel generators and maintain the stability of the frequency and voltage of the ship power system. The dynamic simulation results by using the ship power system model including two diesel generator sets validate the efficiency of the proposed controller. The built simulation platform of the ship power system lay foundations to the control strategy design of the ship power station. The obtained simulation results are useful for system parameter setting and safety strategy selecting of ship power system.

REFERENCES

1. Huang Man-lei, 2013, "Chaos control of diesel-generator set operating in parallel," *Journal of Naval University of Engineering*, vol. 25(2). pp. 5–13.
2. Li Dong-li, Zhao yue-ping, Tang Shi-Iing, Zhu yong-Iuan, 2003, "Effects of the Shipping's big Capability Load upon Power System," *Ship engineering*, vol. 25(6). pp. 47–49.
3. Li Dong-hui, Zhang Jun-dong, He Zhi-bin, 2008, "Modeling and Simulation of the Power System on Practice Teaching Ship," *Journal of Shanghai Jiaotong University*, 42(2), pp. 190–193.
4. Li Hong-lin, 2009, "The study of excitation control on ship multi-generator's parallel operation," M.S dissertation of Harbin engineering university.
5. Shi Wei-feng, Zheng Hua-yao, 2003, "Marine automatic electric power station system simulation," *Journal of System Simulation*, vol. 15(9). pp. 1249–1252.
6. Sun Bin, Zeng Fan-ming, 2010, "Research on frequency and power control based on fuzzy control for ship power plant," *Ship science and technology*, vol. 32(11). pp. 130–133.
7. Yong Chang, Hu Yi-huai. 2010, "Dynamic Modeling and Simulation of the Marine Power Station with Load Disturbance," *Ship building of China*, vol. 51(2). pp. 198–204.

Power and Energy – Kong (Ed.)
© 2015 Taylor & Francis Group, London, ISBN 978-1-138-02782-4

Fuzzy comprehensive evaluation for transmission grid planning based on combination weight

Z.H. Li & Y. Lin
Guangdong Power Grid Development Research Institute Co. Ltd., Guangzhou, Guangdong, China

W.H. Chen, D.H. You, G. Wang, M.L. Dong, Q.R. Chen, Y. Zou & J. Hu
State Key Laboratory of Advanced Electromagnetic Engineering and Technology,
Huazhong University of Science and Technology, Wuhan, Hubei, China

ABSTRACT: In order to remedy the defect that the determination of weight in fuzzy comprehensive evaluation is impacted by subjectivity, the combination weight is introduced. In this paper, the optimization Model combines the subjective weight and objective weight, which is solved by PSO, is built. At the same time, In order to give full consideration to the uncertain and fuzzy factors which involved in the grid planning, this paper combines the combination weight and fuzzy comprehensive evaluation to assess the grid-planning scheme. The calculation results of IEEE Garver-6 show that the proposed fuzzy comprehensive evaluation system is clear in theory and convenient to calculate, the combination weight is objective and reasonable as well as its results are scientific and intuitive.

Keywords: grid planning scheme; fuzzy comprehensive evaluation; combination weight; particle swarm optimization algorithm

1 INTRODUCTION

The grid planning scheme is to use scientific method to determine when or where to build or rebuild what kind of power facilities, in order to achieve the transmission capacity needed in planning cycle, and on the premise of meeting all the technical indicators to minimum the cost[1]. However, With the speeding up of the power grid construction, the traditional goal to minimum the cost can't meet the requirements of the grid planning development, the evaluation of grid planning scheme now need to consider safe, reliable, economic, efficient and other aspects.

At present, studies which about the comprehensive decision-making of the grid planning include the analytic hierarchy process[2], fuzzy comprehensive evaluation[3], etc, and have made certain progress, but these studies for how to determinate the weight has not made in-depth study. Reference [2,6] uses the analytic hierarchy process to determine the attribute weight, which defect is impacted by subjectivity and advantage is take expert experience into account. Reference [4] which uses the order relation method to determine the attribute weight has the same defects and advantages as reference [2,6], Reference [5,6] using the entropy weight method to determine the attribute weight

have the problems that only considering the data information.

The comprehensive decision-making of grid planning scheme is a multi-attribute comprehensive evaluation problem, within it there are many uncertainties and fuzzy factors. In order to consider all these factors and set the attribute weight reasonable, this paper proposes a fuzzy comprehensive evaluation method based on combination weight for the decision-making of grid planning scheme, and the optimization Model making the effect of expert experiences and objective data together combines the subjective weight and objective weight which is solved by PSO is built in this method. The calculation results of IEEE Garver-6 show that the method proposed in this paper has feasibility and effectiveness.

2 THE OPTIMIZATION MODEL FOR COMBINATION WEIGHT

2.1 *The optimization model*

The set of schemes are $S = \{s_1, s_2, ..., s_n\}$, the set of attributes are $I = \{I_1, I_2, ..., I_m\}$, $x_{ij}(i = 1, 2, ..., n; j = 1, 2, ..., m)$ represents the uncertain assessments of the j-th attribute of the i-th alternative scheme s_i, and all the assessments after normalized forms a decision matrix $X = [x_{ij}]_{n*m}$ which dimension is $n \times m$. The weight matrix $\omega = [\omega_1, \omega_2, ..., \omega_p, \omega_{p+1}, ..., \omega_{p+q}]^T$, which is obtained by using subjective weighting methods and objective weighting methods as follows:

$$
\begin{array}{c}
1 \\
\vdots \\
\vdots \\
p \\
p+1 \\
\vdots \\
p+q
\end{array}
\begin{bmatrix}
\omega_{1,1} & \omega_{1,2} & \cdots & \omega_{1,m} \\
\omega_{2,1} & \omega_{2,2} & \cdots & \omega_{2,m} \\
\vdots & \vdots & \ddots & \vdots \\
\omega_{p,1} & \omega_{p,2} & \cdots & \omega_{p,m} \\
\omega_{p+1,1} & \omega_{p+2,1} & \cdots & \omega_{p+1,m} \\
\vdots & \vdots & \ddots & \vdots \\
\omega_{p+q,1} & \omega_{p+q,2} & \cdots & \omega_{p+q,m}
\end{bmatrix}
$$

$$
s.t. \quad \forall i, \sum_{j=1}^{m} \omega_{ij} = 1, \ (i = 1, 2, \dots p, \dots, p+q) \quad (1)
$$

where m is the number of attribute, p is the number of subjective weighting method, q is the number of objective weighting method.

After the above basis, the combination weight $W = (w_1, w_2, ..., w_m)$ is expressed as a linear combination of the various weights as the following equation:

$$
W = \theta_1 \omega_1 + \theta_2 \omega_2 + \cdots + \theta_p \omega_p + \theta_{p+1} \omega_{p+1}
$$
$$
+ \cdots + \theta_{p+q} \omega_{p+q} = \sum_{i=1}^{p+q} \theta_i \omega_i
$$
$$
s.t. \quad \forall i, \theta_i \geq 0; \quad \sum_{i=1}^{p+q} \theta_i = 1 \quad (2)
$$

In order to get θ_i ($i=1,2, ..., p+q$), the optimization model based on the deviation between the combination weight and the original subjective/objective weights as small as possible is built as the following equation:

$$
min \ f_1(W) = \sum_{j=1}^{m} \left(\sum_{i=1}^{p} \alpha_i \left(w_j - \omega_{ij}\right)^2 + \sum_{i=p+1}^{p+q} \beta_i \left(w_j - \omega_{ij}\right)^2 \right)
$$
$$
s.t. \ 0 \leq w_j \leq 1, \sum_{j=1}^{m} w_j = 1, \sum_{i=1}^{p} \alpha_i + \sum_{i=p+1}^{p+q} \beta_i = 1;
$$
$$
w_j = \sum_{i=1}^{p+q} \theta_i \omega_{ij}, \forall i, \theta_i \geq 0, \sum_{i=1}^{p+q} \theta_i = 1
$$
$$
(3)
$$

where $W = (w_1, w_2, ..., w_m)$ is the combination weight, α_i and β_i represents the relative importance of the subjective weights and the objective weights to decision makers respectively. For simplicity, $\alpha_i = 1(i = 1, 2, ..., p)$ and $\beta_i = 1(i = p+1, 2, ...,$

$p+q$) in numerical illustration, and the subjective/objective weighting methods adopted in this paper are analytic hierarchy process[2,6], order relation method[4,6], entropy weight method[3,6], variation coefficient method[6].

2.2 Particle swarm optimization algorithm

In this paper, we refer to the most basic PSO[7] algorithm to solve optimization model. Individuals or particle, are represents by vectors whose length is the number of degrees of freedom of the optimization problem. To start with, a population of N particles is initialized with random positions (L_i^0) and velocities (V_i^0). A cost function, $f_1(W)$, is evaluated for each particle i of the swarm. As time advances, the position and velocities of each particle are updated as a function of its misfit and the misfit of its neighbors. At time-step $k+1$, the algorithm updates positions (L_i^{k+1}) and velocities (V_i^{k+1}) of the individuals as the following equation:

$$
\begin{cases}
v_i^{k+1} = wv_i^k + c_1 r_1 \left(pbest_i^k - x_i^k \right) + c_2 r_2 \left(gbest^k - x_i^k \right) \\
L_i^{k+1} = L_i^k + a * v_i^{k+1}
\end{cases}
$$
$$
s.t. \quad v_{ij} > v_{max}, v_{ij} = v_{max}; v_{ij} < -v_{max}, v_{ij} = -v_{max};
$$
$$
j = 1, 2 \dots m
$$
$$
(4)
$$

where w, c_1, c_2 are called the inertia weight (using linear decreasing strategy) and the local accelerations and the global accelerations, and constitute

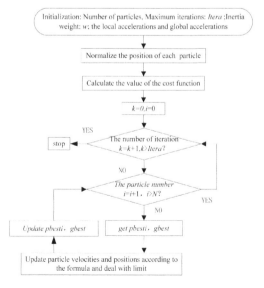

Figure 1. The flow chart of solving the optimization model using PSO.

the PSO tuning parameters. $r_1, r_2 \rightarrow U(0,1)$, $pbest_i^k$ is the best position found so far by the ith particle and $gbest^k$ is the global best position on the whole swarm. k is the iteration. a is the Scaling factor. The flow chart of the optimization model solved by PSO is illustrated by Figure 1.

3 THE FUZZY COMPREHENSIVE EVALUATION

Fuzzy comprehensive evaluation consists of six basic elements, U as the set of evaluation factors (set of all-level-model attributes), V as the evaluation level set (combination of all-level-model evaluation grade), R as the fuzzy relation matrix, A as the weight vector of evaluation factor, "o" as the fuzzy operator and B as the evaluation result vector.

1. Suppose $U = \{u_1, u_2, \ldots, u_m\}$ as the set of evaluation factors, such as safety, reliability, etc.
2. Suppose $V = \{v_1, v_2, \ldots, v_k\}$ as the evaluation level set, and $V = \{$better, good, common, bad, worse, worst$\}$ in this paper.
3. Get fuzzy membership matrix R.

$$R_i = \begin{bmatrix} r_{i1}(v_1) & r_{i1}(v_2) & \cdots & r_{i1}(v_k) \\ r_{i2}(v_1) & r_{i2}(v_2) & \cdots & r_{i2}(v_k) \\ \vdots & \vdots & \ddots & \vdots \\ r_{im}(v_1) & r_{im}(v_2) & \cdots & r_{im}(v_k) \end{bmatrix} \quad (5)$$

where R_i is a single-factor evaluation factor, that is the sub-set in V evaluation set, which corresponds to evaluation factor u_i (all-level-model attributes), $r_{ij}(v_s)$ $(j = 1, 2, \ldots, m; s = 1, 2, \ldots, k)$ is the degree of membership V evaluation set, where $r_{ij}(v_s) \in [0,1]$.
4. Fuzzy Comprehensive Evaluation. In this paper, the formula of fuzzy comprehensive evaluation model can be expressed as follows:

$$B_i = A \circ R_i = (a_1, a_2, \ldots, a_m)$$
$$\times \begin{bmatrix} r_{i1}(v_1) & r_{i1}(v_2) & \cdots & r_{ik}(v_k) \\ r_{i2}(v_1) & r_{i2}(v_2) & \cdots & r_{i2}(v_k) \\ \vdots & \vdots & \ddots & \vdots \\ r_{im}(v_1) & r_{im}(v_2) & \cdots & r_{im}(v_k) \end{bmatrix}$$
$$>> B_i = (b_{i1}, b_{i2}, \ldots, b_{ik}) \quad (6)$$

where $A = (a_1, a_2, \ldots, a_m)$ is the a weight set that is a fuzzy sub-set of U, $B = [b_{i1}, b_{i2}, \ldots, b_{ik}]$ is the sub-set in V evaluation set, Which corresponds to i-th scheme, b_j is the degree of membership V evaluation set. Moreover, if the summation of

$b_j(s = 1, 2, \ldots, k)$ is not equal to 1, take the normalization process: $\hat{b}_{is} = b_{is} / \sum\limits_{s=1}^{k} b_{is}$, and the result of evaluation is $\hat{B}_i = [\hat{b}_{i1}, \hat{b}_{i2}, \ldots, \hat{b}_{ik}]$.
5. Using the results of fuzzy comprehensive evaluation to sort the schemes.

In this paper, the schemes are sorted by the maximum membership degree which is chosen as the evaluation result of the schemes, and if the degree of the maximum membership is the same, then degree of the second largest membership replaces.

4 NUMERICAL ILLUSTRATION

The IEEE Garver-6 system example reference in [8] is tested for verifying the effectiveness of the proposed algorithm in this paper. This example has 2 layers structure, 16 attributes and 5 schemes. The data shows in Table 1.
For cost type attributes, its type is 1:

$$y = \frac{X_{\max} - X}{X_{\max} - X_{\min}} \quad (7)$$

For benefit type attributes, its type is 2:

$$y = \frac{X - X_{\min}}{X_{\max} - X_{\min}} \quad (8)$$

where, X, X_{max}, X_{min} is the actual, maximum and minimum value for all schemes on specified index. The proposed index System has four target index, so it can form 4 normalized decision matrix: $X_1 = [x_{ij}]_{5*8}$, $X_2 = [x_{ij}]_{5*2}$, $X_3 = [x_{ij}]_{5*4}$, $X_4 = [x_{ij}]_{5*2}$. Based on this, the combination weight calculated as follows: Firstly, using the entropy weight method and variation coefficient method to calculate the objective weight of the 4 normalized decision matrix. Then, using the AHP and order relation method to calculate the subjective weight of target layer and Operational layer. At last, solving the optimization model to obtain combination weight by using PSO. The combination weight obtained in this paper and the weight calculated by using DEAHP and AHP in reference [8] show in Table 1.
The fuzzy membership matrix R is obtained by using the membership function of a triangular fuzzy number $r_{ij}(V_j)$ as follows:

$$r_{ij}(V_j) = \begin{cases} [c_{ij} - (P_j - 0.7)]/0.7, & \max(0, P_j - 0.7) \leq c_{ij} \leq P_j \\ [(P_j + 0.7) - c_{ij}]/0.7, & P_j \leq c_{ij} \leq \min(1, P_j + 0.7) \\ 0 & \text{其他} \end{cases}$$
$$(9)$$

Table 1. The index system and data of schemes in IEEE Garver-6.

Target layer	Operating layer	Type	Scheme1	Scheme2	Scheme3	Scheme4	Scheme5	DEAHP	AHP	PSO
Technology X_1	Probability of load curtailments	1	0.0081	0.0131	0.0082	0.0087	0.003	0.0424	0.0573	0.0729
	Expected energy not supplied	1	2634.9	4903	2607	2666.8	487.4	0.0424	0.0573	0.0721
	Safety	2	1	1	1	1	1	0.141	0.1146	0.1263
	Rate of overloaded line	1	0.1333	0.4286	0.1429	0.1333	$1*10^{-7}$	0.0224	0.0152	0.0647
	Rate of light line	1	$1*10^{-7}$	0.1429	0.1429	0.0667	0.0625	0.0087	0.0032	0.0635
	Maximum load rate	1	0.9466	0.9894	0.9345	0.9455	0.8467	0.0204	0.0136	0.0644
	Load unbalance	1	0.2533	0.3804	0.3457	0.3221	0.2497	0.12	0.0061	0.0564
	Short-circuit current rationality	2	0.7994	0.7993	0.7993	0.7993	0.7993	0.0838	0.0382	0.1287
Economic X_2	Line investment costs	1	1468.2	1356.7	1532.8	1574	1585.7	0.0682	0.275	0.1156
	Annual operating costs	1	38.54	43.8	38.11	36.79	39.86	0.0042	0.0306	0.0551
Environment X_3	Power frequency electric field	2	1	1	1	1	1	0.1581	0.1146	0.0033
	Power frequency magnetic field	2	1	1	1	1	1	0.1581	0.1146	0.0033
	Efficiency of using land	2	10.1667	8.1071	7.8321	8.6071	9.8571	0.0351	0.0382	0.0425
	Coordination of landscape	2	0.75	0.9	0.45	0.6	0.5	0.0351	0.0382	0.0231
Adaptation X_4	The maximum power load	2	938.41	908.51	948.48	940.552	985.7	0.13	0.0693	0.044
	Expansion margin of network	2	0.375	0.4167	0.4167	0.375	0.3333	0.0381	0.0139	0.0639

where $j = 1,2, \ldots, 6$; $r_{ij}(V_j)$ is the membership of the j-th set of the i-th attribute. P_j is the parameter of the membership function, in this paper, $P_1 = 0$, $P_2 = 0.2$, $P_3 = 0.4$, $P_4 = 0.6$, $P_5 = 0.8$, $P_6 = 1$.

The fuzzy membership matrix R of scheme1 as follows:

Table 3 shows that the schemes can't be sorted by using the model of C²R, and Table 1 shows that the weight of Line investment costs which is obtained by AHP is very big and this index in scheme2 is best, which results scheme2 is the

$$
R^T = \begin{bmatrix}
0.278642 & 0.305223 & 1 & 0.555697 & 1 & 0 & 0.960651 & 1 & 0.304429 & 0.643367 & 1 & 1 & 1 & 0.52381 & 0.124794 & 0.285714 \\
0.564356 & 0.590937 & 0.714286 & 0.841411 & 0.714286 & 0.285614 & 0.753634 & 0.714286 & 0.590143 & 0.929081 & 0.714286 & 0.714286 & 0.714286 & 0.809524 & 0.410508 & 0.571429 \\
0.850071 & 0.876652 & 0.428571 & 0.872875 & 0.428571 & 0.571328 & 0.46792 & 0.428571 & 0.875858 & 0.785205 & 0.428571 & 0.428571 & 0.428571 & 0.904762 & 0.696223 & 0.857143 \\
0.864215 & 0.837634 & 0.142857 & 0.587161 & 0.142857 & 0.857043 & 0.182206 & 0.142857 & 0.838428 & 0.499491 & 0.142857 & 0.142857 & 0.142857 & 0.619048 & 0.981937 & 0.857143 \\
0.578501 & 0.55192 & 0 & 0.301446 & 0 & 0.857243 & 0 & 0 & 0.552714 & 0.213776 & 0 & 0 & 0 & 0.333333 & 0.732349 & 0.571429 \\
0.292786 & 0.266206 & 0 & 0.015732 & 0 & 0.571529 & 0 & 0 & 0.266999 & 0 & 0 & 0 & 0 & 0.047619 & 0.446634 & 0.285714
\end{bmatrix}
$$

(10)

The membership and sorting results of all schemes are shown in Table 2 and Table 3 respectively.

number one, but considering other indexes, such as Expected energy not supplied, etc, is worst in

Table 2. The membership of all schemes.

	Better	Good	Common	Bad	Worse	Worst
Scheme1	0.21008	0.226637	0.226239	0.178751	0.10695	0.051342
Scheme2	0.145829	0.104633	0.106856	0.147689	0.209355	0.285637
Scheme3	0.119273	0.149067	0.182088	0.199025	0.18835	0.162197
Scheme4	0.110688	0.148018	0.194605	0.215316	0.184487	0.146886
Scheme5	0.23964	0.196005	0.166331	0.132165	0.122414	0.143446

Table 3. The sorting results of all schemes.

	Scheme1	Scheme2	Scheme3	Scheme4	Scheme5
DEAHP	2	5	3	4	1
DEA	*	*	*	*	*
AHP	3	1	4	5	2
PSO	2	5	4	3	1

where "*" indicates that all schemes can't be sorted.

scheme2 and the weights corresponding to this indexes are smaller, so this results that scheme2 is the number one is obviously unreasonable. The method proposed in this paper take the objective weight into account, so the weight vector is more reasonable, such as the weight of Line investment costs is smaller than that in AHP, and the weight of the bad indexes, such as Expected energy not supplied, is bigger than that in AHP. the ranking results of the proposed method is consistent with that in reference [8], and the reason of the different sort between sheme3 and sheme4 is that in some indexes, for example Probability of load curtailments, scheme3 is better than scheme4, but in other indexes, for example rate of overloaded line, scheme4 is better than scheme3, moreover, weights of these indexes are small and the difference between data of scheme3 and scheme4 is little, This shows that the scheme3 and scheme4 are equivalent actually.

5 CONCLUSIONS

A fuzzy comprehensive evaluation method based on combination weight is proposed and the optimization model making the effect of expert experiences and data information together which is solved by PSO is built in this paper, at the same time, the method proposed in this paper can consider uncertainties and fuzzy factors which exists in the decision-making of grid planning, the calculation results show that the method proposed in this paper is clear in theory and convenient to calculate, and its results are scientific and intuitive.

ACKNOWLEDGEMENTS

This work is supported by Guang Dong Power Grid Development Research Institute Co., Ltd (Research on Evaluation System Power Grid Planning Alternatives) (030000QQ00120020).

REFERENCES

1. Cheng H.Z. 2008. Electric power system planning. Beijing: china electric power press.
2. Li X.H. & Zhang L. 2008. The research on the evaluation system for existing network based on analytic hierarchy process and Delphi method. Power System Protection and control 36(14):57–61.
3. Huang W.Z. & Kong D.J. 2001. Application of fuzzy comprehensive judgement in long-term network planning. Distribution & Utilization 5(18):7–9.
4. Zeng M. & Wang L. 2013. Risk Assessment of Smart Grid Based on Rank Correlation Analysis and Gray Triangle Clustering. East China Electric Power, 41(2): 245–248.
5. Lv P. & Qiao Y. & Ge L.T. 2010. Comprehensive decision-making of transmission network planning based on entropy TOPSIS. Journal of North China Electric Power University, 4(37):24–28.
6. Li N.N. & He Z.Y. 2009. Combinatorial weighting method for comprehensive evaluation of power quality. Power System Protection and control, 37(16): 128–134.
7. Kennedy J. & Ebehert R. 1995. Particle Swarm Optimization. Proceeding of the Neural Networks Perth. 1942–1948.
8. Liu Y.L. 2010. Comprehensive Evaluation and Decision-making for Transmission Network Planning Alternatives Based on DEAHP Method. Jinan, Shandong Province: Shandong University.

Power and Energy – Kong (Ed.)
© 2015 Taylor & Francis Group, London, ISBN 978-1-138-02782-4

Assistant decision-making method for distribution network planning based on TOPSIS and combination weight

Z.H. Li, Y. Lin & L. Xu
Guangdong Power Grid Development Research Institution Co. Ltd., Guangzhou, Guangdong, China

X.Y. Yang, J.S. Wu, W.H. Chen, K.M. Zhang & B.H. Zhang
*State Key Laboratory of Advanced Electromagnetic Engineering and Technology,
Huazhong University of Science and Technology, Wuhan, Hubei, China*

ABSTRACT: In order to solve the widespread problems of integrated decisions in grid planning, a comprehensive evaluation method of decision-making is proposed based on composition of TOPSIS method and combination weights. Used in network planning actual case, and describe the method of calculation process step by step. By constructing the optimization model, the subjective weights and objective weights are combined and particle swarm optimization is applied to strike the index weight, and improvement of TOPSIS method is completed. The actual results of the assessment plan cases confirmed the rationality and effectiveness of this method.

Keywords: power system; distribution network planning; TOPSIS; combination weights

1 INTRODUCTION

Network planning is an important basis for the development of the grid, the distribution network planning has generally been carried out currently, but how to scientific evaluate planning is still an urgent problem[1,2].

In the comprehensive evaluation decisions, the current study focused on a single application of subjective evaluation and objective evaluation methods, but the study of combination weight is lacking[4,6]. Reference [3] applies entropy and TOPSIS to evaluate transmission network planning. By constructing positive ideal solution and negative ideal solution, calculate the close degree between programs and positive ideal solution to determine the merits of the proposals. Reference [5] combines the fuzzy set theory and AHP method to evaluate each index of every layer. In these subjective evaluation methods, the calculation of weight all depends on the experts' opinion, so the result has a certain degree of subjectivity. Reference [7] uses IAHP method to calculate attribute weighting of index of every layer and program weighting of every index. Reference [8] combines the advantage of AHP method and select elimination method, and an auxiliary decision method based on distribution network planning is put forward.

Focus on the above problems; subjective weight and objective weight are combined in this paper. AHP, sequence relations are used to obtain subjective weight[5,9], and entropy, variation coefficient are used to obtain objective weight. Use TOPSIS to determine the merits of the proposals.

2 THE OPTIMIZATION MODEL FOR COMBINATION WEIGHT

2.1 Selection of scale system

In practical applications, 1–9 scale and index number scale are most widely used. Either method has its pros and cons, but the advantages of index number scale are particularly evident, when dealing with a small number of indexes. In the comparison of two factors, the weights obtained by 1–9 scale and index number scale method are listed in Table 1.

As Table 1 shows, when A is weakly important than B, the Weight ratio of A to B obtained by 1–9 scale is 0.75:0.25, which is a huge difference to people's expectations. By contrast, index number scale is more reasonable. Hence, index number scale method is selected to construct judgment matrix in this paper.

2.2 Build optimization model to evaluate combination weight

The set of planning schemes are $S = \{s_1, s_2, \ldots, s_n\}$, the set of indexes are $I = \{I_1, I_2, \ldots, I_m\}$, use $x_{ij}(i = 1, 2 \ldots, n; j = 1, 2 \ldots, m)$ to represent the evaluation value of s_i program. $n \times m$ evaluation value of

Table 1. Degree of importance weights at all levels.

Degree of importance at the ratio of A to B	Equally important	Weakly important	Obviously important	Intensely important	Extremely important
Weight ratio of A to B (index number scale)	0.5:0.5	0.63:0.37	0.75:0.25	0.84:0.16	0.9:0.1
Weight ratio of A to B (1–9 scale)	0.5:0.5	0.75:0.25	0.83:0.17	0.87:0.13	0.9:0.1

n planning schemes compose matrix $X = [x_{ij}]_{n*m}$, called normalized decision matrix. Experts use p subjective weighting methods and q objective weighting methods to obtain $p+q$ index weight vectors $\omega = [\omega_1, \omega_2, ..., \omega_p, \omega_{p+1}, ..., \omega_{p+q}]^T$, form a matrix as follows:

$$
\begin{matrix}
1 \\ \vdots \\ \vdots \\ \vdots \\ p \\ p+1 \\ \vdots \\ \vdots \\ p+q
\end{matrix}
\begin{bmatrix}
\omega_{1,1} & \omega_{1,2} & \cdots & \omega_{1,m} \\
\omega_{2,1} & \omega_{2,2} & \cdots & \omega_{2,m} \\
\vdots & \vdots & \ddots & \vdots \\
\omega_{p,1} & \omega_{p,2} & \cdots & \omega_{p,m} \\
\omega_{p+1,1} & \omega_{p+2,1} & \cdots & \omega_{p+1,m} \\
\vdots & \vdots & \ddots & \vdots \\
\omega_{p+q,1} & \omega_{p+q,2} & \cdots & \omega_{p+q,m}
\end{bmatrix}
$$

$$
subject\ to\ \forall i, \sum_{j=1}^{m} \omega_{ij} = 1, (i = 1, 2, ... p, ..., p+q) \quad (1)
$$

m is the number of index, from the first line to p line: p subjective weights of indexes, from $p+1$ line to $p+q$ line: q objective weights of indexes, subjective weights indicates the subjective preferences, objective weights indicates the objective of index data. In order to take the two factors into account and achieve the unity of both sides, combination of the above weights is needed. $W = (w_1, w_2, ..., w_m)$ is combination weight., it is a linear combination of $p+q$ kind of subjective and objective weights.

$$
W = \theta_1 \omega_1 + \theta_2 \omega_2 + \cdots + \theta_p \omega_p + \theta_{p+1} \omega_{p+1}
$$
$$
+ \cdots + \theta_{p+q} \omega_{p+q} = \sum_{i=1}^{p+q} \theta_i \omega_i, \forall i, \theta_i \geq 0; \sum_{i=1}^{p+q} \theta_i = 1
$$
$$(2)$$

In order to make the deviation between W and $p+q$ kind of subjective and objective weight, model is built as follows:

$$
min\ f_1(W) = \sum_{j=1}^{m} \left(\sum_{i=1}^{p} \alpha_i \left(w_j - \omega_{ij} \right)^2 + \sum_{i=p+1}^{p+q} \beta_i \left(w_j - \omega_{ij} \right)^2 \right)
$$

$$
s.t.\ 0 \leq w_j \leq 1, \sum_{j=1}^{m} w_j = 1, \sum_{i=1}^{p} \alpha_i + \sum_{i=p+1}^{p+q} \beta_i = 1;
$$

$$
w_j = \sum_{i=1}^{p+q} \theta_i \omega_{ij}, \forall i, \theta_i \geq 0, \sum_{i=1}^{p+q} \theta_i = 1 \quad (3)
$$

α_i and β_i represent the relative important degree of subjective weights and the objective weights respectively. It can be determined by AHP method. To facilitate the analysis, α_i and β_i is assumed the same in this paper. In this paper, AHP and sequence relations are used to obtain subjective weight, and entropy, variation coefficient are used to obtain objective weight.

In this paper, PSO algorithm is used to optimize the model which is established above. Generate initial particle randomly in the feasible solution space. Determine a value orientation for objective function, and according to the size of the adaptive value to measure the merits of the solution.

3 COMPREHENSIVE ASSESSMENT PROGRAM BASISED ON TOPSIS METHOD

TOPSIS is a kind of method of evaluation scheme of multiple choice based on multiple indices[3]. The basic idea is as follows: calculate the weighted Euclidean distance of solution relative to the positive ideal solution and negative ideal solution and then calculate the close degree of solution relative to the optimal solution on the basis of weighted Euclidean distance. Finally, determine the proposals. Based on this, close degree will be converted into score to make the result more intuitive. TOPSIS integrated evaluation steps are as follows:

Step 1: Measure the value of property of assessment program to determine the matrix, the matrix elements r_{ij} represent the normalized results of the j-th index of the i-th evaluation program.

$$
R = \begin{bmatrix}
r_{11} & r_{12} & \cdots & r_{1m} \\
r_{21} & r_{22} & \cdots & r_{2m} \\
\vdots & \vdots & \ddots & \vdots \\
r_{n1} & r_{n2} & \cdots & r_{nm}
\end{bmatrix} \quad (4)
$$

Step 2: Determine the absolutely positive, negative ideal point of program, positive ideal point $r^+ = [1,1, ... 1]$, negative ideal point $r^- = [0,0, ... 0]$.

Step 3: Get index weight through combination weighting method.

Table 2. Calculated result of single indicator.

First level	Second level	Type	Program 1	Program 2	Program 3	Program 4	Program 5	Weight
Power grid security	N-1 passing rate (A/B/C/D/E/F)	1	99%/99%/93%/ 74%/65%/55%	100%/100%/94%/ 84%/74%/65%	100%/100%/99%/ 94%/96%/74%	100%/100%/93%/ 83%/73%/59%	100%/100%/99%/ 93%/92%/73%	0.102097
	Ratio of single line and single transformer in high voltage distribution network (110/35)	2	3%/3%	0%/5%	2%/0%	0%/0%	0%/0%	0.031520
	Interconnection ratio of medium voltage distribution lines (A/B/C/D/E/F)	3	100%/100% 94%/84%/ 74%/60%	100%/100% 100%/99%/ 91%/89%	100%/100%/89%/ 71%/53%/45%	100%/100% 100%/ 95%/97%	100%/100%/94%/ 75%/66%/56%	0.044703
	Average number of transformer hooked into public line (A/B/C/D/E/F)	1	14.6%/14.8%/ 15.2%/15.6%/ 15.6%/53.2%	10.6%/10.7%/ 10.9%/11%/ 14.85%/42.5%	11.4%/11.5%/ 11.7%/11.8%/ 15.6%/44%	12.2%/12.5%/ 12.8%/13.2%/ 13.2%/48.6%	10.2%/10.28%/ 10.31%/10.%/ 13.3%/41.5%	0.045084
	Average transformer capacity hooked into public line	1	4.15	4.965	4.321	4.624	5.395	0.034254
	Supply radius of high voltage distribution network (A/B/C/D)	1	5.61/6.98/ 11.65/18.86	4.31/5.78/ 9.31/16.21	3.21/4.78/7.94/ 14.1	4.2/6.01/10.12/ 16.89	3.3/4.16/6.15/11.2	0.033112
	Typical connection ratio in high voltage distribution network	3	89%	86%	95%	97%	99%	0.02249
	Average length of medium voltage distribution public lines (A/B/C/D/E/F)	1	3.3/4.3/6.9/ 6.9/11.1/ 11.5	2.5/3.6/5.3/ 5.3/9.5/14.5	2.5/3.3/4.8/ 4.8/8.3/14	3.3/3.9/6.1/6.1/ 10.1/14.5	2.8/3.4/5.6/ 5.6/9.6/14	0.032005
Power grid reliability	Average interruption hours of customer (A/B/C/D/E)	2	0.0654/0.744/ 1.949/4.854/ 15.154	0.08/0.9094/ 2.335/6.094/ 16.354	0.5884/0.62/ 1.7404/4.2568/ 13.9688	0.0463/0.5686/ 1.3/3.496/ 12.898	0.02578/0.332/ 0.9528/2.156/ 10.9648	0.096915
	Electric reliability (A/B/C/D/E)	3	99.97%/99.96%/ 99.95%/ 99.92%/99.79%	99.99%/99.98%/ 99.97%/99.94%/ 99.81%	99.97%/99.96%/ 99.95%/99.94%/ 99.77%	99.99%/99.98%/ 9.97%/99.95%/ 99.80%	100%/99.99%/ 99.98%/99.95%/ 99.92%	0.155492
	Availability factor of transformer, circuit breaker overhead line	3	99.43%	99.91%	99.92%	99.97%	99.98%	0.045845
	Cable rates	3	33.64%	26.75%	34.77%	37.13%	40.53%	0.047748

(Continued)

Table 2. *Continued*

First level	Second level	Type	Program 1	Program 2	Program 3	Program 4	Program 5	Weight
Utilization rate of equipment	Maximum distribution line load rate of each voltage level (110/10)	3	50%/70%	35%/60%	40%/55%	30%/50%	20%/30%	0.064516
	The distribution of maximum distribution line voltage load rate (110/10)	3	60%/40%/ 0%/65%/ 35%/0%	45%/35%/20%/ 35%/40%/25%	65%/35%/0%/40%/ 30%/30%	80%/20%/ 0%/70%/ 30%/0%	60%/30%/ 10%/35%/ 50%/15%	0.037853
	Maximum distribution transformer load rate of each voltage level (110/10)	3	70%/80%	60%/70%	50%/50%	30%/60%	40%/40%	0.074129
	The distribution of distribution transformer load (110/10)	3	45%/35%/20% 65%/35%/0%	60%/40%/0%/35%/ 40%/25%	80%/20%/0%/70%/ 30%/0%	60%/30%/ 10%/35%/ 50%/15%	65%/35%/0%/45%/ 55%/0%	0.036638
Efficiency of investment	Unit cost of distribution capacity (110/10)	2	25/30	20/20	40.16/25	55/45	65/55	0.009572
	Unit cost of line length (110/10)	2	80/35	90/20	100/45	120/45	135.338/50	0.014201
	Distribution line loss rate	2	6%	6.1%	7%	8%	9%	0.013641
	The total investment yields	3	5%	4.5%	4.5%	4%	4%	0.015055
	Project capital of net profit margin	3	14%	13.16%	12%	13.16%	12%	0.015634
	Annual coast of unit load	2	20	40	53.772	30	53.772	0.026761

Step 4: Calculate the weighted Euclidean distance of solution relative to the positive ideal solution and negative ideal solution d_i^+, d_i^-:

$$d_i^+ = \sqrt{\sum_{j=1}^{m} \omega_j \left(r_{ij} - 1 \right)^2} \qquad (5)$$

$$d_i^- = \sqrt{\sum_{j=1}^{m} \omega_j \left(r_{ij} - 0 \right)^2} \qquad (6)$$

Step 5: Calculate the close degree of solution relative to the optimal solution C_i^+:

$$C_i^+ = \frac{d_i^-}{d_i^+ + d_i^-} \qquad (7)$$

Step 6: Convert the close degree into score S, then use S to determine the order of each program.

$$S = \frac{100 \times \left(1 - d_i^+ \right) + 100 \times d_i^-}{2} \qquad (8)$$

4 ANALYSIS OF EXAMPLE

4.1 *Membership function*

Membership function of interval index

$$y = \begin{cases} e^{(k1(x-a))} & x < a \\ 1 & b \leq x \leq a \\ e^{(k2(x-b))} & x > b \end{cases} \qquad (9)$$

where a, b are the upper limit and lower limit of the indicator's reasonable range, k_1, k_2 are the curves parameters.

Membership function of cost type index

$$y = \frac{x_{max} - x}{x_{max} - x_{min}} \qquad (10)$$

where x_{min}, x_{max} are the minimum value and maximum value of the indicator

Membership function of benefit type index

$$y = \frac{x - x_{min}}{x_{max} - x_{min}} \qquad (11)$$

where x_{min}, x_{max} are the minimum value and maximum value of the indicator.

4.2 *Analysis of example*

Take the safe and reliable, economic and efficiency indexes in a city of Guangdong province for example, calculated result of single index is shown as Table 2. Among the index types, 1 indicates interval type index, 2 indicates cost type index, 3 indicates benefit type index.

Firstly, through entropy evaluation method and coefficient of variation method, objective weight of every index in second level is calculated. Secondly, through AHP and Rank Correlation Analysis, subjective weight of every index is calculated. Finally, according to optimization model that constructed above, calculate the combination weight through PSO. Final weight factor of each evaluation indicator is shown in Table 2.

Utilize the calculated weight shown in Table 1; the close degree of every program to the optimal ideal program is obtained based on TOPSIS. The result is shown in Table 3.

The higher the Cable rate is the higher the Safety and reliability is, but it will also raise the Annual cost of unit load so as to lower the Economic efficiency. When the Maximum distribution line/transformer load rate is high, it means Equipment utilization efficiency is high, so the Economic efficiency strategic layer is good but at the same time it increase the Average transformer capacity hooked into pubic lines which influences Power grid security. Based on this, Safety and reliability and Economic efficiency is a contradiction indicator.

It can be concluded from Table 3, the program 1 to program 5, the Economic efficiency of the first program is the highest but the Safety and reliability is at a disadvantage. The Alternative program 2 and 3 have the similar level of safe and reliable, but

Table 3. Result of close degree of every program.

Planning program	Safe and reliability		Economic and efficiency		The total goal	
	Close degree	Score	Close degree	Score	Close degree	Score
Program 1	0.5121	51.4764	0.7448	75.0444	0.5652	57.7155
Program 2	0.6048	62.9269	0.6604	66.4109	0.6175	63.8024
Program 3	0.5768	59.6662	0.5895	59.458	0.5799	59.5801
Program 4	0.6693	69.5221	0.6693	59.5286	0.6477	66.5953
Program 5	0.7326	75.7013	0.5914	48.8611	0.6529	67.1704

the Economic efficiency of program 2 is obviously higher. The Alternative program 3 and 4 have the similar level of Economic efficiency, but the safe and reliable of program 4 is obviously higher. Program 5 is of the highest safe and reliable and the lowest Economic efficiency. Therefore, in the current weighting factor, taking the high safe and reliable requirements for the grid power supply into account, ultimately select program 5 as the optimal solution. The trade-off between the program 4 and 5 will depend on policy makers' balance between safe and reliable and economic efficiency.

5 CONCLUSION

A comprehensive evaluation decision-making method based on TOPSIS and combination-weighting method is proposed in this paper. The combination of subjective weighting method and objective weighting method make the weight more reasonable. It takes both expertise and theoretical support into account. Besides, the close degree of planning program to ideal program and the rank of programs is calculated by TOPSIS. The effectiveness and rationality for proposed method have been verified in actual case.

ACKNOWLEDGEMENTS

This work is supported by Guang Dong Power Grid Development Research Institute Co., Ltd. (Research on Evaluation System Power Grid Planning Alternatives) (030000QQ00120020).

REFERENCES

1. Cao, Y. & Meng, H.H. & Zhao, L. 2007. New rural low voltage distribution network comprehensive evaluation based on analytic hierarchy process. Power system technology 31(8):68–72.
2. Liu, L.G. & Zhao, W.L. & Wang, Z.D. 2013. Research on the construction of large power grids economic evaluation of the content and evaluation. Electric power construction: 22–26.
3. Lv, P. & Qiao, Y. & Ge, L.T. 2010. Comprehensive decision-making of transmission network planning based on entropy TOPSIS. Journal of North China Electric Power University. 4(37):24–28.
4. Lu, Z.G. & Ma, Y.L. 2011. The establishment of the distribution network evaluation index system of economic operation. Power system technology 35(3):108–112.
5. Wang, Q. & Wen, F.Q. & Liu, M. 2009. Comprehensive evaluation on the electricity market based on the fuzzy set theory and analytic hierarchy process method. Automation of electric power systems 33(7):32–37.
6. Wang, W. & Liu, Z.Q. & Zeng, B. 2011. City power grid comprehensive index system and evaluation model. Modern power 28(4):24–28.
7. Xiao, J. & Wang, C.S. & Zhou, M. 2004. Urban power grid planning and comprehensive evaluation decision Based on interval analytic hierarchy process. Proceedings of the CESS 24(4):50–57.
8. Zhang, T.F. & Yuan, J.S. & Kong, Y.H. 2006. Auxiliary decision method of distribution network planning Based on the analytic hierarchy process and select elimination III. Proceedings of the CSEE 26(11):121–127.
9. Zhang, G.H. & Zhang, J.H. & Peng, Q. 2009. Power grid safety evaluation index system and method. Power system technology 33(8):30–34.

Power and Energy – Kong (Ed.)

Short-term wind power completing and forecasting based on Rough Set and BP neural network

Y.F. Zeng, B.H. Zhang, J.L. Wu, B.J. Jin, M. Li & S. Zhao
State Key Laboratory of Advanced Electromagnetic Engineering and Technology,
Huazhong University of Science and Technology, Wuhan, Hubei, China

K. Wang & D. Zeng
China Electric Power Research Institute (Nanjing), Nanjing, Jiangsu, China

ABSTRACT: When we use BP neural network to forecast the wind power, the selection of input variables is the key issue that affects the forecasting results. This paper proposes a method using rough set theory to solve this problem. On the basis of considering the correlation of the wind turbines, we use the Rough-Set analysis methods extract turbines which are strong associated, then use the wind power data of them as the partial inputs of BP neural network and build the RS-BP model (Neural network based on Rough-Set). For the case of wind power data missing, we use the RS-BP to complete the missing data. Then, we use the completed wind power data to forecast the wind power data. This method can select the inputs of neural network reasonably and has the better prediction accuracy. The numerical example shows that this method is effective and feasible.

Keywords: Rough-Set; BP neural network; wind power forecasting; correlation

1 INTRODUCTION

The wind power forecasting has great significance for safe and economical operation of power system. Wind power prediction methods include the Persistence method, ARMA, ANN, and so on. Excessive BPNN (BP Neural Network) inputs can't improve the calculation accuracy, and will reduce computational efficiency. So, there is a problem to select the main factors, when considering various factors that impact the wind power.

Rough-Set theory deals with imprecise ore uncertain mathematical information. By the Rough-Set Algorithm, we can extract major factors among numerous influencing factors. Some scholars used the Rough-Set to select the various weather factors, and then combine with neural networks for forecasting. The RS-BP model proposed on this paper, use the Rough-Set to extract associated turbines from numerous adjacent turbines, then use the data of associated-turbines as inputs. The combination of Rough-Set and BPNN, can exert respective advantages and improve computational efficiency.

2 ROUGH-SET THEORY

2.1 Conception of Rough-Set

The main idea of Rough-Set theory is using the known knowledge to describe the imprecise or

uncertain knowledge approximately. For the information system $S = (U, A, V, f)$, U is the non-empty finite set of objects, called domain; A is the non-empty finite set of attributes; V is the total range of values of attribute a; $f : U \times A \rightarrow V$ is a information function. For the knowledge base $K = (U, C)$, C is the condition attributes, D is the decision attributes. The lower and upper approximations are defined as equation 1.

$$\begin{cases} \underline{C}X = \{x \in U \,|\, [x] \subseteq X\} \\ \overline{C}X = \{x \in U \,|\, [x] \cap X \neq \varnothing\} \end{cases} \quad (1)$$

A is an object in U, $[x]$ is the equivalence class that including x divided according to the C. The approximation-classified quality is as equation 2.

$$\gamma_C(D) = \frac{|\underline{C}X|}{|U|} \quad (2)$$

2.2 Discretization

Rough-Set theory to deal with objects that is discrete values. Discretization methods include equidistance discretization, equal frequency discretization, based on importance of attribute discretization and so on. Here we use equal frequency discretization to discrete the information system, then we get the decision tables. Equal frequency

103

discretization makes each frequency interval has the same number of numerical.

2.3 *Reduction algorithm*

The reduction algorithm aimed to reduce the unimportant knowledge on the condition of keeping the classification ability. Here we use the importance index to extracting the attributes. The importance of attributes subset $c_i \subseteq C$ about D is defined as equation 3.

$$\sigma_{CD}(c_i) = \gamma_C(D) - \gamma_{C-c_i}(D) \qquad (3)$$

We use an example to explain the reduction algorithm. The decision table of example is Table 1. We calculate the equivalence class of each condition attribute as:

$$\begin{cases} U/\{c_1,c_2\} = \{\{e_1,e_2,e_3\},\{e_4,e_6,e_8\},\{e_5,e_7\}\} \\ U/\{c_1,c_3\} = \{\{e_1\},\{e_2\},\{e_3\},\{e_4\},\{e_6,e_8\},\{e_5,e_7\}\} \\ U/\{c_2,c_3\} = \{\{e_1,e_4\},\{e_2\},\{e_5,e_7\},\{e_3,e_6,e_8\}\} \\ U/C = \{\{e_1\},\{e_2\},\{e_3\},\{e_4\},\{e_5,e_7\},\{e_6,e_8\}\} \end{cases}$$

The approximation classified quality and the importance are:

$$\gamma_C(D) = 0.5, \quad \gamma_{C-c_1}(D) = 0.375$$
$$\gamma_{C-c_2}(D) = 0.5, \quad \gamma_{C-c_i}(D) = 0$$

$$\begin{cases} \sigma_{CD}(c_1) = \gamma_C(D) - \gamma_{C-c_1}(D) = 0.125 \\ \sigma_{CD}(c_2) = \gamma_C(D) - \gamma_{C-c_2}(D) = 0 \\ \sigma_{CD}(c_3) = \gamma_C(D) - \gamma_{C-c_3}(D) = 0.5 \end{cases}$$

Form the results we can know that c_3 is the most important condition attribute, then is the c_1, the c_2 is the redundant attribute.

Table 1. Example of the decision table.

| Sample | Condition attributes (C) | | | Decision attributes (D) |
	c_1	c_2	c_3	d
e_1	1	1	0	0
e_2	1	1	1	1
e_3	1	1	2	1
e_4	0	1	0	0
e_5	0	0	1	0
e_6	0	1	2	1
e_7	0	0	1	1
e_8	0	1	2	0

3 RS-BP FOR WIND POWER FORECASTING

First, for the case of missing wind power data, we use the RS-BP (Neural network based on rough sets) to complete the missing data, and then use the completed wind power data for forecasting.

3.1 *RS-BP model*

The BPNN is multi-layer forward neural network based on error back propagation algorithm. It

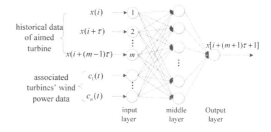

Figure 1. The RS-BP model.

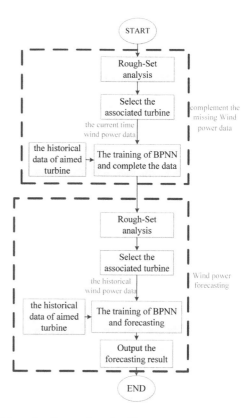

Figure 2. Forecasting algorithm process.

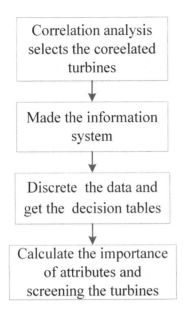

Correlation analysis selects the coreelated turbines

↓

Made the information system

↓

Discrete the data and get the decision tables

↓

Calculate the importance of attributes and screening the turbines

Figure 3.   Rough-Set analysis process.

can be precisely approaching any nonlinear mapping and has the ability of fault tolerance. We use the wind power data of associated turbines and the historical data of aimed turbine as the inputs of BPNN. The RS-BP model is shown in the Figure 1.

Here, we call the turbine for wind power forecasting as aimed turbine, the selected turbines by reduction algorithm as associated turbines.

When we use the RS-BP to complement the missing data, the inputs of BPNN are: 1) the historical data of aimed turbine; 2) the current time wind power data of associated turbines. When we are forecasting, they are: 1) the historical data of aimed turbine; 2) the historical data of associated turbines.

3.2   Algorithm flow

The wind power forecasting algorithm process based on RS-BP model is shown in Figure 2. The process of Rough-Set analysis is shown in Figure 3.

4   NUMERICAL EXAMPLES AND ANALYSIS

The data of numerical examples is provided by United States Renewable Energy Laboratory (NERL). The 66 wind turbines called FJ1-FJ66 come from one farm, the FJ1 called aimed turbine, we will extract associated turbines from another

65 turbines, then build the RS-BP model and complete the wind power and forecasting.

4.1   The determination of inputs

First, we choose the wind power data of 66 turbines within 5 days as sample, analyze the correlation between the turbines in a long time scales. After that, we select 9 turbines which have strong correlation with aimed turbines FJ1, they are FJ3, FJ9, FJ12, FJ23, FJ27, FJ35, FJ37, FJ48, FJ49.

We use the 9 turbines' data that within 150 minutes before the data missing, build the information system. The decision attribute D is the wind power of FJ1, condition attribute $C = \{c_1, c_2, ..., c_9\}$ is the wind power of 9 turbines. For the information system, we discrete it by $n = 5$, then get the discretion table and calculate the importance of each condition attribute as:

$$\begin{cases} \sigma_{CD}(c_2) = \dfrac{2}{15}, & \sigma_{CD}(c_8) = \dfrac{2}{15} \\ \sigma_{CD}(c_1) = \sigma_{CD}(c_3) = \sigma_{CD}(c_4) = 0 \\ \sigma_{CD}(c_5) = \sigma_{CD}(c_6) = \sigma_{CD}(c_7) = \sigma_{CD}(c_9) = 0 \end{cases}$$

From the importance, we can know that determinates are c_2, c_8, that means the associated turbines are FJ3, FJ9.

4.2   Completing of missing data

To complete the wind power data, here we use the current data of FJ3, FJ9 and historical data of FJ1 as the inputs of BPNN, then complete the data within 200 minutes. The result is shown in Figure 4.

The relative error of it is 8.23%, by the correlate analysis, we know that the completion do not change the correlation of turbines, the result of completion is good.

4.3   Results of wind power forecasting

Here we use the completed wind power data to forecasting. Analyzing the wind power data that before forecasting data, we extract FJ3, FJ9, FJ35

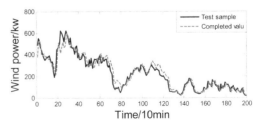

Figure 4.   Completing wind power data using RS-BP.

as the associated turbines. Then using the RS-BP model to forecasting the wind power data for one-step (ten minutes ahead), there-step (thirty minutes ahead), six-step (sixty minutes ahead), we get the prediction wind power data in 24 hours of one day. The results are shown on Figures 5–7.

4.4 *Results analysis*

To analyze the results of forecasting, we adopt the indexes of MAE (mean absolute error), RMSE (RMS error), MAPE (average relative error) to describe the accuracy of forecasting. The calculate formulas of them shown as formulas 4. y' means the wind power prediction value of t moment, y_t is the wind power data of t moment, n is the length of forecasting time. The value of each index is on the Table 2.

$$\begin{cases} MAE = \frac{1}{n}\sum_{t=1}^{n}\left|y_t' - y_t\right| \\ RMSE = \sqrt{\frac{1}{n}\sum_{t=1}^{n}\left(y_t' - y_t\right)^2} \\ MAPE = \frac{1}{n}\sum_{t=1}^{n}\left|\frac{y_t' - y_t}{y_t}\right| \end{cases} \quad (4)$$

From the error analysis table, we can see that the forecasting method based on RS-BP model has higher prediction accuracy; the average relative error is lower than 15%. For predicting step change, this method has stronger adaptability.

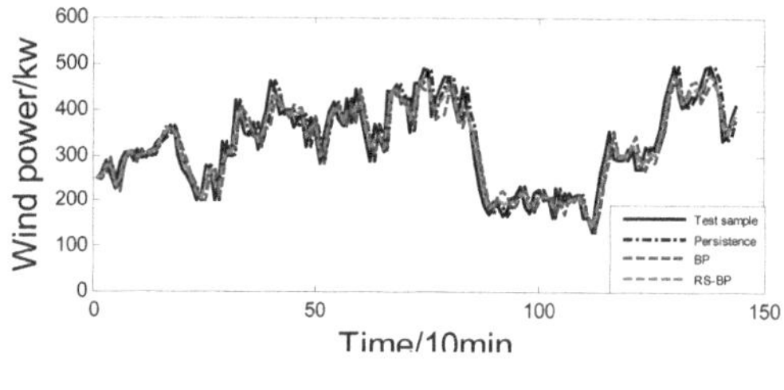

Figure 5. Ten minutes ahead.

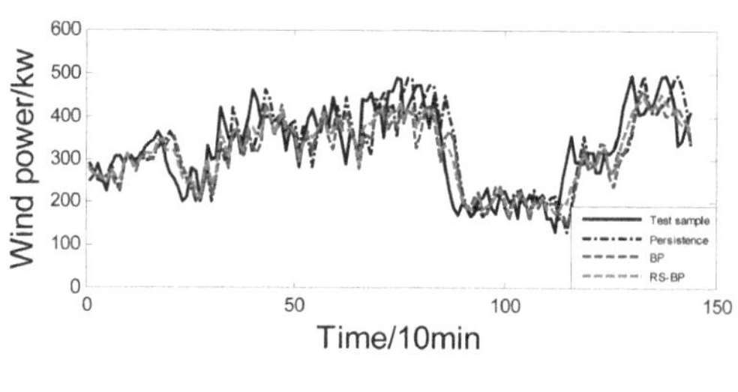

Figure 6. Thirty minutes ahead.

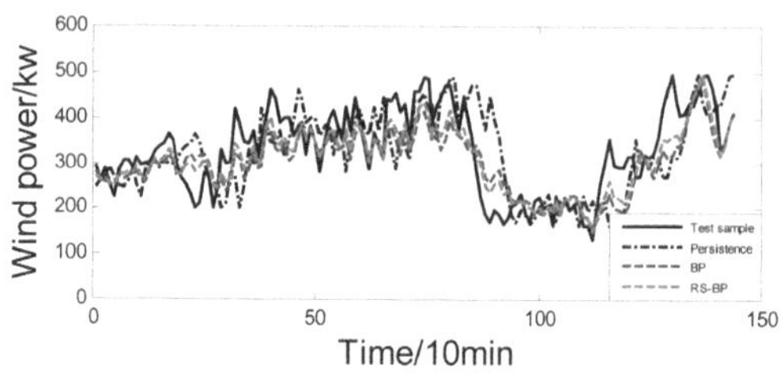

Figure 7. Sixty minutes ahead.

Table 2. Error analysis.

		Evaluation index		
Forecasting step	Methods	MAE (kw)	EMSE (kw)	MAPE (%)
10 min	Persistence	34.11	42.79	10.97
	BPNN	35.29	44.15	11.10
	RS-BP	31.98	39.32	9.54
30 min	Persistence	50.72	66.32	16.47
	BPNN	47.95	62.21	14.42
	RS-BP	41.73	58.28	12.65
60 min	Persistence	63.20	83.05	21.11
	BPNN	49.08	60.67	15.65
	RS-BP	46.87	62.01	14.23

5 CONCLUSIONS

Using the Rough-Set analysis methods based on correlation analysis can extract turbines that are strong associated effectively, then use the wind power data of associated turbines as the inputs of neural and build the RS-BP model. The RS-BP model for wind power completing has good data completing effect. We use the completed wind power data to forecasting; the numerical examples show that the method is effective and feasible. For predicting step change, this method has stronger adaptability.

ACKNOWLEDGEMENTS

This work was supported by State Grid Corporation of China (the Study of Steady-state Analysis Method in the 'Source-Grid-Load' Interactive Environment) (dz71-13-036).

REFERENCES

1. Gao, S. & Dong, L. & Gao, Y. 2012. Mid-long term wind speed prediction based on rough set theory. *Proceedings of the CSEE* 32(1):32–37.
2. Gu, X.K. & Fan, G.F. & Wang, X.R. 2007. Summarization of wind power prediction technology. Power system technology 31(2):335–338.

3. Kariniotakis, G. & Nogaret, E. & Dutton, A.G. 1999. Evaluation of advanced wind power and load forecasting methods for the optimal management of isolated power systems. Proceedings of the European Wind Energy Conference: 1082–1085.

4. Kariniotakis, G. & Stavrakakis, G. & Nogaret, E. 1996. Wind power forecasting using advanced neural network models. IEEE Transactions on Energy Conversion 11(4):762–767.

5. Li, P. & Song, K. & Xiao, B. 2008. Medium and long-term load forecasting using associated clustering based on rough sets. Relay 36(1):43–47.

6. Xie, H. & Cheng, H.Z. & Zhang, G.L. 2003. Applying rough set theory to establish artificial neural networks for short term load forecasting. Proceedings of the CSEE 23(11):1–4.

7. Zhang, W.X. & Wu, W.Z. & Liang, J.Y. 2001. Rough set theory and method. Beijing: Science Press.

8. Zhou, S.X. & Lu, Z.X. 2011. Wind power and power system. Beijing: China Electric Power Press.

Power and Energy – Kong (Ed.)
© 2015 Taylor & Francis Group, London, ISBN 978-1-138-02782-4

Research on SVM-based ice thickness model of transmission lines

Weiwei Yang, Xuemin An, Mingfeng Wu, Huiping Zheng & Shuyong Song
State Grid Shanxi Electric Power Research Institute, Taiyuan, Shanxi, China

ABSTRACT: Ice-covering is a kind of special meteorological condition, and the existence of it seriously affects the safe operation of overhead transmission lines. In recent years, due to frequent occurrence of snow and ice disasters, ice problems and research work has become a symbol to evaluate whether a country's electric industry is able to deal with ice emergency and whether an electric enterprise is modern or not. Thus, ice problem immediately becomes a hot issue. By analyzing the meteorological parameters and related historical data of No. 109 tower located in a 220 kV transmission line in a region where is easier to produce ice covering, this paper studies on the establishment of ice thickness model of transmission lines, which is based on SVM (Support Vector Machines), to predict the ice thickness identification in the future so as to guide the operation and prevent troubles.

1 INTRODUCTION

The world's earliest recorded transmission lines ice accident occurred in the United States in 1932. Ice-covering causes serious damages, such as the power transmission hardware damage, wire stocks breaking, disconnection, line galloping, tower tilting or collapse, or insulator string flashover. So far, ice problem is still threatening and influencing the safe operation of power grid, thus, positive monitoring of ice growth can effectively prevent ice disaster, and reduce the loss of manpower and financial resources.

The ice model in the world is mainly studied from the viewpoint of meteorology, fluid mechanics, and thermodynamics to research on the ice-covering mechanism of the transmission line conductor and insulator. The ice forecast model has reached as many as 20. However, so far, neither empirical model nor theory model was proved to be perfect. The same ice events might be predicted quite differently. Therefore, it is necessary to design a new method for the establishment of the ice model.

2 THE MECHANISM OF ICE FORMATION AND THE INFLUENCING FACTORS

2.1 The mechanism of wire ice formation

Cold rain splashing on the objects with temperature below freezing point (0 °C) forms ice, and if it condenses on transmission line, it is line ice. Ice forms when the following conditions are fulfilled: firstly, the temperature should be below 0 °C; secondly, it must have enough humidity (RH% > 85%); thirdly, water droplets in the air moves with the wind speed above 1 m/s.

2.2 The influencing factors of wire ice

There are many factors that can influence the line ice, mainly including meteorological factors, terrain and geographical conditions, altitude and line suspension height, etc.

a. Meteorological factors: mainly including air humidity, environmental temperature, wind speed, wind direction, the diameter of the cooling water droplets or fog droplets in the air, condensation level, etc.
b. The terrain and geographical conditions: mainly including mountain strike, slope direction and watershed etc.
c. Altitude and line suspension height: the higher the line suspension height is, the easier ice formation will be and the thicker the ice will be.

3 DATA ACQUISITION

The selected data of this article is derived from the ice online monitoring system, which is installed on No. 109 tower of 220 kV transmission line located in the leeward slope of a certain area of mountainous watershed topography. Such topography is easy to form strong winds and severe ice conditions, especially in the top of the mountain and the windward slope. When it is windy, the air mass containing supercold droplet is forced up along the mountainside and then adiabatically expands, which makes the supercold droplet content increase and thus the line ice increases. With this system, it can collect real-time data of the transmission line weight and the inclination angle of the insulator string, and can also measure the environmental temperature and humidity, wind speed, wind direction and other meteorological information. All sorts of relevant data parameters are acquired and packed. With the help of GSM wireless network, the data are transmitted to monitoring center, and the thickness of ice can be calculated at the query terminal using mechanics formula.

Through analyzing the data acquired from the ice online monitoring system, ice quantity model of transmission line at the No. 109 tower was established so that the ice quantity in the future can be identified and predicted. Accordingly, the corresponding measures for removing or preventing the ice were put forward so as to guide the production and reduce the consumption of manpower and financial resources.

4 INTRODUCTION AND ADVANTAGES OF THE MECHANISM VECTOR MACHINES

Learning is a basic human intelligent activity, and learning ability is the fundamental characteristic of human intelligence. However, machine learning refers to the process that machine simulates and realizes varieties of learning behaviour in artificial intelligence system.

SVM (Support Vector Machines) is fundamentally based on SLT (Statistical Learning Theory), which is a kind of machine learning theory specialized on limited sample. SVM exhibits many unique advantages in view of small sample, nonlinear and high dimensional pattern recognition.

Based on structural risk minimization principle rather than the traditional empirical risk minimization principle, SVM have gradually developed into a new type of structured learning method, which is good at establishing high-dimensional model with limited sample and the model performs well in prediction.

5 ESTABLISHMENT OF INTELLIGENT ICE-COVERING IDENTIFICATION MODEL BASED ON SVM

5.1 Data processing

The ice online monitoring system could detect many different types of data. In this paper, the most representative meteorological factors were selected as a basis for data simulation and identification, including the factors like temperature,

humidity, rainfall as input ones, and the ice thickness as output one. Obviously, these data bear different magnitude orders, and the individual data gap is too small so that data normalization must be considered.

5.2 Simulation modelling based on SVM

When modelling based on SVM, half of the sample data are selected for training and the data input and output were normalized to [−1,1] by using MATLAB simulation software so that the training speed could be improved.

Then another half of the data are selected as test data, and the trained SVM model was tested with MATLAB simulation. Then, SVM-based ice thickness model was established. In this model, the input is meteorological parameter (temperature, humidity and rainfall) for vector x, the output is ice thickness for vector y. Specific steps are shown in Figure 1. In addition, the basic operations and

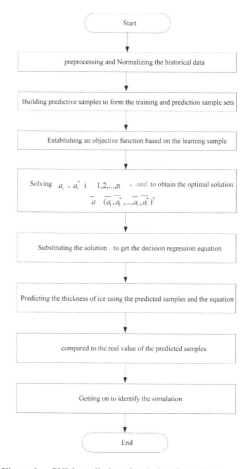

Figure 1. SVM prediction simulation flow chart.

operating results of the programming process are described as follows:

1. First of all, download and install lssvm toolbox under MATLAB directory;
2. Secondarily, to optimize the parameters. To optimize the parameter with tune1 ssvm function came from lssvm. The function format is like this: [gam, sig2, cost] = tunelssvm ({Pn, Tn, type, gam, sig2, kernel, preprocess}). Then the gam which is a regularization parameter to decide the minimalized fitting error, and the sig2 which is used to determine the smoothing degree are obtained. The optimization results are listed as follows: gam = 9976.2996, sig2 = 9.9439, and then to integer: gam = 10000, sig2 = 10. The specific optimization process is shown in Figure 2.
3. Thirdly, to establish the model. Firstly, put the optimization results into the program, and then establish the model for regression prediction and simulation of the training sample; secondly, to test the trained SVM model by using testing sample to get the simulation curve of the ice thickness, the results are shown in the following Figure 3 (c).

From Figure 3(a), we can find that the original training data curve is exactly the same as the regression prediction data curve; from Figure 3(b) we can find that fitting degree of the training sample forecast curve and the actual curve is quite high so that it can be concluded that the training sample's simulation results of ice thickness model coincides with the real value; from Figure 3(c), we can find that the raw data test curve is almost the same as the regression prediction curve; from Figure 3(d), we can find that the fitting degree of the test sample forecast curve and the actual curve is quite high, basically there is no obvious discrete points.

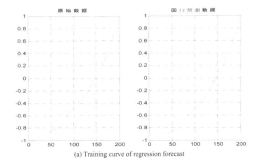

(a) Training curve of regression forecast

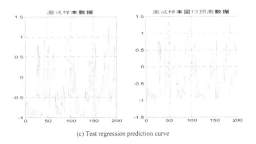

(b) Training fitting curve

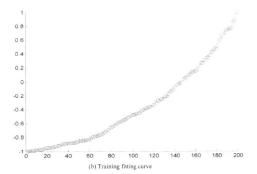

(c) Test regression prediction curve

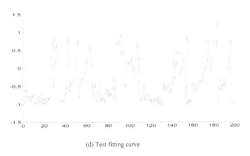

(d) Test fitting curve

Figure 3. SVM model simulation curve of the ice cover thickness.

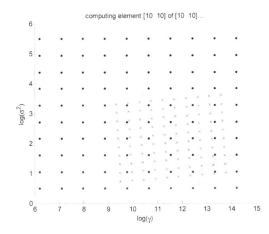

Figure 2. The results of SVM parameter optimization.

In Table 1, the Sum-square Error of the SVM ice thickness model and the Mean Squared Error are presented, to be specific, SSE = 4.6579 e-004, and MSE = 2.3407 e-006. We can see from the performance parameters that SVM has strong nonlinear

Table 1. Comparison table of performance indexes of SVM.

Performance indicators	SVM
SSE	4.6579e-004
MSE	2.3407e-006
The simulation time (s)	3.625360

fitting ability and higher prediction accuracy, and it completely conforms to the requirements of the engineering application, since it's accurate and effective for the ice thickness identification and prediction.

6 CONCLUSION

The extreme weather such as ice and snow disaster occurred frequently, which leads to a series of ice problem to bring a huge threat for the power grid. The changes of ice amount on the transmission lines are not only influenced by fixed factors, but also influenced by many random factors, some of which cannot be identified to have what impaction on ice amount. In addition, the regional difference renders different ice amount. Therefore, particular case should be analyzed particularly. To fully research and analyze on historical data and then establish proper identification model is the key to obtain accurate prediction result. This paper, according to the geographic location and historical data of No. 109 tower located in the 220 kV transmission line in a region where it is easier to form ice-covering, analyzed the data and considered the influences of meteorological parameters on ice amount. SVM was applied to ice thickness identification model and the ice prediction model of a 220 kV transmission line was established. The model identified the thickness of ice through the meteorological parameters, and can predict the ice thickness in the future. In the coming days, it can guide the production, reduce human, material and financial resources, and be ready to respond to disasters and prevention in advance. In addition, it's necessary to continue to study on ice disaster with various measures to reduce ice-covering disaster of transmission lines so as to better and accurately guide the transmission operation.

REFERENCES

1. D. Jiang Xing Liang, "The Present Study on Conductor Icing of Transmission Lines," Chongqing: Chongqing University, 1997.
2. J. Bai Liang, Lao Songyang, and Hu Yanli, "Comparative Research on Support Machine Algrithm," Computer Engineering and Applications, Vol. 38, No. 3, 2005, pp. 79–81.
3. J. Xie Yunhua, The Relation Between Meteorological Elements and Ice Accretion in Three Gorges Region.
4. Electric Power, Vol. 38, No. 3, 2005, pp. 35–39.
5. M. Deng Naichao and Tian Yingjie, "New Methods In Data Mining: SVM," Science Press, 2004.
6. M. Jiang Xingliang and Yi Hui, "Transmission Line Ice-covering and Prevention," China electric power press, 2002R.
7. Wu Wenhui, "Cause and Precaution Measure for Tripping Trouble of Transmission Line Covered with Ice," High Voltage Engineering, Vol. 32, No. 2, 2011, pp. 109–111.

Power and Energy – Kong (Ed.)

A cloud automatic recognition method for PV power prediction

Jianfeng Che, Feng Shuanglei & Bo Wang

China Electric Power Research Institute, Beijing, China

ABSTRACT: Fluctuations of solar irradiance are known to have a significant influence on electric power generation by solar energy systems, and cloud-obscuring sun is the main reason of solar irradiance fluctuations. The traditional PV power prediction methods can't solve this problem effectively. The PV power prediction based on cloud images is a feasible way: Recognize cloud clusters in the consecutive image frames automatically, judge the motion direction of cloud clusters, and estimate the situation of cloud obscuring sun, finally, modify the PV power prediction result. In this method, cloud clusters recognition is important basic. Recognition result is served for PV power prediction, so it needs to be finished automatically, accurately and quickly. In our paper, a cloud clusters recognition method based pattern recognition and image processing was proposed, which was verified by test cloud images. The result indicated that this method had high accuracy and met the need of engineer.

1 INTRODUCTION

Generation of photovoltaic power station has a large fluctuation. The largest fluctuation in a few minutes can reach 80 percent to 90 percent of installed capacity, which may lead to power system frequency fall-off. It is necessary to develop minutes-level ultra-short-term power prediction of PV station.

Cloud clusters blocking sun is the main reason of PV generation fluctuation, conventional PV power prediction methods [1–5] show only limited accuracy due to lack of information on cloud situation. The prediction method based on cloud images is an effective way to solve this task, which uses sequences of cloud images to extract information on clouds motion. Pattern recognition and image processing algorithms for motion detection can be applied. For this method, cloud automatic recognition is one key and difficult point.

Several approaches to cloud automatic recognition have been proposed. Methods based on graph cut model, Logistic regression model and Mumford-Shah model have been used to detect clouds by Wenlong Fei et al. [6–9]. Zhiying Lu et al. applied wavelet transformation and cluster analysis to extract cloud image features [10–12].

The purpose of this paper is to present a cloud automatic recognition method. The approach is based on pattern recognition and image processing algorithms, which performs with high accuracy and has the advantage of short computing times.

2 IMAGE PREPROCESSING

Cloud images are caught by image acquisition equipment automatically, due to factors such as weather conditions, actual images may have the characteristics of low contrast and noisy, large error will be produced if processing the images directly, so it is necessary to preprocess the images. Several filter algorithms are introduced in this paper.

2.1 Median filter

Median filter is non-liner smoothing technique, which sets every pixel value to median value of neighborhood pixels within window. Median filter can eliminate sharp different points and reduce influence of noise. The mathematical expression of median filter is as follows:

$$g(x, y) = \text{med}\{f(x{-}k, y{-}l), (k, l \in W)\} \tag{1}$$

where, $f(x, y)$ is the original image, $g(x, y)$ is the processed image; W is two-dimensional template, which is usually 3×3, 5×5 rectangle, and also can be some different shapes, such as linear, circular, ring etc.

2.2 Mean filter

Mean filter is typical liner filtering algorithm, which mainly uses neighborhood average method to smooth image. Mean filter is a simple, intuitive and easy to implement method of smoothing images, which can reduce the amount of intensity variation between one pixel and the other. It is often used to reduce noise in images.

The idea of mean filter is simply to replace each pixel value in an image with the mean ('average') value of its neighbors, including itself. This has the effect of eliminating pixel values, which are unrepresentative of their surroundings. The pattern of neighbors is called the "window", which represents the shape and size of the neighborhood to be sampled when calculating the mean. Often a 5×5 square template is used, and larger templates (e.g. 7×7 squares) can also be used for more severe smoothing. The mathematical expression of median filter is as follows:

$$g(x, y) = 1/m \sum f(x, y) \qquad (2)$$

where, $f(x, y)$ is the original image, $g(x, y)$ is the processed image; m is total pixel number of the template including current pixel.

3 CLOUD RECOGNITION ALGORITHMS

According to atmospheric science theory, atmospheric molecular scatter is inversely proportional to wave length. Sunny sky is blue, however, in visible light, cloud molecular scatter is not closely tied to wave length, and cloud is white in the sky, therefore, we used the method based on color difference to recognize cloud in the image.

The actual cloud images are RGB color model images, which have three channels: red channel(R), green channel(G) and blue channel(B). The RGB color model is an additive color model in which red, green, and blue light are added together in various ways to reproduce a broad array of colors.

The actual cloud images are 2-D digital images. The digital images are made by pixels. Each pixel corresponds to one position in the image.

3.1 One-channel image

Each pixel has a value. The value ranges between 0 and 255, which shows different gray levels, 0 stands for black color, and 255 stands for white color, as shown in the Figure 1.

3.2 Three-channel RGB image

Each pixel has three values that are R channel value, G channel value and B channel value. R, G, B are added together in different ways to reproduce various colors. Each channel value ranges between 0 and 255, as shown in the Figure 2.

According to color characteristic of cloud and sky in the image, we set a judgment threshold based on red channel and blue channel value to

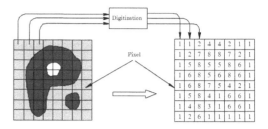

Figure 1.　One-channel image.

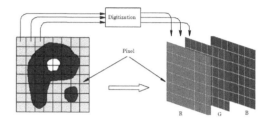

Figure 2.　RGB image.

detect cloud in the image. Assuming that the original image is $img(m,n)$, which has three channels, the processed image is $pimg(m,n)$, which is binary image, the algorithm we proposed is as follows:

$$p\_value(x, y)$$
$$= \begin{cases} 255, b\_value(x,y)/r\_value(x,y) > T; \\ 0, b\_value(x,y)/r\_value(x,y) \leq T. \end{cases} \qquad (3)$$

where,
$p\_value(x,y)$ is the pixel value of $pimg(m,n)$;
$b\_value(x,y)$ is the pixel value of $img(m,n)$ blue channel;
$r\_value(x,y)$ is the pixel value of $img(m,n)$ red channel;
$x \in m$, $y \in n$;
T is the threshold value.

4 COMPUTATIONAL EXAMPLES

Take several cloud images as examples, the images include cloud and sky. We process them in four steps to recognize the clouds and mark them in black (pixel value is 0):

Step 1: (Image preprocessing) Analyze the images; median filter is used to enhance the image. We take 3×3 rectangle as smooth template.

Step 2: (Judgment threshold) According to radiative transfer theory, we simulate the situation by numerical calculation. Based on image sample set, the judgment threshold is determined.

114

Step 3: (Cloud recognition) Based on the judgment threshold, blue and red ratio method is used to detect the clouds, and recognition results are shown in Figures 3–10, Figure 3, Figures 5, 7 and 9 are original images, Figures 4, 6, 8 and 10

Figure 3.   The original image one.

Figure 4.   The processed image one.

Figure 5.   The original image two.

Figure 6.   The processed image two.

Figure 7.   The original image three.

Figure 8.   The processed image three.

Figure 9.   The original image four.

Figure 10.   The processed image four.

are processed images. All examples are ran in Matlab(version 7.10.0), Table 1 shows running time.

Step 4: (Result analysis) From the results, we can see that the results are almost perfect, however,

Table 1. Running time.

ID	Image size	Running time
Figure 3	512×200	1.645529 seconds
Figure 5	512×264	1.976783 seconds
Figure 7	508×226	1.572516 seconds
Figure 9	512×218	1.703879 seconds

in the first image, sun light is detected as cloud too. The reason is that color of sun light is as same as cloud. Overall, the result is very good.

5 CONCLUSIONS

Fluctuations of solar irradiance are known to have a significant influence on electric power generation by solar energy systems, and cloud-obscuring sun is the main reason of solar irradiance fluctuations. The PV power prediction based cloud image is a feasible way.

In this paper, a cloud clusters recognition method based on blue and red channel pixel value ratio was proposed. The result indicated that this method had high accuracy and met the need of engineer. The major advantage of this algorithm was that its running speed met minutes-level PV power prediction need, which provided a basis for irradiance prediction and PV power prediction in the future.

ACKNOWLEDGEMENT

The research work was supported by Science and Technology Project of China Electric Power Research Institute under Grant No. NY83-14-004 and Science and Technology Project of State Grid Corporation of China.

REFERENCES

1. Chen Liu, Feng-xia Li, Yan Zhang. An interactive object cutout algorithm based on Graph-cut and generalized shape prior, Journal of Computer-aided Design & Computer Graphics, 21(12), pp. 1753–1760, 2009.

2. Chang-song Chen, Shan-xu Duan, Jin-jun Yin. Design of photovoltaic array power forecasting model based on neutral network, Transactions of China Electro Technical Society, 24(9), pp. 153–158, 2009.

3. Chun-xiang Shi, Rong-zhang Wu, Xu-kang Xiang. Automatic segmentation of satellite image using hierarchical threshold and neural network, Quarterly Journal of Applied Meteorology, 12(1), pp. 70–78, 2001.

4. Jing Lu, Hai-qing Zhai, Chun Liu, et al. Study on statistical method for predicting photovoltaic generation power, East China Electric Power, 38(4), pp. 563–567, 2010.

5. Jing Lu, Hai-qing Zhai, Shuang-lei Feng, et al. Physical method for photovoltaic power prediction, East China Electric Power, 41(2), pp. 380–384, 2013.

6. Lan Zhang, Yan-xia Zhang, Chang-min Guo, et al. Photovoltaic system power forecasting based on neutral networks, Electric Power, 43(9), pp. 75–78, 2010.

7. Wen-long Fei, Hong Lv, Zhi-hui Wei. A Application of Mumford-Shah Model in Segmentation of Satellite Cloud Image, Journal of Image and Graphics, 14(4), pp. 598–603, 2008.

8. Wen-long Fei, Hong Lv, Zhi-hui Wei. Application of graph cut method in cloud detection in satellite cloud image, Transactions of Atmospheric Sciences, 35(4), pp. 502–507, 2012.

9. Wen-long Fei, Hong Lv, Zhi-hui Wei. Application of Logistic regression method in cloud detection of satellite image. Computer Engineering and Applications, 48(4), pp. 18–21, 2012.

10. Xi Liu, Jian-min Xu, Bing-yu Du. A bi-channel dynamic threshold algorithm used in automatically identifying clouds on GMS-5 imagery, Journal of Applied Meteorological Science, 16(4), pp. 434–444, 2005.

11. Yona A, Senjyu T, Funabashi T. Application of recurrent neutral network to short-term-ahead generating power forecasting for photovoltaic system[C]//IEEE Power Engineering Society General Meeting, Tampa, FL, USA, 2007:1–6.

12. Zhi-ying Lu, Chang-qiao Wang, et al. Image texture feature extraction based on wavelet analysis theory, Pattern Recognition and Artificial Intelligence, 13(4), pp. 434–438, 2000.

Power and Energy – Kong (Ed.)
© 2015 Taylor & Francis Group, London, ISBN 978-1-138-02782-4

A preventive control strategy for voltage stability using Electric Vehicles

Cong Liu, Jiahui Xu & Fujian Chi
State Grid Tianjin Electric Power Company, China

Haifeng Li & Jin Tao
State Grid Jiangsu Electric Power Company, China

Xiaohong Dong & Wang Mingshen
Key Laboratory of Smart Grid of Ministry of Education, Tianjin University, China

ABSTRACT: A novel preventive control strategy using Electric Vehicles (EVs) is proposed to maintain the voltage stability of power system under Vehicle-to-Grid ('V2G') concept. An EV load aggregated model is firstly developed to estimate the EVs charging loads with various State of Charge (SOC) levels. The proposed preventive control strategy is then developed for EVs under charging to respond to the voltage contingencies according to their SOC levels. Two control modes are considered: disconnection of EV charging load ('V1G' mode) and discharge of stored battery energy back to the gird ('V2G' mode). A case study is carried out on the modified IEEE 14-bus system with high penetration levels of wind farms/EVs. The 'dumb' charging scenario of EV is used to analysis the preventive control effect. Simulation results shows that the proposed preventive control strategy can significantly improve the voltage stability of the power system with considerable amount of wind generations, and prevent disturbing the EV owners' daily travel as far as possible.

1 INTRODUCTION

A number of countries have taken specific initiatives to decarbonize power systems and transport sectors by encouraging Electric Vehicles (EVs). In China, the total installed wind capacity reached 44.73 GW at the end of 2010, and it is anticipated to reach 200 GW by 2020 [1]. In order to reduce CO2 emission from the domestic transport sector, the Chinese government has support EV trails with the anticipation that EVs will play a major role in the future transport sector [2].

Wind power is intermittent and non-schedulable while providing clean energy. In addition, wind farms absorb considerable amount of reactive power from system while sending out active power. For these reasons, high penetration of wind generation introduces serious influence on the voltage stability. The capabilities from the economic dispatch of thermal generation units for balancing the intermittent wind power and enhancing the Voltage Stability Margin (VSM) of power system were investigated in [3]. A comprehensive Energy Storage System (ESS) application for regulating wind power variation and improving system voltage stability was developed in [4]. Although static synchronous compensator (STATCOM) can improve the voltage stability of power system with wind farms, large scale adoption of these kinds of devices is not

economical and practical until now [5]. In recent years, the introduction and widespread use of EVs brings large potential of Vehicle-to-Grid ('V2G') [6]. With centralized control, aggregated EV pool can act as a virtual Battery Energy Storage System (BESS) by disconnection charging load from the grid ('V1G' mode) or discharge of stored battery energy back to the grid ('V2G' mode) [7]. The benefits of aggregated EV pool with regards to frequency response, improving system security and relieving transmission congestion were discussed in [8]. However, these control strategies overemphasize on the effectiveness to achieve objectives, while in turn have limited capabilities to consider the EV users' comfort levels (driving habits and requirements) and battery's service life.

In this paper, a preventive control strategy from EVs was proposed to maintain voltage stability of power system under the 'V2G' concept. The preventive control strategy using SOC as an indicator to select EVs for voltage stability depends on two reasons.

2 VOLTAGE STABILITY CONSIDERING WIND FARMS

When large wind farms connect to the power system, e.g. at the receiving end, the transmission mode of power system may be changed. The intermittent power output of wind farm may cause transmission congestion problem. Besides, wind farms usually absorb considerable amount of reactive power from system while sending out active power. As a result, large scale wind farms have serious impacts on the voltage stability of power system. To facilitate the voltage stability phenomena, $S_{GL} = S_G \cup S_L$ is used as the power injection vector, where $S_G = (P_{Gc} \cup Q_{Gc}) \cup (P_{Gw} \cup Q_{Gw})$ corresponds to the power injection vector of the conventional generators and wind farms. $S_L = P_L \cup Q_L$ is the power injection vector of loads. Once S_{GL} is determined, the system's operation status x is determined by the (1).

$$f(x, S_{GL}) = 0$$
$$g(x, S_{GL}) \leq 0 \tag{1}$$

where f is the load flow equation; g is the system operation constraints equation. When x yields to (1) and (2), system is voltage stability; if x satisfies (1) and (3), system is at the critical point of voltage stability. Otherwise, (1) loses equilibrium point and voltage collapse happens.

$$\det(f_x) \neq 0 \tag{2}$$
$$\det(f_x) = 0 \tag{3}$$

where f_x is the Jacobian matrix of the load flow equation. A parameter λ is introduced to obtain the VSM of power system, which can be determined by Continuation Power Flow (CPF) at a fixed load increasing and generation dispatch mode [10]. λ equals to zero at the base case, while λ reaches it max value λ_c at the CP. In post-contingency scenarios, the available λ is usually decreased. In order to prevent voltage stability, it is necessary for system to retain large enough λ under the threat contingencies. A stability criterion was proposed by WECC to maintain a minimum VSM λ_{cr} of 5% [10].

From (1)–(3), λ has close correlation with $(P_{Gw} \cup Q_{Gw})$ which fluctuates following the intermittence of wind speed. A sudden loss of wind generation at the peak load hours may cause λ decreases below λ_{cr}, and lead to voltage collapse. In this situation, preventive controls should be taken to keep the VSM of power system, where load shedding (S_L) was taken as the last measure due to the huge economic losses and inconveniences to end-users.

3 EV LOAD AGGREGATED MODEL

3.1 *EV classification based on transportation*

EV charging load is closely related to the daily driving habits of users. In the UK, three types of

EVs were identified based on the use of transport: HBW, HBO and NHB [13]. Each EV considered in the EV load aggregated model is assigned to one of the above three groups.

The EV classification determines the daily travel distance of each EV. D is the EV's daily travel range which is obtained through the probability density function (*pdf*) in (4) using MCS [12]:

$$h(D; \mu, \sigma) = \frac{1}{D\sqrt{2\pi\sigma^2}} e^{-\frac{(\ln D - \mu)^2}{2\sigma^2}}, D > 0 \tag{4}$$

3.2 *EV battery charging characteristics*

In order to obtain the EV charging load with corresponding SOC profile, Li-ion battery is used for its capability of providing a combination of improved performance, safety, long service life and low cost [13]. The battery-charging profile of Li-ion is shown in Figure 1, whose capacity equals to 29.07 kWh (Cap_{base}) with a maximum travel range (Ran) of 80 miles.

3.3 *Charging load profile of an individual EV*

a. Determine the EV's daily travel range. Equation 4 is used to determine the daily travel range of each EV (HBW, HBO, NHB) by using MCS.
b. Determine the initial state of charge (S_I) when an EV starts to charge. Assuming SOC drops linearly with the travel distance [14], S_I was determined by $S_I = S_0 - D/Ran$, where S_0 is the SOC of an EV before travelling and was generated using MCS assuming S_0 varies uniformly in range [0.8, 0.9] (80%–90% SOC was used in order to maintain the service life of a battery) [11].
c. Determine EV start charging time t_s. The start charging time of an EV battery is determined by people's daily transportation behavior and the charging strategies used. 'Dumb' charging is used to determine t_s. The method to model 'dumb' charging is from [14]. All EVs are

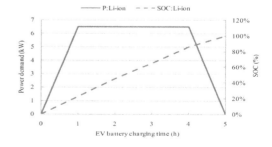

Figure 1.   Battery charging profile of Li-ion.

assumed to start charging just after coming back from their daily trips. The *pdf* of t_s is given by (5):

$$f(t_s;\mu,\sigma) = \frac{1}{\sqrt{2\pi\sigma^2}} e^{-(t_s-\mu)^2/2\sigma^2} \qquad (5)$$

It is assumed that μ is 18:00 and σ is 4 hours. This paper focused on the preventive control strategy from EVs for voltage stability. As a result, only 'dumb' charging is investigated.

d. Obtain the EV charging load with the corresponding SOC profile. The Li-ion battery is used in EV aggregated model to obtain the EV charging load. Once S_0 and S_1 are determined in step (b), S_0 and S_1 are first obtained and marked as points A and B in Figure 3. These points are then projected to the power demand profile (marked as points C and D). Once t_s is obtained in step (c), the power demand profile between points C and D with the corresponding SOC profile between A and B are mapped to the 24-hour coordinate to determine the vehicle charging load along a day, which is shown in Figure 2.

3.4 *Multiple EVs charging load with different SOC levels*

For n EVs, the procedure defined in part 3.3 is repeated n times. Each EV charging load with its SOC profile along a day is recorded. The total vehicle charging load (P_n) for n EVs at a specific time t is defined as (6):

$$P_{nt} = \sum_{i=1}^{n} P_t^i \qquad t = 1, 2, \ldots, 24 \qquad (6)$$

As a result, the total EV charging load available with SOC larger a pre-defined threshold $l\%$ at a specific time t is obtained by (7).

$$P_{nt,SOC \geq l\%} = \sum_{i=1}^{n} P_t^i \cdot \eta_t^i \qquad (7)$$

η_t^i is defined in (8):

$$\eta_t^j = \begin{cases} 1, S_t^i \geq l\% \\ 0, S_t^i < l\% \end{cases} \qquad (8)$$

4 PREVENTIVE CONTROL STRATEGY USING EVS

4.1 *Framework of preventive control from EVs*

As the power capacity of an EV is rather small, their participation in the electricity markets requires a new entity: the EV Management Module (EMM) integrated in the Virtual Power Plant (VPP). The VPP serves as an intermediary between a large number of EVs and market players and/ or system operators. The role of the EMM is to cluster geographically dispersed EVs, and manage their generation and demand portfolios as a single entity.

Figure 3 shows how the VPP interacts with the system operators (DSO and TSO) [8]. Upon recognizing a voltage contingency, the system operator could instruct the VPP to shed part or all of the charging EVs or discharge the stored battery energy back to the gird according to the operation state of each EV. SOC is an indicator for VPP to select the EVs to respond to the instructions for preventing disturbing the users' normal travel and reducing the service lives of batteries. Charging Point Management (CPM) combined with Smart Meter (SM) is used to obtain the SOC information of EVs.

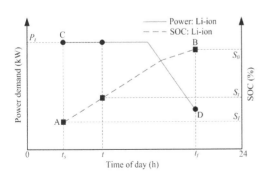

Figure 2. EV charging load from a Li-ion battery with charging starts at t_s.

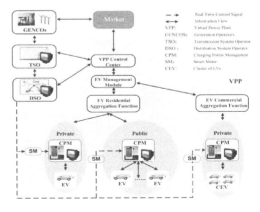

Figure 3. Integration of EVs for preventive control under VPP concept.

4.2 Modal analysis

Modal analysis is used to determine the critical buses for preventive control, which is depicted by (9).

$$\begin{bmatrix} \Delta P \\ \Delta Q \end{bmatrix} = \begin{bmatrix} \mathbf{J}_{P\theta} & \mathbf{J}_{PV} \\ \mathbf{J}_{Q\theta} & \mathbf{J}_{QV} \end{bmatrix} \begin{bmatrix} \Delta \theta \\ \Delta V \end{bmatrix} \quad (9)$$

where ΔP: active power variations of the buses; ΔQ: reactive power variations of buses; $\Delta \theta$: voltage angle variations of the buses; ΔV: voltage magnitude variations of the buses. $\mathbf{J}_{P\theta}$, $\mathbf{J}_{PV}$, $\mathbf{J}_{Q\theta}$ and $\mathbf{J}_{QV}$ are jacobian sub-matrices representing the sensitivities of active and reactive power with respect to voltage angles and magnitudes. The matrix is obtained near the critical point of power system by using CPF, which is written as (10).

$$\mathbf{J}_{2NPQ+NFV} = \mathbf{\Phi} \cdot \mathbf{\Lambda} \cdot \mathbf{\Gamma} \quad (10)$$

where $\mathbf{\Phi}$ is the right eigenvectors of matrix $\mathbf{J}$; $\mathbf{\Gamma}$ is the left eigenvectors of matrix $\mathbf{J}$; $\mathbf{\Lambda}$ is eigenvalue matrix of $\mathbf{J}$. NPQ and NPV are the numbers of PV and PQ bus respectively.

The first (NPV+NPQ) elements of the right and left eigenvectors associated to the critical eigenvalue of $\mathbf{J}$, define the Active Participation Factor (APF), associated with each generator and load bus of system. In this paper, APF is used to identify the best buses for the preventive control to alleviate the voltage stability risk in case of the sudden loses of wind generation.

4.3 Preventive control strategy from EVs

The preventive control strategy using SOC as an indicator to select EVs for voltage stability depends on two reasons. Firstly, the EV batteries need enough energy for user's daily travel; secondly, when SOC reaches a low level, discharge of the stored battery back to the grid damages the performance of EV battery. The following steps consist of the preventive control strategy.

Step 1: Determine the VSM λ_t of power system at current time (t) by using CPF;

Step 2: If $\lambda_t \geq \lambda_{cr}$, system is stable; compute λ for next time, and go to step 1; If $\lambda_t < \lambda_{cr}$ (system enters emergency state), preventive control from EVs is needed and go to step 3.

Step 3: Obtain the APFs by modal analysis technique. Queue the load buses based on the APFs from high to low, and determine the available responsive EV loads from both 'V1G' and 'V2G' prospective by virtual EV load aggregated model developed in section 3.

Step 4: If system exists available EVs for preventive control under 'V1G', go to Step 5; otherwise go

to step 6 for 'V2G'. If the EVs under the above two control modes are unavailable, other voltage emergency control strategies should be taken.

Step 5: The EVs under charging with SOC $\geq l\%$ and connecting to the load bus with the highest APF (among the load buses having EVs with SOC $\geq l\%$) stop charging from the grid for preventive control.

Step 6: The EVs under charging with SOC $\geq l\%$ connecting to the load bus with the highest APF (among the load buses having EVs with SOC $\geq l\%$) discharge the stored battery energy back to the gird for preventive control.

5 CASE STUDY

A modified IEEE 14-bus system with large wind farms (WG3, WG6 and WG8) depicted in Figure 4 is used as a test system to illustrate the preventive control strategy.

According to the functional orientation of the studied area, the system was divided into three functional zones. The normalized load patterns (without EVs charging load) of the above three zone functions supplied by UK GDS shown were used in this example [15].

It is assumed that the wind farms WG3, WG6 and WG8 are composed of doubly fed induction generators, whose typical power outputs along a day is shown in Figure 5 ($S_B = 100$ MW) [15].

In the modified IEEE-14 bus system, the load peak load is 397.05MW, which is 0.58% of the reference UK peak electricity demand. The EV numbers used for different functional zones were given in Table 1, which is also 0.58% of the reference UK national data [14]. The available capabilities from EVs for different functional zones obtained

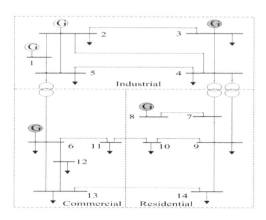

Figure 4. The modified IEEE-14 bus system with large scale wind farms.

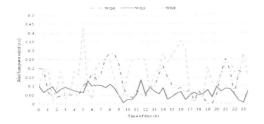

Figure 5. Wind farm output along a day.

Table 1. EV numbers for different functional zones.

	HBW	HBO	NHB	All EV groups
Residential	100650	49500	0	150150
Commercial	0	0	2500	2500
Industrial	0	0	12400	12400
Total no.	100650	49500	14900	165050

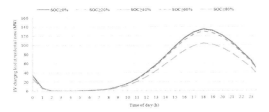

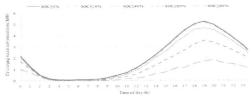

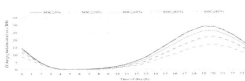

Figure 6. Available EV charging load at different functional zones.
a. Available EV charging load at residential zone.
b. Available EV charging load at commercial zone.
c. Available EV charging load at industrial zone.

by EV load aggregated model are given in Figure 6. Assuming each load bus was equipped with EV charging facilities, the available EV charging

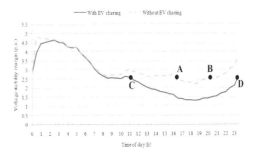

Figure 7. Voltage stability margin along a day (without preventive control).

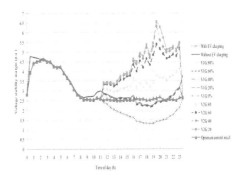

Figure 8. Preventive control results.

load at each load bus was then determined in proportion to its initial load (without EV charging load).

The λ values with and without EV charging along a typical day are shown in Figure 7 (without preventive control). λ_{cr} is defined to be 2.5 in this paper. As depicted in Figure 7, due to the loss of wind generation, systems become unstable between 16:30 to 20:00 (between A and B); when EVs were integrated, system become unstable between 11:30 to 23:00.

The preventive control strategy was then used to alleviate the voltage contingency, where $l_{min}\% = 20\%$, $\Delta l\% = 5\%$ and $l_{max}\% = 80\%$. At point C, the APFs of the load buses (B) from high to low is listed as follows: $B_{14}(0.321) > B_{10}(0.239) > B_{9}(0.1999) > B_{11}(0.108) > B_{13}(0.0317) > B_{12}(0.0180) > B_{4}(0.0084) > B_{5}(0.004) > B_{6}(0.002) > B_{2}(0.001) > B_{3}(0.000)$, which was found unchanged during the whole control process. In order to illustrate effect of preventive control, the control process continued until the total capability of EVs was used up. The preventive control results with different SOC levels and the optimum preventive control result are shown in Figure 8.

6 CONCLUSION

In this paper, a preventive control strategy from EVs was proposed to maintain voltage stability of power system under the 'V2G' concept. An EV load aggregated model was developed to simulate the function of VPP. Modal analysis was used to determine the best buses for preventive control, and SOC was introduced as an indicator for selecting the EV groups to respond to voltage contingencies. Simulation results show that the available capability of EVs for voltage stability control has evident spatial-temporal distributions along a day. The preventive control strategy not only can enhance the voltage stability margin of power system effectively but also avoid disturbing the normal travels of EV users and eliminate the over discharge on EV batteries as far as possible.

REFERENCES

1. GWEC, Global Wind Report, 2010.
2. Y.H. Song, X. Yang and Z.X. Lu, "Integration of plug-in hybrid and electric vehicles: Experience from China," in *Proc. IEEE Power Energy Soc. Gen. Meeting*, 2010, pp. 1–6.
3. B.C. Ummels, M. Gibescu, E. Pelgrum, W.L. Kling, and A.J. Brand, "Impacts of wind power on thermal generation unit commitment and dispatch," *IEEE Trans. Energy Convers.*, vol. 22, no. 1, pp. 44–51, Mar. 2007.
4. Y.F. Mu and H.J. Jia, "An Approach to determining the local boundaries of voltage stability region with wind farms in power injection space," Sci China Ser E-Tech Sci, vol. 53, no. 12, pp. 3232–3240, Dec. 2010.
5. H.T. Le, S. Santoso and T.Q. Nguyen, "Augmenting wind power penetration and grid voltage stability limits using ESS: application design, sizing, and a case study," *IEEE Trans. Power Syst.*, vol. 27, no. 1, pp. 161–171, Feb. 2012.
6. A. Arulampalam, M. Barnes, N. Jenkins, and J.B. Ekanayake, "Power quality and stability improvement of a wind farm using STATCOM supported with hybrid battery energy storage," *Proc. Inst. Electr. Eng. Gen., Transmiss. Distrib.*, vol. 153, no. 6, pp. 701–710, Nov. 2006.
7. J.A.P. Lopes, F.J. Soares, and P.M.R. Almeida, "Integration of electric vehicles in the electric power system," *Proc. IEEE*, vol. 99, no. 1, pp. 168–183, Jan. 2011.
8. A.F. Raab, M. Ferdowsi, E. Karfopoulos, et al., "Virtual power plant control concepts with electric vehicles," in *ISAP 2011, Hersonisos, Crete, Greece*, 2011, pp. 1–7.
9. D.J. Burke and M. O'Malley, "Maximizing firm wind connection to security constrained transmission networks," *IEEE Trans. Power Syst.*, vol. 25, no. 2, pp. 749–759, May 2010.
10. "Guide to WECC/NERC planning standards I.D: voltage support and reactive power," Mar. 06, 2006, WECC.
11. EU Merge Project, Deliverable 2.1: "Modeling Electric Storage devices for electric vehicles," Jan. 2010, Task. Rep.
12. National Statistics, Department for Transport, Transport Statistics Bulletin-Vehicle Licensing Statistics: 2010, Apr. 9, 2011.
13. S. Bruno and G. Jurgen, "Lithium batteries: Status, prospects and future," *J. Power Sources*, vol. 195, no. Issue 9, pp. 2419–2430, May 2010.
14. K. Qian, C. Zhou, M. Allan, and Y. Yuan, "Modeling of load demand due to EV battery charging in distribution systems," *IEEE Trans. Power Syst.*, vol. 26, no. 2, pp. 802–810, May 2011.
15. "Typical load patterns, DG output data," United Kingdom Generic Distribution Systems (UKGDS).

Power and Energy – Kong (Ed.)
© 2015 Taylor & Francis Group, London, ISBN 978-1-138-02782-4

Condition evaluation and analysis of 110 kV and above SF6 circuit breakers

Han Zhang, An Chang, Jun Deng & Qi Wang
Maintenance and Test Center, EHV, China Southern Power Grid, Guangzhou, China

ABSTRACT: Planned maintenance or blackout preventive tests are used universality in judging the operational state of the high voltage circuit breakers. For the sake of solving the lack of regular maintenance, a new concept of condition-based maintenance springs up in the modern society. In this background, this paper evaluates the SF6 circuit breakers of EHV. Three types of circuit breakers are referred, i.e. 110 kV and above AC circuit breakers, DC circuit breakers and series compensation bypass circuit breakers. We make an overall statistics and analysis of all the breakers in multiple aspects—the run time, operator type and the manufacture factory. It comes a result of the condition evaluation based on the condition guide rule. What's more, we put forward some corresponding analysis and useful suggestions of the abnormal condition and the attention condition circuit breakers. As simple and rapid, the evaluation methodology has a good guidance for the condition maintenance of the circuit breakers.

Keywords: planned maintenance; condition maintenance; SF6 circuit breakers; the condition evaluation guide rule; analysis and suggestions

1 INTRODUCTION

The high voltage circuit breakers have control and protection functions and they play a vital role to the safe operation of the power system. For a long time, the operational state of the high voltage circuit breakers were judged by planned maintenance or blackout preventive tests. The planned maintenance mode can eliminate hidden dangers to a certain extent and prevent the accident. But periodic overhaul consumes large quantities of supplies, and the frequently disconnect and restore lead also makes wear to the equipment.

To solve the lack of regular maintenance, modern society presents the new concept of condition maintenance. Condition maintenance is a maintenance mode, which develops maintenance plans based on the current state of the equipment. Based on "110 kV to 500 kV SF6 circuit breaker status assessment guidelines (Trial)", this paper evaluates the SF6 circuit breakers of EHV. The evaluated circuit breakers include 110 kV and above AC circuit breakers, DC circuit breakers and series compensation bypass circuit breakers. It makes the statistics and analysis of basic information and state evaluation situation of SF6 circuit breakers. And it puts forward some corresponding analysis and suggestions of the abnormal condition and the attention condition circuit breakers.

2 MAIN CONTENTS

2.1 *Analysis of basic information breaker*

Breakers involved in this evaluation include AC circuit, DC circuit, and series compensation bypass circuit breakers, which are shown in Table 1.

2.1.1 *Statistics of the run time of the circuit breakers*

Among 606 circuit breakers, there were 342 units in operation life within five years, 148 ones with 5 years to 10 years, 116 ones with more than 10 years, as shown in Figure 1.

As shown in Figure 1, in all 606 sets of SF6 circuit breakers, 56.44% of which are within 5 years

Table 1. Classification statistics of the Sf6 circuit breakers.

Type	Voltage level	Quantity (units)
AC circuit breakers	500 kV	340
	220 kV	234
	110 kV	5
DC circuit breakers	800 kV	4
	400 kV	4
Series compensation bypass circuit breakers	500 kV	19
Total		606

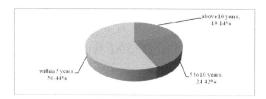

Figure 1. Statistical graph of the run time of the circuit breakers.

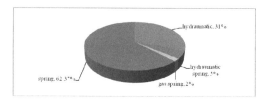

Figure 2. Statistical graph of the operator type of the circuit breakers.

of operation, whose device are relatively stable. Another 19.14% with more than 10 years, which need attention.

2.1.2 Statistics of the operator type of the circuit breakers

The operating mechanism of the circuit breaker is the most important device components, SF6 circuit breaker operating mechanism in this article mainly include four categories, i.e. springs, hydraulic, hydraulic and pneumatic spring. As the statistical analysis of the evaluation report shows, of all 606 units in the breaker, there are 378 units using a spring mechanism, 187 ones with hydraulic mechanism, 30 ones with hydraulic spring mechanism, and the remaining 11 ones with pneumatic spring mechanism, which was shown in Figure 2.

Through the statistic analysis of the operating mechanism, SF6 circuit breaker mechanism is still based on spring and hydraulic mechanism. The ones who use hydraulic spring mechanism are from Western companies, and pneumatic spring mechanism is only used by Mitsubishi in 550-SFM-50 series circuit breaker.

2.1.3 Statistics of the manufacture factory of the circuit breakers

Among all 606 circuit breakers, 239 sets are imported overseas, 214 units are from the joint manufacturers, the remaining domestic 153 ones are from manufacturers, as shown in detail in Figure 3.

As shown in statistics above, SF6 circuit breakers are form foreign manufacturers and joint venture companies almost equally, slightly more than the domestic. To the contrary, DC breakers, breaker converter station group and series compensation filter bypass breakers are all imported overseas.

2.2 Evaluation method outlined

2.2.1 Evaluation method

In accordance with the requirements of the guidelines, state SF6 circuit breakers can be divided into four states: normal, attentive, severe and abnormal. State assessment and evaluation include

Figure 3. Statistical graph of the manufacture factory of the circuit breakers.

components and overall ones. The formal one is based on the evaluation of the independence of the components of SF6 circuit breakers, which can be divided into five parts: Overview, body, operating mechanism, shunt capacitors and resistors closing. Each part was marked by the state deduction standards, simultaneously considering the individual states and total f points deduction of all parts. The corresponding component state evaluation deduction form is shown in Table 2. When any member state and the total points of individual meet the requirements of the normal state in Table 2, they are considered as normal state; in a similar way.

When reach the attention of state in Table 2, considered as attention state; when reach abnormal or critical state in Table 2, considered as abnormal state or critical state. An amount of state is mainly from major of raw data, operational data, maintenance test data and other information, which can directly and effectively reflect the equipment operating conditions, and the trend of failure.

2.2.2. Evaluation results

As shown in Table 3, a station Pinggao LW6-220 circuit breakers exist the following issues: a). the body has family defects, which will lose if lifted, so it has been transformed against loosening; b). reflation interval is less than a year and more than half a year; c). the operating mechanism of the terminal block inside exists corrosion. Among these problems, a) belong to the "Profile" entry points, deducted 12 points (attention state), b) belong to the "circuit breaker body" entry points, deducted 12 points

Table 2. Status and evaluation points for the circuit breakers.

No.	Description and component	Marking and state					
		Normal state (meet one of the following two conditions)		Attention state (meet one of the following two conditions)		Abnormal state	Serious state
		Total points	Individual points	Total points	Individual points	Individual points	Individual points
1	Description	30	<10	≥30	12~16	24	≥30
2	Body	<30	<10	≥30	12~16	20~24	≥30
3	The operating mechanism	<20	<10	≥20	12~16	20~24	≥30
4	Shunt capacitor	<12	<10	≥12	12~16	20~24	≥30
5	Closing resistors	<12	<10	≥12	12~16	20~24	≥30

Table 3. Condition evaluation report of 110 Kv and above Sf6 circuit breakers.

Company name. —

Equipment data	Installation site	–	Run No.	–	Model	LW6–220
	Rated voltage	220 kV	Manufacturer	Pinggao	Rated current	3150 kA
	Rated short-circuit breaking current	50 kA	Agency type	Hydraulic mechanism	Serial number	–
	Date of manufacture	1989	Commissioning date	1992	Last inspection date	2006

Component evaluation results

Evaluation index	Overview	Body	Operating mechanism	Shunt capacitor	Closing resistors
Evaluation of the number of items	2	6	5	4	3
Number of points	1	1	1	0	0
The maximum value of the individual points	12	12	24	0	0
Total points value	12	12	24	0	0
Status classification	Attention state	Attention state	Abnormal state	Normal state	Normal state

Overall results of the evaluation

Evaluation results:	Abnormal state	Total points	48
Device problem description	a) The body has family defects, which will lose if lifted, so it has been transformed against loosening; b) Reflation interval is less than a year and more than half a year; c) The operating mechanism of the terminal block inside exists corrosion.		
Treatment recommendations	Step up inspections, check the location marked on the equipment during power outages timely leak repair, and timely investigation of the causes of abnormal conditions, and it has been included in 2012 to replace.		

(attention the state), c) are "operating mechanism" button itemized deduction of 24 points (abnormal state), "shunt capacitor" and "closing resistance" were evaluated as normal. So overall evaluation of the results of the breaker taking the most severe status of each component is an abnormal state.

2.3 Evaluation results and analysis

2.3.1 Analysis of the general evaluation of the circuit breakers

As shown in Figure 4, in 606 SF6 breakers none was evaluated as a critical state sets; 33 units as

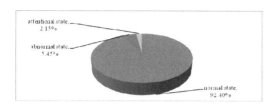

Figure 4. Analysis graph of the general evaluation of the circuit breakers.

abnormal state, taking up 5.45% of the total; 13 units as attentional state, representing 2.15%; 560 units as normal state.

2.3.2 *Statistical evaluation of the abnormal state of the circuit breakers*

The number of the Overall rating of SF6 circuit breakers as abnormal state was 33, whose sets of equipment defects were analyzed as follows:

a. the design of LW6-220 circuit breaker body from Pinggao company exist flaws, the insulated rod of which use rotating docking structure, prone to failure rod loose, belonging to familial defect, and needing to be replaced gradually. Focusing on strengthening LW6-220 circuit breaker insulated rod inspection and maintenance before the replacement. Marking in the circuit breaker switch indication position, after the event, check the mark's displacement to determine whether the insulated rod looses; rotate the insulated rod after decompression, and check the rod to see whether it looses off.

b. Mitsubishi's 500-SFM-50E-breaker shunt capacitance is flawed, because the model breaker uses shunt capacitance ceramic structure whose capacity is too small. It prones to discharge and breaks down after a long hot standby. Therefore, it needs to be replaced by the oil capacitors as soon as possible. Relevant measures should be strictly implemented. Before replacing the circuit breakers, hot standby time of the controlling circuit breakers cannot be more than four hours, simultaneously taking shorten pre-test cycle, tracking changes in capacitance value, periodic testing, for example, SF6 gas decomposition conduct, and other useful means to closely monitor the capacitors' electrical performance.

c. Alstom FX32D-550 circuit breakers and Pinggao Henan Electric Company's LW6-220 Ontology, both of which have run nearly 20 years have body and operating mechanism leakage and frequent defects, which need strengthened surveillance, special patrol and complete overhaul of the operating mechanism as soon as possible. Circuit breaker terminal box component aging,

terminal block corrosion and other defects should be replaced promptly.

2.3.3 *Statistical evaluation of the attention state of the circuit breakers*

According to the evaluation report, the overall evaluation of the 13 sets of SF6 circuit breakers of our company is the state of attention, of which 10 ones in the "Profile" were designated as the note states, 3 ones with note states in "operating mechanism". Defects of the circuit breakers in a note state were analyzed separately as follows:

a. In 10 SB6 m-T 550-type circuit breakers of Vital Power Transmission and Distribution Company., the design of their SF6 density relay have flaws, which easily occurs low-pressure alarm in direct sunlight.

b. The hydraulic mechanism of 2 Alstom FX22D-CIH breakers occurs leakage repeatedly, which was put into operation in 1998.

c. The opening two low voltage jump of a HPL550B2 circuit breaker of Beijing ABB Switchgear Company is 62 V, which is less than the prescribed 66 V procedures.

From the above assessment of the state of operation of the circuit breakers, it comes the conclusion that we should make more strengthened surveillance to them, and meanwhile make analysis of the defects and eliminate the defects timely.

3 CONCLUSION

For a long time, planned maintenance wasted a lot of manpower, material and financial resources, and it also caused the power loss and reduction of the equipment life. This paper evaluates the SF6 circuit breakers of EHV. The evaluated circuit breakers include 110 kV and above AC circuit breakers, DC circuit breakers and series compensation bypass circuit breakers. Evaluation results show that the evaluation method are simple and quick, and it has good guidance to condition maintenance.

ACKNOWLEDGEMENT

Project is supported by National high technology research and development program (863 Program) (2012 AA050209).

REFERENCES

1. Chen Jiabin. SF6 circuit Breakers Technical. Beijing: China Water Power Press, 2004.

2. He Xuliang, Su Bo. Analysis of preventive test methods for electrical equipment. Electric Test. 2014. 91–92.
3. Schlabbach R, Berka T. Reliability-centred maintenance of MV circuit-breakers. Power Tech Proceedings, 2001 IEEE Porto. 541–545.
4. Zhu Yubi. Application of New Technology to Condition Based Maintenance of Electric Equipment. High Voltage Apparatus, 2013, 39(2). 68–70.
5. Yang Liangjun, Xiong Xiaofu, Zhang Yuan. Research on condition-based maintenance policy of electric power equipment based on grey correlation degree and TOPSIS[J]. Power System Protection and Control, 2009, 37(18). 74–78.
6. Han Fuchun, Zhang Hai-long. The multi-level fuzzy evaluation of high-voltage circuit breakers' running. Power System Protection and Control, 2009(37). 60–63.
7. Q/CSG 11099–2010, 110 kV to 500 kV SF6 circuit breaker status assessment guidelines (Trial).

Power and Energy – Kong (Ed.)
© *2015 Taylor & Francis Group, London, ISBN 978-1-138-02782-4*

Exact acquisition of load dynamic characteristic data based on ELM

Zhenshu Wang, Kai Yu, Xiaoyu Liu & Yunpeng Shi
Key Laboratory of Power System Intelligent Dispatch and Control of Ministry of Education,
Shandong University, Jinan, Shandong Province, China

Shaorun Bian
Power Supply Company of Jinan, Jinan, Shandong Province, China

ABSTRACT: Exact acquisition of load dynamic characteristic data is a problem to be settled urgently in load modeling for electric power system, since the precision of the load modeling data affects the veracity of the load model directly. Considering the traditional modeling data acquisition method can't extract the load dynamic characteristic data accurately, a method of exact acquisition of load dynamic characteristic data based on Extreme Learning Machine (ELM) is proposed in this paper. At first, virtual value of voltage, current, active power and reactive power of power fault wave record data are calculated by Discrete Fourier Transform (DFT), then the relative change rate of variables are chosen as the feature vectors. At last, ELM is introduced to the classification of the load dynamic characteristic data. The simulation results show that ELM can be used to acquire load dynamic characteristic data and it has good classification accuracy. Besides, the training speed and generalization ability of ELM are superior to that of BP neural network. The method presented in this paper can extract the load dynamic characteristics data quickly and accurately, which provides a new research thinking for load dynamic characteristic data acquisition.

1 INTRODUCTION

There are many problems and technology involved in load modeling, the exact acquisition of load dynamic characteristics data is a problem to be settled urgently, which is still recognized bottleneck problem in load modeling for electric power system (Li, W. 2011, Ping Ju & Daqiang Ma 2008). The accuracy of the load modeling largely depends on the precision of the modeling data, thus acquiring the exact data which reflect the load dynamic characteristics is the premise of establishing accurate load model.

At present, the threshold test method is still used in load modeling data acquisition (Ping Ju et al. 2011, Li Xiaofang et al. 2012). However, the precision and the speed of the traditional data acquisition method are difficult to meet the needs of the engineering practice with the age of big data advent in electric power system. It is an effective way to solve the above problem by adopting the intelligent algorithm in load modeling data acquisition. BP neural network and Support Vector Machine (SVM) have been widely used in many fields as the classification algorithms (A.Y. Abdelaziz et al. 2007, Yanxi Liu 2010, Zhijun Gao 2011, Shaoning Pang et al. 2005, Mathur, A et al. 2008). However, the traditional BP neural network

is easy to fall into local values, and SVM is sensitive to parameters which are determined through experiment. Besides, the learning speed of both algorithms is slow, the generalization ability is weak and the calculation process is complex.

ELM is a novel algorithm for Single-Hidden Layer Feed-Ward Neural Network (SLFN) (G. Huang et al. 2006). It is clear that the learning speed of both algorithms mentioned above is in general far slower than required, while ELM provides good generalization performance at extremely fast learning speed (G. Huang et al. 2006). In this paper, ELM is used to acquire load dynamic characteristic data. ELM network model of data classification can be built by training using the typical disturbance data as training set. Then the load dynamic characteristic data can be acquired quickly and accurately with the application of this model.

2 THE THEORY OF ELM

There are a large amount of network training parameters need to be set in the traditional neural network learning algorithms (such as BP algorithm), and it's easy to produce the local optimal solution. The number of hidden layer nodes of networks is the only parameter need to set in

ELM. Besides, the weights of input values of the network and the bias of hidden layer do not need to adjust in execution of the algorithm, and ELM can produce unique optimal solution. Therefore ELM has advantages of fast learning speed and strong generalization capability (G. Huang et al. 2006, Guang-Bin Huang et al. 2004). ELM has been widely used in many fields, such as large data analysis, image classification and function approximation (Jiaoyan Chen et al. 2013, Bazi, Y. Alajlan et al. 2007, Y. Chang et al. 2007).

Structure of a typical SLFN is shown in Figure 1. The network is composed of input layer, hidden layer and output layer.

Set numbers of the input layer, hidden layer and output layer to be m, L and n respectively. The weights vector w connecting the input layer and the hidden layer is

$$w = \begin{bmatrix} w_{11} & w_{12} & \cdots & w_{1n} \\ \cdots & \cdots & \cdots & \cdots \\ w_{L1} & w_{L2} & \cdots & w_{Ln} \end{bmatrix}_{L \times n} \tag{1}$$

where w_{ij} is the link weight between the ith input layer and the jth hidden layer.

The weights vector connecting the output layer and the hidden layer is

$$\beta = \begin{bmatrix} \beta_{11} & \beta_{12} & \cdots & \beta_{1m} \\ \cdots & \cdots & \cdots & \cdots \\ \beta_{L1} & \beta_{L2} & \cdots & \beta_{Lm} \end{bmatrix}_{L \times m} \tag{2}$$

where β_{jk} is the link weight between the jth hidden layer and the kth output layer.

The bias of the hidden layer nodes is

$$b = \begin{bmatrix} b_1 \\ \bullet \\ \bullet \\ \bullet \\ b_L \end{bmatrix}_{L \times 1} \tag{3}$$

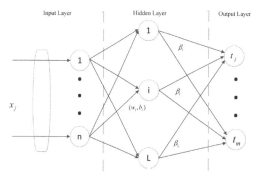

Figure 1. Structure of typical SLFN.

SLFNs with L hidden nodes are mathematically modeled as

$$f_L(x) = \sum_{i=1}^{L} \beta_i G(w_i \cdot x + b_i) = O_j \quad (j = 1, 2, 3 \ldots N.) \tag{4}$$

where $\beta_i = [\beta_{i1}, \beta_{i2} \ldots \beta_{im}]^T$ is the weights vector connecting the ith hidden layer node with output nodes; $w_i = [w_{i1}, w_{i2} \ldots w_{in}]^T$ is the weights vector connecting the ith hidden node with input nodes; $G(w_i \cdot x + b_i)$ is hidden layer feature mapping; $x = [x_1, x_2, \ldots, x_n]^T$ is input vector; b_i is the bias of the ith hidden layer node; O_j is the output of function.

That SLFNs can approximate these N samples with zero error means that

$$\sum_{j=1}^{N} \left\| O_j - t_j \right\| = 0 \tag{5}$$

i.e., there exist w_i, b_i and β_i such that

$$\sum_{i=1}^{L} \beta_i G(w_i \cdot x + b_i) = t_j \quad j = 1, 2, 3 \ldots N. \tag{6}$$

The above N equations can be written compactly as:

$$H\beta = \mathrm{T} \tag{7}$$

where

$$H = \begin{bmatrix} h(x_1) \\ \vdots \\ h(x_N) \end{bmatrix} = \begin{bmatrix} G(\omega_1 \bullet x_1 + b_1) & \cdots & G(\omega_L \bullet x_1 + b_L) \\ \vdots & \cdots & \vdots \\ G(\omega_1 \bullet x_N + b_1) & \cdots & G(\omega_L \bullet x_1 + b_L) \end{bmatrix}_{N \times L} \tag{8}$$

$$T = \begin{bmatrix} t_1^T \\ \vdots \\ t_N^T \end{bmatrix}_{N \times m} \tag{9}$$

H is called the hidden layer output matrix of the SLFN (Huang G-B. 2003). T is network outputs.

The output weights and deviation can be given randomly in ELM, and then the hidden layer matrix H becomes a definite matrix. To train an SLFN is simply equivalent to finding a least-squares solution $\hat{\beta}$ of the linear system $H\beta = \mathrm{T}$:

$$\left\| H\hat{\beta} - T \right\| = \min_{\beta} \left\| H\beta - T \right\| \tag{10}$$

The smallest norm least-squares solution of the above linear system is:

$$\hat{\beta} = H^{\dagger}T \tag{11}$$

where $H^{\dagger}$ is the Moore–Penrose generalized inverse of matrix H (Serre D. 2002). The least-squares solution is unique, it makes the network model has minimum trained error and has the strongest generalization ability. Solving the input weights of least-square solutions can complete the network training. Thus, ELM can be summarized as follows:

Step 1: Randomly assign hidden node parameters (ω_i and b_i);
Step 2: Calculate the hidden layer output matrix H;
Step 3: Calculate the least-square solutions $\hat{\beta}$.

3 EXACT ACQUISITION OF LOAD DYNAMIC CHARACTERISTIC DATA BASED ON ELM

3.1 Load modeling data

Load dynamic characteristics data used in load modeling must be fluctuations data of voltage, active power and reactive power caused by large disturbance outside the area. Generally, the form of the load disturbance has the following performance: active power and reactive power decline during voltage drop; active power and reactive power restore to the steady state in a simple form of oscillation when voltage return to normal value. The Power Fault Recording and Monitoring System (PFRMS) collecting transient data, the above data contain the load dynamic characteristic data. Exploiting load dynamic characteristic data from massive data is the primary task of load modeling. PFRMS can automatically record the information of electrical quantities when fault occur. In this paper, the PFRMS is used to obtain the fault data for load dynamic characteristic data acquisition.

3.2 Data processing

Data used in load modeling is the effective value of voltage, active power and reactive power, while the PFRMS record the instantaneous value of the electrical quantity. Therefore the effective values of the electrical quantity need to be calculated by DFT.

Steps of calculating the effective value by DFT are as follows:

1. Extract the effective value and phase angle of fundamental waves;

Expressions of real part and imaginary part of fundamental are as follows:

$$X_{1s}(n) = \frac{2}{N}\sum_{k=0}^{N-1} x[n-N+k] \times \sin[\omega k T_s]$$

$$= \frac{2}{N}\sum_{k=0}^{N-1} x[n-N+k] \times \sin\left[\frac{k}{N}2\pi\right] \tag{12}$$

$$X_{1c}(n) = \frac{2}{N}\sum_{k=0}^{N-1} x[n-N+k] \times \cos[\omega k T_s]$$

$$= \frac{2}{N}\sum_{k=0}^{N-1} x[n-N+k] \times \cos\left[\frac{k}{N}2\pi\right] \tag{13}$$

where N is sampling number of sine period, ω is angular velocity.

2. Calculate the effective value and phase angle;

$$X_1 = \sqrt{X_{1c}^2 + X_{1c}^2}\Big/\sqrt{2} \tag{14}$$

$$\alpha_1 = arctg\frac{X_{1c}}{X_{1s}} \tag{15}$$

where X_{1s} is the real part of fundamental, X_{1c} is the imaginary part of fundamental, α_1 is the phase angle.

3. Calculate the active power and reactive power.

$$\varphi = \alpha_u - \alpha_i \tag{16}$$

$$\dot{S} = \dot{U}\overset{*}{I} = UI\angle\varphi = P + jQ \tag{17}$$

where α_u is phase angle of voltage, α_i is phase angle of current.

The effective value of voltage, current, active power and reactive power are obtained by above calculations. The variables mentioned above occurs obvious fluctuations during electrical disturbance. Data characteristics are extracted by following functions:

$$\Delta u = u(t) - u(t_0) \tag{18}$$

$$\Delta i = i(t) - i(t_0) \tag{19}$$

$$\Delta P = P(t) - P(t_0) \tag{20}$$

$$\Delta Q = Q(t) - Q(t_0) \tag{21}$$

where $u(t), i(t), P(t), Q(t)$ is the value of variable at time t, $u(t_0), i(t_0), P(t_0), Q(t_0)$ is the value of variable at time t_0.

3.3 Parameter setting of ELM

The number of hidden layer nodes is key parameter of ELM. Generally speaking, the more hidden layer nodes are used, the better generalization ability ELM has, and however the network model is more complex at the same time. Therefore, consideration should be given to both classification accuracy and complexity of the ELM model. In this paper, ELM with 12 hidden layer nodes is applied to acquire load dynamic characteristic data, which is verified to be suitable in section 4.

There are many functions can be chosen as activation function, such as Sigmoid function, Gaussian radial function and RBF function. Sigmoid function is chosen as action function in this paper, that is

$$f(x) = \frac{1}{1 + e^{-x}} \tag{22}$$

3.4 *Flow chart of exact acquisition of load dynamic characteristic data based on ELM*

Data acquisition can be roughly divided into three steps:

1. Preparation stage. The data feature extraction should be completed firstly, then select training set and testing set of ELM. At last, set the number of hidden layer nodes and choose activation function.
2. Training stage. Training ELM and obtain the network model of ELM.
3. Application stage. Acquire load dynamic characteristic data using the network model obtained in training stage.

The whole flow chart of load dynamic characteristic data acquisition based on ELM is shown in Figure 2.

4 SIMULATION ANALYSIS

4.1 *ELM load model building based on typical disturbance data*

A three-phase short current fault occurred to Yangqiu line On February 27, 2012. PFRMS recorded fault waveforms of five substation (Dongcheng substation, Guangli substation, Jinqiu substation, Yangjia substation and Yongan substation). The disturbance data of Dongcheng substation is chosen as training data in this paper. Figure 3 shows the off-line data analysis software of PFRMS.

The PFRMS recorded the instantaneous value of the electrical quantity, and the effective value of the electrical quantity can be calculated by DFT. The curves of voltage, current, active power and reactive power are shown in Figure 4, Figure 5, Figure 6 and Figure 7 respectively.

Data characteristics can be extracted by (18) to (21), and part of the data characteristics are shown in Table 1.

As can be seen from Table 1, output = 1 means that data belong to the steady data, output = 2 means that data belong to disturbance data.

4.2 *Select the number of hidden layer nodes of ELM*

The influence of hidden layer nodes of ELM on classification accuracy of load dynamic

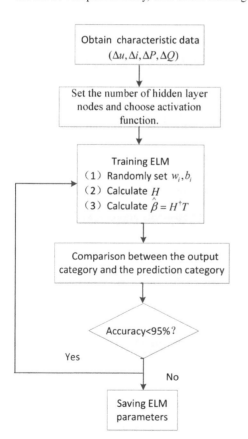

Figure 2. Flow chart of exact acquisition of load dynamic characteristic data based on ELM.

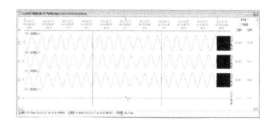

Figure 3. Off-line data analysis software of PFRMS.

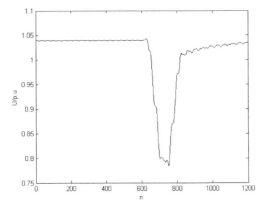

Figure 4.   Curve of voltage.

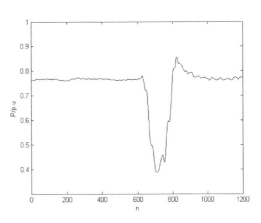

Figure 6.   Curve of active power.

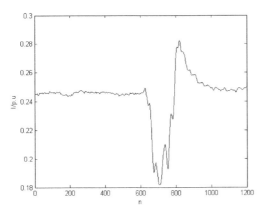

Figure 5.   Curve of current.

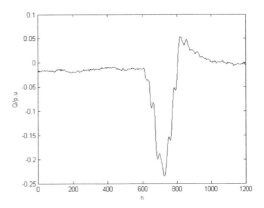

Figure 7.   Curve of reactive power.

Table 1.   Part of the data characteristics.

Classification types	Δu	Δi	ΔP	ΔQ	Output
Steady data	8.00E-05	0.00142	0.00459	0.0039	1
Steady data	0.015092	0.00292	0.00192	0.0188	1
Steady data	0.009830	0.00419	0.00592	0.0176	1
Disturbance data	0.22507	0.05834	0.34868	0.1716	2
Disturbance data	0.06835	0.03253	0.04468	0.0258	2
Disturbance data	0.01606	0.00558	0.00546	0.0255	2

characteristic data acquisition are studied in this part. ELM models are established with N = 2, 4, 6 ... 40, and each classification accuracy is calculated. The relationship between N and classification accuracy is shown in Figure 8.

Figure 8 shows that the classification accuracy becomes higher when larger number of hidden layer nodes is set, however the model becomes more complex at the same time. As observed form Figure 8, the classification accuracy of ELM becomes relatively stable when the number of hidden nodes is set to be 12. As mentioned above, a simple model is desirable. Thus, establishing ELM model with 12 hidden layer nodes is suitable for load dynamic characteristic data acquisition.

4.3 *Performance analysis of ELM model*

According to the procedures shown in section 3, ELM model is established based on typical disturbance data. The prediction of the train set and the test set are shown in Figure 9 and Figure 10 respectively.

As is shown in Figure 9, there is only one sample, in which the prediction category isn't consistent with the output category. The accuracy of the training results is 99.1667%, and as can be seen from Figure 10, the accuracy of the test results is 96.67%.

4.4 *Performance comparison between ELM and BP neural network*

The classification results of BP neural network with the same training set and test set are shown in Figure 11 and Figure 12 respectively.

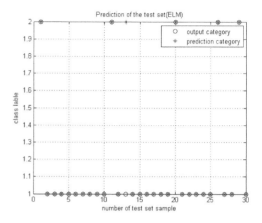

Figure 10.   Prediction results of test set of ELM.

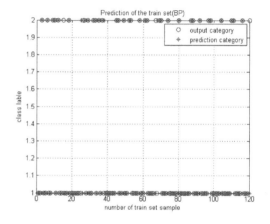

Figure 11.   Prediction results of training set of BP.

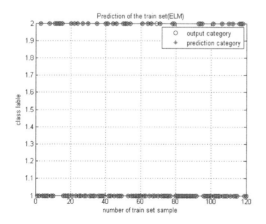

Figure 8.   The relationship between N and classification accuracy.

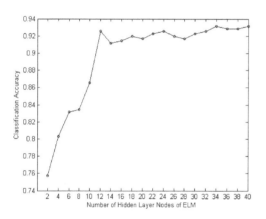

Figure 9.   Prediction results of training set of ELM.

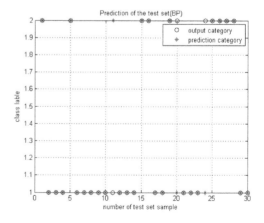

Figure 12.   Prediction results of test set of BP.

Table 2. Performance comparison between ELM and BP neural network.

Algorithm	Training time/s	Training precision	Test precision
ELM	0.48474	99.1667%	96.67%
BP neural network	12.765	94.1667%	90%

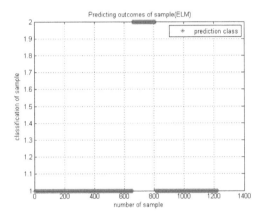

Figure 13. Classification results of Guangli substation.

As can be seen from Figure 11, there are more samples, in which the prediction category isn't consistent with the output category than that in Figure 9. The performance comparison between ELM and BP neural network is shown in Table 2.

As can be seen from Table 2, ELM algorithm has faster learning speed and higher classification accuracy.

4.5 Application of ELM

The data used for simulation is Guangli substation data, and the classification results of the simulation data using ELM is shown in Figure 13.

As can be seen from the classification results, ELM can acquire load dynamic characteristic data accurately.

5 CONCLUSION

Exploiting load dynamic characteristic data from massive data becomes a popular concerned problem of load modeling with the age of big data advent in electric power system. ELM is used to acquire load dynamic characteristic data in this paper. The PFRMS is used to obtain the fault data for dynamic data acquisition. The simulation results verify that the ELM model provides good generalization performance at extremely fast learning speed. The method proposed in this paper provides a new research thinking for load dynamic characteristic data acquisition.

ACKNOWLEDGMENT

This research was supported in part by the National Natural Science Foundation of China (51377099) and the National Natural Science Foundation of Shandong Province, China (ZR2011EEM017).

REFERENCES

1. A.Y. Abdelaziz, M.A.L. Badr, A.H. Younes. "Dynamic load modeling of an Egyptian primary distribution system using neural networks," International Journal of Electrical Power and Energy Systems, vol. 29, no. 9, pp. 637–649, 2007.
2. Bazi, Y. Alajlan, N. Melgani, F. AlHichri, H. "Differential Evolution Extreme Learning Machine for the Classification of Hyperspectral Images," Geoscience and Remote Sensing Letters, IEEE. vol. 11, no. 6, pp. 1066–1070, 2014.
3. G. Huang, Q. Zhu, et al. "Extreme Learning Machine: Theory and applications," NEUROCOMPUTER, vol. 71, no. 13, pp. 489–501, 2006.
4. Guang-Bin Huang, Qin-Yu Zhu, and Chee-kheong Siew. "Extreme Learning Machine: A New Learning Scheme of Feedforward Neural Networks," 2004 IEEE International Joint Conference on Neural Networks, 2004. Proceedings, vol. 2, pp. 985–990, July. 2004.
5. Huang G-B. "Learning capability and storage capacity of two-hidden-layer feedforward networks," IEEE Trans Neural Netw, vol. 14, no. 2, pp. 274–281, 2003.
6. Jiaoyan Chen, Guozhou Zheng, Huajun Chen. "ELM-MapReduce: MapReduce Accelerated Extreme Learning MaChine for Big Spatial Data Analysis," 2013 10th IEEE International Conference on Control and Automation (ICCA) Hangzhou, China. pp. 400–405, June. 2013.
7. Li, W. Load Modeling. New York: Wiley-IEEE Press, 2011, pp. 21–47.
8. Li Xiaofang, Peng Minfang, He Hao, Liu Tao. "Dynamic Load Modeling for Power System Based on GD-FNN," Proceedings of the 3nd International Conference on Digital Manufacturing & Automation, pp. 339–342, 2012. Mathur, A, Foody, G.M. "Multiclass and Binary SVM Classification: Implications for Training and Classification Users," Geoscience and Remote Sensing Letters, IEEE. vol. 5, no. 2, pp. 241–245, 2008.
9. Ping Ju, Chuan Qin, Feng Wu, Huiling Xie, Yan Ning. "Load modeling for wide area power system," International Journal of Electrical Power and Energy Systems, vol. 33, no. 4, pp. 909–917, 2011.

10. Ping Ju, Daqiang Ma. Power system load modeling. Beijing: china Electric Power Press, 2008, pp. 1–34.

11. Shaoning Pang, Daijin Kim, Sung Yang Bang. "Face membership authentication using SVM classification tree generated by membership-based LLE data partition," IEEE Transactions on Neural Networks, vol. 16, no. 2, pp. 436–446, 2005.

12. Serre D. "Matrices: theory and applications," Springer, New York, 2002.

13. Y. Chang, F. Wang, et al. "The Comparative Study between ELM and SVM on Predicting the Post-operative Survival Time of NSCLC Patients," Pattern Recognition and Artificial Intelligence, vol. 18, no. 5, pp. 636–640, 2007.

14. Yanxi Liu. "The BP neural network classification method under Linex loss function and the application to face recognition," 2010 The 2nd International Conference on Computer and Automation Engineering (ICCAE). Singapore, pp. 592–595, Feb. 2010.

15. Zhijun Gao. "Application of the BP neural network classification model in the value-added services of telecom customers," 2011 3rd International Conference on Advanced Computer Control (ICACC) Harbin, pp. 620–623, Jan. 2011.

Power and Energy – Kong (Ed.)
© 2015 Taylor & Francis Group, London, ISBN 978-1-138-02782-4

Overview of the Automatic Test System of relay protection

Jingjing Huang, Li-an Chen & Yan Liu
School of Electrical Engineering and Automation, Xiamen University of Technology, Xiamen, China

Gang Li
ABB Ltd., PPMV Technology Center China, Xiamen, China

ABSTRACT: The summary to the relay protection testing technology is mainly carried on in this paper. It introduces the development history and current situation of relay protection testing technology, features of hardware and software of automatic test system and the present stage of relay protection, and with the ABB automatic test system of ATS as an example to illustrate the working process of the automatic test system of relay protection. The significance and requirements of relay protection testing technology are expounded and the importance of relay protection and automatic test system are also pointed out in this paper.

1 INTRODUCTION

The importance of the safe and stable operation in power protection system is self-evident[1]. With the development of society, modern society's increasing demand for electricity, power grids around the world are rapidly developing, with the massive use of power transmission equipment, large numbers of various types of protection devices are emerged. In order to break through the traditional tests with tuning function modules and individual tests based on test mode, the relay automatic detection system gradually developed can greatly improve the efficiency of the testing work.

2 THE HISTORY OF THE DEVELOPMENT OF THE RELAY PROTECTION TESTING

In a fairly long period of time before the 1990s, in order to inspect and debug a variety of protection devices, generally when checking and adjusting the amplitude and phase angle of three-phase voltages and currents, usually use voltage regulator, high precision phase shifter, varistor, and accurate voltmeter, ammeter and phase meter, or other meters with high physics precision, to record various physical parameters which is needed for test during the process of measuring. As the most traditional method of testing, the operation is complex and requires testers to record the test data repeatedly during the test, it's inefficient and has low accuracy.

The development process of relay protection testing technic is mainly evolution through following several typical stages: the combination type, integrated circuit, digital control and microcomputer relay protection test device. Among them, test device with combination type is mainly used to measure various physical parameters with different measuring devices, and combine different measuring datum, finally to obtain the measurement results of relay protection. As a traditional measurement device, this method has few debugging functions, cannot carry out complex test and the efficiency is low; integrated circuit testing device is based on the basic of the hardware test device, adopts an integrated circuit chip to control test device, and complete digital display in order to read the test data more accurately and simple; the digital control type test device is an important development of relay protection measuring technology, mainly using the computing and storage functions of SCM.

With the development of microcomputer technology, the strong performance of microcomputer control is also applied in relay protection technology, microcomputer relay protection test device can fully meet the performance requirements of microcomputer protection test, enriching and improving the performance of relaying protection system. Microcomputer relay protection test device has gone through four developing stages: 1) as an intelligent controller with single-chip period; 2) with PC as the intelligent controller period; 3) using Windows operating system as platform period; 4) using high performance DSP core controller and embedded industrial control computer period. In this process, the relay protection testing technology is heading to mature[2].

3 CHARACTERISTICS AND APPLICATION OF THE AUTOMATIC TESTING SYSTEM OF RELAY PROTECTION

3.1 *Characteristics of relay protection testing*

The introduction of automated tools in the process of automated tests is very efficient and practical method[3]. The software system as a part of relay protection test system is able to formulate test method scientifically according to the test requirements of the work and the setting parameters of automatic test of relay protection, complete the test and manage the test report automatically and scientifically, the development of test technology transits from conventional, steady state test to the accurate dynamic test successfully, in order to make testing more efficient.

In addition to the rational design of hardware modules and structure, the software, as a part of automatic test system, need to have simple, powerful features. In present stage, the relay protection system of the automatic test software system can not only storage the test data, setting calculate, but also can generate and output standard test reports automatically, and what is more important is that it has the real-time and multi-tasking features, which can greatly improve the efficiency of the program. At present, domestic relay protection products are mainly based on the IEC60255 series of standards and national standards GB/T14047[4]. In general, protection devices go through the testing process in four phases: research and development testing, network testing, production testing and on-site verification.

The target of R&D testing is to test whether the hardware and software design are reliable and comply with protection product standards, due to the security and stability of the power system is particularly important, therefore testing requirements for protection products in the testing stage of R&D is extremely stringent. It is generally based on the product model and test equipment should be in a very high degree of specialization and automation, which need many test projects and the project cycle is long. In the development phase, the test content of protection products includes:

– The test of stability of the software program in device and the reliability of the hardware operation;
– Device communication protocol testing, device information communication function detection;
– Device action properties, action principle, functional testing and time parameter test;
– Devices fast transient, electrostatic discharge, radiating magnetic field testing;
– Device temperature rise, insulation, mechanical test.

The test projects in R&D produce process should be standardized and need a heavy workload, so introduction an automatic test system into R&D test procedure is largely a revolutionary breakthrough for R&D test.

The goal of production test is to ensure the quality of manufactured products meet the factory requirement and customer can use it normally, so the test project is simple and light. In the delivery inspection and on-site testing, some conventional test of the hardware and the software related to the specific device will be carried on, ensuring that the device hardware is reliable, action performance parameters are reasonable, the protection function is reliable and time synchronization, etc.

R&D and production testing together constitute the complete process of relay protection test, both of them are very important and indispensable, ensuring the reliability of the product, or it will endanger the stability of power grid.

3.2 *Application of automatic test system of relay protection*

The ATS (Automatic Test system) is researched and designed by ABB independently, and it is a highly automated system which is mainly used in development stages to test the reliability of relay protection products according to international specification and standard. ATS system is mainly used for function module test, including evaluating the function performance and verifying the functionalities, function performance evaluating includes static accuracy test (e.g. operate value & reset ratio, operate time & reset time, service value); dynamic accuracy (e.g. transient value, critical impulse time, impulse margin time, recovery time); dependency tests (power frequency, harmonics). Verifying the functionalities is to test the effectiveness of one function.

The main task of ATS is the development of Test case for new AFL's (application function library), and as the resource of product technical data (e.g. time accuracy, measurement accuracy of resistance to disturbance), and it is also used for product project-regression test of the old AFL's.

3.2.1 *The hardware structure of ATS*

Figure 1 for the system structure of ATS unit, the figure including the relay protection test device, the network data server which is used for storing the test case and test reports, testing software for editing test cases and test execution in computer, and Austria company OMICRON's OMC 256, CMA 156 and CMB IO-7 for outputting the analog signal and digital signal.

The measured relay protection device must choose the corresponding testing card, including hardware and software version.

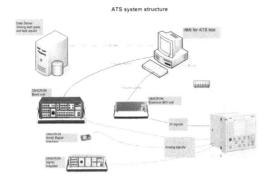

Figure 1. ATS unit system structure (picture comes from ABB).

CMC256 as a kind of relay protection testing device, but also can be used as a universal calibration scheme for electric energy measurement device, transmitter. CMC256 has the advantages of wide application range and the high accuracy in the entire output amplitude range. In the ATS, CMC256 is mainly used to output analog voltage or current signal and digital signal to test object.

CMA 156 is a current amplifier and CMC256 tester connected by a cable to control the equipment. The output and input of the amplifier and the ground is the electrical isolation, the outputs of CMC are independent with each other, but the output signal can be also used with CMC tester at the same time. The CMA amplifier 156 is divided into two independent groups (A, B), each group containing 6 current channels, each channel can output 25A current independently, the maximum output of each channel can be up to 150A. The signal output can be configured as a variety of ways (parallel connection, series connection, etc.).

CMB IO-7 is a simple serial development product, which can be expanded to 96 outputs with high accuracy.

3.2.2 Testing software of ATS

Testing Software system of ATS adopts IEC61850 standard to achieve GOOSE real-time communication, and support data sampling simulation and receive. The system applies advanced hardware equipment and friendly user interface, which can be used in the WINDOWS operating system environment.

The ATS software mainly includes test editing software, test execution software, Web server, and some relevant program importing software, etc.

The test editor is mainly used for editing test cases, which is divided into several modules by different relay protection functions, each functional module is used for testing a independent relay

protection function, such as protection, monitoring, control, and has its own test parameter list, related configuration files, several input and output files, tolerance setting. Once created, the TO (test object) could not be deleted, and the name could not be changed. A new revision of a TO could be created based on the present revision. Only the totally new TO could be used to import the data of an existing TO.

Every function module consists of dozens of or hundreds of test cases, defining all the related datum related to a test (e.g. settings, I/O mapping, configuration file, test procedure, expected result, criteria to evaluate the test result, curve for the test result, etc.). The development of a test case is based on a specific test template (e.g. operate value, operate time, test sequence, etc.). A test case could be modified or deleted and could be created based on an existing test case. Modification of related setting (e.g. parameter value setting, tolerance) might affect the other test cases, which also use the same setting.

Test executive is client software for ATS database and it's used to execute test cases in ATS database. Database is source of test data and it stores test cases, files, test project data and test hardware information. And it also gives the link to run the test engine. It's like a bridge to connect the database to the testing hardware. The test executive part is the highest component in ATS. It includes: user interface, test sequencer, database processor, testing result evaluation program, and test project processor. The test execution interface can display the test schedule and test results of the assessment, so it is easy for the tester to learn about the current status of the test.

ATS webpage server is an application which is used to browse the test project and report generated by the test system. It allows users with permissions to modify the project or test state. Webpage server can be created, changed or close the item. User can not only see the test project, but also view the obtained test information previously saved in the project which has been closed.

3.2.3 The test process of ATS

The test was divided into four stages:

Firstly, Test preparation stage: create a test project in web server, select test module and test cases. And then prepare for relay protection device making the device's test board and CPU version meet the test requirements. Gets the test program package and import it into the relay protection device by software, complete preparation for hardware equipment. The hardware connection, connect the I/O mapping of OMICRON according to the relay protection device card types. At last, set test parameters according to

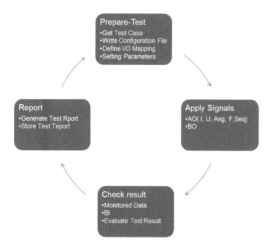

Figure 2.   ATS test flow chart.

the purpose of the test and check the protection setting value.

Secondly, apply test signals: test execution software begins to execute test cases. Download a setting package of the particular test case according to the test editor, the test execution software call the OMICRON 256 to output analog signals or digital signals to device.

Thirdly, check test result: relay protection device generate action after receiving signals, OMICRON can detect and record any output signals from the I/O of the relay protection test device, then transfer the information to test execution software, the software analysis the datum, then compare with expectation results, and obtain the test result.

Finally, share test report: test editor will upload test report to the server, save and share in Web server. The whole test process is executed by test software automatically.

If the test is failure, tester need to analysis the reason, to ensure the test process is normal and parameters is reasonable, finally draw the conclusion that whether the relay protection function program package has any design flaw, tester and summarize the problems in testing and feedback to the program developer, which is probably one of the original intentions to develop the ATS, to find any possible flaw in relay protection module program which is not in conformity with the norms and standards of relay protection part, ensuring the reliability of relay protection products. The whole test process is finished automatically, and testers just select corresponding hardware and connect test device, reducing the professional requirements of testers, and greatly improving the test efficiency.

3.3 The functions of automatic relay test system which has been implemented

According to the above process development of test system and report template, the following benefits can be achieved:

1. Organize the test items and test processes by inspection procedures to ensure the integrity and accuracy;
2. Reusable, once the test instance is created, it can be called at any time to repeat;
3. Test template gives a lot of tips of help information during the execution of the test; testers can do the test without checking drawings or instructions;
4. Function calls, parameter setting, and the evaluation of test results are achieved automatically during the testing process. Testers just import, check the protection setting, follow the steps to complete all the testing process automatically. A standardized and automation test method is truly completed.

4   CONCLUSIONS

The importance of relay protection for the stable operation of power system is self-evident, but to improve the level of automation of relay protection testing technology is even more crucial. Highly automated relay automatic test system is the development trend of current relay test technology, it can not only greatly improve the efficiency of the relay test work, improving the accuracy of the test, what is more important is that relay protection automatic test technology enables automatic processing of the test results, reducing the cost of the testing process cost. The research and the application of relay protection automatic test technology made a useful demonstration for reforming and improving the management of relay test work, which will greatly promote the development of the standardization for relay automatic test technology, and fully tap and play the effectiveness of relay automatic test systems. At the same time take a positive role in reducing the labor cost and improving the efficiency of the inspection work and standardized management and other aspects.

REFERENCES

1. G. Liu, "Research of relay protection test technology ". Silicon Valley Silicon Valley, Vol. 3, 2013, pp. 26–27(in Chinese).
2. J. Gao, "About the automatic test system of relay protection". Wireless Internet technology. Vol. 2, 2013, pp. 55(in Chinese).

3. Z.-A. Li, Q.-R. Shen, Y.-G. Wang, et al. "Electrical power system intelligence installment automated test system design". Automation of Electric Power Systems, Vol. 33, No. 8, 2009, pp. 77–79(in Chinese).

4. Z.-Q. Yao, "Some thought about development target of testing for relay protection equipment". Relay, Vol. 36, No. 11, 2008, pp. 76–78(in Chinese).

5. Z.-H, Ying, J.-B. Hu, R.-D. Zhao, et al. "Research and design of relay protection equipment automated test system". Power system protection and control, Vol.38, No. 17, 2010, pp. 142–146(in Chinese).

6. Q. Lai, J.-W. Hua, Y. Lu, Y.-F. Chen, J. Xu. "Research on general relay protection auto-test system software". power system protection and control, Vol.38, No. 3, 2010, pp. 90–94(in Chinese).

7. X.-C. Zheng, W.-H. Ding, X. Han, Z.-Y. Wang, H.-L. Zheng, "Research and realization of autotest technology for protection relays based on test template". Power system protection and control, Vol. 38, No. 12, 2010, pp. 69–73(in Chinese).

8. J.-L. He, C.-J. Song. The theory of power system protection. Beijing: China electric power press, 2004(in Chinese).

9. G. He, B. Hu, C.-L. Chen, L.-L. Zheng. "OMICRON tester in digital protection device test". Power system protection and control, Vol. 38, No. 12, 2010, pp. 132–134(in Chinese).

Power and Energy – Kong (Ed.)
© 2015 Taylor & Francis Group, London, ISBN 978-1-138-02782-4

A design of smart charger for ultracapacitors on electric vehicle

Ying Deng, Guang-bing Jiang, Lin-feng Wang & Meng Jiang
College of Engineering and Technology, Southwest University, Chongqing, China

ABSTRACT: Compared with traditional energy storage battery, ultracapacitor is widely used for electric vehicle because of its friendly charge-discharge properties, while the key and foundation of this kind of vehicle is how to get ultracapacitor batteries charged. This paper adopts modular design method to design a hardware system of smart charger for ultracapacitors which uses single chip PIC with CAN function as control core and IGBT power components to make up DC/DC converter. The paper also makes a series of protection design in detail including under-voltage and over-voltage protection, default phase protection, over-current and short circuit protection, overheat protection and so on. CAN interface is designed in charging system for real-time data exchange between the charger and ultracapacitors, which provides a basis for closed-loop and smart control. Finally, we test the voltage stabilization precision of smart charger in condition of constant current, the change of IGBT temperature working in condition of different output power and charging time and work efficiency of charger. The results show that its performance index meets design requirement and it can operate reliably.

1 INTRODUCTION

In recent years, electric vehicle has become an important green vehicle in our life because of energy deficiency and request for protecting environment. Power supply is the heart and key technology of electric vehicle. Compared with traditional power like lead-acid battery, NI-MH battery, lithium battery and so on, ultracapacitors have great advantage of fast charging, higher specific power, higher specific energy, long cycle life and so on, it is widely used in electric vehicle applications.

The key of developing electric vehicles with ultracapacitors is the charging technology of ultracapacitor batteries. On the one hand, each ultracapacitor should be in the same charging state to avoid under-voltage or over-voltage that can reduce the life of ultracapacitor batteries in series and increase driving range. On the other hand, there are so many charging methods applicable to the basic charging properties of ultracapacitors, such as constant-voltage charging, constant-current charging, constant-current—constant-voltage charging, constant-power charging and so on. Additional, ultracapacitor batteries can get fully charged with heavy currents in a short time, the great changes of transient state have a big effect on power network. So the charging device requires a high reliability. Aiming at above-mentioned characteristics, this paper adopts modular design method to design a hardware system of smart charger for the ultracapacitors which uses single chip PIC with CAN function as control core and IGBT power components to make up DC/DC converter. The

paper also makes a series of protection design in detail including under-voltage and over-voltage protection, default phase protection, over-current and short circuit protection, overheat protection and so on. CAN interface is designed in charging system for real-time data exchange between the charger and ultracapacitors, it provides a basis for closed-loop and smart control.

2 HARDWARE STRUCTURE OF SYSTEM AND OPERATING PRINCIPLE

The hardware structure of this charger specially designed for ultracapacitors consists of three subsystems which are main charging circuit, control circuit of charging and output module for display and alarm. The first subsystem called main charging circuit includes three-phase bridge rectification circuit, filter circuitfull-bridge converter circuit using IGBT, high frequency transformer and output rectification and filter circuit. The second subsystem called control circuit of charging is made up of single chip PIC18F4580 as control unit, data acquisition circuit, protection circuit, IGBT drive circuit, communication interface circuit and man-machine interaction circuit. The system structure block diagram is available in Figure 1.

The smart charger can operate in one of two modes: intelligent mode and manual mode. The following sequence of events will be executed when the system is working in intelligent mode: first step is self-diagnosis taken by control system, meanwhile reading basic information about the ultracapacitors

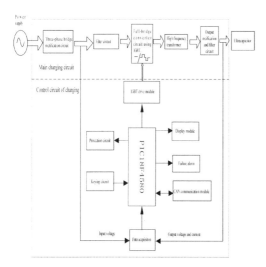

Figure 1. The system structure block diagram.

before charging such as its terminal voltage, capacitance and so on via CAN bus interface. Then control system sends the instruction to switch on the three-phase Solid-State Relay (SSR) in main charging circuit. Thus three-phase Alternating Current (AC) as input power passes through three-phase bridge rectification circuit and filter circuit to generate more smooth direct current (DC). Finally, DC/DC conversion will be finished by full-bridge converter circuit using IGBT, high frequency transformer and output rectification and filter circuit. The main charging circuit actually implements a change of AC—DC—AC—DC. The entire charging process is controlled by single chip PIC18F4580 which can get some data about charging voltage, charging current, the ultracapacitors' terminal voltage, capacitance and so on as feedback single from data acquisition module and CAN bus. PIC18F4580 can make a decision to choose the best reasonable charge mode based on its analysis and judgment in accordance with a certain mathematic model. After drive single is sent to IGBT drive module by PIC18F4580, the full-bridge converter circuit using IGBT begins to work while the ultracapacitors get charged. The whole working process is totally controllable.

In intelligent mode, charging voltage, charging current and charging method are automatically determined by system. When ultracapacitor batteries get fully charged, system stops work automatically. While in manual mode, all these can be set at will.

Many detection circuits are designed in Charging control system in order to avoid phase default, over-voltage, under-voltage, over-current, short or overheat. Once a failure occurs, PIC18F4580 sends a command for stop immediately to IGBT drive

module, the three-phase Solid-State Relay (SSR) is switched off automatically. Thus charger stops charging and gives audible and visible alarm, also fault location is displayed on LCD so as to eliminate fault in time. Charging process and battery parameter can be displayed on LCD in output module.

3 THE DESIGN OF MAIN CHARGING CIRCUIT

The main charging circuit diagram is available in Figure 2. It contains three-phase solid-state relay,

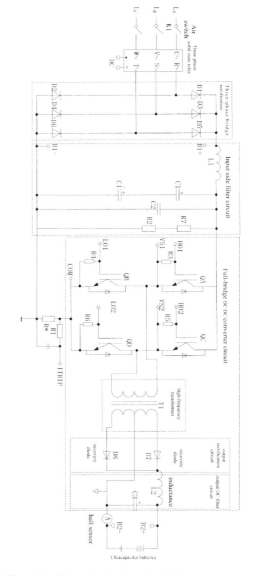

Figure 2. The main charging circuit diagram.

three-phase bridge rectification circuit, input side filter circuit, full-bridge DC/DC converter circuit, output high frequency transformer and output rectification and filter circuit.

When the three-phase solid-state relay gets power on, the three-phase bridge rectification circuit made up of six diodes transforms 380V AC into about 530V DC, then passing through filter circuit which consists of L(L1) and C(C1~C3) to be transformed into more smooth Direct Current (DC). Full-bridge DC/DC converter circuit consists of four IGBT each two of which in series are parallel, high frequency transformer and low side rectification circuit. By controlling the condition of four IGBT, 530V DC is transformed into AC and input to high side of high frequency transformer, thus low side generates induced AC, then it is transformed to DC by recovery diodes (D7~D8) and passes through filter circuit consisting of L (L2) and C (C3) so as to be transformed into smooth DC to get ultracapacitors charged. Charging methods such as constant-voltage charging, constant-current charging, constant-power charging and so on can be realized by controlling the four IGBT.

As shown in Figure 2, A represents a hall sensor which is cascaded in the charging circuit and it is used for measuring the charging current of ultracapacitors. R* represents a sample resistance and it is cascaded in full-bridge DC/DC converter circuit. So the current of IGBT can be measured to avoid over-current damage.

4 THE DESIGN OF CHARGING CONTROL SYSTEM

Control system is the core and control center of the charger, it plays a critical role in guaranteeing security, reliability and stable operation of charger. It contains smallest control unit, input/output voltage and current detection module, environment/IGBT temperature detection module, IGBT drive module, man-machine interaction module, CAN communication module, audible and visible alarm drive module and display drive module. The schematic diagram of structure is available in Figure 3.

4.1 *Voltage/current detection circuit design*

Voltage/current detection is the foundation that determines the precision and reliability of the charging system. This section contains primary side input voltage of main charging circuit detection circuit, output voltage of charger detection circuit and output current detection circuit.

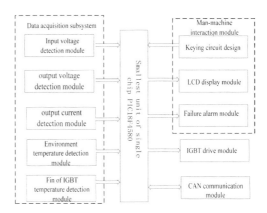

Figure 3. The schematic diagram of structure.

4.1.1 *Primary side input voltage of main charging circuit detection*

When charging system operates normally, primary side input voltage of main charging circuit is about 530V DC. In order to prevent this kind of strong electricity disturbing the weak electricity in control circuit, the circuit adopts high-linearity analog optocoupler HCNR201 to provide electrical isolation. As shown in Figure 4, high-voltage (VB1) is reduced to less than 5V by divider resistance (R1~R3) and followed by a voltage-follower consisting of LM324. Then it is input to HCNR201 through the reverse input terminal of the secondary operational amplifier LM324 which forms feedback on HCNR201 at the same time. Finally, an operational amplifier (LM324) and resistor are connected to make up integral circuit as wave converter and PIC18F4580 gets analog voltage signal for A/D from the resistor. Input voltage detection of main charging circuit has finished.

4.1.2 *Output voltage of charger detection*

As shown in Figure 5, it's the same difference between input and output voltage of main charging circuit detection.

4.1.3 *Output current of charger detection*

As shown in Figure 6, output current detection circuit consists of a hall sensor and A/D converter ADS7825. The ADS7825 is a 16-bit sampling A/D converter which has a higher precision than PIC18F4580.

4.2 *The design of IGBT full-bridge drive circuit*

The four IGBT in full-bridge converter circuit are driven by driver IR2133. Drive circuit is available in Figure 7. It integrates a Current

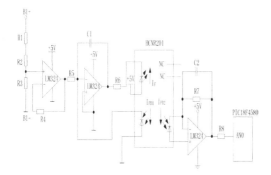

Figure 4. Primary side input voltage of main charging circuit detection circuit.

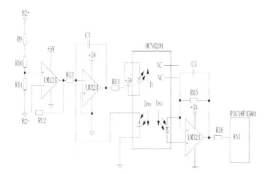

Figure 5. Output voltage of charger detection circuit.

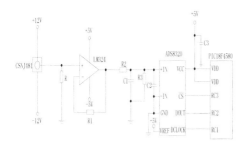

Figure 6. Output current of charger detection circuit.

Amplifier (CA), a Current Comparator (CC), a Under-Voltage Detector (UVD) of its own power supply, a fault logic processing unit (FL), three under-voltage detectors of high side drive signal (UVDR), six low resistance output drivers (DR), three input latches and a OR gate circuit. A shut-down function is available to terminate all six outputs. An open drain fault signal is provided to indicate that an over-current or under-voltage shutdown has occurred.

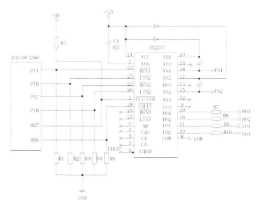

Figure 7. IGBT full-bridge drive circuit.

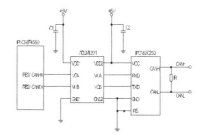

Figure 8. CAN communication module circuit.

4.3 The design of CAN communication module circuit

CAN communication module consists of control IC PIC18F4580, digital isolator ADUM1201 and CAN controller interface PCA82C250. CAN communication module circuit is available in Figure 8.

5 THE DESIGN OF PROTECTION SYSTEM OF CHARGER

We design the protection system to make sure the security and reliability of charger, including under-voltage and over-voltage protection, fault phase protection, over-current and short circuit protection, overheat protection.

5.1 Under-voltage and over-voltage protection design

There is some voltage amplitude fluctuation when adopting 380V AC as power supply. So the charger may not operate in preset mode even damaged because of under-voltage or over-voltage. Thus under-voltage and over-voltage protection have to be designed. The input voltage of main charging circuit can only vary from 486V to 590V. If not, the relay switches off immediately to cut down the power.

5.2 Fault phase protection design

Fault phase can aggravate the burden of the circuit and degrade charging performance. The protection circuit is available in Figure 9. In the circuit, 4N25 is an industry standard single channel phototransistor coupler. Each optocoupler consists of gallium arsenide infrared LED and a silicon NPN phototransistor. If the power supply phase is connected, the LED is lighted and the silicon NPN phototransistor is 'on', otherwise, the NPN is 'off'. The DM7411 contains three independent gates each of which performs the logic NAND function. The DM7411 is a logic gate which produces an output that is true only if all its inputs are true.

5.3 Over-current and short circuit protection design

Over-current and short circuit protection is designed to prevent over-current and short circuit of IGBT and charger output. Over-current protection of IGBT is realized by over-current protection function of IR2233. ITRIP pin of IR2233 can detect operating current, once surpassing the largest operating current set by 570 mV, IR2233 stops automatically the drive output of IGBT.

Short circuit protection of IGBT is realized by a comparator that compares sample resistance voltage with set voltage, if the former is bigger, external interrupt occurs in INT pin of single chip, the controller cuts down the power to protect circuit.

Over-current and short circuit protection circuit of IGBT is available in Figure 10.

Output over-current protection circuit of charger is available in Figure 6. The current is measured by a hall sensor, once it exceeds 5 percent of largest output current, single chip sends stop signal to make IGBT stop working, meanwhile displaying fault message on LCD and giving an alarm.

Output short protection circuit of charger is available in Figure 11. It uses a zero-crossing comparator and a gate circuit to control the relay. When short circuit occurs, the relay switches off and power is cut down, meanwhile fault message is displayed on LCD and an alarm is given.

5.4 Overheat protection design

The temperature of environment and fin of IGBT is both detected. Temperature sensor DS18b20 is applied in the circuit. Once the detected temperature exceeds set control value, charging current begins to reduce to avoid heat fault. Overheat protection design is available in Figure 12.

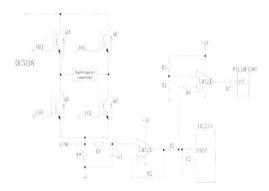

Figure 10. Over-current and short protection circuit of IGBT.

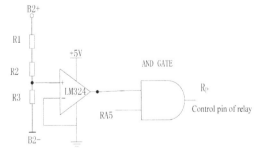

Figure 11. Output short circuit protection circuit of charger.

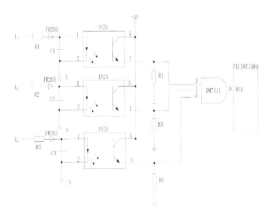

Figure 9. Fault phase protection circuit.

147

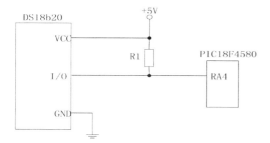

Figure 12. Overheat protection circuit.

6 EXPERIMENT AND CONCLUSION

We test the voltage stabilization precision of smart charger in condition of constant current, the change of IGBT temperature working in different condition of output power and charging time and work efficiency of charger. The experiment results indicate that:

1. Waveform of output ripple voltage of charger is detected. Peak voltage is less than 100 mV, waveform of voltage is less than 0.5 percent. The design meets requirement.
2. The difference between measured value and set value of the voltage stabilization precision of smart charger in constant-voltage, constant-current and constant-power charging method changes slightly over time, but all the precision detected is less than 0.5 percent.
3. The detection of work efficiency of charger working in different condition of output power shows that charging efficiency exceeds 90 percent, the design meets requirement.

ACKNOWLEDGEMENTS

This work was supported by "Fundamental Research Funds for the Central Universities" (XDJK2015D007) and Chongqing Science & Technology Commission Application and Development Project (cstc2013yykfA80003). The corresponding author of this paper is Meng Jang (driver1964@swu.edu.cn).

REFERENCES

1. Gualous H, Bouquain D, Berthon A, et al. Experimental study of super capacitor serial resistance and capacitance variations with temperature [J], Power Sources, 2003.123(1): 86~93.
2. Jeong JU, Lee HD, Kim CS, et al. A Development of an Energy Storage System for Hybrid Electric Vehicles Using Super capacitor [J]. IEEE Industrial Electronics Society, 2009(9): 10~15.
3. Kalan B.A, Lovatt H.C, Brothers M. System design and development of hybrid electric vehicles [C]. Power Electronics Specialists Conference, 2002(2):23.
4. Kotza R, Carlen M. Principles and applications of electrochemical capacitors [J]. Electro chimica Acta, 2000, (45):15~16.
5. Simpson AG, Lifecycle costs of ultracapacitors in electric vehicle applications [J]. IEEE Annual Power Electronics Specialists Conference, 2001(2):1015~1020.
6. Stuart R, Beale Roger Gerson, The Use of Ultracapacitors as the Sole Power Plant in an Autonomous [J]. Electric Rail-Guided Vehicle, IEEE, 2004(4):1178~11.

Power and Energy – Kong (Ed.)
© *2015 Taylor & Francis Group, London, ISBN 978-1-138-02782-4*

DG output forecasting based on phase space reconstruction and one-rank local-region method

Kun Peng & Xingying Chen
College of Energy and Electrical Engineering, Hohai University, Nanjing, China

Chunning Wang
Nanjing Power Supply Company, Nanjing, Jiangsu, China

Yingchen Liao & Kun Yu
Nanjing Engineering Research Center of Smart Distribution Grid and Utilization, Nanjing, China

ABSTRACT: Distributed generation output is influenced by the factor of weather and environment, which means it has the high degree of volatility and uncertainty. This paper proposes an distributed generation output forecasting approach based on phase space reconstruction and one-rank local-region method, which can reconstruct the historical data of distributed generation into the phrase point of the phase space, it can avoid the prediction error caused by uncertainty of environmental forecasting, improve the accuracy of distributed generation output prediction. At end of this paper, a simulation analysis was carried out by using the historical data of a wind generation.

1 INTRODUCTION

Distributed Generation (DG) is an intermittent power supply, environmental factors change has great influence on its power output, so the output power of the distributed generation is inconclusive. It is not beneficial for the power grid dispatchers to dispatch the conventional power supply with the distributed generation. Therefore, we need to forecast output power of the DG and obtain the power output development curve, then, coordinated controlling the conventional power supply with the DG to reduce its influence on power system and improve security and stability of the operation of power system. There are two main methods used for DG output forecasting: indirect method and direct method. The indirect method, is through the meteorological station or the information of the weather forecast, according to the change of relative parameters of environment. Through the calculation of a certain mathematical model to obtain the prediction of the DG output. There are commonly used methods, such as neural network algorithm. Due to the sun light characteristics, S.-X. Wang and N. Zhang proposed a combination-forecasting model based on the gray neural network in the literature. Atsushi Yona, Tomonobu Senjyu and Ahmed Yousuf Saber used three different methods of artificial neural networks: feedforward neural network, radial basis function neural network and recurrent neural network, to predict distributed

photovoltaic power output by predicting the solar light intensity. However, indirect method is too heavily reliant on the meteorological environment factors, which can lead to larger deviation.

Direct method used the history data of DG's power output, starting from the data itself, to find some kind of implicit regularity rules in the data, and predict DG power output. It mainly build DG power output history model based on Markov chain, then predict distributed generation power output. The trainable intelligent algorithm can provide effective and accurate methods for this direct prediction. However, this kind of method often uses the historical data and historical environmental data as input to predict distributed generation power output.

Therefore, this paper selects phase space reconstruction method based on historical data, which implies the influence of environmental factors in the recursive relationship of phase points that in the reconstruction phase space and avoids the prediction error caused by uncertainty of environmental forecast, to predict DG output.

2 THE BASIC PRINCIPLE OF PHASE SPACE RECONSTRUCTION

Phase space reconstruction first proposed by Packard, Takens etc, is a new kind of time series data processing method, by finding the inner relationship between the data, and building prediction model. The phase space reconstruction is also called dynamic system reconstruction. Through the one-dimensional time series, reverse constructed the phase space structure of original system. Delay vector method is a commonly used method, its basic idea is the evolution of the arbitrary component in the system is determined by other components which interact with it, thus, the information of these relevant components are implying in the process of development of arbitrary component.

The main mechanism of phase space reconstruction is as follow. For time series $x[i]$, in the phase space, the state vector which reconstruction dimension is m and time delay is τ can be described as $X[i] = [x(i), x(i + \tau), ..., x(i + (m-1)\tau)]$. Takens has proved when reconstruction dimension m is big enough, the reconstruction algorithm is an embedding mapping, phase space reconstruction can keep down many characteristics in power system, and it can restore the dynamic characteristics of the system in the sense of the topological equivalence. Then, there are a variety of calculation methods to calculate reconstruction dimension m and time delay τ. Based on the view that the value of reconstruction dimension m and time delay τ is not related, it can use autocorrelation method, mutual information algorithm, average displacement method and depolarization complex self-correlation algorithm to obtain τ, then it can use G-P algorithm or false nearest neighbor points algorithm etc.

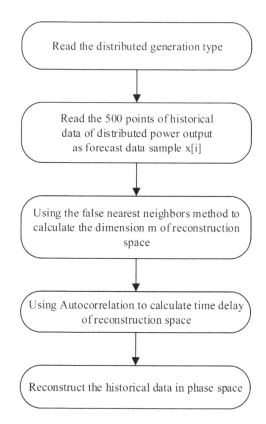

Figure 1.   Flow chart of phase space reconstruction.

3   PHASE SPACE RECONSTRUCTION OF DISTRIBUTED GENERATION OUTPUT POWER DATA

Firstly, select n distributed generation historical output data points which in time series space to form distributed generation output temporal sequence $\{g(i), i = 1, 2, 3 ... N\}$. Then, using autocorrelation method to calculate the value of time delay τ. Because of the view that the value of reconstruction dimension m and time delay τ is not related, this paper use false nearest neighbor algorithm to calculate reconstruction dimension m. According to the value of the reconstruction dimension m and time delay τ, we can reconstruct temporal sequence $\{g(i), i \in (1, N)\}$ to the phase points sequence in m dimensional space$\{G(i), i \in (1, 2, 3, ..., N-(m-1)\tau)\}$, where $G(i) = \{g(i), g(i + \tau), g(i+2\tau),..., g(i + (m-1)\tau)\}$.

After we obtain the phase points by reconstructing phase space, this paper use the one-rank local-region method to predict the size of next phase point. The flow chart of phase space reconstruction data processing is shown in Figure 1.

4   DISTRIBUTED GENERATION OUTPUT FORECASTING BASED ON ONE-RANK LOCAL-REGION METHOD

When the time series develop to a m dimensional phase space, Only a certain neighbor range of the phase points have great effect on the pre-diction value, and the influence of long-relative distance phase points can be negligible. Therefore, this paper selects one-rank local-region method to do the phase point forecasting. In a sense, the influence of neighbor points to forecast value are the same, but the phase points in the reconstructed phase space are the multi-dimensional structure of time series, so we need to join the phase point influence weight.

Firstly, according to the phase points $G(i)$ in the reconstructed phase space which we have already reconstructed in Part 3 by using the distributed generation historical output data, then we select some core point as follows.

$$Gc(1 + j * x), j = 0, 1, 2, ... \frac{N - (m-1)\tau}{x} - 1 \quad (1)$$

where x is neighbor radius, the neighbor points are the points, which distance to core point is smaller than x. To the set that consist of each core point and its relatedly neighbor points, we use the following formula to calculate the distance of each phase point to the core point:

$$d_i = \|Gc(1+jx), G(1+jx+i)\|, i = 1, 2, \ldots x \quad (2)$$

Then select the nearest distance d_{min} to the G_c to calculate each neighbor points weight as the following formula.

$$P_i = \frac{\exp(-c(d_i - d_{min}))}{\sum_{i=1}^{q} \exp(-c(d_i - d_{min}))} \quad (3)$$

c is the calculation parameter, in most cases, we use $c = 1$ to calculate.

m order matrix and linear fitting phase points evolution to:

$$G_{ci+1} = ae + bG_{ci}, i = 1, 2, \ldots x \quad (4)$$

a, b are fitting coefficients, e is a m dimension i vector

By using weighted least square method, we can get formula as follows:

$$\sum_{i=1}^{x} P_i(G_{ci+1} - ae - bG_{ci}) = \min \quad (5)$$

This formula can be seen as a two-variable linear equation about a, b, then calculate partial derivative respectively and simplify to obtain a two-variable linear equation contains matrix:

$$\begin{cases} ae\sum_{i=1}^{q} P_iG_{ci} + b\sum_{i=1}^{q} P_iG_{ci}^2 = \sum_{i=1}^{q} P_iG_{ci}G_{ci+1} \\ ae + b\sum_{i=1}^{q} P_iG_{ci} = \sum_{i=1}^{q} P_iG_{ci+1} \end{cases} \quad (6)$$

Linked with a and b, $x+1$ phase points exist in the phase point sequence, and to the next phase point we predict, the closer sequence to the phase point, the closer value of a, b. Therefore, we calculate the distance of each core point to the core point of final step, and obtain its weight. Choosing the different weight and different value of fitting coefficients to calculate fitting coefficients a_y, b_y of next step forecasting phase point. After that, taking a_y, b_y into the fitting formula can obtain the G_{M+1} phase point:

$$G_{M+1} = (g_{M+1}, g_{M+1+\tau}, \ldots, g_{M+1+(m-1)\tau}) \quad (7)$$

$$M = N - (m-1)\tau \quad (8)$$

where $g_{M+1+(m-1)\tau}$ is the No. m element in G_{M+1}, the value of the $g_{M+1+(m-1)\tau}$ is equal to the predicted value of distributed generation output that linked to next forecasting time points in time series.

Then we join the predicted values as known data to the original time series, after repeated the above steps, we can obtain multi step prediction. Its logical structure is shown as Figure 2.

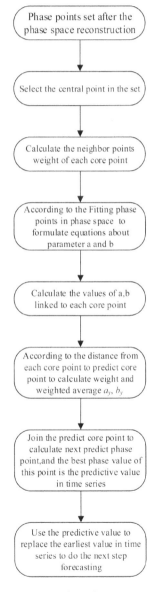

Figure 2. Flow chart of distributed generation output forecasting based on one-rank local-region method.

5 CASE STUDY

To the methods of research in this paper, its prediction method is starting from historical data itself. For wind power generation and photovoltaic power generation is using a similar process and method, it is not influenced by the environmental factors. This paper uses a wind power plant as an example to simulate and verify the forecasting method based on phase space reconstruction and one-rank local-region in this paper.

This paper uses 550 sets of historical data from a wind power generation. Firstly, the former 500 sets of historical data are used as the training data. Then the internal relations have been obtained by the method of phase space reconstruction. After that, by using the one-rank local-region method repeatedly, we finally obtain output forecasting of the 50 follow-up time point, and compare predict result with the real historical data.

The former 500 groups of data are normalized, and the output curve is shown in Figure 3.

According to the phase space reconstruction combine with one-rank local-region method which have been proposed in this paper, the core point radius is selected as 4, then compare the prediction value of later 50 time points with 50 sets of historical data following the former 500 sets of historical data. The results are shown as Figure 4.

As a comparison, the traditional one rank local region method without phase space reconstruction is used to predict distributed generation output. Choosing the same former 500 sets of historical data as training data. The later 50 sets of predict data are compared with later 50 sets of historical data.

From Figure 4 and Figure 5, it is obvious that the proposed method in this method achieves a higher accuracy.

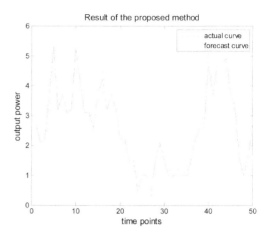

Figure 4. Result of the proposed method.

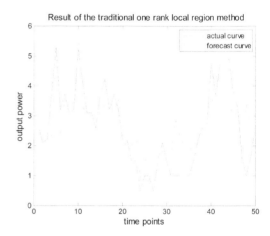

Figure 5. Result of the traditional on-rank local-region without phase space reconstruction.

6 CONCLUSION

Simulation analysis is carried out by using real historical data of a distributed wind power generation, verified the feasibility of distributed generation output prediction method based on phase space reconstruction and one-rank local-region method based mainly discussed in this paper. Considering the randomness distributed generation output under the comprehensive weather and environment factor, it will keep down many effective characteristics of the system by the method of phase space reconstruction, and restore the original characteristics of system in the sense of topological equivalence, imply the influence of environmental factors in the recursive relationship of phase points that

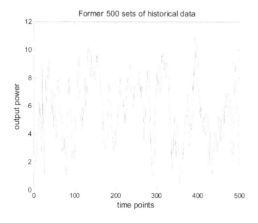

Figure 3. Former 500 sets of historical data from a wind power generation.

in the reconstruction phase space and avoids the prediction error caused by uncertainty of environmental forecast. The proposed method in this paper just need the historical data of the distributed generation, taking no account of the change of weather and environment factors, it can reduce the cost of distributed generation output forecasting. Compare with one-rank local-region method without using phase space reconstruction, we can see the proposed method in this paper achieves a higher accuracy.

ACKNOWLEDGEMENTS

This research is supported by the National High Technology Research and Development of China 863 Program (2012AA050214), National Natural Science Foundation of China (51207047) and State Grid Corporation of China (key technology research, equipment development and demonstration of demand side's optimal operation, 2013-1209-6998).

REFERENCES

1. Yona, T. Senjyu and A.-Y. Saber, "Application of neural network to one-day-ahead 24 hours generating power forecasting for photovoltaic system," ISAP International Conference on Intelligent Systems Applications to Power Systems, 8 April 2007.
2. B.-J. Li, Y.-J. Fang and W.-D. Yang, "Problems and Countermeasures for Large Power Grids in Connection with Photovoltaic Power," Power System and Clean Energy, Vol. 26, No. 4, 2010, pp. 52–59.
3. Fang, Y.-H. Lv and Y.-M. Zhang "Research and design of high frequency link inverter in stand-alone PV system," Power System Protection and Control, Vol. 36, No. 16, 2008, pp. 61–65.
4. G.-F. Fan, W.-S. Wang and C. Liu, "Wind Power Prediction Based on Artificial Neural Network," Proceedings of the CSEE, Vol. 28, No. 34, 2008, pp. 118–123.
5. G.-F. Fan, W.-S. Wang and C. Liu, "Artificial Neural Network Based Wind Power Short Term Prediction System," Power System Technology, Vol. 32, No. 22, 2008, pp. 72–76.
6. I. Abouzahr and R. Ramakumar, "An approach to assess the performance of utility: interactive wind electric conversion systems," IEEE Trans on Energy Conversion, Vol. 6, No. 4, 1991, pp. 627–638.
7. J. Li, H.-H. Xu and H.-X. Zhao, "Dynamic Modeling and Simulation of the Grid-connected PV Power Station," Automation of Electric Power Systems, Vol. 32, No. 24, 2008, pp. 83–87.
8. J.-Z. Li, L. Ru and J.-C. Niu, "Forecast of power generation for grid-connected photovoltaic system based on grey model and Markov chain," 3rd IEEE Conference on Industrial Electronics and Applications, 17 May 2008.
9. J.-Z. Li, J.-C. Niu, and L. Ru, "Research of multi-power structure optimization for grid-connected photovoltaic system based on Markov decision-making model," International Conference on Electrical Machines and Systems, 15 June 2008.
10. M. Ding and N.-Z. Xu, "A Method to Forecast Short-Term Output Power of Photovoltaic Generation System Based on Markov Chain," Power System Technology, Vol. 01, No. 6, 2011, pp. 152–157.
11. M. Ding, L. Wang and R. Bi, "A short-term prediction model to forecast output power of photovoltaic system based on improved BP neural network," Power System Protection and Control, Vol. 40, No. 11, 2012, pp. 93–99.
12. S.-X. Wang and N. Zhang, "Short-term Output Power Forecast of Photovoltaic Based on a Grey and Neural Network Hybrid Model," Automation of Electric Power Systems, Vol. 36, No. 19, 2012, pp. 1–5.
13. W.-N. Ling, N.-S. Hang and R.-Q. Li, "Short-term wind power forecasting based on cloud SVM model," Electric Power Automation Equipment, Vol. 33, No. 7, 2013, pp. 34–38.
14. X.-L. Yuan, J.-H. Shi and J.-Y. Xu, "Short-term Power Forecasting for Photovoltaic Generation Considering Weather Type Index," Proceedings of the CSEE, Vol. 34, No. 34, 2013, pp. 57–64.
15. X.-L. Wang and P.-J. Ge, "PV array output power forecasting based on similar day and RBFNN," Electric Power Automation Equipment, Vol. 33, No. 1, 2013, pp. 100–103.

Power and Energy – Kong (Ed.)
© 2015 Taylor & Francis Group, London, ISBN 978-1-138-02782-4

Modeling and simulation of ship power system

Chunhua Li
Department of China Satellite Maritime Tracking and Controlling, Jiangyin, Jiangsu, China

Yuewei Dai
Jiangsu University of Science and Technology, Zhenjiang, Jiangsu, China

Jiahong Chen & Yang Lu
Department of China Satellite Maritime Tracking and Controlling, Jiangyin, Jiangsu, China

ABSTRACT: The ship power system model is indispensable to evaluate its operating performance and design its controlling strategy. The models of the diesel, synchronous generator, and ship loads are integrated into the ship power system simulation platform in this work. The speed governor of the diesel is regulated by a PID controller to maintain the system frequency steady. The excitation device of the generator is controlled by a PID controller to prevent the system voltage from fluctuating. The simulation results validate the correctness of the proposed ship power system model.

1 INTRODUCTION

As a key part of the ship, the ship power system has an important effect on the safety and reliability of the ship operation. The ship power system is mainly composed of the ship station and ship grid (Xiao 1998; Shi 2003; Huang 2004). The ship power system with limited capacity is independent, and the capacity of some loads is sizable to the capacity of a diesel generator. The change of the power load will lead to the big fluctuation of the voltage and frequency of ship power grid (Li et al. 2004) Because of the various ship loads, complex operation conditions and poor working environment, the problems in ship power system are characterized of the random and concealment. In addition, upgrading transformation of the large power equipment makes the structure of the ship grid more complex and its anti impact ability more vulnerable. More important is that the traditional ship power monitoring method only detects manually several points and the performance of the ship power system at a whole can't be evaluated and studied. Therefore, an accurate system simulation model of the ship power system need be developed to lay the foundation for the stability analysis, fault diagnosis and performance evaluation of the whole ship power system.

In this work, the Models of the diesel, synchronous generator, and ship load are developed and integrated into the ship power system simulation system. There are the diesel rotation speed controlling subsystem and generator excitation device controlling subsystem in the ship station model. A PID controller is designed to regulate the speed governor of the diesel to maintain the system frequency steady. Another PID controller is proposed to control the excitation device of the generator to prevent the system voltage from fluctuating. The simulation platform of ship power system is built in Matlab/Simulink environment. The dynamic simulation results verify the correctness of the ship power system model under different load disturbances.

2 SHIP POWER SYSTEM MODEL

The ship power system mainly includes two parts: the ship station and the ship grid as shown in Figure 1. The ship station is composed mainly of the diesel, the synchronous generator, the diesel rotation speed controlling subsystem and generator excitation device controlling subsystem. ω is the measurement value of the rotation speed of the generator stator; ω_{ref}(pu) is the set value of the rotation speed, V_{ref} is the set value of the output voltage of the generator, m are the comprehensive measurement signals of the generator, V_d is the d-axis voltage of the generator, V_q is the q-axis voltage of the generator, and V_{stab} is the ground zero voltage. The models of the speed governor and diesel engine output the mechanical power P_{mec} of the diesel engine to the generator model, and the excitation device outputs the excitation voltage V_f to the generator model.

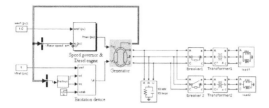

Figure 1. Ship power system model.

2.1 Synchronous generator model

The model of the synchronous generator is composed of the electromechanical transient process and the electromagnetic transient process. The electromechanical transient process is described by the rotor motion of the generator (Bo 2004):

$$T_J \frac{d^2\delta}{dt^2} = P_{mec} - P_e - P_D \qquad (1)$$

$$\frac{d\delta}{dt} = \Delta\omega \qquad (2)$$

where: T_J is the inertia time constant of the generator; P_e is the electromagnetic torque of the generator; P_D is the damping torque of the generator; δ is the power angle of the generator.

The electromagnetic transient model of the synchronous generator utilizes the five order state equations in the Matlab/Simulink environment.

The park transformation changes the three-phase currents (A, B, and C phase) of the stator to the d-axis current, q-axis current and 0-axis current of the rotor.

$$\begin{bmatrix} i_d \\ i_q \\ i_0 \end{bmatrix} = \frac{2}{3} \begin{bmatrix} \cos\theta & \cos\left(\theta - \frac{2\pi}{3}\right) & \cos\left(\theta + \frac{2\pi}{3}\right) \\ -\sin\theta & -\sin\left(\theta - \frac{2\pi}{3}\right) & -\sin\left(\theta + \frac{2\pi}{3}\right) \\ \frac{1}{2} & \frac{1}{2} & \frac{1}{2} \end{bmatrix} \begin{bmatrix} i_a \\ i_b \\ i_c \end{bmatrix} \qquad (3)$$

The stator terminal voltage is defined as:

$$V_t = \sqrt{V_d^2 + V_q^2} \qquad (4)$$

$$V_d = -Ri_d + \omega X_q i_q \qquad (5)$$

$$V_q = -Ri_q - \omega X_d i_d + \omega E_q \qquad (6)$$

where: $X_d(X_q)$ is the d(q)-axis winding reactance; R is the resistance of the stator winding; $i_d(i_q)$ is the d(q)-axis winding current; E_q is the q-axis ancient potential. The above equations ignore the 0 axis current of the Park equation.

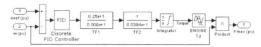

Figure 2. Rotation speed controlling subsystem model.

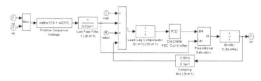

Figure 3. Excitation device controlling subsystem model.

2.2 Rotation speed controlling subsystem

Figure 2 shows the model of the diesel and its speed governor. The PID controller is used to regulate the rotation speed of diesel by adjusting engine throttle. The diesel and its throttle actuator are modelled by using a second order transfer function. Output torque of the diesel is obtained by integrating and delaying the output of the second order system. The mechanical power is obtained by multiplying the output torque and rotation speed measurement value, and then output to drive generator.

2.3 Excitation device controlling subsystem

Figure 3 shows the excitation subsystem model of the generator. The voltage deviation value are input into the PID controller so as to output the compensating signal. The excitation voltage is attained after the compensating signal through the saturation limit and exciting parts. The differential part in the feedback loop plays a dampling and stabilizing role on the system.

2.4 Ship load model

According to the load characteristic, the ship loads can be divided into three categories: induction motor load, static load and reactive power compensation device (Yong 2010). Generally, there are three modelling method of ship load: the constant impedance model, dynamic model of motor considering its static characteristic, and static model which power varies with the voltage and frequency of the ship grid. In this work, the big disturbance is modelled by the sudden loading and unloading, and the static model of the ship load is adopted.

3 SYSTEM SIMULATION

Figure 1 shows the simulation structure of the ship power system. The ship power grid which is

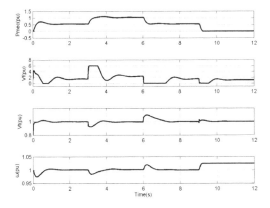

Figure 4. Dynamic responses of the ship power system.

a AC three-phase AC400 V, 50 Hz power system, is equipped with one diesel synchronous generators (AC400 V, 770 kW, 1389 A, 750 r/min). Two 400 kW, 50 kvar loads are connected to the ship power grid through two isolation transformer, respectively.

Figure 4 shows the dynamic responses of the ship power system. At the beginning, one load is connected to the ship grid. The generator speed ω becomes steady with the output mechanical power P_{mec} increasing. The excitation voltage V_f decreases from a big value to a steady value, and the terminal voltage V_t and the rotation speed ω of the generator need 2 s to reach their stable states. At the 3rd second, another load is connected with the ship grid. At the 6th second, the second load is disconnected. At the 9th second, the first load also is disconnected. The above simulation results show: the ship power controller can quickly and stably ensure the operating frequency and terminal voltage of the generator steady under the big disturbance from the ship load.

4 CONCLUSIONS

The dynamic model of the ship power system is developed in this work. The diesel engine, the synchronous generator, the diesel rotation speed control subsystem, the generator excitation subsystem and the ship load are modelled and integrated into the system simulation platform. The simulation results validate the correctness of the proposed system model and show that the system frequency and voltage under big disturbance can keep steady. The ship power system model will be useful for the stability analysis, fault diagnosis and performance evaluation of the whole ship power system.

ACKNOWLEDGEMENT

This work is supported by the National Natural Science Foundation (NNSF) of China (Grant 51307074), the natural science foundation of Jiangsu Province of China (Grant BK20130466) and the China postdoctoral science foundation (No. 2014m562615).

REFERENCES

1. Bo Hong-guang, 2004, "Marine power station simulation software system," M.S dissertation of Dalian Maritime University.
2. Huang Man-lei, Li Dian-pu, 2004, "Mathematical model of synchronous generator voltage regulation system on ship power station," *Journal of Harbin Engineering University*, vol. 25(3). pp. 305–317.
3. Li Dong-li, Zhao yue-ping, Tang Shi-Iing, Zhu yong-Iuan, 2003, "Effects of the Shipping's big Capability Load upon Power System. Ship engineering, vol. 25(6). pp. 47–49.
4. Shi Wei-feng, Zheng Hua-yao, 2003, "Marine automatic electric power station system simulation," *Journal of System Simulation*, vol. 15(9). pp. 1249–1252.
5. Xiao Yong-ming, Zhang Wei-zhong, 1998, "Mathematical model on marine electric station simulation system," *Journal of Shanghai Marine University*, vol. 19(4). pp. 52–58.
6. Yong Chang, Hu Yi-huai, 2010, "Dynamic Modeling and Simulation of the Marine Power Station with Load Disturbance," *Shipbuilding of China*, vol. 51(2). pp. 198–204.

Power and Energy – Kong (Ed.)
© 2015 Taylor & Francis Group, London, ISBN 978-1-138-02782-4

An influence analysis method for renewable DG to distribution system

Qun Yang, Xiangyu Kong & Yunfei Mu
Key Laboratory of Smart Grid of Ministry of Education, Tianjin University, China

Songling Pang
Electric Power Research Institute of Hainan Power Grid Corporation, China

Ning Lu & Jiahong Yan
Future Renewable Electric Energy Delivery and Management (FREEDM) Systems Center, North Carolina State University, USA

ABSTRACT: More renewable Distributed Generations (DG) will integrate in distribution system, it is important to analyze the DGs influence. In this paper, a method to analysis the influence of renewable DGs to distribution system was provided, which not only takes the technology but the forecasting of the associated meteorological parameter into considering. Some models of DG sources and load were studied, and a flow chart to analyze the influence of DGs integrated was given. At last, a distribution system optimization formulation considering the DGs Some case studies show the validity of the proposed method.

Keywords: power system; distributed generation integration; distribution system

1 INTRODUCTION

The utilization of renewable energy technologies, due to their suitability for remote applications, cost effectiveness and environmental impact, is a vastly increasing practice in the present and future power grid [1]. Distributed Generation (DG) is any type of electrical generator or static inverter producing alternating current that (a) has the capability of parallel operation with the utility distribution system, or (b) is designed to operate separately from the utility system and can feed a load that can also be fed by the utility electrical system via a change-over switch [2].

A distributed generation is sometimes referred to simply as "generator". As the sustainable energy technologies available, the present trends indicate a thrust for the large-scale integration of PVs and Wind technologies particularly into the power system to be utilized as distributed generation, which is in favor of costly central generation expansions within the power system [3]. There have been drives towards the development of components for the realization of new planning and operational tools that are geared to the study, monitoring and operation of the new renewable energy resources integrated grid.

This paper aims at presenting a method to analysis the influence of renewable DG to distributed network. It is propelled by the recent increase in the integration of renewable energy resources in today's power system. The integration plays a major role in the Smart Grid development. Within the objectives of the Smart Grid, the facilitation of remote utilization of renewable energy resources and enhancement of functionality of associated technologies are key components.

2 MODELING OF DG SOURCES AND LOAD

Technologies for distributed electricity generation include wind, solar, biomass, fuel cells, gas microturbines, hydrogen, combined heat and power, and hybrid power systems [5]. Most wind generators and PV systems use inverters. An inverter converts DC voltage and current into AC voltage and current, via power electronics and microprocessors. The inverter system also provides most of the protective relay functions and automatically synchronizes with the voltage and frequency from the electric grid, eliminating the need for discrete relays for voltage and frequency protection. The power electronics can be used for power factor correction and provide greater flexibility than non-inverter systems.

As the host of technologies presently being utilized as renewable energy DGs, solar and wind technologies capacities have been consistently trending upwards. The two sustainable energy sources have the characteristic of variability on account of climatic phenomena, namely solar insolation and wind speed. As such modeling of these resources requires consideration of not only the technology but the forecasting of the associated meteorological parameter.

2.1 Wind technology modeling

For wind power technology, the wind turbine is connected to an asynchronous/inductive machine that consumes reactive power and produces real power. The electric output of Wind Turbine Generation (WTG) depends on the wind characteristics as well as on the aero-turbine performance and the efficiency of the electric generator. These factors must be combined to obtain a probabilistic profile of the WTG output. The quantification of the capacity/real power output is given by equation (1):

$$P_w = \frac{1}{2}\rho\pi R^2 V^3 C_P \tag{1}$$

where ρ is the air density (kg/m^3), and R is the turbine radius (m); C_p is the turbine power co-efficient power conversion efficiency of a wind turbine and V is the wind speed (m/s).

The electrical power output as,

$$P_{w,m} = \eta_m \eta_g P_w \tag{2}$$

where, η_m and η_g are the efficiency of the turbine and the generator respectively.

2.2 PV technology modeling

As in the case of the wind technology, there exist a host of PV models in existence. A direct method for calculating the output of the entire PV array, without extensive knowledge of the particulars of the individual PV cells, was provide in reference [6]. It depends on available meteorological data and allow for ready integration of the solar insolation model as in Equation (3), which presents a model to be used for the modeling of the output of the PV panels.

$$P_{pv,m} = \frac{G}{G_{ref}} P_{m,ref}[1 + \gamma(T - T_{ref})] \tag{3}$$

where, G is the incident irradiance; P_m and $P_{m,ref}$ are the maximum power output and the maximum power output under standard testing conditions, respectively; T stands for the temperature and T_{ref}

stands for the temperature for standard testing conditions; G_{ref} is reference incidence, and γ is the maximum power correction for temperature.

2.3 Load modeling

Load modeling is very challenging when implemented in power analysis, because the electrical system has a large number of devices whose characteristics vary continuously. Therefore, it is not useful to exactly represent each individual component. To increase the performance of the dynamic analysis, load models have been implemented.

$$P = \frac{k_p}{100}\left(\frac{V}{V_0}\right)^{\alpha_p}(1 + \Delta w)^{\beta_p} \tag{4}$$

Power system analysis refers loads as real and reactive power demand. Generally, the load is represented by a static model. The model simulates the characteristics of the load at any given time as algebraic function of voltage and frequency [7]. Active and reactive powers are considered separately.

3 THE IMPACTION OF DGS TO DISTRIBUTION SYSTEM

3.1 Distribution system analysis method considering DGs

First lines of paragraphs are indented 5 mm (0.2″) except for paragraphs after a heading or a blank line (First paragraph tag). Different operation modes for DG have different impacts on the power supply reliability of distribution network system [8], mainly including three cases: 1) DG acts as the standby power for the system; 2) System power acts as the standby power for the DG; and 3) DG supports the load points in parallel to system power.

After more and more DGs, like solar and wind, connect to the distribution system, the intermittency and uncontrollability will increase the difficulty of reliability analysis [9]. To analyze a distribution system, different kinds of fault need to be assembled at different nodes and in different timing sequence. The computation will be massive, so the fault status is sampled by Monte-Carlo Method for analysis in this paper.

First, divide the load points in the cut set into two parts according to the point number, one part is the same branch-group as some DG, the other is the different branch-group from that DG. Then, based on the judgment of feeder the cut set belongs to, analyze the quantity of DG in the cut set and read the data of corresponding line. After that, use the Monte-Carlo Method to sample the operational status to form a state sequence of the system and

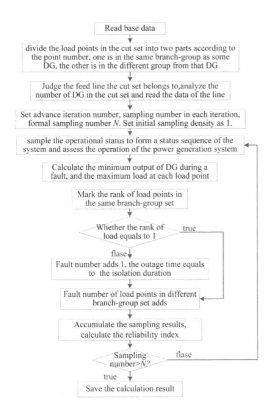

```
Read base data
        ↓
divide the load points in the cut set into two parts according to
the point number, one is in the same branch-group as some
DG, the other is in the different group from that DG
        ↓
Judge the feed line the cut set belongs to,analyze the
number of DG in the cut set and read the data of the line
        ↓
Set advance iteration number, sampling number in each iteration,
formal sampling number N. Set initial sampling density as 1.
        ↓
sample the operational status to form a status sequence of the
system and assess the operation of the power generation system
        ↓
Calculate the minimum output of DG during a
fault, and the maximum load at each load point
        ↓
Mark the rank of load points in
the same branch-group set
        ↓
Whether the rank of     true
load equals to 1
        ↓ flase
Fault number adds 1, the outage time equals
to the isolation duration
        ↓
Fault number of load points in different
branch-group set adds
        ↓
Accumulate the sampling results,
calculate the reliability index
        ↓
Sampling        flase
number>N?
        ↓ true
Save the calculation result
```

Figure 1. Distribution system analysis flow chart considering DGs.

assess the operation of the power generation system. For a fault on the line, analyze the load restoration according to the comparison between the sizing of DG and the amount of planned loads as well as the importance of load in forward and backward regions. Last, get the reliability index of distribution system. The algorithm flow is shown in Figure 1.

Using this method, the intermittence and uncontrollability of solar and wind due to climate changing need to be considered. In the reliability analysis, the climate factor affects the distribution system reliability a lot. It can be solved by multiplying the device reliability in different time and different climatic situations by a weight coefficient, and the weight coefficient is used to simulate different climatic status also.

3.2 Optimization formulation of DG integrated system

The distribution system has become a multiple power network and customers can gain power from different power supply. DG may submit hourly energy bids and their corresponding prices to the Microgrid actors, while consumers submit

hourly energy demands and their respective maximum buying prices.

The classical economic dispatch problem optimizes the power scheduling of generators to meet their load demand and the transmission system losses, while maintaining their acceptable performances such as operating the generators within their thermal and capability limits and maintaining the reliability of the transmission grid. In a market with m_w wind turbines and m_{pV} PV banks that supply power to loads, the economic dispatch problem can be stated as:

To minimize

$$F = \sum_{j=1}^{m_w} C_{pv}(P_{pv,j}) + \sum_{j=1}^{m_{pv}} C_w(P_{w,j}) \quad (5)$$

Subject to:

$$\sum_{j=1}^{n} P_{pv,j} + \sum_{j=1}^{n} P_{w,j} + P_{co} = P_{Load} + P_{Losses} \quad (6)$$

$$P_{pv,j} + P_{w,j} \le P_{DG,j,\max} \quad (7)$$

$$P_j^{\min} \le P_{w,j} \le P_j^{\max} \quad j = 1, 2, ..., m_w$$
$$P_j^{\min} \le P_{pv,j} \le P_j^{\max} \quad j = 1, 2, ..., m_{pv} \quad (8)$$

$$V_j^{\min} \le V_j \le V_j^{\max} \quad j = 1, 2, ..., n \quad (9)$$

where, $P_{w,j}$ and $P_{pv,j}$ is real power output of wind and PV units installed at bus j; $C(P_{w,j})$ and $C(P_{pv,j})$ are the cost per kW of generated by wind and PV per time interval based on rated capacity of unit; P_{Load} is assumed to be a constant representing all the loads in system for a specific time period and P_{co} is the power from distributed power system; $P_{DG,j,\max}$ is the maximum permissible penetration at bus j. $V_j^{\min}$ and $V_j^{\max}$ are the voltages limits. Equation (6) describes real power balance equation and equation (7) is maximum penetration limit of DGs per bus, equation (8) shows the operational limits on the real power output of PV and wind units, and equation (9) is the upper/ Lower limits of node voltages.

Being different from transmission losses, the generators' real power limits are not affected by other transactions in the network, and it could be dealt with easily. In this paper only real power losses are considered, which is equal to the difference of the line real nodal power flow at different end of the line [10], that is:

$$P_{Losses} = \sum_{i=1}^{n} I_i^2 r_i = \sum_{i=1}^{n} \frac{(P_i^2 + Q_i^2)}{V_i^2} r_i \quad (10)$$

where P_i and Q_i are the active and reactive power flow of node I, respectively, and V_i is the node voltage.

Loss of Load Probability (LOLP) is calculated by dividing the number of simulations with at least one shortfall event by the total number of simulations. If one case represents a potential "future" for customers, then LOLP is the probability that customers will face an interruption at some point in the future. LOLP does not represent the probability of a single shortfall event occurring. In some cases, a utility or region may only want to focus on its most critical part of the year, such as winter. In this case, hourly simulations only need to be done over the winter months and the resulting LOLP is thus a winter reliability measure only.

The following equation is a generalization of how LOLP is calculated:

$$LOLP = \frac{1}{N} \sum_{i=1}^{N} S_e \qquad (11)$$

where: S_e is a case in which at least one significant event occurs, a significant event occurs when load and operating reserve obligations exceed resources including contingency operations (or event threshold limits). N is the number of cases for the period, which is typically one year using hourly level of granularity.

4 CASE STUDY

A grid with wind and PV units is used to modify the impact of DGs in paper [2]. This simplistic system has been augmented to reflect the excessive loading on a radial distribution system. The DGs are connected to main medium voltage grid and under test system includes the following components: 1 battery units and 4 renewable energy generators: PV and wind, and several loads with different priority levels.

The seasonal variability of load for a particular network or location is usually calculated based on statistical data. The load is assumed to vary as a percentage of the base/peak load value over the periods of hours, weeks, months and seasons of the period of a year. This variation is presented as a percentage of base/peak loads. A profile is then generated based on this data. Sample load profiles for the variation of the load are presented in Figure 2.

Tables 1 shows sample system performance results for three cases studied. Here only the PV units are utilized to meet the demands of the system. The voltage improvement is demonstrated in the cases. For this operation, it is evident that

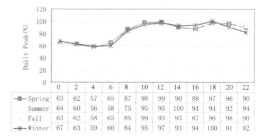

	0	2	4	6	8	10	12	14	16	18	20	22
Spring	63	62	57	65	87	98	99	90	88	97	96	90
Summer	64	60	56	58	75	95	95	100	94	94	92	94
Fall	63	62	58	65	85	99	93	93	87	96	98	90
Winter	67	63	59	60	84	95	97	93	94	100	91	82

Figure 2.   Hourly peak load by seasons.

Table 1.   System performance measures for winter study at load max.

Time	0:00	6:00	12:00	18:00
Bus voltage (mean)				
Bus 1	0.99745	0.99743	0.99741	0.99733
Bus 2	0.99964	0.99956	0.99865	0.99826
Bus 3	0.99694	0.99643	0.99611	0.99562
Bus 4	0.99693	0.99684	0.99655	0.99614
Bus 5	0.99865	0.99831	0.99798	0.99732
Bus 6	0.99932	0.99874	0.99854	0.99843

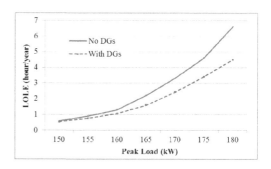

Figure 3.   Variation of the loss of load expectation with the peak load.

the PV only installation cannot meet the required demands.

In order to quantitatively assess the benefits of DGs, the LOLE indices of the distribution system before and after adding some DGs are plotted as functions of the peak load in Figure 3. It can be seen that there is a load carrying capability benefit from the DGs addition.

An argument for using a seasonal Load and DG balance as opposed to an annual balance is the idea of perception. Figure 4 illustrates a case in which the annual balance is about 1,000 average kilowatts but the winter months are deficit. If the region were same DG output surplus on an annual basis, the deficits in the winter months should be of no concern partially.

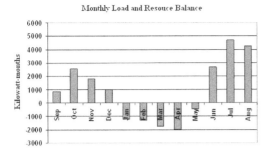

Figure 4. Monthly load and resource balance in test system.

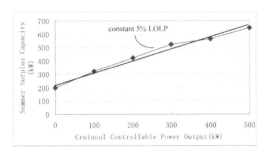

Figure 5. Relationship between other supply and load balance.

Because it would be difficult for individual utilities to plan to a regional load-resource balance deficit, an alternative would be to plan to a zero load-resource balance with critical controllable power, such as water or diesel units. Figure 5 shows the relationship between out-of-region supply and adverse diesel unit for a constant 5% LOLP. Average conditions mark on the x-axis, if 400 kilowatts of out-of-region capacity of controllable power were available, the region should plan to approximately 200 kilowatts to achieve a Pre-set LOLP.

5 CONCLUSION

This paper aims at presenting a heuristic method for the optimal placement and power level of the different types of generation units. The recent increase in the integration of renewable energy technologies into the electric power system, has led to the development of a methodology for the optimal use of these resources. Though there is presently, integration at the transmission, sub-transmission and distribution level of the system, the use of distributed generators at the customer/load-end of the distribution system is of particular concern.

ACKNOWLEDGMENTS

The authors acknowledge the National Science and Technology Support Program of China (2013BAA01B03).

REFERENCES

1. Pengwei Du. Operation of Power System with distributed Generation [D], 2006, Electrical, Computer and System Engineering, Rensselaer Polytechnic Institute, New York.
2. Wang kai, Kong Xiangyu, Zhao Shuai. A heuristic method for optimal placement and monitoring of renewable energy DG [C]. 5th International Conference on Electricity Distribution, shanghai, 06–8. Oct, 2012. pp:1–6.
3. E. Martinot, J. Swan, Renewable Global Status Report: 2009 Update [R]. Renewable Energy Policy Network for the 21st Century (REN21), 2009.
4. V. Piccolo A, Siano P. Clderaro. Maximizing DG Penetration in distribution netowrk by means of GA based reconfiguration [C]. Future Power Systems, International Conference on, November 2005.
5. N. Mahat P, Mithulanathan N Acharaya. An analytical approach for DG Allocation in Primary Distribution Networks [J]. Electrical Power and Energy Systems, 2006, 28(6): 669–678.
6. M.A.Masoum, M. Ladjevardi, A. Jafarian, et al. Optimal Placement, replacement and Sizing of Capacitor Banks in Distorted Distribution Networks by Genetic Algorithms [J], IEEE Transactions on Power Delivery, 2004, 19(4): 122–125..
7. J.C Awodele, K. Johannes. The impact of Distributed Generation on Distribution Voltage and Network Power Losses [C]. 8th International Conference in Power System Operation and Planning (ICPSOP), January, 2010, pp: 18–21.
8. Xiangyu Kong, Qun Yang, Wen Sun. Reliability Analysis Method for Distribution System with Distributed Generation. Applied Mechanics and Materials Vols. 448–453 (2014) pp 2649–2653.
9. Xu Shengyou, Chen Minyou, Wade Neal, et al. Reliability evaluation of electric power system containing distribution generation [J]. Advanced Materials Research, 2012, (383): 3472–3478.
10. C.L Borges, Djalma M. Falcao. Planning and Operation of Distribution Systems in the Presence of Dispersed Generation [C], CIGRE Symposium, Athens, Greece, April 2005.

Power and Energy – Kong (Ed.)
© 2015 Taylor & Francis Group, London, ISBN 978-1-138-02782-4

Study on security benefits of power network index hierarchy and evaluation methods

Yongzheng Mu, Zongxiang Lu & Ying Qiao
State Key Laboratory of Power Systems, Department of Electrical Engineering, Tsinghua University, Beijing, China

Han Huang & Yang Wang
State Grid Energy Research Institute, Beijing, China

ABSTRACT: An index hierarchy for security benefits of power network evaluation is proposed, which includes three levels of indices such as module level, factor level and index level. Three module level indices which are direct benefit indices, indirect benefit indices and benefit coefficient indices reflect all aspect of power network security benefits. The index hierarchy using economic evaluation method and benefit coefficients indices to avoid the deviation cause by the subjective favor and index evaluation results. The social and environmental benefits are brought into the index system as the result of the development of the power network and the integration of large amount of new energy. The example shows the rationality of the index hierarchy and the environmental benefit can compensate the security impact of the new energy integration in certain circumstances.

1 INTRODUCTION

With the development of the power industry deregulation, the power grid and power generation units belong to different economic bodies, respectively. To evaluate the comprehensive properties of power system, including power grid and units, two aspects are proposed separately, i.e. benefit evaluation and security assessment.

While the development of the new energy to power industries, the continuously growing investment becomes a new challenge for power network, and the available benefit evaluation method cannot meet the requirement. Considering the social responsibility of the power grids as utilities, security is a key concept that cannot be ignored in benefit evaluation. The traditional security assessment of the power system focuses on single aspect evaluation such as: voltage stability, angle stability, topology vulnerability and reliability. By using comprehensive evaluation methods (Wang, 1998,

Chen et al., 2004), several index systems of power system security are proposed (Zhang et al., 2009, Zhang et al., 2009, Li et al., 2008, Yang et al., 2009, Zhao et al., 2013), but the security evaluation results of these index system do not considered the cost of security. Enterprise security benefits theory is studied in the lecture (Zhang et al., 2009), but the index system does not aim at power network and the social influence is not considered.

For the power network improvement project such as smart grid (Wu, 2010), fixed assets investment (Liu et al., 2007) and sustainably energy integration (Brenna et al., 2007), the index system gives the comprehensive benefit result, but evaluation methods of projects can only evaluation few aspects of power network. The planning and operation of distribution network pay more attention to the economic benefit. Some evaluation methods such as Analytic Hierarchy Process (AHP) method (Xiao et al., 2008), dynamic weighting method (Juan et al., 2012) and gray related degree model (Zhang and Meng, 2011) are used in the comprehensive evaluation of the security benefit of distribution network. The economic benefit evaluation of power system is studied with gray related degree model (Ma et al., 2011) and AHP method (Gao et al., 2013), which focuses on the profit and does not consider the security risk. AHP method (Fan and Niu, 2007, Yue et al., 2014) is

This work was supported by The National High Technology Research and Development Program of China (2011 AA05 A103) and the State Grid Shandong Company Economic Research Institute ('Research of Power System Technical and Economic Evaluation Method, Model and Application Adapting to Strong and Smart Grid Development').

used to evaluation power system comprehensively, but the social and environmental benefit caused by security is not included in research and the method causes some subjective diversion in the result. Back Propagation (BP) neural network (Wang and An, 2010) requires a large number of samples to training which are hard to obtain.

An index hierarchy for security benefits of power network evaluation is proposed, which includes three levels of indices such as module level, factor level and index level. By using the economic evaluation method and benefit coefficients, the index hierarchy avoid the deviation cause by the subjective favor and index evaluation results while using traditional methods such as AHP method, dynamic weighting method, gray related degree method and BP neural network method. Three module level indices which are direct benefit indices, indirect benefit indices and benefit coefficient indices reflect all aspect of power network security benefits. The benefit coefficient indices show the indirect benefit participation factors of power networks, and they are different from each other because of the economic and geographic characteristics of power network. The example of two region power network security benefits evaluation shows that the index hierarchy is reasonable and suit for power network security benefit evaluation.

2 SECURITY BENEFIT EVALUATION INDEX HIERARCHY

2.1 Principle of constructing power network security benefits evaluation index system

The evaluation of power network security benefit is complicated. To ensure the practicability and validity of the index hierarchy, the construction principle of the index hierarchy should consist of following rules:

a. Comprehensive characteristic: the index system should systemically cover all the aspect of network security benefit.
b. Objective characteristic: the index system should focus on the evaluation object and ignore the irrelevant index.
c. Simplify characteristic: the index system should prevent redundancy indices, which are repeated and strong correlated with the exist index.
d. Hierarchy characteristic: the index system should have several hierarchies of indices to express the connection of indices and simplify the calculation.
e. Utility characteristic: the index system should be easy to operate and realize.

2.2 Security benefit evaluation index hierarchy

According to the principle of constructing index system, the 3-level hierarchy is proposed. The three levels of evaluation index hierarchy are module level, factor level and index level. According to different effect patterns of security benefits, the index hierarchy can be divided into two module level indices: the direct benefit and indirect benefit.

3 INDEX EVALUATION MODEL AND CALCULATION METHOD

3.1 Direct benefit evaluation indices

The index hierarchy uses the traditional safety evaluation concept to study the direct security benefit of the power network, which is the power network company lost due to the security event. The direct benefit includes two kinds of factor level indices: equipment failure risk and staff accident risk.

The equipment failure risk index is the expect cost of repairing and replacing the fail equipment of power network, which has three affiliated indices: the number of equipment, the equipment failure rate and equipment repair price. The equipment of power network mainly includes transformer, transmission line, bus, disconnector and circuit breaker. The average failure rate and repair price of these kinds of equipment can be estimated by the historical data.

$$R_{equipment} = \sum_{i=1}^{m} N_i P_{ei} C_{ei} \qquad (1)$$

where, i is the equipment type serial number, N_i is the number of the ith kind of equipment, P_{ei} is the average failure rate and C_{ei} is the average repair price.

The staff accident risk index is the expect loss of the power network accidents which causes deaths of the employee. The compensation and casualty are two affiliated index which can be obtained by using the historical data, laws and regulations.

$$R_{staff} = N_{pd} C_{pd} \qquad (2)$$

where, N_{pd} is the number of deaths caused by accident during the evaluation period and C_{pd} is the compensation.

3.2 Indirect benefit evaluation indices

The indirect benefit of power network security is that caused by increasing power supply and power quality, which includes two kinds of factor level indices: risk index and benefit coefficient.

3.2.1 Power network stability risk

The power network stability risk indicate the expect loss of two kinds of power system unstable events: transient unstable and static unstable. The transient stability reflects the dynamic characteristics of power system and mainly considers three kinds of unstable event after large disturbances: voltage stability, angle stability and frequency stability. Because most of the static unstable events are not driven by the power network component or device, the static stability indices should not be include in the hierarchy.

The power network stability risk have three affiliated indices: operation mode, instability event probability and load curtailment, which can be obtained by simulation of power system in each operation mode.

$$R_{stability} = \sum_{m=1}^{M} \eta_m P_{sm} E_{sm} \qquad (3)$$

where, M is the number of operation mode and η_m is the mth kind of operation mode proportion, P_{sm} is the unstable event probability in mth operation mode and E_{sm} is the average load curtailment energy.

The disturbances in simulation are set in following rules:

Single fault: the failure of single lines, transformers and substations or any double circuit line, the adjacent components in electrical connections.
Double fault: the protect device failed to act correctly while the single opponent failure happens.
Multiple fault is ignored while the evaluation of power network stability risk because of the low probability, and it will be evaluated in following index.

3.2.2 Power supply risk

The power supply risk represents the power network reliability while all kinds of uncertainty of power network are considered. The power supply risk has two affiliated indices: Expected Energy Not Supplied (EENS) and power failure penalty. By summing up the severity of power supply failure into energy shortage, EENS index is proposed in index level. The power supply failure penalty index shows the imposed additional fine according to the regulation when the load curtailment exceeds certain value. Using sequential Monte-Carlo simulation, the indices of power supply risk can be calculated.

$$R_{EENS} = \sum_{i=1}^{m} P_{ri} E_{ri} \qquad (4)$$

$$R'_{reliability} = \sum_{i=1}^{m} P_{ri} k_p (E_i) \qquad (5)$$

where, m is the total amount of events, P_{ri} and E_{ri} are the probability and severity of the ith load curtailment event, k_p is the penalty function which is determined by the regulation and $R'_{reliability}$ is the total penalty.

3.2.3 High influence low probability event risk

The extreme event, which can be defined as High Impact Low Probability event (HILP event), includes floods, earthquakes, typhoons and severe sleets. This event may cause multiple fault and blackout in power system with a great quantity of load curtailment and economic loss.

The HILP risk index has two affiliated index: the HILP event probability and HILP event load curtailment. The HILP risk can be evaluated by the historical data.

$$R_{HILP} = \sum_{i=1}^{n} P_{hi} E_{hi} \qquad (6)$$

where, P_{hi} and E_{hi} are the probability and load curtailment energy of nth type extreme event.

3.2.4 New energy application

The new energy developed rapidly nowadays because of the threat of global warming and climate change, and power network plays important role in the integration of centralized developed wind and solar power. The new energy power centers are often far from the load center and the utilization factor of new energy integration network is relatively low because of the fluctuation of the new energy. The power network security benefits of carbon emission reduction can be evaluated by three affiliate indices: delivered new energy and new energy curtailment, unit electric energy carbon emission reduction income.

3.2.5 Power benefit coefficient and environmental benefit coefficient

By using the benefit coefficients, the indirect benefit participation factors of power network can be calculated. To prevent the influence of index evaluation result above, the coefficient is based on the analysis of power network character.

The power benefit coefficient indicates the social benefit of continuous power supply which is strongly affected by the economic development level and the industrial structure. The coefficient of different power networks can vary significantly because of the area type such as hi-technology garden, commercial center, smelting plant and farmland.

Three index level indices are proposed such as ratio of output value to unit electric energy, net income per unit electric energy, power network delivery energy.

The environmental benefit coefficient indicates the power network contribution to the new energy integration and the carbon emission, which can be evaluated by comparing the reference power network without new energy integration. Two index level indices are proposed: new energy integration cost and new energy generator cost.

Replace the new energy generators with conventional ones by the rules of same annual energy output, the reference power network can be simplified because of the downsizing of generators capacity. Define the cost of new energy integration and new energy generators as the difference network and generator cost between two scenarios respectively, we can get the environmental benefit factor.

$$r_e = \frac{C_{nn}}{C_{ng} + C_{nn}} \times 100\% \qquad (7)$$

where, C_{nn} is the cost of new energy integration, C_{ng} is the cost of new energy generators.

The full sketch map of the index hierarchy is shown in Figure 1.

3.2.6 Economic evaluation method

The economic evaluation method is a comprehensive evaluation method which uses price index to normalize the index and gives comprehensive result.

Using net income per unit electric energy and ratio of output value to unit electric energy, the economic and social benefits of power network security can be calculated.

$$B_c = -(R_{stability} + R_{EENS} + R_{HILP})p - R'_{reliability} \qquad (8)$$

$$B_s = -(R_{stability} + R_{EENS} + R_{HILP})(r_v - p) \qquad (9)$$

where, r_v is the ratio of output value to unit electric energy, p is the net income per unit electric energy.

The environmental benefit is

$$B_e = (r_e E_{ne} - E_{nc}) \times C_c \qquad (10)$$

where, E_{ne} and E_{nc} are delivered and curtailed new energy respectively, C_c is the unit energy carbon emission reduction income.

After normalize the risk by the annual delivered, the security benefit of power network is:

$$\begin{aligned} B &= (B_{direct} + B_{indirect})/E_a \\ &= [(-R_{equipment} - R_{staff}) + (B_c + B_s + B_e)]/E_a \end{aligned} \qquad (11)$$

4 EXAMPLE

Two regional power networks are used as examples and the proposed index hierarchy is calculated.

4.1 Profile of the power network

The two power network cases are located in North China and East China, case A is a power network with annual delivered energy 1.949×10^8 MWh and case B is a power network with annual delivered energy 4.957×10^8 MWh.

4.2 Results and analysis of the index hierarchy

4.2.1 Direct benefit

The equipment failure risk is calculated by the historical failure rate data and planning equipment price and number, the result is shown in Table 1.

The staff accident index can be evaluated by the historical data; the result is shown in Table 2.

4.2.2 Benefit coefficient and indirect benefit

To evaluate the indirect benefit, the benefit coefficient is calculated. The social benefit coefficient and environmental benefit coefficient result is shown in Table 3. The power network in Case B doesn't have new energy integration, so the environmental factor is zero.

Using the simulation result of the power networks, the power network stability index and power supply index are calculated. The HILP risk index is obtained by the historical events list and

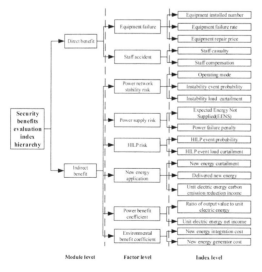

Figure 1.   Security benefits evaluation index hierarchy.

Table 1. Equipment failure risk index result.

Voltage	Equipment type	Case A $R_{equipment}$ (10^4 Yuan)	Case B $R_{equipment}$ (10^4 Yuan)
220 kV	Transformer	9.66	10.752
	Transmission line	0.5196	1.56
	Bus	0.1143	0.0978
	Disconnector	0.04785	0.01083
	Circuit breaker	0.5895	0.5985
500 kV	Transformer	35.7	86.4
	Transmission line	0.547	0.5352
	Bus	0.00549	0.034
	Disconnector	0.027	0.02214
	Circuit breaker	1.1625	3.22
Equipment failure risk index		48.37324	103.2305

Table 2. Staff accident risk index result.

	Staff casualty (death per year)	Compensation (10^4 Yuan)	Staff accident risk index (10^4 Yuan)
Case A R_{staff}	0.4	43.6	17.44
Case B R_{staff}	1.1	49.1	54.01

Table 3. Benefit coefficient index result.

	Case A	Case B
r_v	$0.864*10^4$ Yuan/MWh	$1.347*10^4$ Yuan/MWh
p	0.4 Yuan/KWh	0.45 Yuan/KWh
C_{nn}	$1*10^3$ Yuan/KWh	0
C_{ng}	$7*10^3$ Yuan/KWh	0
r_e	0.125	0

the meteorology data, the probability event defined by 'once in one hundred years' is 0.01. The result is shown in Table 4.

The evaluation result shows that the expect loss of power network in both case is relatively small, which indicate that the power networks are in good condition. If the environmental benefit is neglected as traditional evaluation method does, the benefit index of case A is −0.0245 Yuan/MWh, which is worse than the index in case B. The reason is that the power network security level is dropping while the large amount new energy integration.

After taking environmental index into the evaluation, the power network comprehensive benefit result shows that the security benefit of power system in case A is better, the environmental benefit has compensate the security impact of the new energy integration.

Table 4. Indirect benefit index result.

	Case A	Case B
M	1	1
P_{sm}	0.944%	0.366%
E_{sm}	31500 MWh	63200 MWh
$R_{stability}$	297.36 MWh	231.31 MWh
R_{EENS}	17.58 MWh	31.18 MWh
$R'_{reliability}$	$2.846*10^4$ Yuan	$1.982*10^4$ Yuan
P_{hi}	0.01	0.01
E_{ei}	15750 MWh	18300 MWh
R_{HILP}	157.5 MWh	183 MWh
B_c	$-21.74*10^4$ Yuan	$-22.029*10^4$ Yuan
B_s	$-389.29*10^4$ Yuan	$-580.03*10^4$ Yuan
E_{ne}	$1.9*10^6$ MWh	–
E_{nc}	$2*10^5$ MWh	–
C_c	0.054 Yuan/KWh	–
B_e	$235.58*10^4$ Yuan	–
B	−0.0124 Yuan/MWh	−0.0153 Yuan/MWh

5 CONCLUSION

This paper proposed an index hierarchy for security benefits of power network evaluation. By using the economic evaluation method and benefit coefficients, the index hierarchy avoids the deviation cause by the subjective favor and index evaluation results. Three module level indices which are direct benefit indices, indirect benefit indices and benefit coefficient indices reflect all aspect of power network security benefits. The social and environmental benefits are brought into the index system while the power network is developing and large amount of new energy is integrated. The index hierarchy can help the operation and planning of power network in areas with different economic and geographic characteristic, while the environmental benefit can trade of with the increase risk caused by the new energy integration.

REFERENCES

1. Brenna, M., Foiadelli, F. & Roscia, M. (2007) Sustainability Energy Indicators by means of Fuzzy Logic. *Power Engineering Society General Meeting, 2007. IEEE.* Tampa, FL.
2. Chen, Y., Chen, G. & Li, M. (2004) Classification & research advancement of comprehensive evaluation methods. *Journal of Management Sciences in China*, 69–79.
3. Fan, L. & Niu, D. (2007) An IAHP-based MADM method in transmission network planning. *RELAY*, 47–51.
4. Gao, C., Wu, Q., Chen, X. & Chen, S. (2013) Assessment Method of Operational Efficiency in the Transmission Network. *East China Electric Power*, 485–490.

5. Juan, L., Shuyong, L., Hongfang, N., Rui, H. & Ming, Z. (2012) Investment Benefit Evaluation of Distribution Network Based on Dynamic Weighting Comprehensive Evaluation. *Power and Energy Engineering Conference (APPEEC), 2012 Asia-Pacific.* Shanghai.

6. Li, W., Zhou, J., Xie, K. & Xiong, X. (2008) Power System Risk Assessment Using a Hybrid Method of Fuzzy Set and Monte Carlo Simulation. *Power Systems, IEEE Transactions on,* 23, 336–343.

7. Liu, L., Wang, G., Tian, Y. & Li, R. (2007) Research on Evaluation of Benefits on Fixed Assets Investment Based on Gray Related Degree Model. *Machine Learning and Cybernetics, 2007 International Conference on.* Hong Kong.

8. Ma, L., Lu, Z., Chang, L. & Qu, C. (2011) Study of index system of economic operation of grid based on gray correlation degree. *Power System Protection and Control,* 39, 22–26.

9. Wang, J. & An, X. (2010) Evaluation on economic benefit for electric power enterprise based on BP neural network and DEA. *Networking and Digital Society (ICNDS), 2010 2nd International Conference on.* Wenzhou.

10. Wang, Z. (1998) On the Methods, Problems and Research Trends of Comprehensive Evaluation. *Journal of Management Sciences in China,* 75–81.

11. Wu, J. (2010) China smart grid benefit evaluation and policy research. *Journal of Electric Power Science and Tehchnology,* 42–46.

12. Xiao, J., Cui, Y., Wang, J., Luo, F., Li, Y., Wang, S. & Wang, H. (2008) A Hierarchical Performance Assessment Method on the Distribution Network Planning. *Automation of Electric Power Systems,* 36–40.

13. Yang, Z., Mao, Z., Yang, B. & Xiao, T. (2009) Safety Assessments of Transmission Grid Based on Factor Analysis and Neural Network. *Power System Technology,* 26–30.

14. Yue, Y., Huang, Y. & Han, R. (2014) Construction of a Comprehensive Index System for Evaluating Power Transmission Network Planning. *Smart Grid,* 41–45.

15. Zhang, G., Zhang, J., Peng, Q. & Duan, M. (2009) Index System and Methods for Power Grid Security Assessment. *Power System Technology,* 30–34.

16. Zhang, J., Miao, J., Zhou, X. & Huang, X. (2009) Analysis and Evaluation on the Effect of Work Safety and Its Relation with Safety Investment. *China Safety Science Journal,* 19, 49–54.

17. Zhang, J., Zhang, G., Duan, M. & Yang, J. (2009) Indices System and Methods for Urban Power Grid Security Assessment. *Power and Energy Engineering Conference, 2009. APPEEC 2009. Asia-Pacific.* Wuhan.

18. Zhang, Y. & Meng, X. (2011) Research on economic evaluation model for distribution network operation based on grey relation method. *Electric Utility Deregulation and Restructuring and Power Technologies (DRPT), 2011 4th International Conference on.* Weihai, Shandong.

19. Zhao, J., Wang, K., Yang, S., Liu, J., Li, F. & Feng, S. (2013) Analysis for evaluation indexes of complex structural security in transmission networks. *Power System Protection and Control,* 41, 48–53.

Energy

Power and Energy – Kong (Ed.)
© 2015 Taylor & Francis Group, London, ISBN 978-1-138-02782-4

Study on the questionnaire's design of mine safety signs

Shibo Li

*Department of Resources and Safety Engineering, China University of Mining and Technology (Beijing),
Beijing, China*
Jizhong Energy Group Handan Co. Ltd., Yunjialing Coal Mine, China

Wei Jiang & Zhiming Zhu

*Department of Resources and Safety Engineering, China University of Mining and Technology (Beijing),
Beijing, China*

ABSTRACT: As an important means of the safety management, safety signs play a significant role in the production and behavior control. In this paper, interviews, questionnaires tests, and mathematical statistical analysis are used to study the effectiveness of mine safety signs and evaluation questionnaire, and then the paper obtains 14 indicators to evaluate the effectiveness of mine safety signs, including the position, color, brightness, size, pattern, text messages, typography, attention, hazard perception, safety tips, safety knowledge, safety awareness, and safety habits. Then the author got a questionnaire, the mine safety signs and designed the questionnaire type, the form, number, content and type of the questions on the basis of the above work.

1 INTRODUCTION

Studies on the risk management have shown that[1] the safety signs which transfer and warn the dangerous information can help people to avoid the risk of hazard effectively. The safety signs are used widely at the mine site to avoid accidents happen. But some safety signs exist as many problems[2], for example, non-standard design lead hazard information is not transferred clearly enough, so that the actual effects are poor. The failure of safety signs, in the very great degree, directly leads to the occurrence of accidents in production. Therefore, the evaluation indicators of mine safety signs and questionnaire are needed to be designed in order to improve the effectiveness of safety signs at mine site.

2 EVALUATION INDICATIONS OF SAFETY SIGNS, EFFECTIVENESS

Scholars have done a lot of research to study the safety sign-itself feathers which can affect the effectiveness of safety signs. To determine the evaluation indications of the effectiveness of safety signs itself feathers, this paper now offers some achievements obtained by some scholars in the domestic and abroad, which are shown in Table 1.

By contrast to the seven kinds of safety signs above and analyze their feathers, position, color, brightness, size, pattern, and text messages that six feathers are widely recognized as the feathers which affect the effectiveness of safety signs greatly. So, the six feathers are chosen as the evaluation index when evaluating safety signs. Mine safety signs sometimes are several safety signs being set together, so the typesetting should be used as one of evaluation. Therefore, seven indications are chosen to measure the effectiveness of safety sign-itself feathers, namely position, color, brightness, size, pattern, text messages, and typesetting.

The concept explanation of the indicators is as following:

1. Position: the place that safety signs are set in the workplace;
2. Color: the color of safety signs;
3. Brightness: the level of lighting that safety signs are in the workplace;
4. Size: the size of safety signs;
5. Pattern: the main part of the safety signs graphics;
6. Text messages: the auxiliary information of safety signs;
7. Typesetting: the order of several signs being set together;

The indications of (1)~(4) determine the significance of safety signs, namely safety signs need to have enough staff to attract significant attention; the indications of (5)~(7) determine the comprehensibility of safety signs, and the comprehensibility is the key of safety signs to be identified.

Table 1. The safety sign-itself feathers achieved by researched n the domestic and abroad.

No.	Researchers	The feathers of safety sign-itself
1	Kenneth R. Laughery[3]	The size, position, color, information words, information content is clear or not
2	Braun C.C[4]	The size, shape, product packaging, signal words, icons, color
3	Emília Duarte[5]	Simple, shape, color
4	Hancock, Rogers[6]	Size, shape, color contrast, information words
5	Tam, C.M., Fung[7]	Text auxiliary flag, arrangement, signage models
6	Huang yi[8]	Size, position, color contrast, brightness
7	Jingpeng Yuan[9]	Color, information words, shape
8	YichengHu, Xiao Zhou[10]	Position, environmental contrast, signage models, logos, text auxiliary flag
9	Kong jian, Shu yu[11]	The brightness, text size, information expression, information presentation,
10	Xianwei Tang[12]	Color, shape, words

The safety signs are carrier which can transfer safety information, and reminder or alert the target audience to product attention and safe behavior. When evaluating the effective of safety signs, whether they can achieve the desired effect is also needed to be researched.

The safety signs are set in the workplace, which can effectively work for a long time. The audience is exposed to coverage under safety sign, and to see and focus on the safety signs more frequently, they will remember the context better[3] and the function of the safety signs will be greater.

Studies have shown that, compliant behavior of safety signs is regarded as the function of dangerous (risk) perception, and the higher the risk perceived, the more likely people to comply with safety signs[4]. There is a consistent finding in researches: consumers' reaction to the warning information is impacted by the degree of risk perceived, and the degree of risk perceived is related to the clarity of the warning messages[13].

In addition, safety signs are distinguished by focusing on safety or operation. Focusing safety simply refers to remind danger, accident or injury type, and focusing operation refers to reaching the goal of task and reminding danger at the same time. So, the safety signs not only make people perceive danger, but also give people some information related to safety tips.

The safety signs help people to avoid unsafety behavior by visual reminder, which can be set in the workplace and work effectively for a long time. Its function is widely, and have some effects on the habitual behavior of the "2–4" model in behavioral safety, namely safety knowledge, safety awareness, safety habits[14].

Through the above analysis, the other evaluation indications of (8)~(13) can be obtained, namely attention, hazard perception, safety note, safety knowledge, safety awareness, and safety habits.

The concept explanation of the indicators is as following:

8. Attention: attention people paid to the safety signs;
9. Hazard perception: the color of safety signs;
10. Safety note: the effect of the safety tips;
11. Safety knowledge: how much the safety signs increase person's safety knowledge;
12. Safety awareness: how much the safety signs improve safety awareness of human;
13. Safety habits: how much the safety signs improve people's safety habits.

3 THE DESIGN OF QUESTIONNAIRE

3.1 *Type of questionnaires*

Depending on the method of investigation, questionnaires can be divided into different types: send interviewers visit questionnaire, telephone survey questionnaire, mailed questionnaires, online questionnaires and forum questionnaires[15]. Due to the limitations of the questionnaire, survey was conducted among the coal mine workers, who often carry out their duties during working hours and almost have no extra time to fill out the questionnaires, and it is even harder to fill out the questionnaires together. Therefore, the questionnaires were sent to workers so that they could fill out the questionnaires after work time.

3.2 *Questions*

The questionnaire has three parts: questionnaire name, introduction, notes, and main part of the questionnaire. The main part of the question-naire has two parts: the basic personal information and questions. The basic personal information: name, age, education level, staff categories, years of working experience, safety training or not, and experienced safety accident or not. The design of this part is to analyze the distribution of the sample when analysis the result. The question-part of the questionnaire is consisted of several questions which covering some evaluation indications of safety sign-itself feathers.

The question consists of number, stem, options. The stem is interrogative sentence, needing every participator choose the most appropriate answer depending on there own will. The answer is no right or wrong and every question must be answered.

3.3 The number of questions

The number of questions in an excellent questionnaire is not the more the better. If too many questions, participators may feel fidgety which will affect the quality of the answer, and too litter will not reflect the actual situation[16].

14 questions are selected in the questionnaire, covering 13 evaluation indications about the effectiveness of safety signs. This not only ensures the effectiveness of the questionnaire but also reduce the number of question at most, and the quality is improved as well.

3.4 Content of questions

The 13 evaluation indications about the effectiveness of safety signs and the content of every indication are regarded as a theoretical basis for the design of the questions.

Table 2. The content of test.

No.	Indication	The content of test
1	Position	Significance of environmental safety signs, position
2		Significance of partial safety signs, position
3	Size	Significance of safety signs, size
4	Color	Significance of safety signs, color
5	Brightness	Significance of safety signs, brightness
6	Pattern	Comprehensibility safety signs, pattern
7	Text messages	Comprehensibility safety signs, text messages
8	Typesetting	Comprehensibility safety signs, typesetting
9	Attention	Employee, attention for safety signs
10	Safety note	Effect of safety signs, safety note
11	Hazard perception	Effect of safety signs, hazard perception
12	Safety knowledge	Effect of safety signs improve workers, safety knowledge
13	Safety awareness	Effect of safety signs improve workers, safety awareness
14	Safety habits	Effect of safety signs improve workers, safety habits

The content of test is shown in Table 2 as following:

3.5 Type of questions

As the purpose of this questionnaire is to survey staff's views on the effectiveness of the new set safety signs, the "consciousness-type" questions are selected in the questionnaire. That means participants have to choose the option based on the views they obtained in the workplace, and the answer is reflected to the subjective perception of reviewers.

4 RELIABILITY ANALYSIS OF THE QUESTIONNAIRE

Before using questionnaire to evaluate those mine accidents, reliability and validity must be examined so that the reliability and rationality can be guaranteed.

4.1 Reliability text

Reliability means to examine the stability and consistency of the results, and the lager the stability is, the smaller the standard error. Continue to analyze the reliability of all levels of scale and the total scale after analyze the indications. Reliability means the reliability or stability of the scale[17], and the coefficient created by L.J. Cronbach is used widely to text the reliability in Attitude Scale; the formula is as following (K is the total number of questions):

$$\alpha = \frac{K}{K-1}\left(1 - \frac{\sum S_i^2}{S_T^2}\right) \qquad (1)$$

α is between 0 and 1, and the probability of the two extreme value of 0 or 1 is very low (but it is possible); Scholars take different views on how much of α means high reliability; Nunnally (1978) considered that 0.7 is a low but acceptable boundary scale. DeVellis (1991) has his opinions: if α is between 0.60 and 0.65, it should be discarded; α between 0.65 and 0.7 is the minimum acceptable value; α between 0.70 and 0.80 is pretty good, and between 0.8 and 0.9 is completely good. Therefore we can set the reliability evaluation form as shown in Table 3.

Analyzed by SPSS 18.0 for all valid questionnaire survey results, we can get the reliability evaluation form of Cronbach's Alpha as shown in Table 4 and Table 5.

Above all, the Cronbach's Alpha of all the questions is between 0.8 and 0.9, except the title 5 and title 10 whose value is lower than 0.8.

Table 3. The scale of reliability evaluation.

α	Reliability evaluation	Scale evaluation
$\alpha<0.60$	Lower	Abandon and revised it
$0.60\leq\alpha<0.70$	Low	Need to be modified
$0.70\leq\alpha<0.80$	Medium	Acceptable
$0.80\leq\alpha<0.90$	High	Ideal
$\alpha\geq0.90$	Higher	Perfect

Table 4. The reliability evaluation form of whole questionnaire, Cronbach's Alpha.

Cronbach's Alpha	Cronbach's Alpha based on standardized items	Number
0.807	0.829	14

Table 5. The reliability evaluation form of all questions, Cronbach's Alpha.

No.	The deleted Cronbach's Alpha	No.	The deleted Cronbach's Alpha
1	0.838	8	0.816
2	0.828	9	0.809
3	0.870	10	0.796
4	0.830	11	0.813
5	0.797	12	0.821
6	0.831	13	0.825
7	0.829	14	0.811

4.2 Validity test

Validity means the correctness of measurement that refers to the degree of measuring tool can measure the constructs correctly. According to measurements and related forms of perspective, validity can be divided into three main types: content validity, criterion-related validity and construct validity. Content validity, also known as logical validity or face validity, reflects the range of measurement itself. Related validity, known as the empirical validity or statistical validity, indicates the level of effectiveness measuring tools. The construct validity is used to measure the degree of theoretical structure or qualities[18,19].

In this paper, the related validity is used to analyze the validity of measuring tools. In order to reflect the criterion-related validity of the questionnaire accurately, this paper analyzes the related validity between all questions and the total sample

Table 6. The correlation coefficient of the questions and the total sample.

No.	The total sample	No.	The total sample
1	0.442**	8	0.642**
2	0.463**	9	0.711**
3	0.489**	10	0.845**
4	0.451**	11	0.686**
5	0.826**	12	0.601**
6	0.374**	13	0.486**
7	0.395**	14	0.686**

Note: ** means significant correlation on the level of 0.01 (bilateral); * means significant correlation on the level of 0.05 (bilateral).

in the questionnaire and then obtains the correlation coefficient. If the correlation coefficient is significant, that means the guesstimate of this measuring entry is high, so it is better to reserve it; if the correlation coefficient is not significant, that means the guesstimate of this measuring entry is low, and it should be abandoned. The higher the correlation coefficient of the questions and the total sample is, the higher the validity of this questionnaire is. The result of analysis shows that the correlation coefficient of the questions and the total sample is significant on the level of 0.01, which proves that the criterion-related validity of the questionnaire is perfect.

5 CONCLUSION

In summary, this paper obtains some conclusions as following:

1. This paper gives 14 indicators to evaluate the effectiveness of mine safety signs: the position, color, brightness, size, pattern, text messages, typography, attention, hazard perception, safety tips, safety knowledge, safety awareness, and safety habits.
2. This paper obtains the questionnaire of mine safety signs based on the above 14 indicators and designs the type of questionnaire, questions, the number and content.
3. Analyze the reliability and validity of the questionnaire, and prove that the reliability the questionnaire is high, and the criterion-related validity is favorable.

ABOUT THE AUTHORS

Shibo Li is a doctoral students of Department of Resources and Safety Engineering in China

University of Mining & Technology (Beijing), studying on safety behavior and safety signs.

Contact: Wei Jiang, Tel:13426323576, E-mail: jiangwei678@126.com.

Address: the Department of Resources and Safety Engineering in China University of Mining & Technology (Beijing), No. Ding 11, Xueyuan Road in Haidian District of Beijing, Post Code:100083.

REFERENCES

1. Braun, C.C, Kline, P.B., Silver, N.C. The influence of color on warning label perceptions [J]. International Journal of Industrial Ergonomics, 1995,15(3):179–187.
2. Chuanyu Zou, Xiaozhe Li. Current Situation of safety signs in china [J]. World Standard Information, 2008,52(6):123–124.
3. Emília Duarte, Francisco Rebelo, Júlia Teles, Michael S. Wogalter, Safety sign comprehension by students, adult workers and disabled persons with cerebral palsy, Safety Science, 2014,62:175–186.
4. Enying Jing. Questionnaire design process and attention problems [J]. Hubei Institute for Nationalities, 2009,06.
5. Gui Fu. Safety Management—accident prevention behavior control method [M]. Beijing: Science Press, 2013.
6. Guoyi Lai, Chao Chen. The common features and application examples of spss 17.0 [M]. Electronic Industry Press, 2010.
7. Hancock, H.E., Rogers, W.A., Schroeder, D., Fisk, A.D., 2004. Safety symbol comprehension: effects of symbol type, familiarity, and age. Human Factors 46, 183–195.
8. Jingpeng Yuan. Factors affecting the effectiveness of safety signs empirical research [D]. Hangzhou: Zhengjiang Uiversity, 2009.
9. Jian Feng, Yu Shu, Shi Ma. Comprehensive Evaluation System of subway station service marks [J]. Journal of Tongji University, 2007,35(8):1064–1068.
10. Kenneth, R. Laughery. Safety communications: Warnings [J]. Applied Ergonomics, 2006, (37):467–478.
11. Laughery, K.R. Safety communications: Warnings [J]. Applied Ergonomics, 2006,37(1):467–478.
12. Leece, J., Parker, F. Use and misuse of SPSS. Statistical Package for the Social Sciences[J]. Software Practice and Experience, 1978,3(8):301–311.
13. Monica Trommelen. Effectiveness of explicit warnings [J]. Safety Science, 1997,25(1–3):79–88.
14. Minglong Wu. Statistical analysis of survey operations and application practices-spss [M]. Chongqing University Press, 2010.
15. Tam, C.M., Fung, I.W.H., Yeung, T.C.L., Tung, K.C.F. Relationship between construction safety signs and symbols recognition and characteristics of construction personnel, Construction Management and Economics, Vol. 21 No. 7, pp. 745–53.
16. Wenwang Yang. Behavioral training methods of preventing coal mine gas explosion [D]. China University of Mining & Technology (Beijing), 2013.
17. Xianwei Tang. Factors affecting the effectiveness of safety signs [J]. Modern Business. 2010,(32):283.
18. Yi Huang. Research on mine Safety advertising and the promotion for safe production [D]. Shandong University of Science and Technology, 2011.
19. Yiwei Cheng, Xiaohong Zhou, Liang Wang. Evaluate the effectiveness of the project site safety signs [J]. China Safety Science Journal, 2012,(08):37–42.

Power and Energy – Kong (Ed.)
© 2015 Taylor & Francis Group, London, ISBN 978-1-138-02782-4

Research on the policy guarantee for development of China's low-carbon electricity

Ying Li
Research Center for Beijing Energy Development, North China Electric Power University, Beijing, China

Di Yu
School of Humanities and Social Sciences, North China Electric Power University, Beijing, China

ABSTRACT: At present, China is the world's largest emitter of carbon. It becomes an inevitable choice for China to develop low carbon electric industry. From this sight, a deep study of low-carbon electricity policy guarantee system will be a typical model to promote China's low-carbon economy strategic adjustment. The current China's policy guarantee system of low-carbon electricity has many problems, such as lack of long-term planning; the insufficient implementation of power industrial structure policy to reach the designated position, the absence of government guidance as to science and technology innovation, unsound electricity market and defective electrical power regulation system, etc. Therefore, it is necessary to further study China's low-carbon electricity policy guarantee system and discuss the development countermeasure in order to come up with the corresponding countermeasures.

1 INTRODUCTION ON LOW-CARBON ELECTRICITY

1.1 The concept of low-carbon electricity

Low-carbon electricity refers to an idea of adopting the method of comprehensive resource strategy planning to achieve energy conservation and emissions reduction from the whole link of power production and power consumption. Developing low-carbon electricity demands the changing of traditional planning system of electricity industry. That is, clean energy generation is encouraged, and new technology and strategy are adopted to reduce pollutant emissions in terms of supply; and at the demanding side, carbon emissions in the process of power consumption will be cut down by the market mechanism function, law regulation and policy guidance.

1.2 The significance of developing low-carbon electricity

In September 2013, China enterprise management board of low carbon committee and International low carbon cooperation alliance jointly issued Chinese public company constraint report in carbon. The report presented that electric power, steel and chemical engineering have been the first three energy-hungry industries in China, and thermal power industry produces 50% of the total emissions of the country, most of which come from coal-burning electricity. In 2012, thermal power industry consumes average standard coal of 324 g/kWh, equivalent to carbon emission of 8.4 tons/m kWh. In such a situation, developing low-carbon electricity becomes a task of great urgency and practical significance. Specifically speaking, the development of low-carbon electricity shows three benefits as follows.

1.2.1 Improvement for the structure of the power supply

At present, with coal as main part of the power structure, coal power-generation accounted for more than 70% the total installed capacity, and carbon dioxide emissions of coal power accounted for more than 95%, suggesting that the power structure is very irrational with quite low proportion of low-carbon power. Therefore, it will optimize the current terrible power structure and reduce the power coefficient of carbon emissions to develop low-carbon electricity in China.

1.2.2 Environmental protection

The intensive utilization of power generated mainly by coal gave rise to great pressure to our environment. Large amount of land has been occupied and polluted by piles of coal gangue, which made lots of water resource consumed or polluted, and areas affected by acid rain amounted to 1,200,000 sq. km. The main pollutants and greenhouse gas emissions rank first in the world. Now it is hard for domestic ecological environment to carry such an extensive development mode. With increasing pressure responding to international climate change, it becomes very urgent for low-carbon electricity.

1.2.3 *Promotion for national economy*

The twelfth five-year plan for national economy and social development program demonstrates that in the face of increasingly constraining resources and environment, we must strengthen consciousness of crisis and establish the concept of developing green low carbon to improve sustainable development and ecological civilization. All of those can be reached through energy conservation and emissions reduction, efficient incentive and constraint mechanism, resource saving, and environmentally, friendly mode of production and consumption. It will help drive national economy to a new round of growth through developing low-carbon electricity, exploiting and utilizing renewable energy such water energy, solar energy, wind energy and nuclear power, and improving the level of energy saving and efficiency of electric power facilities.

2 THE CURRENT SITUATION OF POLICY GUARANTEE FOR CHINA'S LOW-CARBON ELECTRICITY

In May 2005, China's national energy-leading group was founded and it aimed at studying the national energy development strategy and planning, and great policies of energy developing and saving, energy safety and urgency, foreign cooperation to give suggestions to the State Council. The establishment of the national energy-leading group, to a certain extent, promoted the subsequent low-carbon electricity policy in our country. In the same year, the State Council issued several opinions about promoting healthy development of the coal industry, by which the national development and reform commission was responsible for planning and improving policies, organizing and constructing demonstration project, and financial support was given to promote the development of clean coal technology and industrialization.

In January 2006, renewable energy law was enforced, in which the protection of green energy in Rule 9 was matched to power generation management, price share, and technology specification, etc. At the same time, the law also provided that the higher part of renewable energy power generation price to the conventional energy price would be shared by all the electricity users, so that less than 0.001 Yuan was shared by every end-user

per KWh to make renewable energy companies gain a foothold in the market competitions.

In April 2007, the national development and reform commission issued the energy development of the eleventh five-year plan, according to this plan, by 2010, total energy consumption of coal in China accounted for the proportion fell to 66.1%, while natural gas, nuclear and water energy and other renewable energy for the proportion increased slightly. In August the same year, the national development and reform commission issued the planning of long-term renewable energy development, pointing out that the amount of clean and renewable energy in energy structure should be gradually promoted about 10% in 2010 and 15% in 2020. In October the same year, the national development and reform commission issued the long-term nuclear power development planning (2005–2020). The plan aims at installing nuclear power capacity of 40 million kilowatts, and 18 million kilowatts project under construction to carry forward after 2020, which marks the development China's nuclear power, has entered a new stage.

In March 2008, the national development and reform commission issued again renewable energy development of the eleventh five-year plan. The 11th five-year plan points out, by 2010, the proportion of renewable energy in the energy consumption in China shall reach 10%, and the national renewable energy consumption reach 300 million tons of standard coal, which is nearly double higher than that in 2005.

In March 2009, the ministry of finance issued fiscal subsidy funds management interim measures for solar photovoltaic building application. The implement of this measure helped to carry out the energy conservation and emissions reduction strategy by the state council, speed up the application for solar photovoltaic technologies in urban and rural construction, promote clean coal technology industrialization with clean energy such as nuclear power, wind power and solar power development, etc.

In January 2013, the state council issued energy development of the 12th five-year plan, and pointed out to realize double control of both energy consumption and intensity. It is hoped that by 2015 the total energy consumption is expected to reach 4 billion tons of coal, 6,150 billion KWh of electricity use, energy consumption per unit of GDP 16% lower than that of in 2010. In addition, the comprehensive energy utilization efficiency increased to 38%, thermal power supply of standard coal consumption dropped to 323 g/KWh, and comprehensive processing energy consumption dropped to 63 kg/t standard oil.

3 PROBLEMS IN POLICY GUARANTEE FOR CHINA'S LOW-CARBON ELECTRICITY

3.1 *The policy of power industry structure did not be implemented to the right place*

In some parts of our country, the government officials only focus on the pure GDP Numbers and power projects and engineering in short time, but often ignore the long-term investment and low efficiency of low carbon power projects and measures. Moreover, some local governments continue to support and protect quite a number of medium and small electric power enterprises whose outdated technologies and equipments result in low resources utilization and terrible pollution, just to alleviate the pressure of the local tax and unemployment. The above local governments disobey the central policy, making low carbon power measures very difficult to implement locally. Especially the power structure in rural areas around the country, coal is still the main resources; low utilization rate, serious waste and terrible pollution are quite common.

3.2 *Science and technology innovation lacks of government policy guidance*

When electric power enterprises are carrying on scientific and technological innovation, they further intensify the structure adjustment for energy conservation and emissions, arrange reasonable capital investment, strengthen the equipment maintenance and management, optimize the power dispatching operation and improve the efficiency of generating units. Science and technology innovations of low-carbon electricity not only need the principal role of the electric power enterprises, but also require the government's policy guidance. However, government policy as the role of guidance did not play well. Firstly, the government did not organize efficient low-carbon electricity leading group, the existing division of labor in management and organization was not clear, incentive mechanism for low-carbon electricity innovation was not sound enough to encourage enterprises to better proceed scientific innovations for low carbon power industry through energy conservation and emission reduction.

3.3 *Imperfect electricity market*

Firstly, the current policy is poorly implemented in the management market. Since 2004, China has made the desulfuration electrovalence policy that the grid purchase price raises 1.5 fen per KWII, which has a positive effect on improving the motivation of installing desulfurization facilities for power generation enterprises, and reducing sulfur dioxide emissions. However, some problems still exist in the desulfuration electrovalence implementation. First, the desulfuration electrovalence compensation cannot be in place in time; second, the desulfuration electrovalence is not deducted as required strictly for those power plants whose desulfurization facilities commissioning rate does not reach the standard; third, the current desulfuration electrovalence policy is difficult to solve the cost of sulphur coal units desulfuration, old power plant desulfuration technical innovation, small units below 300 thousand KW and heating units desulfuration; fourth, China has not made relevant policies to encourage power plant desulfuration yet, which brings great management pressure, and seriously dampens the initiative and enthusiasm of enterprise desulfuration. According to the Ministry of Environmental Protection, about one third of the built De-NOx units can operate normally, one third of those operate abnormally, and one third of those do not operate. Secondly, local government policy is not in agreement with the central government policy. Some places do not carry out differential electricity price policy, and some introduce preferential price without authorization, some do not carry out punitive electrovalence in accordance with the regulations, and some do not impose governmental funds for self-owned power plants, resulting in the lack of orderly management and unbalanced development in the nationwide electricity market.

3.4 *Defective electricity supervision system*

In the function division of power regulator, the functions of government departments and electrical supervisor are not clear. In terms of regulatory system, the relevant regulators functions are relatively dispersed, at the same time, it makes the responsibility unclear among various departments, and lack mutual coordination, resulting in repeated regulation. At the same time, China electricity installing capacity is rising, thus the service ability of regulators can't satisfy the current development demand of low-carbon electricity in China.

4 ADVICES ABOUT IMPROVING CHINA LOW CARBON ELECTRICITY POLICY GUARANTEE

4.1 *The establishing of framework document for China low carbon electricity policy guarantee*

For the aspect of planning, we should research scientifically the low-carbon electricity demand for medium and long term economic and social development based on China's national conditions, formulate the macroscopic planning and adjust and improve flexibly according to the environment, ensuring that the low-carbon electricity strategy implementation can be put into practice in a planed and orderly way. Two relations should be focused on as follows: first, the short, medium and long term goals for the low-carbon electricity strategy matching with social and economic development situation, excessively fast or slow development may cause severe imbalance for electricity supply and demand, which has a negative impact on the electric power industry and national economy; second, the comprehensive coordination of each link in the power industry, namely electricity generation, transmission, power supply and power utilization, is different from the adjustment of inventory for general merchandise, while the electricity generation, transmission, power supply and power utilization can be achieved instantaneously, and the resource mismatch in any link will cause a huge waste in upstream and downstream. Based on the policy documents, "National Assessment Report on Climate Change", "National Climate Change Program", the white paper of "China's Energy Situation and Policy", the white paper of "China's Climate Change Policy and Action", "the pilot work implementation of pilot provinces and low carbon city", "the Eleventh Five-year Plan Outline" and "the Eleventh Five-year Plan Outline" and so on, we should carry out integrative construction from a macroscopic view, combining with the "the Twelfth Five-year Plan for National Economic and Social Development Program", "the Twelfth Five-year Plan for National Environmental Protection" by the state council and "the Medium and Long-term Planning for Renewable Energy" by the national development and reform commission, and establish the programmatic document of China's low-carbon electricity policy guarantee— China's low carbon electricity development plan, and guide the development of China's low-carbon electricity on this basis.

4.2 Formulate security system of sustainable development policy for China low carbon electricity

4.2.1 Policy guarantee of construction and operation of low carbon electricity

Further optimize the power structure in power generation side and adjust the industrial structure in demand side to raise the utilization ratio of electricity. First, from power generation side, China's electric power industry should: (1) give priority to develop hydropower; (2) optimize the thermal power equipment; strive to develop the high efficient and clean high-capacity machine units,

shut down the inefficient small units; (3) develop nuclear power on the basis of safety; promote solar power in a planned way; (4) strengthen the grid connection of new energy, especially the wind power, optimize the energy utilization, promote the comprehensive, coordinated and sustainable development of power industry in China. Second, we should strengthen the demand-side management. Electricity demand-side management is a way of electricity usage that the electrical network enterprises guide users based on the comprehensive planning of energy, which is established in accordance with the mature power market mechanism. The promotion in the construction of smart grid and simultaneous demand-side management of electrical network enterprises can not only make the power grid operate smoothly, but also reduce the waste of energy power between peak load and valley load. According to Lawrence Berkeley lab in the United States, if we further improve the efficiency of terminal electricity in the future 30 years, about 40% of the energy investment can be reduced by the developing countries, so actively carrying out electricity demand side management in power grid has a realistic significance.

4.2.2 Policy guarantee of technological innovation for low carbon electricity

In order to develop the low-carbon electricity, the traditional electric power development mode should be changed through energy efficiency and clean energy development, pollution emissions reduction and industrial structure adjustment, and system and mechanism innovation, achieving the clean, efficient and sustainable development of electric power industry. [1] from the actual situation of China's electric power development, we should strive to complete the following tasks to realize the low carbon electric power development: (1) pay attention to the effective development and clean utilization of coal resources, vigorously promote the coal-fired power generation, including the supercritical (ultra-supercritical) units of 600 thousand KW or above, low NOx combustion technology, cogeneration units of 300–600 thousand KW, IGCC, supercritical (ultra supercritical) circulating fluidized bed, air cooling power technology, cogeneration, cogeneration cooling heating and power, and thermoelectric gas multi-generation; (2) carry on the overall planning of and comprehensive development of hydropower resources, and the orderly exploitation and healthy development of renewable energy sources such as wind power and solar energy, promote the scientific planning and rational operation of power grid construction, and achieve the electric power safety and optimal allocation of resources.

In addition, China should actively promote the popularity of the smart grid. The development of smart grid should comply with the world electric power development tendency, and the internal requirement, such as the stage characteristics of domestic economic and social development, energy resources endowment, electric power development stage, sustainable development of energy, etc. According to the "Advices on Strengthening the Construction of Smart Grid" issued by SGCC, we should study the policy guarantee, such as large-scale batch of new energy grid and large-scale batch of new energy grid and storage, intelligent power utilization, intelligent scheduling and control of large power grid, intelligent equipment, etc. Improve the level of power grid development and independent innovation ability through the application of advanced information communication, automatic control, energy conservation, etc.

4.3 Policy guarantee of low carbon electricity market

The price mechanism is the core of electric power market mechanism. In the face of increasingly severe energy saving situation, we should promote the reform of power price system actively and steadily, gradually establish the price mechanism that can reflect the degree of resource scarcity, relation of market supply and demand and environmental costs, correctly guide the energy consumption and production to develop into low carbon, especially increase the support of clean energy, achieve the development of low-carbon electricity by economic leverage.

In addition, China can also establish a special low-carbon electricity incentive mechanism. According to the regulations issued by the Ministry of Finance, such as the "Interim Procedures for Special Funds Management of Power Generation Equipment Industrialization", "Interim Procedures for Subsidy Funds Management of Straw Energy-oriented Utilization", "Interim Procedures for Fiscal Subsidy Funds Management of Solar Photovoltaic Building Application", "Interim Procedures for Fiscal Subsidy Funds Management of Golden-sun Demonstration Project", "Interim Procedures for Special Funds Management of Marine Renewable Energy", "Interim Procedures for Fiscal Incentive Funds Management of Contract Energy Management Project", etc, introduce the systematic low-carbon electricity incentive mechanism, vigorously support the long-term development of low-carbon electricity financially.

4.4 Policy guarantee of low carbon electricity regulation

Electricity regulators should regularly summarize and analyze the performance of low carbon and emissions reduction for electric power enterprises, strengthen the responsibility assessment of low carbon and emissions reduction for electric power enterprises, implement supervision in accordance with the law. First of all, the electricity regulators should strengthen the energy conservation supervision and improve the law enforcement level of energy conservation. Establish the municipal, district (county) law enforcement team of energy conservation as soon as possible; carry out the energy conservation law-enforcement monitoring in accordance with the law. As the energy conservation administrative department, the development and reform department is responsible for unified management, guidance and coordination of the energy conservation supervision work in the whole city. Next, establish the perfect supervisory process. Perfect supervisory process is the precondition of supervision of regulators in accordance with the law, the possibility of errors occurring in the supervisory process lows only according to the supervisory process strictly. In order to realize the fairness and rationality of law enforcement, a set of strict supervisory process must be developed, and regulators must enforce the law within the procedural framework, and must not change the law enforcement procedures optionally. Especially improve the decision-making procedures, consultation and approval procedures, realize high and positive cycle of electricity regulation. Finally, improve the diaphaneity and public participation of electricity regulatory. The supervision should take full advantage of the characteristics of the current Internet popularization, report the regulatory information to the society through several media means, such as the news media and the Internet, at the same time, broaden the problem reaction channels, improve the application proportion of the Internet in public participation, and accept the public supervision constantly, thus forming the positive supervision cycle.

REFERENCES

1. Guan Yingzhe, Gao Song, Qi Guohong i. Development Countermeasure for Power Enterprises in Low-carbon Economy Context. Cooperative Economy and Technology. Late Jan. 2014 (Issue 481).
2. Sun Yaowei, Path Selection of Low-carbon Electricity Strategy. China Power Enterprises Management. May 2010.
3. Wang Wei, Guo Weiyu, the Policy Research of China Energy Development in Low-carbon Time. China Economic Publishing House. Jan. 2011. P214.

Power and Energy – Kong (Ed.)
© 2015 Taylor & Francis Group, London, ISBN 978-1-138-02782-4

Investigating on supercritical heat cooling in transcritical CO_2 cycle

Zhi Li
Hebei Electric Power Design and Research Institute, Shijiazhuang, China

Weigang Shi
Hebei Xibaipo Generation Co. Ltd., Shijiazhuan, China

ABSTRACT: In gas cooler, a key component of transcritical heat pump system, CO_2 is always in supercritical status. A well design gas cooler should get the full informations of supercritcal CO_2 in-tube cooling heat transfer characteristic. An EES program is developed to calculate the supercritical CO_2 cooling heat transfer performances based on 4 selected correlation equations. According to the results and the thermopgysical parameters of supercritical CO_2, such as specific heat, density, thermal conductivity and dynamic viscosity, two region, the starting area of supercritical region and the main supercritical region, is suggested for supercritical CO_2 cooling heat transfer by temperature at 80°C. In the staring area, the variations of heat transfer and other parameters are intensive, while in the main supercritical region, the variations are clined to be stable.

1 INTRODUCTION

Carbon dioxide, as a kind of natural refrigerant, is making great contribution to reducing the ozone depletion, reducing the Green House Effect and promoting the sustainable development of environment. This refrigerant is widely used in automobile air conditioning, water heater, and so on. System adopt CO_2 are usually transcritical cycle, and gas cooler is a key component in the system. For the CO_2 is in supercritical gas status, the heat transfer characteristic is not very clearly.

At present, many scholars have contributed a lot work on the flow and heat transfer characteristics of CO_2 at supercritical conditions. Commonly used empirical correlations are summarized in Table 1. The heat transfer inside gas cooler becomes more complex because of the supercritical carbon dioxide fluid has dual physical properties both liquid and gas, in the supercritical initial region especially, density, specific heat, heat conducting coefficient and dynamic viscosity have tremendous changes with the temperature and pressure changing. General mathematics physics model of condensation heat transfer for traditional refrigerant is not applied for supercritical CO_2 in transcritical carbon dioxide cycle system.

Therefore, establishing appropriate mathematic physical model for supercritical CO_2 refrigerant in gas cooler, through modeling and simulation the process of carbon dioxide cooling heat transfer, is necessary and helpful to optimize the performance of gas cooler. In this paper, an EES program is developed to calculate the supercritical CO_2 cooling heat transfer performances based on 5 selected correlation equations. The research on the flow and heat transfer characteristics of CO_2 at supercritical conditions is to provide reasonable theoretical basis for optimize the system and improve heat transfer performance.

2 THERMODYNOMIC MODEL

Figure 1 is a flow chart of transcritical CO_2 cycle. The system consists of a compressor, evaporator, reservoir, gas cooler, regenerator and throttle valve, to form a closed loop. Figure 2 shows the pressure enthalpy diagram of transcritical CO_2 cycle.

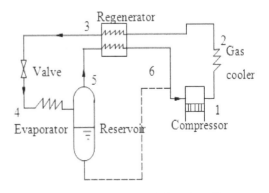

Figure 1. Transcritical CO_2 cycle flow chart diagram.

Figure 2. Transcritical CO_2 cycle p-h diagram.

The transcritical CO_2 cycle has the following characteristics: 1) cooler outlet temperature is higher than the CO_2 critical temperature, cooling pressure is higher than CO_2 critical pressure, which means that the cooling of CO_2 in the gas cooler under supercritical condition; 2) the evaporator outlet state is in the two phase region or steam line.

3 CORRELATION

Many scholars have done a lot of research on the flow and heat transfer characteristics of CO_2 at supercritical conditions, obtained some new empirical formula or the association made some modification (Table 1). But because of experimental device and working conditions is different, the correlations have itself uncertainty, limitations and application conditions. In this paper, we choosed four widely application range and commonly used correlation (Equations 1, 2, 7 and 9).

1. Dittus-Boelter[12]

$$Nu_{DB} = 0.023 Re_b^{0.8} Pr_b^{\;n}$$

$10^4 \leq Re_b \leq 1.2 \times 10^5$ $L/D \geq 10$, $0.7 \leq Pr_b \leq 120$

2. Petukhov et al[7]

$$Nu_{PK} = \frac{(f/8)(Re_b Pr_b)}{1.07 + 12.7\sqrt{f/8}(Pr_b^{2/3} - 1)}$$

$10^4 \leq Re_b \leq 10^6$ $0.7 \leq Pr_b \leq 200$

3. Krasnoshchekov et al[1]

$$Nu_{KKP} = \frac{(f/8)(Re Pr_b)}{1.07 + 12.7\sqrt{f/8}(Pr_b^{2/3} - 1)}$$

$$\times \left(\frac{\rho_w}{\rho_b}\right)^{0.3} \left(\frac{\bar{c}_p}{c_{p,w}}\right)^n$$

$9 \times 10^4 \leq Re_b \leq 3.2 \times 10^5$ $6.3 \times 10^4 \leq Re_w \leq 2.9 \times 10^5$

4. Petrov-Popov[6]

$$Nu_{PP} = \frac{(f_w/8)(Re Pr_b)}{A + 12.7\sqrt{f_w/8}(Pr_w^{2/3} - 1)}$$

$$\times \left(1 - 0.001\frac{q_w}{G}\right)\left(\frac{\bar{c}_p}{c_{p,w}}\right)$$

$3000 \leq Re_w \leq 10^6$ -350 J/kg $\leq q_w/G \leq 0$ J/kg

5. Son-Park[9]

$$Nu_{SP} = Re_b^{\;a} Pr_b^{\;c} \left(\frac{\rho_b}{\rho_w}\right)^d \left(\frac{c_{p,b}}{c_{p,w}}\right)^e$$

When $T_b/T_{pc} > 1$, $a = 0.55$, $c = 0.23$, $d = 0$, $e = 0.15$
When $T_b/T_{pc} \leq 1$, $a = 0.35$, $c = 1.9$, $d = -1.6$, $e = -3.4$

6. Fang[3]

$$Nu_F = \frac{(f/8)(Re - 1000) Pr_b}{A + 12.7\sqrt{f/8}(Pr_b^{2/3} - 1)}$$

$$\left(1 - 0.001\frac{q_w}{G}\right)\left(\frac{\bar{c}_p}{c_{p,w}}\right)^n$$

$3.1 \times 10^4 \leq Re_b \leq 8 \times 10^5$ $1.4 \times 10^4 \leq Re_w \leq 7.9 \times 10^5$
-350 J/kg $\leq q_w/G \leq 29$ J/kg

7. Gnielinski[4]

$$Nu_{G,M} = \frac{f/8(Re_b - 1000)Pr_b}{1 + 12.7\sqrt{f/8}(Pr_b^{2/3} - 1)}[1 + (D/L)^{2/3}]$$

$2300 < Re_b < 10^6$ $0.05 < Pr_b/Pr_w < 20$

8. Huai[5]

$$Nu_H = Re_w^{0.8} Pr_w^{0.3} (\rho_r/\rho_w)^{-1.47} (\bar{c}_p/c_{pw})^{0.083}$$

9. Yoon et al[8]

$$Nu_b = a Re_b^{\;b} Pr_b^{\;c} \left(\frac{r_{pc}}{r_b}\right)^n$$

When $T_b/T_{pc} \leq 1$, $a = 0.014$, $b = 1.0$, $c = -0.05$, $n = 1.6$
When $T_b/T_{pc} > 1$, $a = 0.14$, $b = 1$, $c = 0.69$, $n = 0.66$.

4 PROGRAMMING AND SIMULATION IN EES

The distributed-parameter method is used to establish model and programme for the sleeve cooler in this paper. First, ascertaining the procedure inlet parameters: sleeve cooler structure parameters, refrigerant inlet temperature inside tube and assuming outlet temperature, cooling water inlet and outlet temperature and pressure outside tube, the mass flow of refrigerant and cooling water.

Then the outlet parameters of procedure include the outlet parameters of refrigerant and cooling water and heat exchange amount in each tube section.

Dichotomy of iteration algorithm is used in the program to make the application simple and reliable.

1. Assume refrigerant outler temperature. According to the assumption, the value range of refrigerant can be ensured. Cooling water temperature is the lower limit; discharge temperature of compressor is the upper limit. Take the limit value's arithmetic mean as initial iterative algorithm temperature difference value Tro.
2. According to outlet temperature assumption of refrigerant, calculate the heat exchange amount Qr of refrigerant side basis on flowing enthalpy.
3. Calculate heat transfer amount of cooling water side basis on heat exchange temperature difference.
4. Calculate heat transfer control error condition:

$$\delta = \frac{2|Q_r - Q_w|}{Q_r + Q_w} \qquad (1)$$

5. Estimate whether the δ less than assigned error.
6. If the δ less than the assigned error, export results, on the contrary repeat the above steps.

5 THE ANALYSIS OF SIMULATION RESULTS

Supercritical fluid has many unique thermal physical properties. The density of supercritical CO_2 gas is larger than normal gas, it also has the liquid and the gas physical properties: (1) the density of supercritical CO_2 is higher than ordinary gas, but its viscosity is close to ordinary gas and less than the general the viscosity of liquid; (2) the diffusion coefficient of supercritical CO_2 is higher than general liquid diffusion coefficient, and close to the gas diffusion coefficient.

This paper, by using the NIST company Refprop property database, compiling program calls density, specific heat, coefficient of thermal conductivity and physical properties in the range of supercritical CO_2 such as dynamic viscosity parameters, pressure and temperature change rule. Figure 3 the physical parameters change curve with temperature is studied respectively from the density, thermal conductivity, specific heat and dynamic viscosity under different pressure.

It can be observed from the above curves that: the temperature and pressure have a huge impact on the various thermal physical properties of CO_2 as the temperature rises, the change of properties is very extreme. After the critical point, the change of properties decreases gradually.

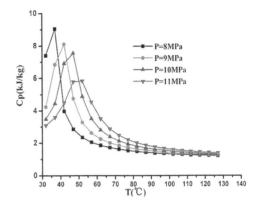

Figure 3A. The change of thermal properties with temperature and pressure.

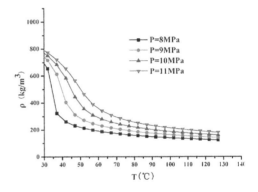

Figure 3B. The change of thermal properties with temperature and pressure.

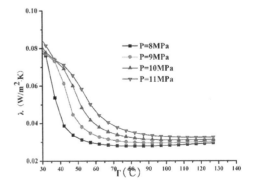

Figure 3C. The change of thermal properties with temperature and pressure.

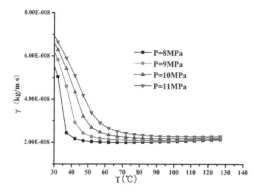

Figure 3D. The change of thermal properties with temperature and pressure.

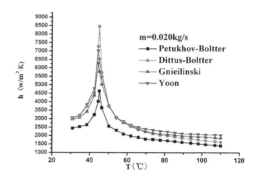

Figure 4B. The heat transfer coefficient calculated by different correlation equations changes with temperature of CO_2 under different CO_2 mass flow.

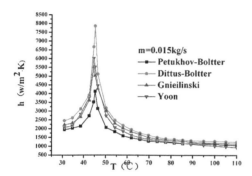

Figure 4A. The heat transfer coefficient calculated by different correlation equations changes with temperature of CO_2 under different CO_2 mass flow.

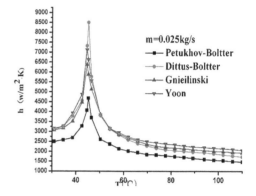

Figure 4C. The heat transfer coefficient calculated by different correlation equations changes with temperature of CO_2 under different CO_2 mass flow.

Be different from constant physical property fluid, the lumped parameter method is not suitable to obtain the state parameters in gas cooler due to carbon dioxide fluid thermal physical properties waving remarkably with the change of temperature and pressure, especially in the initial region of supercritical condition, the waving is tremendous. In this paper, using distributed-parameter method to establish the mathematic physical model of gas cooler and the heat exchanging tubulation is divided into 6 sections, the sections lengths are 500 mm, 400 mm, 400 mm, 400 mm, 400 mm, and 500 mm.

Modeling conditions:carbon dioxide mass flow m respectively 0.015 kg/s, 0.020 kg/s, 0.025 kg/s, cooling water mass flow respectively 300 g/s, 400 g/s, 500 g/s, optimum operating pressure selected 8.5 MPa on high pressure side. As Figure 4 shown, the heat transfer coefficient's change trend with the change of temperature of CO_2 calculated through different correlations is consistent. From 30°C, heat transfer coefficient

increases rapidly with the CO_2 temperature increasing, reached a peak at the critical point, then gradually reduced, Reaches about 80°C, the heat transfer coefficient changes gradually slow and the rate of change tends to zero.

The simulation results show that CO_2 transforms from the "gas-like" to the "liquid-like" in the supercritical heat transfer process and the complex cooling heat transfer performance internal pipe. In consideration of complex characteristics in supercritical CO_2 area, it is necessary to divide the region close to the CO_2 critical point in order to put the study of the CO_2 heat transfer characteristics and flow characteristics in a better position. According to the range of pressure and temperature, the supercritical region can be roughly divided into two parts: the starting area of supercritical region close to the critical point (31.1°C < T < 80°C) and the main supercritical region away from the critical point (T > 80°C).

6 CONCLUSION

In this paper a modeling method of the gas cooler in theory is introduced and the results of programming and calculation in the EES software are given. The following conclusion is given based on the above works:

1. In order to put the study of the CO_2 heat transfer characteristics and flow characteristics in a better position. According to the range of pressure and temperature, the supercritical region can be roughly divided into two parts: the starting area of supercritical region close to the critical point ($31.1°C < T < 80°C$) and the main supercritical region away from the critical point ($T > 80°C$).

2. The calculation results for different heat transfer correlations show that cooling heat transfer coefficient internal tube with temperature change varies very extreme within the supercritical initial area and achieves the crest value at the critical point, but the change of heat transfer coefficient varies very slow in the supercritical main region.

REFERENCES

1. Chen, N., Kwon, H., Ray, A. & Wong, R. Flow in the Pipes: Correlation between Fanning Friction Factor and Reynolds Number. Transport Process Laboratory, Carnegie Mellon University, Pittsburgh, Pennsylvania. 2005.
2. Dittus F.W, Boelter L.M.K. Heat transfer in automobileradiators of the tubular type [J]. University of CaliforniaPublications of Engineering, 1930, 2:443–61.
3. Fang X., Bullard C.W, Hrnjak P.S. Heat transfer and pressure drop of gas cooling. ASHRAE Trans, 2001, 107 (Part1):255–66.
4. Gnielinski V. New equations for heat and mass transfer in turbulent pipe and channel flow [J]. International Chemical Engineering 1976, 16(2):359–68.
5. Huai, X.L., Koyama, S. & Zhao, T.S. An experimental study of flow and heat transfer of supercritical carbon dioxide in multiport mini channels under cooling conditions [J]. Chemical Engineering Science, 2005, 60:3337–3345.
6. Petrov N.E, Popov V.N. Heat transfer and resistance of carbon dioxide being cooled in the supercritical region [J]. Thermal Engineering, 1985, 32(3):131–4.
7. Petukhov B.S, A.V Polyakov, Boundaries of Regimes with "worsened" Heat Transfer for Supercritical Pressure Coolant [J]. Teplofizika Vysokikh Temperature, 1974, 12(1): 221–224.
8. Seok Ho Yoon, Ju Hyok Kim, Yun Wook Hwang. Heat transfer and pressure drop characteristics during the in-tube cooling process of carbon dioxide in thesupercritical reign [J]. International Journal of Refrigeration, 2003,(26):857–864.
9. Son, C.H. & Park, S.J. An experimental study on heat transfer and pressure drop characteristics of carbon dioxide during gas cooling process in a horizontal tube [J]. International Journal of Refrigeration, 2006, 29:539–546.

Power and Energy – Kong (Ed.)
© 2015 Taylor & Francis Group, London, ISBN 978-1-138-02782-4

Study on VSC based MTDC for offshore wind farms

Hongyuan Huang, Feijin Peng & Xiaoyun Huang
Foshan Power Supply Bureau, Guangdong Power Grid Co. Ltd., Foshan, Guangdong, China

Aidong Xu, Jinyong Lei, Lei Yu & Zhan Shen
Electric Power Research Institute of China Southern Power Grid, Guangzhou, Guangdong, China

ABSTRACT: In view of the drawbacks of traditional HVDC transmission technology in the offshore wind farms, this paper studied on the basis of VSC-MTDC offshore wind power grid technology, and proposed a new type of VSC-MTDC system structure with energy storage device. This paper designed a controller that can adjust the receiving-end power adaptivity by using the State of Charge (SOC) of the energy storage device. Simulation results show that the VSC-MTDC system with energy storage device can greatly improve the utilization rate of the offshore wind and ensure the stability and smoothness of the system output.

1 INTRODUCTION

Problems of energy depletion and environment pollution have being becoming more and more serious in recent years, so offshore wind power, which is clean, safe, stable, and sustainable, is drawing more and more attentions (Lei 2005, Yang 2012). However, a regular offshore wind farm is always far away from land, and how to transmit power from offshore wind farms to the onshore load centers is a key problem. On the other hand, since the power of the offshore wind farm has the characteristics of randomness and volatility, the power quality of public power grid will be impacted greatly if large scale of wind power injected (Bao 2011).

In Giddani et al. (2010), VSC-HVDC system was used for large-scale offshore wind farm, the flexibility of control and ability of riding through fault is studied. Prabha (2001) studied the offshore wind farm VSC-HVDC system and its protection with power grid, and proposed an offshore wind power transmission network topology for the study of the relay device under different fault conditions and for the wind power reliability. In Huang (2011), parameters of VSC-HVDC controller were analyzed and calculated, and based on the coordinate system of d-q-0 frame, and a closed-loop vector controller that allows offshore wind farm VSC-HVDC transmission system is designed for the connection of off-shore wind power. Wen et al. (2013) proposed the usage of hybrid multi-terminal HVDC, that is use VSC converters to connect an offshore wind farm which is a weak terminal, and use LCC converter to connect AC power grid which

is a strong terminal. A hybrid five terminal HVDC systems are studied in details. Lu et al. (2003) discussed the issue of offshore wind farm development, in order to maximize wind capture, proposes multi-terminal VSC-HVDC system connected to the AC side of fan. Chen et al. (2014) proposed optimal control strategy of local wind storage systems based on very short-term wind power prediction. Yuan et al. (2014) presented a policy based on storage to stabilize fluctuations in wind power output. In Ruan et al. (2008), the free wheel energy storage unit as the energy buffering means of the grid-connected wind farms to stabilize the power output of wind farms and wind designs a control strategy with output active power and reactive power of wind turbines as control signals. The Rao et al. (2014) proposed topology of two flexible terminals HVDC connected grid transmission system based on energy storage devices.

Currently VSC-HVDC system, which is used for offshore wind farm, mainly focuses on two terminal HVDC systems. There are few studies with multi-terminal MTDC for offshore wind farm. However, multi-terminal MTDC system has more advantages compared with the two terminal HVDC systems in the economy and flexibility, so this paper, which is based on the in-depth analysis of the two terminals VSC-HVDC system, applies the multi-terminal HVDC system (VSC-MTDC) to the offshore wind farm. It establishes a new topology of multi-terminal HVDC system that contains energy storage device, and designs the control strategy of converter station and energy storage device, and puts forward a new controller that is adapted by the end of the power by oneself. Simulation results show that with energy storage device, MTDC will have better operating characteristics against with the volatility of wind power. At the same time, it also greatly improves the utilization rate of the offshore wind power.

2 SYSTEM TOPOLOGIES

This paper presents a multi-terminal MTDC system which is used for offshore wind farm. Topology of this system is shown in Figure 1.

Figure 1 uses the direct driving wind generator; the electricity energy is transformed into DC through the controlled rectifier. The DC which several direct drive wind generators generated is aggregated to the DC aggregation platform through the DC short line, then transfer the electricity energy to the land converter through the total seafloor DC cable, which connect to the power network. At the same time, in order to support DC voltage for the DC aggregation platform and ensure the stability of the dc voltage, the energy storage device with the DC aggregation platform is connected through the DC/DC converter.

This paper uses the ideal super capacitor as the ideal energy storage device. First-order Linear RC model is selected as equivalent model of the super capacitor. The interface circuit of energy storage system uses DC/DC converter. Energy storage system in wind power generation system need to have the ability of low reservoir and high incidence, that is to say, energy storage system store energy when wind power output is excess, energy storage system release energy when wind power output is shortage, so bidirectional DC/DC converter is used. Bidirectional DC/DC converter can double-quadrant operation, thereby achieving two-way flow of energy through regulating the energy transfer direction.

Topology of bidirectional DC/DC converter circuit is shown in Figure 2, where L is the energy storage inductor, C is filter capacitor, S_1 and S_2 are IGBT switch, D_1 and D_2 are diodes (integrated in

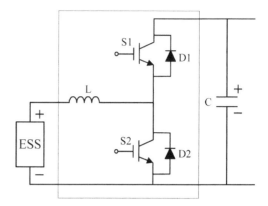

Figure 2. The bidirectional DC/DC converter structure.

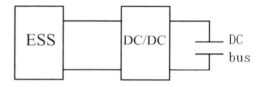

Figure 3. Energy storage system schematic diagram.

the corresponding switch tube). The circuit that is also called bidirectional Buck-Boost DC converter normally has three working modes, including Buck, Boost, and Stop.

Through power electronic interface technology and the appropriate control method, energy storage system can meet the requirements of wind power generation system in the volume density and power density at the same time. Energy storage system is composed of the ideal energy storage device and the super capacitor of power electronic power converter circuit, and the working principle diagram is shown in Figure 3.

3 CONTROL STRATEGY OF MULTI-TERMINAL HVDC SYSTEM

3.1 Control strategy of offshore converter station

MTDC control system, which this paper studies, has a converter which uses the direct current control by the inner and outer loop. The converter of the offshore wind turbines operates in a controlled rectifier state. The outer loop of the converter adopts tracking active power and fixed AC reactive power control. Inner and outer ring control is shown in Figure 4.

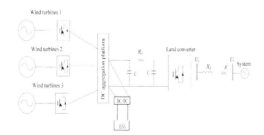

Figure 1. System structure of VSC-MTDC.

3.2 Control strategy of energy storage device

Ideal super capacitor is used in energy storage device of this paper, which provides the support to the collection platform of DC voltage. In order to maintain a stable DC voltage, and keep the voltage of super capacitor unchanged, we use a constant voltage source. Working principle of constant voltage source is shown in Figure 5.

According to the principle shown in Figure 5, mathematical model that contains charge and discharge of super capacitor can be obtained:

$$\begin{cases} L_{DC}\dfrac{dI_{SC}}{dt} = -U_{SC} + U_{PWM} \\ C_{SC}\dfrac{d(U_{SC} - I_{SC}R_{SC})}{dt} = I_{SC} \\ C_{DC}\dfrac{dU_{DC}}{dt} = -I_{DC} + I_{PWM} \end{cases} \quad (1)$$

By Equation (1), we can obtain DC/DC transfer function block diagram, as shown in Figure 6.

We can get the super capacitor reference current i_{scref} by the subtraction value of the super capacitor DC reference voltage U_{dcref} and the actual voltage measurements of the collection platform U_{dc} after

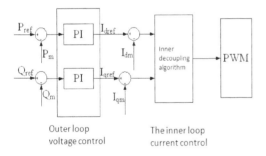

Figure 4. The inverter control block diagram rings.

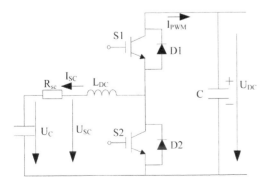

Figure 5. Working principle of the super capacitor.

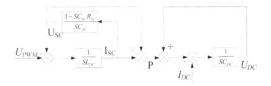

Figure 6. DC/DC transfer function block diagram.

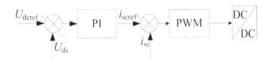

Figure 7. Super capacitor dc voltage control block diagram.

PI regulation. By the subtraction value of i_{scref} and the super capacitor current measurements i_{sc}, we can get the PWM control signal after PI controller. The signal drives the DC/DC switch to make the DC voltage meet the requirements.

3.3 Control strategy the converter station on land

Converter station on land operating in inversion state, and it takes the inner and outer loop control mode of the direct current control. The outer ring adopts the active power and constant AC reactive power control mode, the integral outer ring control as shown in Figure 4. Moreover, the constant active power reference value of the converter on land outer ring is variable, and the value of constant active power on the receiving terminal mainly depends on the State of Charge (SOC) of energy storage device. The SOC of energy storage device is used to describe the storage capacity of the energy storage device, which is usually expressed as a percentage. Its range is 0–1. When SOC = 0, the power of energy storage device is completely exhausted. When SOC = 1, the energy storage device is fully charged. SOC can be expressed as follows:

$$SOC = \frac{W_{re}}{W_{full}} \quad (2)$$

where W_{re} represents the current stored energy, and W_{full} represents the maximum stored energy. In this paper, we talk about the ideal super capacitor. First we choose a W_{full}, and take the initial SOC is 0.5. We will get the value of the SOC through the energy absorbed or released by the energy storage device. The SOC is sent to the receiving terminal through the wireless communication mode.

The active power reference increases or decreases according to the SOC. Power adjustment steps proposed in this paper are as follows:

1. When SOC ≥ m (0.5 < m < 1), the active power reference will increase until SOC < m;
2. When SOC ≤ n (0 < n < 0.5), the active power reference will decrease until SOC > n;
3. When n < SOC < m, the active power reference keeps unchanged.

The specific control structure diagram is shown in Figure 8. In the diagram, ∇P is the adjustment of active power reference value; P_m is the actual measured active power. We can obtain I_{sdref} through the difference between the given reference power P_{ref} and ∇P as well as P_m next after through PI regulator and limiting links. Then the I_{sdref} enters the inner ring current control links.

When there is a error in the storage system, the DC voltage of the DC transmission line will rise (or fall), even make the whole transmission system collapse. So we should set upper and lower limit of the receiver DC voltage. When the receiver DC voltage U_{dc} in the range, we can adopt the constant active power and reactive power control mode whose outer ring reference value of the receiver converter station is variable. When the voltage U_{dc} is out of the range, we will adopt the outer ring constant DC voltage and reactive power control mode of the receiving terminal station. The approach is illustrated by Figure 9. Interval 2 in this figure represents a control method, by which the outer loop's reference value of the receive side can be changed,

with both active power and reactive power being set. Interval 1 and interval 3 manifest another control manner, with the DC voltage and reactive power of the outer loop being set in purpose of ensuring the stability of the system.

4 SIMULATION ANALYSIS

To validate the reliability of the multi-terminal HVDC system configured with energy storage devices and applied in offshore wind power, simulation model, shown by Figure 10, is realized with the Matlab. The control of each converter and energy

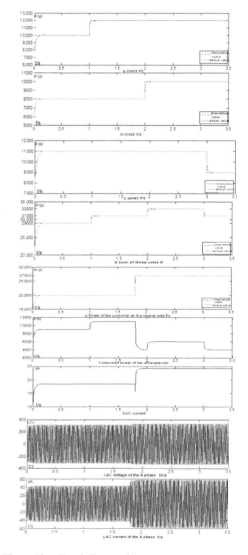

Figure 10. Simulation results.

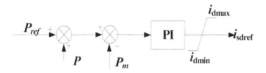

Figure 8. Receiver converter loop control block diagram.

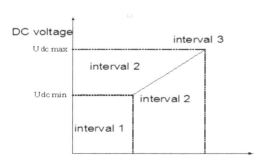

Figure 9. Receiver control mode.

storage device is mentioned above. Parameters are as follows: $R_1 = R_2 = R_3 = 0.5\Omega$, $L_1 = L_2 = L_3 = 0.01\text{H}$; $R_d = 2\Omega$, $C = 1\mu\text{f}$; $R = 0.5\Omega$, $L = 0.01\text{H}$. The ideal super capacitor is utilized as compensation devices, with its inertia constant being regarded as extremely large and hence the USC is invariable. m is 0.8, n is 0.3, and W_{full} is 40000 J. Energy storage devices carry 20000 J energy initially.

Simulation results are as follows. From Figure 10a to Figure.10c, we can observe that the power step to the three units in different time. Change of the whole power converged into the offshore energy convergence platform is shown by Figure 10d. Figure 10e presents us with the power of the converter on the receive side. Figure 10f represents the power absorbed by the energy storage devices. Figure 10f represents the DC current of the transmission line. Figure 10i to Figure 10j represents the AC voltage and current concerning the A phase on the AC side of the receive one. At the time of 1.8 or so, the switch signal of the reference of power is sent to the receive side by communication and then, the reference alters from 20000 W to 27000 W. Also, I_{dc} and I_{ca} are both changed. At the time of approximate 3.1 s, a new stable status is reached, when the total power generated by the wind facility and that from the converters on land and the super capacitor exactly reach a new balance.

From the simulation results, we can find that this approach can adjust the values of the receive side automatically. The system is able to track wind farm's power by the SOC value of the energy storage device, which enhances the efficiency of offshore wind power's utilization and enables the energy storage devices to be in normal working condition.

5 CONCLUSION

A novel multi-terminal HVDC system configured with energy storage devices is proposed in this paper is characterized as follows:

1. The multi-terminal HVDC system makes it possible that through the DC transmission and convergence, the wind farm realized connection to the grid on land, and the transmission has the feature of flexibility, economy and efficiency.
2. In terms of the fluctuation of wind farm's output power, the multi-terminal HVDC has a better output characteristic with the configured of energy storage devices. Meanwhile, the utilization rate of the wind resources is enhanced and the output stability is ensured in the case that the receive side reference of active power can be

adjusted automatically via the SOC of the storage device.

ACKNOWLEDGMENT

This paper is supported by Science and Technology Project of Guangdong Power Grid Co. Ltd (K-GD2013-044) and the National High Technology Research and Development Program of China (863 Program) (2011 AA05114).

REFERENCES

1. C. Huang. "Offshore wind farm VSC-HVDC converter of flexible study,". ShangHai: ShangHai Jiaotong University, 2011.
2. C.C. Rao, H.Y. Wang, W.Q. Wang, "Increase research capacity and stable operation of the wind power system large-scale HVDC technology based on flexible energy storage decice," Power system protection and control, 2014.
3. J.P. Ruan, J.C. Zhang, J.H. Wang, "Flywheel energy storage system to improve the research network stability and wind farms," Electric Power Science and Engineering, Vol. 18, No. 3, 2008, PP.5~8.
4. J.Q. Bao, "Application of VSC-HVDC system in offshore wind power transmission," Journal of Shenyang Institute of Engineering (Natural Science Edition), Vol.1, No. 7, 2011, pp. 5~8.
5. J.Y. Wen, X. Chen, "Research on hybrid multiterminal HVDC technology for offshore wind farms and networks," Power System Protection and Control System, Vol. 2, No. 41, 2013, pp. 55–61.
6. K. Prabha. "Power system stability and control,". Beijing: Chinese Power Press, 2001.
7. Giddani, P. Adam, O. Anaya. "Grid integration of offshore wind farms using multi-terminal DC transmission systems (MTDC)," 5th IET International Conference on Power Electronics, Machines and Drives (PEMD 2010), 2010.
8. Q.Y. Chen, W. Hu, Y. Min, "Wind Cluster storage system wide coordination of joint optimization of control," Chinese Society for Electrical Engineering, 2014.
9. T.J. Yuan, C. Hao, "Energy storage system to improve the strategy of large-scale wind farm output fluctuations," Power system protection and control ISTIC EI PKU, 2014.
10. W. Lu, O. Teck, "Optimal acquisition and aggregation of offshore wind power by multiterminal voltage-source HVDC," IEEE Transactions on Power Delivery, Vol. 18, No. 1, 2003, pp. 201–206.
11. W.L. Zhang, M. Qiu, X.K. Lai, "Energy Storage Technology in Power System," Grid Technology, Vol. 32, No. 7, 2008, PP. 1–9.
12. Y.Z. Lei, "Wind power generation and electricity market," Automation of Electric Power Systems, Vol. 29, No. 10, 2005, pp. 1–5.
13. Y.L. Yang, H.M. Wang, "Analysis of China's offshore wind power development situation," China Waterway, Vol. 12, No. 12, 2012, pp. 45–46.

Power and Energy – Kong (Ed.)
© 2015 Taylor & Francis Group, London, ISBN 978-1-138-02782-4

Investigation on performance for the two stage transcritical CO₂ cycle with injector

Y.B. Xie, T.J. Wang & H.B. Ji
Department of Power Engineering, North China Electric Power University, Baoding, Hebei, China

ABSTRACT: The use of two-stage compression or replace throttle valve by ejector can improve the Coefficient of Performance (COP) of CO_2 trans-critical refrigeration cycle effectively. This paper analyzed the performance of CO_2 trans-critical two-stage compression/ejection refrigeration cycle, and contrasted it with the CO_2 trans-critical two-stage compression refrigeration cycle. Under stable operation conditions, effects of heat rejection pressure, outlet temperature of gas-cooler, evaporation temperature and super-heat of compressor on the COP are investigated. Results indicate that at given condition, the COP of two-stage compression/injection cycle is better than the two-stage compression cycle obviously; the COP reduces 50.4% and 41.5% respectively with the increase of outlet temperature of gas-cooler and decrease of evaporation temperature; increasing of super-heat of low pressure and high pressure compressor also results in decreasing of COP.

1 INTRODUCTION

The natural working fluid CO_2 is paid more and more attention as a powerful refrigerant replacement of HCFCs and CFCs in refrigeration field around the world [1,4,10,12]. CO_2 has good flow and heat transfer characteristics, ODP = 0, GDP = 1. Moreover, it is safe and non-poisonous, large volumetric refrigeration capacities and compact structure systems. But its critical temperature (31.1°C) is below the typical environmental temperature at the summer working condition so that systems usually work in the trans-critical condition. The key problems to popularize and apply CO_2 trans-critical refrigeration cycle are CO_2 has much higher operating pressure, larger throttling loss and lower system efficiency than conventional refrigerants. Adopting two stage compression cycle is able to decrease throttling loss and improve Coefficient Of Performance (COP) [5,9,13].

Using injector to replace the throttle valve can not only reduce the loss of throttling, recycle pressure energy, but also improve the compressor suction pressure, and reduce the consumption of the compressor. Moreover, it has lots of advantage: simple structure, low cost, no moving parts and not easy to damage. Theoretical analysis results show that the COP of trans-critical CO_2 ejection refrigeration cycle is 18.6% higher than the regenerative cycle, 22.0% higher than the conventional cycle [2]. Experimental researching results show that the refrigerating capacity and COP of transcritical CO_2 injection refrigeration cycle system

improves 8% and 7%, compared with the conventional cycle, respectively. The injector can recycle about 14.5% of the loss of throttling [11]. The optimization analysis on several different forms trans-critical CO_2 injection refrigeration cycle has carried out [6,7].

This paper analyzed the performance of CO_2 trans-critical two-stage compression/ejection refrigeration cycle, and contrasted it with the CO_2 trans-critical two-stage compression refrigeration cycle. Under stable operation conditions, effects of heat rejection pressure, outlet temperature of gas-cooler, evaporation temperature and super-heat of compressor on the COP are investigated. Aimed at providing the reference for the optimization design of trans-critical CO_2 refrigeration system.

2 CO₂ TRANS-CRITICAL TWO-STAGE COMPRESSION/EJECTION CYCLE SYSTEM

Figure 1 is the flow diagram and P-H diagram of the injection cycle. The gas liquid two phase refrigerant out of injector is divided into two in the gas-liquid separator: saturated steam after a pipeline overheating into the low pressure compressor, compressed to the middle pressure p_2 into the inter-cooler, cooled in the inter-cooler into high-pressure compressor, compressed to high-pressure steam into the gas cooler, gas cooler outlet CO_2 gas through the nozzle of injector becomes low pressure high velocity flow to inject low pressure

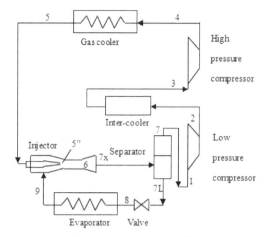

Figure 1a.   Flow description.

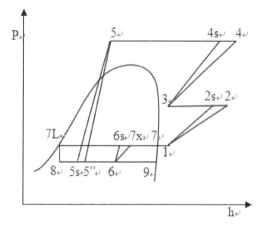

Figure 1b.   Pressure-enthalpy diagram.

saturated steam from evaporator, the two flow mixed into state 6 in mixing chamber, then through diffusion chamber pressurization deceleration out injector; saturated liquid after throttle valve throttling action into the evaporator, evaporating and producing refrigerating capacity, the low pressure steam after evaporating ejected into the injector.

3   ANALYSIS OF CIRCULATION PERFORMANCE

3.1   Model hypothesis

Given following hypothesis: (1) ignore the inlet and outlet kinetic energy of injector; (2) mixing process is constant pressure in the injector; (3) refrigerant is one dimensional steady flow in the ejector;

(4) except heat exchangers, heat transfer between refrigerant and the ambient is negligible; (5) compression process is adiabatic and isentropic process; (6) pressure drop in the connecting pipes and heat exchangers are negligible; (7) evaporator outlet steam, gas liquid separator outlet vapor and liquid are in saturated state.

3.2   Calculation formula

Take unit working fluid for an example to calculate. Injection coefficient is defined as:

$$\mu = \frac{\text{Injection flow quality}}{\text{Working flow quality}} \quad (1)$$

Then the injection flow quality and working flow quality respectively are $\mu/(1 + \mu)$ and $1/(1 + \mu)$.
Nozzle efficiency in ejector is:

$$\eta_n = \frac{h_5 - h_{5''}}{h_5 - h_{5s}} \quad (2)$$

Mixing and diffusion chamber comprehensive efficiency is:

$$\eta_d = \frac{h_{6s} - h_6}{h_{7x} - h_6} \quad (3)$$

Energy conservation equation is:

$$\frac{1}{1+\mu}h_5 + \frac{\mu}{1+\mu}h_9 = h_{7x} \quad (4)$$

$$\frac{1}{1+\mu}h_5 = \frac{1}{2}\frac{1}{1+\mu}u_{5''}^2 + \frac{1}{1+\mu}h_{5''} \quad (5)$$

$$h_6 + \frac{1}{2}u_6^2 = h_{7x} \quad (6)$$

Momentum conservation equation is:

$$\frac{1}{1+\mu}u_{5''} = u_6 \quad (7)$$

Coefficient of performance of the system is:

$$COP = \frac{Q}{W} \quad (8)$$

$$Q = \frac{\mu}{1+\mu}(h_9 - h_8) \quad (9)$$

$$W_1 = \frac{1}{1+\mu}\left(\frac{h_{2s} - h_1}{\eta_1}\right) = \frac{1}{1+\mu}(h_2 - h_1) \quad (10)$$

$$W_2 = \frac{1}{1+\mu}\left(\frac{h_{4s}-h_3}{\eta_2}\right) = \frac{1}{1+\mu}(h_4 - h_3) \quad (11)$$

$$W = W_1 + W_2 \quad (12)$$

where, h is specific enthalpy; Q is refrigerating capacity; W_1, W_2 and W are respectively energy consumption of low and high-pressure compressor and total consumption of compressor. η_1, η_2 are respectively isoentropic efficiency of low and high compressors. They are determined by the follow formulas [8]:

$$\eta_i = 1.003 - 0.121 r_i \quad (13)$$

$$r_1 = \frac{p_2}{p_1} \quad (14)$$

$$r_2 = \frac{p_4}{p_3} = \frac{p_4}{p_2} \quad (15)$$

$$p_2 = \sqrt{p_1 p_4} \quad (16)$$

4 RESULTS ANALYSIS

The vapor compression/ejection refrigeration cycle system is stable operating state when the ejection efficiency meets the injector outlet dryness [3]. It means that:

$$x_{7x} = \frac{1}{1+\mu} \quad (17)$$

In the thermodynamic calculation, the system always meets the stability condition. Nozzle efficiency in injector η_n is 0.9, comprehensive efficiency η_d is 0.85; evaporation temperature T_e is –20°C~0°C; gas cooler exit temperature T_c is 30°C~50°C; the heat discharge pressure P_c is 8 MPa~13 MPa; the suction gas superheat of low and high pressure compressor T_{sh1} and T_{sh2} are 5°C.

The relation between COP and high pressure of two-stage compression/ejection refrigeration cycle and two-stage compression cycle at evaporating temperature Te = 0°C and gas cooler outlet temperature Tc = 35°C is shown in Figure 2. And from Figure 2, the tendencies of COP changing with high pressure of the two cycles are basically the same. At high pressure of 9 MPa, the maximum of COP (COP$_{max}$) of both cycles can be found at the pressure named optimum high pressure (P$_{c,opt}$). The performance coefficient of two-stage compression/ejection refrigeration cycle is higher than that of conventional two-stage compression cycle evidently. And at the pressure of P$_{c,opt}$, the COP of two-stage compression/ejection refrigeration cycle is 27.2% higher than that of two-stage

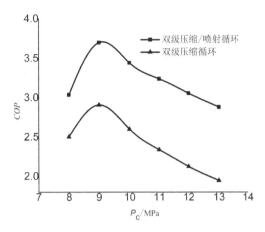

Figure 2.   Variation of COP with heat rejection pressure.

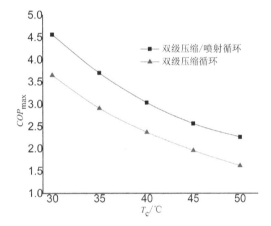

Figure 3.   Effect of gas cooler exit temperature on COP$_{max}$.

compression cycle, and this is because the throttle was replaced with an ejector which decreases the throttling loss, increases the suction pressure of the compressor and so decreases the compressor power consumption.

The COP$_{max}$ of the two cycles changing with the gas cooler outlet temperature Tc is shown in Figure 3 at evaporating temperature Te = 0°C. With the rising of gas cooler outlet temperature, the COP$_{max}$ of the two cycles decline with a similar speed. Within the whole range of the gas cooler outlet temperature, the COP of two-stage compression/ejection refrigeration cycle is better than that of conventional two-stage compression cycle evidently. When Tc increased from 30°C to 50°C, the COP$_{max}$ of two-stage compression/ejection refrigeration cycle declined 50.4%.

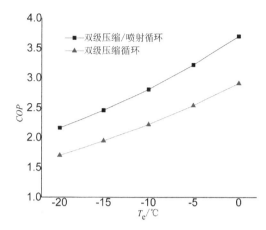

Figure 4.   Effect of evaporation temperature on COP$_{max}$.

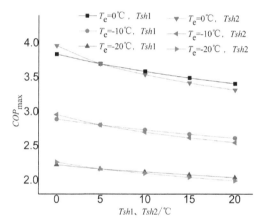

Figure 5.   Effect of Tsh1 and Tsh2 on COP$_{max}$.

The variation trend of the max performance coefficient COP$_{max}$ of the two cycles with respect to evaporating temperature Te at gas cooler outlet temperature Tc = 35°C is shown in Figure 4. With the rising of evaporating temperature, COP$_{max}$ of the two cycles will rise too. When Te declined from 0°C to −20°C, the COP$_{max}$ of two-stage compression/ejection refrigeration cycle declined 41.5%.

The relationship between COP$_{max}$ and T$_{sh1}$ or T$_{sh2}$ of two-stage compression/ejection refrigeration cycle at different evaporating temperature with temperature of 35°C at gas cooler outlet was shown in Figure 5. The performance coefficient of the cycle will decrease with the increase of T$_{sh1}$ and T$_{sh2}$ as can be observed from Figure 5. The performance coefficient decline caused by the increase of T$_{sh1}$ and T$_{sh2}$ will be less with the decrease of evaporating temperature.

At the conditions of unchanged evaporating temperature and unchanged gas cooler outlet temperature, the decrease rate of performance coefficient caused by the increase of T$_{sh2}$ is faster than that caused by the increase of T$_{sh1}$. At Te = 0°C, the rising of T$_{sh1}$ and T$_{sh2}$ from 0°C to 20°C caused a performance coefficient decline of 11.9% and 16.3%, respectively. And At Te = −20°C, the rising of T$_{sh1}$ and T$_{sh2}$ from 0°C to 20°C caused a performance coefficient decline of 11.9% and 16.3%, respectively.

5   CONCLUSIONS

1. A optimum high pressure was found where CO$_2$ trans-critical two-stage compression/ejection refrigeration cycle can obtain the highest performance coefficient, and the performance coefficient of two-stage compression/ejection refrigeration cycle is higher than that of conventional two-stage compression cycle evidently at the calculation conditions of this article.
2. In the given range of temperature, the performance coefficient of two-stage compression/ejection refrigeration cycle declined obviously with the temperature increase at the outlet of the gas cooler and the decrease of the evaporating temperature.
3. The rising of superheat degree in the suction process of both high-pressure compressor and low-pressure compressor can reduce the performance coefficient of two-stage compression/ejection refrigeration cycle. The performance coefficient decline caused by T$_{sh2}$ exceeded that caused by T$_{sh1}$, and the performance coefficient decline caused by T$_{sh1}$ and T$_{sh2}$ will be less with the decrease of evaporating temperature.

ACKNOWLEDGMENTS

The Project supported by Natural Science Foundation of Hebei province (number E2014502085).

REFERENCES

1. Andy Pearson. Carbon dioxide-new uses for an old refrigerant [J]. Int J. Refrig, 2005, 28(8):1140–1148.
2. Deng Jianqiang, Jiang Peixue, Lu Tao, Wang Guoliang, Luwei. Theoretical analysis of a transcritical CO2 apor compression/ejection refrigeration cycle [J]. Tsinghua Univ (Sci & Tech) 2006, Vol. 46, No. 5.
3. Deng Jian-Qiang, Jiang Pei-xue, Lu Tao, Lu Wei. Performance comparison of trans-critical CO2 vapor-compression/ejection refrigeration cycle [J]. Journal of Engineering Thermophysics, 2006, Vol. 27, No. 3.

4. G. Lorentzen. Revival of carbon dioxide as a refrigerant [J]. Int J. Refrig, 1994, 17(5):292–301.
5. Jun Lan Yang, Yi Tai Ma, Sheng Chun Liu. Performance investigation of trans-critical carbon dioxide two-stage compression cycle with expander [J]. Energy, 2007, 32(3):237–345.
6. Jahar Sarkar. Optimization of ejector-expansion transcritical CO2 heat pump cycle [J]. Energy, 2008, 33(9):1399–1406.
7. K. Cizungu a, M. Groll a, Z.G. Ling. Modeling and optimization of two-phase ejectors for cooling systems [J]. Applied Thermal Engineering, 2005, 25(13):1979–1994.
8. Liao S.M., Zhao T.S., Jakobsen A. A correlation of optimal heat rejection pressure in transcritical carbon dioxide cycles [J]. Applied Thermal Engineering, 2000, 20(9):831–834.
9. M. Kim, J. Pettersen, C.W. Bullard, Fundamental process and system design issues in CO2 vapour compression systems [J]. Progress in Energy and Combustion. Science, 2004, 30(2):119–174.

10. Pettersen J, Hafner A, Skaugen G. Development of compact heat exchangers for CO2 air-conditioning systems [J]. Int J. Refrig, 1998, 21(3):180–193.
11. Stefan Elbel, Pega Hrnjak. Experimental validation of a prototype ejector designed to reduce throttling losses encountered in transcritical R744 system operation [J]. Int J. Refrig, 2008, 31(3):411–422.
12. Yitai Ma, Junlan Yang, Shengchun Liu. Thermodynamic analysis for transcrutical CO2 cycle and conventional refrigeration cycle [J]. Acta Energiae Solaris Sinica, 2005, 26(6):836–841.
13. Yingbai Xie, Ganglei Sun, Chuntao Liu. Thermodynamic analysis of CO2 trans-critical two-stage compression refrigeration cycle [J]. Journal of Chemical Industry and Engineering, 2008, 59(12):2985–2989.

Power and Energy – Kong (Ed.)
© 2015 Taylor & Francis Group, London, ISBN 978-1-138-02782-4

Research on the impacts of DC-bias on the loss and temperature rise of the transformer

Chang-yun Li & Ya-kui Liu

School of Electrical Engineering and Automation, Qilu University of Technology, Shandong, China

ABSTRACT: The phenomenon of DC-bias refers to a DC current flowing through the neutral point into power transformer winding, resulting in the abnormal working state in the transformer. This paper first analyses the mechanism of DC-bias and several kinds of loss types, then establishes the model of iron loss and makes a numerically solve, finally discusses the influence of temperature rise to transformer.

1 GENERATING MECHANISM OF DC-BIAS

There are a variety of reasons produce a large DC component in transformer winding, mainly can be divided into two cases: one is the monopolar ground circuit operation mode or the bipolar asymmetrically operation mode of the HVDC transmission system.

As shown in Figure 1, when using monopolar ground return operating mode, the dc current flowing through grounding electrodes into the earth, the earth becomes a part of the circuit, the ground current is the current of HVDC transmission, so as to produce DC potential difference between two grounding. Under this situation, DC will pass by the windings of the transformers through the earthed neutral point.

The other is Geomagnetically Induced Current (GIC) which is caused by the interaction of the geomagnetic field and the dynamic movement of the ionic wind. The change of the geomagnetic field produces electric potential gradient, which give rise to low-frequency induction current. It is DC because its frequency is very low.

The mechanism of transformer DC-bias phenomenon is shown in Figure 2. Full line in figure (a) says no DC component of the magnetic flux curve, the typical Φ-i curve for transformer is shown in Figure (b), full line in Figure (c) says excitation current curve with no DC component, as the dotted line stands on the opposite. By analyzing Figure 3, when a DC current flowing in the winding of transformer, the core produces a DC magnetic flow which named Φ0. Under the action of Φ0, the cure of magnetic flow, which named Φ, begins to migrate, so that the working point of the transformer from linear area into the saturated zone, as the exciting current increase and makes a distortion. Form that, the excitation current distortion with DC component caused by the nonlinear parts in Φ-i curve for transformer, to put the transformer into a saturated state. So the size of the exciting current in addition to related to transformer design, also is closely related to the size of the DC current.

The work point of the transformer under DC bias will shifts, and its iron will be saturated which affect the normal work of the transformer seriously. This gives rise to lots of bad influence, for example, the peak value of the exciting current increases significantly, and the waveform of the current is seriously distorted with a lot of harmonics; the heavy noise; the serious vibration; the increases of transformer's leakage and losses and temperature build-up; the local heating of metal structures; the malfunction of relay protection devices of electric power system; which endanger the operation safety of the transformers and electric power system. So the DC bias phenomenon of the transformer is concerned.

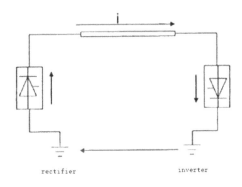

Figure 1. The photograph of monopole earth loop way.

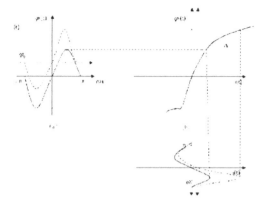

Figure 2. Mechanism of DC bias.

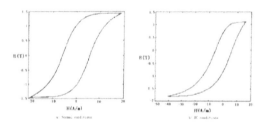

Figure 3. Hysteresis loop of core material.

2 THE MECHANISM OF LOSSES AND TEMPERATURE RISE

Transformer will produce a certain amount of loss in operation process. On the one hand, the loss will influence the operational efficiency of transformer; on the other hand, it will lead to raise the temperature of transformer components. The loss including winding loss (copper loss) and core loss (iron loss).

2.1 Copper loss under DC-bias

Copper loss for transformer includes basic copper loss and the additional copper loss. Under the effect of direct current, excitation current could significantly increase, leads to basic copper consumption increase dramatically. But as a result of the main magnetic flux are sine wave, and the flux density change is relatively modest, so the flow of DC-bias current on the additional copper loss of relatively small, copper consumption is mainly basic copper loss.

2.2 Iron loss under DC-bias

Iron loss for transformer includes basic iron loss (hysteresis loss and eddy current loss) and the

additional iron loss (loss of magnetic flux leakage). Hysteresis loss is proportional with the frequency of magnetic flux and core material of the hysteresis loop area, and hysteresis loop associated with core flux waveforms and size. In the normal operation of the transformer, core has the positive and negative half wave symmetrical magnetic flux, and core material flux hysteresis loop as the core in the DC-bias of translational overall upward or downward. This paper selects 30QG120 orientation of silicon steel sheet. As Figure 3 shown is hysteresis loop operation of the transformer core material under normal situation and DC-bias, it can be observed that the DC-bias magnet core material hysteresis loop area, and changes hysteresis losses. Relationship between the rate of change of the core magnetic flux core eddy current loss. DC bias magnetic flux will only make the core with sine wave form the overall upward or downward shift, and it will not change the waveform of magnetic flux. The transformer will not change core eddy current loss because of the small rate of flux change under DC-bias. So the core hysteresis losses constitute the main effect of core loss under DC-bias situation.

2.2.1 Loss function model

In the bias magnetic field, considering the relationship among the field current and the magnetic flux, the induction electromotive force, exciting current can be divided into two parts, part of phase with magnetic flux, part of the phase with the induction electromotive force. So, we can introduce the following function in the relationship with Φ-i

$$i(t) = i_1(\varphi) + i_2(\varphi') \tag{1}$$

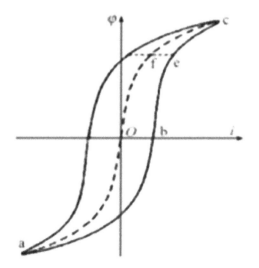

Figure 4. Hysteresis model based on consuming function.

In Figure 5, i_1 corresponding to the middle magnetization curve "aOc"; any point in hysteresis loop with the distance between the middle magnetization curve reflects the hysteresis effect of iron core, with i_2 corresponding. When the magnetization curve rises to the highest point, i_2 drops to minimum value zero; when the magnetization curve drops to zero, i_2 rises to maximum value.

Based on the Φ-I relationship between the dc bias magnetic hysteresis loop, Φ_{ac} magnetic flux ac corresponding excitation current is zero, there are two values, both the absolute value of the average and record for the I_{ob}, loss function relationship with flux is as follows:

$$\begin{cases} \varphi = \varphi_m \cos \omega t + \varphi_{dc} \\ i_2(\varphi') = -I_{ob} \sin \omega t \end{cases} \qquad (2)$$

where φ_m = amplitude of the Ac magnetic flux; φ_{ac} = Dc magnetic flux.

Loss function can be written as further

$$i_2(\varphi') = D\frac{d\varphi}{dt} = D\varphi_m \omega \cos(\omega t) = I_{ob} \cos(\omega t) \qquad (3)$$

D is a function loss coefficient, according to the properties of the material is different, can be more groups of measuring data into the formula, using the least squares method to find out.

At this point, we can draw core loss model of DC-bias condition.

$$P = f\int_0^T (i \cdot u)\, dt = f \cdot n \int_0^T \left(i \cdot \frac{d\varphi}{dt} \right) dt \qquad (4)$$

where n = coil number of turns; f = frequency.

Both the nondestructive function i_1 and loss function i_2 constitute i in equations (4), and i_1 corresponding core loss is zero, so can be directly to introduce i_2 to build core loss model of bias field.

$$P = f \cdot n \int_0^T \left(i_2 \cdot \frac{d\varphi}{dt} \right) dt \qquad (5)$$

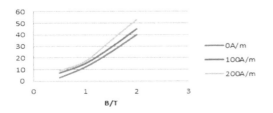

Figure 5. Core loss under different DC bias condition.

2.2.2 Numerical calculation

Calculation of DC magnetic field strength of 0, 100, and 200 A/m core loss, the result is shown in Figure 5.

It can be observed from the table, with the increase of core partial magnetic strength, core loss also increases.

2.3 The temperature rise of the plate and support plate under DC-bias

Core transformer iron core of the plate or shell type transformer iron core of the support plate is usually adopts magnetic materials, in order to obtain enough mechanical strength, is located in the core of table and core board or support plate, the same as the magnetic field strength of core silicon steel sheet and its thickness is more than the thickness of the silicon steel sheet thickness.

Operation of the transformer, there will be a part of the magnetic flux leakage into the casing wall, core plate and a structure, such as fire and eddy current loss in these structures, which leads to its temperature rise. In order to obtain enough mechanical strength, the transformer enclosure wall generally has the guide magnetic steel materials (e.g., A3 steel), normal operation of transformer, the permeability of core material is far greater than enclosure wall structures such as permeability, so the magnetic flux is mainly to flow inside the core, only a small amount of magnetic flux leakage. Dc-bias of transformer core half wave after saturation, core permeability has fallen sharply, thus a lot of leakage magnetic flux through the transformer oil, casing wall and core plate and pressure plate structure form a loop, resulting in the increase of the eddy current loss components, and then caused the increase of the temperature rise of metal components. If the iron core of the board or support plate with nonmagnetic materials, temperature can be greatly reduced.

3 THE DANGERS OF LOSS AND TEMPERATURE

Under DC-bias magnetic transformer losses and the increase of the temperature rise of the unfavorable influence on the operation of the transformer itself, its harm is mainly manifested in the following party:

1. The efficiency of transformer. The efficiency of the transformer is the ratio of transformer output active power and active power input. Dc bias condition, the transformer loss increase, and therefore need to input more active power to make the output of the same active power, thereby lowering the efficiency of transformer.

2. Transformer internal fault. Transformer winding wires, enclosure wall core pulling, and so on. Parts are made by metal material. Metal materials strength will reduce because the temperature was too high rigidity, metal material structure deformation, even causing the internal fault of transformer, such as winding temperature rise is too large to cause the deformation, winding string of melting it happen, even leading to the failure of large power transformer winding, serious when can burn transformer; meanwhile transformer local temperature is too high, and may damage the insulation board or insulating paper, causing the internal fault of transformer.

3. The insulation of the transformer accelerated aging so as to shorten the life of the transformer. The life of the insulating material determines the service life of the transformer, and the service life of all kinds of insulation materials correspond to a specific temperature. In oil-immersed transformer, transformer oil and insulation board, insulation paper constitutes the main insulation materials, and so on. Insulation aging is under the action of heat or other physical chemical gradually lose electrical insulating material strength and mechanical strength. Temperature, the oxidation degradation of the material, moisture and oil decomposition influence caused the transformer insulation aging, such as temperature mainly determines the speed of aging. Under dc bias, the temperature rise of the transformer winding and core plate increases, the higher the working temperature of the insulating materials and its contact, accelerated the speed of oxidation, led to the transformer insulation aging is accelerated, reduce the intensity of the mechanical strength and electrical insulation materials, thus shortening the service life of the transformer.

4. Gas content in transformer oil. Normal operation of the transformer, because of transformer insulation aging, solid organic insulating materials and transformer oil will release a small amount of hydrocarbon, carbon dioxide and carbon monoxide, most of these gases dissolved in transformer oil. Dc bias can lead to raise the temperature of transformer insulation material and metal parts of local overheating, from solid organic insulating material into the content of carbon dioxide, carbon monoxide and transformer oil decomposition of hydrocarbon can obviously increase, when the gas reaches a certain number of gas relay in transformer in into the part of the gas, causing gas protection action.

REFERENCES

1. Atbertson V.D., Boroki B., Fccro W.E., et al. Geomagnetic disturbance effects on power systems [J]. IEEE transactions on power delivery, 1993, 8(3): 1206–1216.
2. Boteler D. II, Shier R.M., Watanabc T., et al. Effects of geomagnetically induced currents in the BC hydro 500 kV system [J]. Power Delivery, IEEE Transactions on, 1989, 4(1):818–823.
3. Kappennman J.G., Albertson V.D., Mohan N. Current transformer and relay performance in the presence of geomagnetically-induced currents [J]. Power Apparatus and Systems, IEEE Transactions on, 1981, (3):1078–1088.
4. Meliopoulos A.P.S., Christoforidis G. Effects of DC ground electrode on converter transformers [J]. Power Delivery, IEEE Transactions on, 1989, 4(2): 995–1002.
5. Meliopoulos A.P.S., Glytsis C.N., Cokkinides G.J., et al. Comparison of SS-GIC and MHD-EMP-GIC effects on power systems [J]. Power Delivery, IEEE Transactions on, 1994, 9(1):194–207.
6. Picher P., Bolduc I., Dutil A., et al. Study of the acceptable DC current limit in core-form power transformers [J]. Power Delivery, IEEE Transactions on, 1997, 12(1):257–265.
7. Towle J.N., Prabhakara F.S., Ponder J.Z. Geomagnetic effects modeling for the PJM interconnection system. 1. Earth surface potentials computation [J]. Power Systems, IEEE Transactions on, 1992, 7(3): 949–955.

Power and Energy – Kong (Ed.)
© *2015 Taylor & Francis Group, London, ISBN 978-1-138-02782-4*

Numerical simulation of pressure fluctuations in an axial-flow pump

X.J. Yu, C. Kang, G.F. Zhang & L.T. Li
School of Energy and Power Engineering, Jiangsu University, Zhenjiang, China

W.F. Gong & C.J. Li
Shanghai Marine Equipment Research Institute, Shanghai, China

ABSTRACT: To investigate unsteady flow characteristics in an axial-flow pump, numerical simulation is performed using commercial ANSYS-CFX code. Stator vane numbers of 5, 7 and 9 are examined and a comparison is thereby accomplished. Stress is placed on small flow rate conditions. Pump performance is assessed for the three schemes involving different stators. While the variation of pump head is slightly influenced by stator vane number, pressure distributions between the impeller and the stator show evident periodicity associated with stator vane number. In the impeller-stator interaction zone, large stator vane number contributes to the suppression of the discrepancy among characteristic pressure fluctuation frequencies. With the increase of flow rate, low-frequency components are stimulated.

1 INTRODUCTION

Axial-flow pumps play an important role in the operation of energy-conversion systems. Of interest are diverse flow characteristics inside the flow passage of the axial-flow pump with its high specific speed (A. Furukawa and T. Shigenmitsu, 2007). In most cases, vertical installation manner is adopted for the axial-flow pump. Consequently, both fluid dynamics and structural aspects manifest distinct features (G.L. Wen, 2010). As for the performance of the axial-flow pump, the contribution of hydraulic factors is always highlighted. In addition, experiments for measuring flows in the axial-flow pump, particularly between the impeller and the stator, are problematic in the presence of current flow measurement techniques. Alternatively, numerical simulation can lend its support in this context. Steady simulation results are not sufficient in terms of treating complex flow phenomena residing between the impeller and the stator, thus unsteady simulation has gained much popularity in recent years (M. Vahdati et al. 2000). Here, an axial-flow pump is investigated with special attention paid to the interaction of the impeller and the stator. Different stator vane numbers are adopted to match the same impeller. Numerical simulation is utilized to reveal flow features between the impeller and the stator. Furthermore, unsteady pressure fluctuations between the impeller and the stator are extracted and analyzed. It is anticipated to cast light on hydraulic factors affecting the performance of the axial-flow pump.

2 NUMERICAL PROCEDURES

2.1 Geometrical model

The nominal flow rate Q_0 and pump head H_0 of the axial-flow pump considered are 1000 m³/h and 5.8 m, respectively. Under nominal conditions, the impeller rotates at a rotational speed of 1450 r/min. The outlet diameter of the impeller is 300 mm and the blade number is 4. Three stators with 5, 7 and 9 vanes were designed with streamline method to match the same impeller. Based on geometrical parameters of the hydraulic components involved, a computational domain is established. A sectional view of the pump with the 7-vane stator is shown in Figure 1.

Unstructured grids were used to discretize the computation subzones associated with the impeller and the stator in view of geometrical complexity of the blades and vanes. Other subzones were discretized with structured grids. Local grids were refined near the impeller blades and in the subzone between the impeller and the stator. With the grid-independence examination, the total grid number of 6732182 was determined for the 7-vane scheme. And the 5-vane and 9-vane schemes employed total grid numbers of 6325085 and 6941962, respectively.

2.2 Numerical methods

Three-dimensional unsteady governing equations were solved in the simulation. Various turbulent models have been attempted in the simulation of flows in the impeller pump (F.J. Shi and

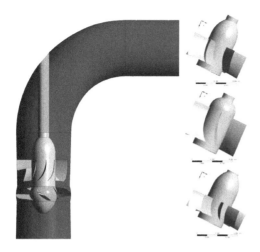

Figure 1. A sectional view of the axial-flow pump with the 7-vane stator. Geometrical models of the three stators are arranged on the right.

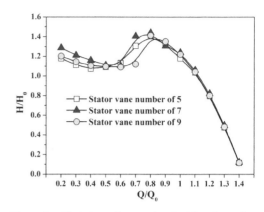

Figure 2. Variation of pump head with volume flow rate.

H.X. Chen, 2006). Here, the Renormalization Group (RNG) k-ε turbulence model, recognized as an improvement of the standard k-ε turbulence model, was used. This model was established based on principles of fuzzy mathematics, and the parameters involved in this model were defined according to relevant formulae instead of empiricism or experiments. Cavitation was not taken into account in view of the submerged operation conditions.

Present numerical work was carried out with the commercial code ANSYS-CFX, which is flexible in the treatment of complex turbulent flows and rotating flows. Central finite difference scheme was adopted to treat those advection terms. Discretization of momentum and turbulent kinetic energy equations was accomplished using second-order upwind scheme. Velocity inlet boundary conditions were defined at the inlet of the whole computational domain and thus volumetric flow rate could be modified accordingly. Free outflow conditions embedded in ANSYS-CFX were used at the outlet of the whole computational domain. Non-slip condition was applied for all solid boundaries. Scalable wall functions are fulfilled in near-wall flow regions.

3 PUMP PERFORMANCE

Energy delivery performance of the pump was assessed through statistically calculating total pressure at the inlet and outlet sections. Cases corresponding to flow rate values ranging from $0.2Q_0$ to $1.4Q_0$ were examined. And the results are plotted in Figure 2.

Under the nominal operation conditions, the pump head of the pump with the 7-vane stator is the largest. At high flow rates, the three curves apparently overlap each other. In this context, the effect of stator vane number is weak. Under low-flow-rate conditions, a pump head valley is presented for the three schemes. Such a shape is in accordance with the characteristic saddle-shape performance curve. Pump head depends largely upon the impeller, thus it can be extrapolated from Figure 2 that different stators do not cause a significant effect on pump head. However, hydraulic losses are directly related to pump head and low flow rates contribute to irregular flow structures in the impeller passage. Thus low flow rate conditions are stressed in subsequent analysis.

4 PRESSURE DISTRIBUTIONS

Turbulent flows inside the impeller and the stator are featured by distinct pressure distributions (Z.D. Qian and Y. Wang, 2010). For the three stator schemes, pressure distributions corresponding to the flow rates of $0.7Q_0$ and $0.5Q_0$ at which the variation of pump head is evidently seen. Steady simulation results at $0.7Q_0$ are presented in Figure 3 where two cross sections at impeller outlet and vane outlet are monitored.

Fluid discharged from the impeller clearly bears the effect of the rotating impeller. For the three schemes, the pressure distributions immediately downstream of the impeller contain circumferential periodicity corresponding well to the four blades. For the 9-vane scheme, flows near the hub are seemingly uniform. For the three cross sections near the stator, they are represented by periodicity in accordance with the stator vane number. Although the subzone between the impeller and

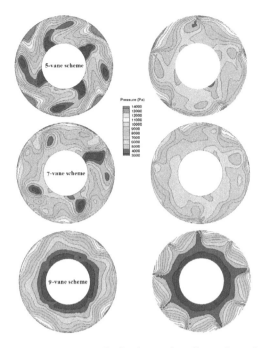

Figure 3. Pressure distributions at impeller outlet and stator inlet at flow rate of $0.7Q_0$. The three left-hand contours are associated with impeller outlet, and the three right-hand contours are associated with stator inlet.

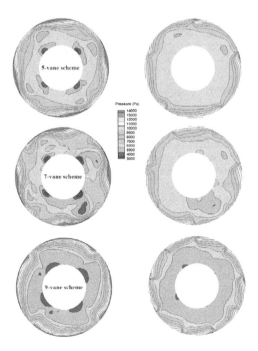

Figure 4. Pressure distributions at impeller outlet and stator inlet at flow rate of $0.5Q_0$.

the stator is rather small, axial variation of pressure distribution is remarkable. At $0.5Q_0$, pressure distributions are shown in the same fashion in Figure 4.

Relative to Figure 3, Figure 4 indicates that flow uniformity is improved as flow rate increases, which is particularly obvious near the stator. Still, pressure distributions in Figure 4 are closely related to the rotating impeller. For pressure contours near the stator vane, the effect of stator vane number can still be recognized.

5 PRESSURE FLUCTUATION

Pressure fluctuations between the impeller and the stator prove to be an influential factor underlying the hydraulically induced pump vibration (F.J. Wang et al. 2007). Nevertheless, only some preliminary conclusions have been recognized since that characteristic pressure fluctuations are dependent upon not just geometrical properties of the pump but also the operation parameters. Here, pressure fluctuations at flow rates of $0.5Q_0$ and $0.8Q_0$ are obtained and further processed with fast Fourier transformation method. The monitored points between the impeller and the stator are shown in Figure 5.

Characteristic frequencies at flow rate of 0.5Q are shown in Figure 6. Here, high frequencies components are not displayed in view of the comparatively low pressure fluctuation magnitudes corresponding to frequencies higher than 300 Hz. For the four points near the stator, shaft frequency (24.17 Hz) is not evident. In comparison, both shaft frequency and blade passing frequency (96.68 Hz) can be discerned in the spectra associated with the four points near the impeller. Although the axial distance between the two groups of monitored points is rather small, the pressure fluctuation characteristics are considerably different. At such a low flow rate, similarity among individual spectrum is weak. As for the 7-vane scheme, from the hub to the pump casing, the variation tendency of characteristic

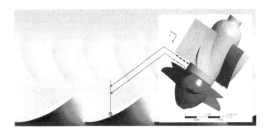

Figure 5. Monitored points between the impeller and the stator.

209

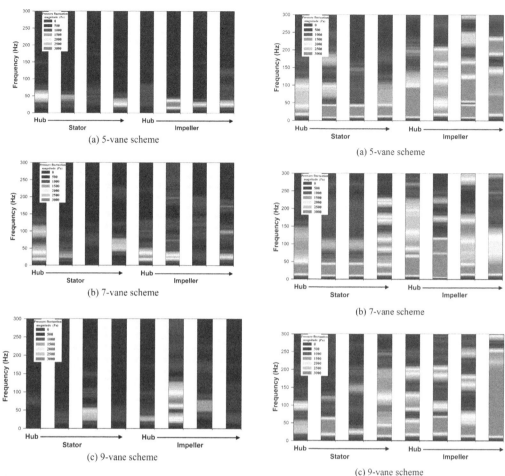

Figure 6. Pressure fluctuation spectra at flow rate of $0.5Q_0$. The four left-hand spectra are associated with the four points near the stator, while the four right-hand spectra correspond to the four points near the impeller.

Figure 7. Pressure fluctuation spectra at flow rate of $0.8Q$.

frequency distribution is explicit and the shaft frequency is influential, particularly near the hub.

As flow rate is increased to $0.8Q_0$, characteristic frequencies are shown in Figure 7 in the same manner. An apparent change relative to Figure 6 lies in that overall pressure fluctuation magnitudes in Figure 7 are improved.

For the 5-vane and 7-vane schemes, near the stator, shaft frequency is predominant in the eight spectra. In this connection, the 9-vane scheme yields relatively uniform distributions of characteristic frequencies. Regarding the four points near the impeller, several frequencies lower than the blade passing frequency is stimulated. The three schemes are analogous in terms of the variation from the hub to the pump casing. In comparison, characteristic frequencies associated with the 9-vane scheme are

distinct due to several identifiable frequencies higher than the blade passing frequency. It is difficult to relate the pressure fluctuation spectra with the pump vibration since the effect of hydraulic factors on the entire pump demands experimental verification.

6 CONCLUDING REMARKS

At low flow rates, the performance curve of the axial-flow pump is characterized by a pump head valley which is closely related to complex flow structures in the flow passage. Another important factor affecting pump performance is impeller-stator interaction. Pressure distribution between the impeller and the stator serves as an indicator of the effect of stator vane number. Large stator vane number causes uniform pressure distributions near stator inlet. Based upon unsteady simulation results, the increase of flow rate contributes to the stimulation of low-frequency components between the impeller and the stator. Meanwhile, the dominance of shaft frequency prevails at large flow rates.

ACKNOWLEDGEMENTS

This study is financially supported by the College Industrialization Project of Jiangsu Province (Grant No. JHB2011-37). The authors would extend their gratitude to the project funded by the Priority Academic Program Development of Jiangsu Higher Education Institutions.

REFERENCES

1. Furukawa and T. Shigenmitsu, "Performance Test and Flow Measurement of Contra Rotating Axial Flow Pump," Journal of Thermal Science, Vol. 16, No. 1, 2007, pp. 7–13.

2. F.J. Shi and H.X. Chen, "Applicability of Turbulent Models in Simulation of Internal Flow within Axial Flow Pump," Journal of Shanghai University (Natural Science Edition), Vol. 12, No. 3, 2006, pp. 273–276.

3. F.J. Wang, L. Zhang and Z.M. Zhang, "Analysis on Pressure Fluctuation of Unsteady Flow in Axial-flow Pump," Journal of Hydraulic Engineering, Vol. 38, No. 8, 2007, pp. 1003–1009.

4. G.L. Wen, "Verifying Performance of Axial-flow Pump Impeller with Low NPSHr by Using CFD," Engineering Computations, Vol. 28, No. 5, 2010, pp. 557–577.

5. M. Vahdati, A.I. Satma and M. Imregun, "An Integrated Nonlinear Approach for Turbomachinery Forced Response Prediction. Part II: Case Studies. Journal of Fluid and Structures, Vol. 14, 2000, pp. 103–125.

6. Z.D. Qian and Y. Wang, "Numerical Simulation of Water Flow in an Axial Flow Pump with Adjustable Guide Vanes," Journal of Mechanical Science and Technology, Vol. 24, No. 4, 2010, pp. 971–976.

Power and Energy – Kong (Ed.)
© 2015 Taylor & Francis Group, London, ISBN 978-1-138-02782-4

Stability analysis of wind farm connected power grid

Y.D. Zhang
State Grid Sichuan Electric Power Research Institute, Chengdu, China

H.C. Xu, T.Y. Wang & L.Y. Li
School of Electrical Engineering, Southwest Jiaotong University, Chengdu, China

ABSTRACT: Wind energy is renewable and green energy, and developing wind power is demand of the world energy development strategy. Due to the randomness of wind power, a large-scale wind energy injection means a great challenge for security and stability of the power grid. Based on PSASP, this paper first established an electromechanical transient model of direct driven permanent magnetic wind farm. Then, static and transient security for wind farm connected Sichuan power grid is analyzed. Also, the impacts of the capacity of the planned wind farm on the system bus voltage level are analyzed. Finally, the impacts that whether the wind farm has the potential of low voltage ride through has on system stability are analyzed.

1 INTRODUCTION

As the world energy crisis broke out, the countries all over the world have to seek technology breakthrough of new energy. As wind energy is a kind of renewable, friendly, green energy, to develop the wind power generation is the global energy development strategy. China is rich in wind energy resources. The 10 GW class wind power base was established in Jiuquan, Gansu in 2011. Also, Sichuan province will build GW class wind farm. According to the national 12th five-year relevant requirements in the planning of wind power, by 2015, cumulative parallel operation wind capacity will reach 100 GW and total annual capacity will be more than 190 TWh. And total installed capacity is required to be 180 GW in the long-term 2020 planning. Wind power of our country has entered the large-scale development stage, and will further extend along the route of large-scale development, centralized construction, and long-distance transmission.

Wind power is environmentally, friendly energy source. But after wind power integration, the natural wind randomness and uncontrollability will lead to volatility of wind turbine generator's output. Meanwhile, the operating characteristics and control methods differences between wind turbine generator and general generator may affect the safe and stable operation of the system. Once the wind permeability increased, these effects will be more obvious. After interconnection, a large-scale wind farm will bring a certain impact to the power flow, reactive voltage level, static security analysis, transient stability of the power system. So, combining the national planning requirements for wind power development in Sichuan province, according to the actual situation of power grid in Sichuan, carry out stability study on wind power connecting to Sichuan power grid has theoretical and realistic significance.

Before analyzing the stability and security of the wind farm connected power grid, it's necessary to establish an appropriate model of the wind turbine. The Power System Analysis Software Package (PSASP), popular in China, is convenient for analyzing the power grid security and safety. What's important, it has a plat called UD for users defining their own models. So it becomes feasible to develop a reasonable electro—mechanical transient model of wind turbine using PSASP UD plat.

The paper is organized as follows. An electro—mechanical transient model of direct-driven permanent magnet synchronous generator (D-PMSG) is established on PSASP UD. Then it takes Sichuan power grid as the research object. After the wind farm connecting to power grid, static security of power system is analyzed. Planning conservative access capacity of wind farm is given from a conservative point of voltage level. Transient stability is also analyzed. The simulation is taken to analyze the influence of wind power low voltage through ability on the stability of the system.

2 ELECTROMECHANICAL TRANSIENT MODEL OF PMSG

An electromechanical transient model of PMSG is established on PSASP UD as shown in Figure 1.

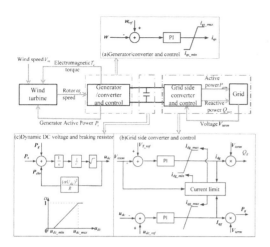

Figure 1. Electromechanical transient model of D-PMSG.

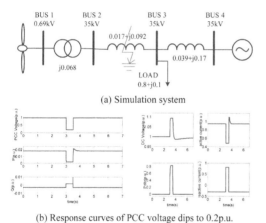

(a) Simulation system

(b) Response curves of PCC voltage dips to 0.2p.u.

Figure 2. Response curves of UD model demonstration.

Both the generator with its converter control system and the grid side converter with control system are greatly simplified. The model also consists of braking resister and considers the dynamic process of DC voltage, which can reveal the operation characteristic of Low Voltage Ride Through (LVRT) process. Finally, the order of the proposed electromechanical transient model is greatly decreased into eight order.

The UD model of PMSG is implemented in the single machine infinite bus system for simulation in PSASP. Considering short circuit fault, the PCC (point of common connection) voltage, active power and reactive power at PCC are shown to verify the correctness of the model. Simulation results in Figure 2 demonstrate that the proposed model in

PSASP UD can reflect the electromechanical transient characteristics of the wind power system.

3 STATIC SAFETY ANALYSIS OF WIND FARM CONNECTED GRID

3.1 Power flow analysis

Wind power is concentrated in Sichuan power grid, mainly distributed in southwest Sichuan province. As shown in Figure 3, buses of YC—YL—HD—HL—DJW—PG—ZJB—YC constitute a ring grid, where the wind power is concentrated, like DC wind farm (90 MW), LM wind farm (49.5 MW), and so on. The total capacity of all these ten wind farms is 530.5 MW. So this area is a good research object to analyze the stability of large-scale wind power connected power grid.

The results of power flow calculation show that the wind power transmission lines can meet the requirements for wind farm to send out power. There is no line power flow in the network exceeds the limit, and all the bus voltages are in the reasonable range of deviation.

3.2 N-1 static safety analysis

The N-1 static security analysis is carried out on the wind farm connected Sichuan power grid. This paper only lists the analysis results of several lines which are closer to wind farm, as shown in Table 1.

It can be concluded from Table 1, that voltages of all nodes are in the normal operation. No relevant line appears overload phenomenon.

3.3 Influence of wind power capacity to bus voltage

We should pay attention to the impact of wind power change on the power grid voltages of all levels, for ensuring the security of the power grid and the wind turbines. As the requirements for the

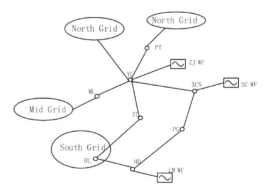

Figure 3. Wind farm accessing to power grid.

214

Table 1. Results of N-1 static safety analysis.

Fault line	Observation line	Result
ZJB-PG 220 kV	ZJB-PG line	No overload
ZJB-PG 220 kV	HD-HL line	No overload
ZJB-PG 220 kV	PG-DJW line	No overload
HL-HD 220 kV	HL-HD line	No overload
HL-HD 220 kV	PG-DJW line	No overload
HL-HD 220 kV	YC-ZJB line	No overload
YC-ZJB 220 kV	YC-ZJB line	No overload
YC-ZJB 220 kV	PG-DJW line	No overload
YC-ZJB 220 kV	HL-HD line	No overload

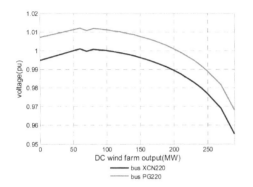

Figure 4. Voltage changes with changed output of wind farm.

technology standards of wind power connecting to grid, the wind farm should be able to control the voltage deviation of the PCC point in a range from −3% to +7% of the rated voltage. On the premise of unchanged other wind power capacity, gradually increase the capacity of DC wind power, and determine the response of the bus voltages by using conventional flow algorithm. Figure 4 shows the influence to each bus voltage when DC wind farm are in different wind power output.

As shown in Figure 4, when the capacity of DC wind farm rises to 240 MW, the voltage of XCN bus reduces to the limit value. Using the same analytical method, we gradually increase the capacity of LM wind farm respectively. When the capacity of LM wind farm rises to 230 MW, the voltage of XCN bus reduces to the limit value.

4 TRANSIENT STABILITY ANALYSIS OF POWER GRID WITH WIND FARM

4.1 Transient stability analysis after three phase fault

Take a main 220 kv YL-IIL line which occurs three phase short circuit fault as an example to analyze

transient stability of Sichuan grid which integrates with wind farm. HL side occurs three phase fault at 1 s, and double ends jump at 1.12 s. Figure 5 shows curves of generator relative power angles and voltages within the gird.

It can be observed from Figure 5(1), relative power angles between the main generators have a trend of damped oscillation, and eventually return to a stable state with a first swing angle much smaller than 180. As to Figure 5(2), voltages of system main buses restore stability after a short-term oscillations in the stability criterion. It can be observed from the above analysis, when AC line occurs three-phase permanent short-circuit fault, it can restore a steady state after a short-time adjustment.

The results of transient stability analysis of main 220 kv and 500 kv lines when faults occur have been shown in Table 2.

4.2 Transient stability analysis when wind farm are out of service due to fault

In addition to the normal shutdown, wind farm could be cut by the action of system protection due to short circuit fault of bus line. The worst shutdown situation is cutting wind farm when its output is large power or even full power.

XCN-YC line is the outgoing line of AY wind farm. The simulation example is set as XCN line occurs three phase permanent fault at 220 kv side of XCN substation. Curves of generator relative power angles and voltages when wind farm has been cut are simulated. The three phase fault occurs at 1s, and double ends jump at 1.12s. Figure 6 shows curves of relative power angles and voltages within the gird.

It can be observed from Figure 6(1), relative power angles between the main generators have a trend of damped oscillation, and eventually return to a stable state with a first swing angle much smaller than 180. From Figure 6(2), voltages of system main buses restore stability after a short-term oscillations in the stability criterion. It can be observed from the above analysis, when AC line occurs three-phase permanent short-circuit faul, it can restore a steady state after a short-time adjustment.

The results of transient stability analysis of other lines occurs different faults has been shown in Table 3.

4.3 Effects caused by wind farm with LVRT on improving system stability

HL–YL line is in first single-phase transient fault accident. When the wind farm doesn't have the low voltage ride through capability, because of bus voltage of the wind farm drop down under 0.9 p.u.,

215

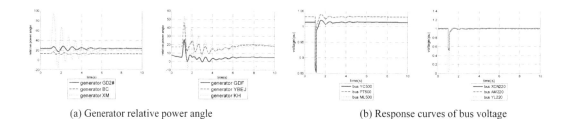

(a) Generator relative power angle   (b) Response curves of bus voltage

Figure 5.   YL-HL three phase short circuit fault.

Table 2.   Results of transient stability analysis.

Voltage level	Fault line	Fault type	Stability
220 kV	PG-ZJB	Three phase short circuit	Stable
220 kV	YC-XCN	Three phase short circuit	Stable
220 kV	YC-YL	Three phase short circuit	Stable
220 kV	YL-HL	Three phase short circuit	Stable
220 kV	ML-YC	Three phase short circuit	Stable
500 kV	YC-GD	Three phase short circuit	Stable
500 kV	YC-PT	Three phase short circuit	Stable
500 kV	GL-PT	Three phase short circuit	Stable

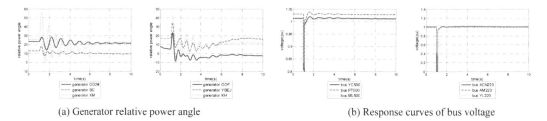

(a) Generator relative power angle   (b) Response curves of bus voltage

Figure 6.   XCN-YC three phase short circuit fault.

Table 3.   Transient stability analysis when wind farm are out of service due to fault.

Line	Fault side	Fault type	State of power system	State of wind farm
XCN-YC	XCN	Three phase short circuit	Stable state	Out of service
220 kv line	XCN	Single-phase instantaneous fault	Stable state	Out of service
	XCN	Single-phase sustained fault	Stable state	Out of service
	YC	Three phase short circuit	Stable state	Out of service
	YC	Single-phase instantaneous fault	Stable state	Out of service
	YC	Single-phase sustained fault	Stable state	Out of service
HD-NN	HD	Three phase short circuit	Stable state	Out of service
220 kv line	HD	Single-phase instantaneous fault	Stable state	Out of service
	HD	Single-phase sustained fault	Stable state	Out of service
	NN	Three phase short circuit	Stable state	Out of service
	NN	Single-phase instantaneous fault	Stable state	Out of service
	NN	Single-phase sustained fault	Stable state	Out of service

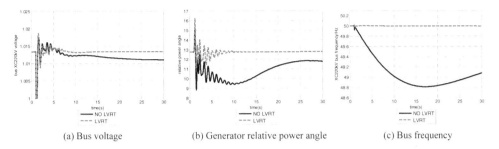

(a) Bus voltage   (b) Generator relative power angle   (c) Bus frequency

Figure 7. Influence of LVRT.

all the wind farm turbine is cut off, out of run. On the contrary, all the wind farm turbine will not be cut off when wind farm has the low voltage ride through capability. Figure 7 (1), (2), (3) respectively shows the effects on YC Station bus voltage, generator relative power angle and frequency caused whether the wind farm has the low voltage ride through capability.

When the wind farm doesn't have the low voltage ride through capability, if three-phase short circuit fault happens in the main lines of the system, effects would be caused on system bus voltage and frequency. System bus voltage will be a little lower after the fault. Generator relative power angle needs a long time adjustment to a new steady state after the fault. System frequency needs a long time adjustment, and will be a little lower after the fault. When wind farm turbine has the low voltage ride through capability, the simulation figures show that bus voltage adjustment process is short, and it can reach the level before the fault.

Generator relative power angle adjustment process is not large, it recovers rapidly to the steady state before the fault. System frequency does not have large fluctuation. Therefore, when wind farm has the low voltage ride through capability, it can help recover system bus voltage and frequency level, and improve system stability. It's advised that installing the wind farm turbine with the low voltage ride through capability.

5 CONCLUSION

As clean energy, wind power has wide prospect of application. In this paper, the stability of Sichuan power grid with the planned wind power integration is analyzed, mainly from the influence that wind power integration has on power flow, static security analysis, transient stability, and so on. The conclusions are as follows: (1) the transient stability of power grid with the planned wind power integration is more critical than static security. (2) With the increase of wind power, node voltage slightly increases and then decreases. (3) When AC line three-phase permanent short circuit occurs in Sichuan power grid with the wind power integration, the voltage curve returns to a stable state after short-term oscillations, and the system has no instability phenomenon. (4) Wind farm with the capability of low voltage ride through, can help the system recover the bus voltage and frequency level, and improve the system stability.

REFERENCES

1. S. Neris, N.A. Vovos, and G.B. Ginnakaopoulos, "A Variable Speed Wind Energy Conversion Scheme for Connection to Weak AC System," IEEE Transactions on Energy Conversion, vol. 14(l), pp. 122–127, 1999.
2. Zhe, J.M. Guerrero, and F. Blaabjerg, "A Review of the State of the Art of Power Electronics for Wind Turbines," IEEE Trans on Power Electronics, vol. 24(8), pp. 1859–1875, 2009.
3. N. Kosterev, C.W. Taylor, and W.A. Mittelstadt, "Model validation for the August 10, 1996 WSCC system outage," IEEE Transactions on Power Systems, vol. 14(3), pp. 967–979, 1999.
4. H.J. Pan, and X.F. Xu, "Research Overviewing of Influence on Stability Owing to Wind Power Integration," Journal of Shenyang Institute of Engineer (Natural Science), vol. 1, pp. 54–57, 2013.
5. J.K. Kaldellis, "The wind potential impact on the maximum wind energy penetration in autonomous electrical grids," Renewable Energy, vol. 33, pp. 1665–1677, 2008.
6. R.Y. Luo, Y. Lin, and Y. Qian, "The Development and Prospects of World Wind Power Industry," Renewable Energy Resources, vol. 2, pp. 14–17, 2010.
7. T.Y. Wang, L.Y. Li, and X.R. Wang, "An Electromechanical Transient Model of Direct-Driven Permanent Magnet Synchronous Generator Based Wind Power System," Proceedings of the 2014 International Conference on Power System Technology, 2014.

8. T. Zhu, "Review of Some Problems Related to Wind Farm Integration," Southern Power System Technology, vol. 3, pp. 58–63, 2009.

9. X. Fu, H.W. Li, and B.H. Li, "Review on Influences of Large-scale Wind Farms Power System and Countermeasures," Shaanxi Electric Power, vol. 1, pp. 53–57, 2010.

10. Y.N. Chi, "Studies on the Stabiliy Issues about Large Scale Wind Farm Grid Integration," China Electric Power Research Institute, 2005.

11. Z.J. Cao, "Study on Control Strategy of DFIG wind turbine for the Participation of System Primary Frequency Regulation," Southwest Jiaotong University, 2012.

Power and Energy – Kong (Ed.)
© 2015 Taylor & Francis Group, London, ISBN 978-1-138-02782-4

Emission characteristics of particulate matters from coal-fired power plant equipped with WFGD

Yi Zhang, Zhe Liu & Zhanjun Xie
San-He Power Generation Limited Liability Company, Langfang, Hebei, China

Wenbo Jia
School of Environmental Science and Engineering, North China Electric Power University, Baoding, Hebei, China

ABSTRACT: Wet Flue Gas Desulfurization system (WFGD) has an obvious trapping effect on flue gas particulate matters. In this paper, particulate sampling was conducted on a coal-fired power boiler equipped with WFGD. The size distributions, element composition and microstructure of particles before and after WFGD were tested. The results show that WFGD system had a significant removal effect on particles greater than 2 μm, but the effect was not obvious for particles that size below 2 μm. Compared with the entrance of WFGD, the element content of S and Mg in particles was significantly increased after WFGD, and element F was found in outlet particles. Particles from entrance are uniform spherical particles with smooth surface, while in outlet they are irregularly shaped particles formed with fine particles agglomerated with rough surface. The dust removal efficiency of WFGD system is 62.22% and the flue gas purified by WFGD meets the emission standard.

1 INTRODUCTION

Particulate Matters (PM) pollution has caused an extensive concern of the society from all walks of life. Coal-fired power plant is one of the main sources of PM in the atmosphere (Hao & Wang & Shen 2007). At present, the Wet Flue Gas Desulfurization (WFGD) system is generally installed after the dust removal device in the large coal-fired power plants. The wet desulfurization device use the limestone slurry as the absorbent, slurry through spray device countercurrent contact with flue gas to removal SO_2 in the desulfurization tower. On the one hand, this system has a high SO_2 removal efficiency and the advantages of a strong adaptability; on the other hand, the lime stone slurry of desulfurization tower spray layer has a washing effect on the flue gas. It can capture the fine particles escaped from the electrostatic precipitators, which increasing the possibility of flue gas emission meeting the standard (Xu & Guo 2002, He & Yao 2001, Xu & Chen & Qi 2001).

A series of study about the effects of WFGD system for particles removal had been launched by the domestic and foreign scholar. Wang Hui et al. (Wang & Song 2008) found that the limestone/gypsum wet desulfurization system can effectively remove SO_2 and the larger particles, but has a lower efficiency to the smaller particulate matters, and the removal efficiency is decreased significantly with the particle size decrease. The study also found that S and Ca element in the fine particle increased significantly after WFGD system. There is not only the coal fly ash in the outlet of the desulfurization tower, but also contains about 7.9% of the gypsum particles and 47.5% of the limestone particles. Meij et al. (Meij & Winkel 2004) found that fine particulate matters in the outlet of the limestone-gypsum desulfurization system mainly includes 40% of coal fly ash, 10% of desulfurization gypsum components and 50% of solid particles formed by droplet evaporation. The study showed that the wet flue gas desulfurization system has a significant capture effect on the larger particles, but a low effect on the fine particulate matters such as the $PM_{2.5}$. How to promote the dust removal efficiency of WFGD and prevent the solid particles of desulfurization slurry carried by the flue gas remain to be further researched.

2 THE MECHANISM FOR COLLECTING DUST BY WFGD SYSTEM

The three available mechanisms for collecting dust by wet flue gas desulfurization system may be classified as inertial impaction, interception and Brownian diffusion. In desulfurization spray tower, flue gas contacts with desulfurization slurry, then the purified flue gas expelled from the top.

2.1 Inertia collision

When the dust-containing flue gas encounter the droplets during the process of upward motion in the spray tower, the flue gas is captured by the droplets due to its inertia. The quality of the dust particle plays a decisive role for trapping by inertia, so that the dust particles were assumed only quality without the size during our analysis. Among various kinds of mechanism of the dust particles capture, the inertial impaction effect is the most common and important, which show a better effect on the capture of dust particles to the diameter exceed 1 µm. The inertial collision can be described by the Stokes number as follow: (Hao & Chen 2005)

$$S_t = \frac{Cd_p^2 \rho_p u}{9\mu_g D_c} \tag{1}$$

where C is the Cunningham correction factor, d_p is the diameter of dust; ρ_p is the dust density; u is the relative speed of the gas and the liquid, μ_g is the gas viscosity coefficient; D_c is the diameter of the liquid drop.

2.2 The interception

The dust particles can be captured by the droplets when it moves along the gas flow line directly to the droplets within the distance of D/2. The size of dust particles plays an important role on the intercept, regardless of the dust mass and gas velocity. The smaller particles shown a stronger intercept effect. The intercept effect can be described by the direct interception ratio as follow: (Hao & Chen 2005)

$$R = d_p/D_c \tag{2}$$

In which, d_p is the diameter of dust and D_c is the diameter of the liquid drop.

2.3 Brownian diffusion

Due to the effect of Brownian diffusion, the fine dust particles were captured when they moved around the droplets. The smaller dust particles have a more intense Brownian diffusion. Brownian diffusion should be taken into account when the diameter of dust particles is smaller than 2 µm. (Yue & Yuan & Hong 2006).

Typically, there is not only one of the three principles but all of them have effect on the dust particles when the flue gas countercurrent contact with desulfurization slurry. In addition, it is also conducive to trap dust particles separated that the dust humidity is increased and the effect of cohesion is raised among the dust particles.

3 EXPERIMENTAL

3.1 Sampling methods and equipments

The particulate matters in the flue gas was sampled at the selected points depending on the isokinetic sampling theory, which pumped a certain amount of flue gas that contained dust using sampling tube by the way of isokinetic sampling. The dust was captured with trapping devices and the concentration of smoke in the flue gas was analyzed according to the quality of the trapped soot and the volume of the smoke. The sampling equipment was Qingdao Laoying 3012H automatic smoke/gas tester.

3.2 The selection of sampling points

Particulate sampling points were selected at the rectangular flue before and after Wet Flue Gas Desulfurization system (WFGD), which is used to test the size distributions, element composition and microstructure of particulate matters. The trapping effect of WFGD system for particulate matters was researched by the test result. During the test, the boiler stay at 90% load. Sampling positions are observed in Figure 1.

3.3 The field tests

a. The numbered filter cartridges were dried and weighted before the test in the laboratory. After the test, the dust concentrations were calculated by differential method.
b. The field tests: First, we should choose the suitable sampling mouth by predicting the flue gas velocity, then start the isokinetic sampling after putting the filter cartridges into the sampling tube. After getting a certain amount of flue gas, the filter cartridges should be sealed and reserved for subsequent experiments analysis.
c. The operation data of boiler was recorded during the test.

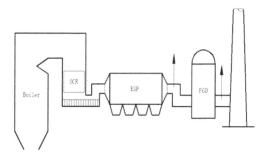

Figure 1. Sampling positions of flue gas.

4 THE RESULTS AND ANALYSIS

4.1 The operation parameters of WFGD

The test was conducted on a boiler under a 90% load of a 350MW power plant at Hebei province China. Flue gas pollution controlling devices include Selective Catalytic Reduction (SCR), electrostatic precipitation (ESP) and Wet Flue Gas Desulfurization (WFGD). The multilayer sprinkling was taken in the desulfurization tower, and the flue gas contact counter-currently with desulfurization slurry. In the above process, the washing effect of the slurry can effectively remove particulate matter in the flue gas. The main design parameters of the absorption tower within WFGD system are listed as Table 1.

4.2 The size distribution of particles

Size distributions of particulate matters at entrance and outlet of WFGD were tested by the Malvern Mastersizer2000 particle size analyzer. The test results show that the peak of particle size before WFGD system is 2.139 μm, while the outlet is 1.753 μm. Particles with the size more than 2.5 μm are sparse. According to the results of particle size analysis, a comparison of particle size distribution was made between entrance and outlet of WFGD, shown as Figure 2. It can be observed from the Figure 2, large particles take a priority at the inlet of WFGD with diameters greater than 2 μm, and fine particles have a high content at the outlet of WFGD with diameters 1~2 μm, where bulky particles were effectively removed. After ESP, the flue gas was washed by desulfurization slurry and bulky particles were basically removed. This phenomenon shows that WFGD system has a significant removal effect on particles greater than 2 μm,

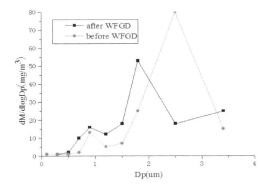

Figure 2. The particle size distribution of PM at entrance and outlet of WFGD.

but the effect on particles little than 2 μm is not obvious.

4.3 Element compositions of particulate matters

Elemental compositions of particulate matters were tested by X Ray Diffraction (XRD). The element compositions of particulate matters at the entrance and outlet of WFGD are listed as Table 2. As it can be observed from the table that the differences of element types of particulate matters before and after WFGD system are not obvious. The element of F was found in outlet particles, which may be carried by the gas from desulfurization slurry. Each element content is not very different, but the content of S at outlet increased significantly than inlet. It is generally believed that the flue gas at outlet carry the solid particles formed by desulfurization droplets, which increased the content of S. In addition, the content of Mg also increase obviously. It is generally believed that the desulfurization slurry containing Mg^{2+}, which attached on the fine particulate matters during the process of desulfurization slurry countercurrent contacting with flue gas.

4.4 The microstructure of particulate matters

Particulate matters before and after WFGD were analyzed by Scanning Electron Microscopy (SEM). Figure 3a stands for SEM scanning images of particulate matters before WFGD. Particulate matters seeing from the Figure 3a are uniform spherical particles with smooth surface, which are independent individuals and relatively simple. In addition to spherical particles, there are also some fine particles attached on the spherical particles. X-ray Energy Dispersive Spectroscopy (EDS) analysis shows that the main component of these fine particles is Si, Al, and so on. And the element content of S in particulate matters at outlet is higher (Figs. 4a and 4b).

Table 1. The main design parameters of WFGD absorber system.

Projects	Units	Parameters
Inlet flue gas volume	Nm³/h	1251943
Inlet flue gas temperature	°C	≤160
Sulfur content of coal	%	0.7
SO₂ concentration of inlet flue gas	mg/Nm³	1651
Inlet dust content	mg/Nm³	≤55
Desulfurization efficiency	%	≥95
Ca/S	Mol/mol	1.03
Liquid-gas ratio	L/m³	16.47
Outlet gas temperature	°C	≥81
Outlet dust content	mg/Nm³	≤28
SO₂ concentration of outlet flue gas	mg/Nm³	≤83

Table 2. Elemental compositions of particulate matters before and after WFGD.

Element weight percentage %	Before WFGD	After WFGD
C	7.72	8.47
O	38.95	36.79
Na	0.81	1.32
Mg	0.66	3.68
Al	15.7	12.23
Si	19.75	12.14
Ca	5.52	3.12
Fe	5.32	2.62
S	3.72	14.64
K	0.88	0.95
Ti	0.98	0.81
F	–	3.21

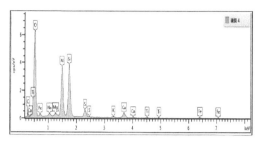

Figure 4a. The energy spectrum of particulate matters before WFGD.

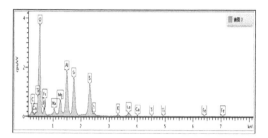

Figure 4b. The energy spectrum of particulate matters after WFGD.

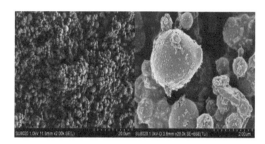

Figure 3a. SEM scanning images of particles before WFGD.

Figure 3b. SEM scanning images of particles after WFGD.

These particulate matters may mainly contain aluminum silicate and pyrite. Figure 3b stands for SEM scanning images of particulate matters at the outlet of WFGD. Particulate matters in Figure 3b are spherical particles without rule. They are irregularly shaped particles formed with fine particles agglomerate and flocculent with roughness surface. In Figure 3b, the mesh-liked particles also were observed. It is generally believed that should be nuclear particles from desulfurization droplet condensation.

4.5 Dust removal efficiency

The volume of flue gas and the quality of particulate matters were used to calculate the mass concentration of particulate matters in flue gas at inlet and outlet of WFGD. Mass concentration of particulate matters in flue gas before WFGD was 45 mg/m^3, while outlet of WFGD was 17 mg/m^3. The dust removal efficiency of WFGD system was 62.22%. Particles after WFGD system in flue gas meet the national standard that is 20 mg/m^3, and the dust removal efficiency of WFGD system meets the requirement of design.

5 CONCLUSIONS

Size distributions of particulate matters at the entrance and outlet of WFGD system were tested by the Malvern Mastersizer2000 and the analysis of element composition and morphology was completed using SEM and XRD. The study shows that the WFGD system has a good effect on the removal of particles with the diameters greater than 2 μm, but not effective for the particles under 2 μm. Elements of S and Mg in particles increased greatly and element of F appeared after WFGD. It is generally believed that the flue gas after WFGD carries the fine particles formed by desulfurization droplets that contain elements of S, Mg and F, and then the content of these elements increased obviously after WFGD. The results of SEM show that particles before WFGD are uniform spherical particles with smooth surface which mainly consisted of aluminum silicate and pyrite, while the particles after WFGD are irregularly shaped particles formed by fine particles with roughness surface. These fine particles should be nuclear particles formed by desulfurization droplets. The study revealed that the dust removal efficiency of WFGD is 62.22%. After the ESP and WFGD, the concentration of particles decreases apparently, basically meeting the standard of discharge.

REFERENCES

1. Hao J., Wang L., Shen M., et al. 2007. Air quality impacts of power plant emissions in Beijing. Environmental Pollution 147:401–408.
2. He S.H., Xiang G.M., Yao Q., et al. 2001. Comparison of a few different $CaCO_3/CaO-CaSO_4$ wet FGD towers. Electric Power Environmental Protection 17(3):5–8.
3. Yue H.L., Yuan Y.T., Hong Z. 2006. Analysis of dust removal mechanism in sprayer of lime stone/gypsum wet flue gas desulfurization system. Electric Power Environmental Protection 22(6):13–15.
4. Meij R., Winkel H. 2004. The emissions and environmental impact of PM_{10} and trace elements from a modern coal-fired power plant equipped with ESP and wet FGD. Fuel Processing Technology 85(6–7): 641–656.
5. Hao Q., Chen J. 2005. The factors affecting dust removing characteristics in lime stone/gypsum wet FGD system. Electric power environmental protection 21(2):33–34.
6. Wieprecht W., Lutz M., John A., et al. 2004. PM_{10} aerosol mass and composition in and around Berlin (Germany). Journal of Aerosol Science 35:453–464.
7. Wang H., Song Q., Yao Q., et al. 2008. Experimental study on removal effect of wet flue gas desulfurization system on fine particles from a coal-fired power plant. Proceedings of the CSEE 28(5):1–7.
8. Xu X.C., Chen C.H., Qi H.Y., et al. 2003. A Study on Coal-smoke Air Pollution Control and Its Technology. Journal of University of Science and Technology of Suzhou (Engineering and Technology) 16(1): 8–15.
9. Xu J., Guo J., Guo B., et al. 2002. Integrated wet dust-removal and desulfurization equipment with high efficiency. Urban Environment and Urban Ecology 15(4):51–52.

Power and Energy – Kong (Ed.)
© 2015 Taylor & Francis Group, London, ISBN 978-1-138-02782-4

Forecast output power of photovoltaic system based on grey system and BP neural network

Weiping Zhu
Jiangsu Electric Power Company Research Institute, Jiangsu, China

Zhicheng Wang
State Grid Jiangsu Electric Power Company, Jiangsu, China

Xiaodong Yuan
Jiangsu Electric Power Company Research Institute, Jiangsu, China

ABSTRACT: This paper presents a output power prediction model of PV system, based on grey system and Back-Propagation (BP) artificial neural network; this model is used to solve the impact on power system, brought by the increasing of PV system penetration. Forecast total daily output power of PV system by improved grey system; use BP artificial neural network based on similar day algorithm to acquire one hour rated output power of PV system; combine the total daily output power and one hour rated output power, and obtain the correction values of one hour rated output power. By comparing the predicted result, verify the accuracy of the algorithm, and decrease the forecast deviation when using BP artificial neural network lonely.

1 INTRODUCTION

In recent years, the application of distributed photovoltaic power generation technology not only can improve the reliability of power system and prevent blackouts, but also reduces the impact of natural disasters on the electric power system and has great significances for the economy and safety of the country [1].

However, PV has characteristics of fluctuations, intermittent and periodicity. When the grid-connected PV permeability increases, it has an important effect on grid quality, such as voltage fluctuation and flicker, harmonic pollution and reactive power imbalance [2–3]. Thus, accurate prediction of PV output is very important and the electricity sector can take advantage of prediction data of power to dispatch and plan the power reasonably.

Existing PV output prediction methods are divided into direct and indirect [4], direct refers to the use of history data of the PV stations power output and weather information for the prediction of power output, and indirect means to predict effects of each individual impact factors, these factors are then used as inputs by PV power models to obtain the method of predicting the outcome.

In the direct prediction method, a single BP forecast is predicted only by similar historical day data and ignores the closed daily variation of

output power and grey prediction only uses closed the forecast data and ignore similar historical variation of output power.

Based on the above reasons, this paper starts from improving PV output power prediction accuracy, then uses similar historical data through BP neural network to predict one hour rated output power, then uses the historical data through grey system to predict total daily output power, finally adjusts one hour rated output power according to the total daily output power. Comparing the experimental results with the actual PV output, this paper verifies the accuracy of the method that can improve the prediction.

2 BP NEURAL NETWORK AND GREY SYSTEM

2.1 BP neural network

Since the BP neural network has the following advantages: first, nonlinear mapping ability. Second, good at learning useful knowledge from the input and output data. Third, it is easy to use the software and hardware implementation. For the above reasons, BP neural network can accurately predict the trends of PV power output per hour in different weather conditions. [5–8] The structure is shown in Figure 1.

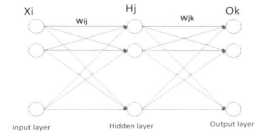

Figure 1. Three layers structure of BP neural network.

Wherein the input layer $X = \{x_1, \ldots x_i, \ldots x_n\}$, Hidden layer $H = \{h_1, \ldots h_i, \ldots h_m\}$ Output layer $O = \{o_1, \ldots o_k, \ldots o_l\}$.

And w_{ij} is the weights from the input layer to the hidden layer, w_{jk} is the weights from the hidden layer to the output layer.

When the network output is not equal to expectations, there is an error we named E. Mathematics is defined as Eq.(1)

$$E = \frac{1}{2}\sum_{l=1}^{l}(d_k - o_k)^2 \qquad (1)$$

According to the definition of the error, extended to the hidden layer.

$$E = \frac{1}{2}\sum_{l=1}^{k}\left(d_l - f\left(\sum_{j=0}^{m}w_{jl}h_j\right)\right)^2 \qquad (2)$$

And further to the output layer.

$$EE = \frac{1}{2}\sum_{l=1}^{l}\left(d_k - f\left(\sum_{j=0}^{m}w_{jk}f\left(\sum_{i=0}^{n}w_{ij}x_i\right)\right)\right)^2 \qquad (3)$$

where, Transfer function $f(x)$ is unipolar sigmoid function

$$f(x) = \frac{1}{1+e^{-x}} \qquad (4)$$

According to Eqs. (2) and (3), network output error is the function of w_{jk} and w_{ij}, Thus adjust the weights can change the error E.

Weight adjustment equations are given below:

$$\begin{cases} \Delta w_{jl} = \mu(d_l - o_l)o_l(1-o_l)h_j \\ \Delta w_{ij} = \mu\left\{\sum_{l=1}^{k}\left[(d_l - o_l)o_l(1-o_l)w_{jl}\right]\right\}h_j(1-h_j)x_i \end{cases} \qquad (5)$$

where, constant $\mu \in (0,1)$, it represents a proportionality constant, reaction training speed.

2.2 Grey System

Grey system theory is a method to study small amounts of data and problem of uncertainty. Grey system theory treats uncertain systems that has less information as the research object, the main method is that to realize the system runtime behavior, the correct description of the evolution and effective monitoring forecast by developing part of the known information and Extracting valuable information [9–13].

Grey system has three main basic principles:

a. principle of non-unique answer: the information is not entirely and uncertain, the answer is non-unique.
b. minimum information principle: the characteristic of grey system theory is full development and utilization of existing "minimal information."
c. new message priority principle: the learning effects of new information is greater than the old information.

Based on the above principles, especially the third principle, model DGM(1,1) (discrete grey model) with metabolic to predict the total daily output of photovoltaic power. That is, enter the latest $x^{(0)}(n+1)$, to remove the oldest message $x^{(0)}(1)$ to predict the next set of data. Grey model change with time change.

For the model DGM (1,1):
Set up the original sequence:

$$X^{(0)} = \left(x^{(0)}(1), x^{(0)}(2), \ldots x^{(0)}(n)\right) \qquad (6)$$

Claimed that:

$$\propto^{(1)} x^{(0)}(k) + a_1 x^{(0)}(k) + a_2 z^{(0)}(k) = b \qquad (7)$$

Eq. (7) is named model DGM(1,1).
Among them:

$$\propto^{(1)} x^{(0)}(k) = x^{(0)}(k) - x^{(0)}(k-1), k = 2, \ldots n; \qquad (8)$$

$$x^{(1)}(k) = \sum_{i=1}^{k}x^{(0)}(i), \ k = 1,2,\ldots n; \qquad (9)$$

Claimed that:

$$\frac{d^2 x^{(1)}}{dt^2} + a\frac{dx^{(1)}}{dt} = b \qquad (10)$$

Eq.(10) is named Whitenization equation of model DGM(1,1).

For the Matrix:

$$B = \begin{bmatrix} -x^{(0)}(2)\,1 \\ -x^{(0)}(3)\,1 \\ \vdots \quad \vdots \\ -x^{(0)}(n)\,1 \end{bmatrix},\ Y = \begin{bmatrix} \propto^{(1)} x^{(0)}(2) \\ \propto^{(1)} x^{(0)}(3) \\ \vdots \\ \propto^{(1)} x^{(0)}(n) \end{bmatrix} \quad (11)$$

The least squares estimation of parameters column $\vec{a} = [a\ b]^T$ is:

$$\vec{a} = (B^T B)^{-1} B^T Y \quad (12)$$

Solving differential equations:

$$x^{(1)}(k+1) = \left[\frac{b}{a^2} - \frac{x^0(1)}{a} \right] e^{-ak} + \frac{b}{a}(k+1)$$
$$+ \left[x^{(0)}(1) - \frac{b}{a} \right] \frac{1+a}{a} \quad (13)$$

Ultimately, the mathematical model of the original sequence is:

$$x^{(0)}(k+1) = x^{(1)}(k+1) - x^{(1)}(k) \quad (14)$$

Combine BP neural networks and grey system, adopt the method that is shown in Figure 2.

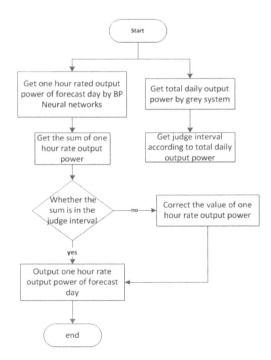

Figure 2.   The flowchart of prediction.

3   JUDGED INTERVAL

The selection of judged interval is very important. Because it is directly related to the power forecast accuracy. The judged interval can be determined by grey system model [14]. Build the Mathematical Model of the sample data by grey system modeling prediction method. According to the principle of new message priority, establish model DGM (1,1) with metabolic.

Select 50 groups of consecutive historical values about one hour rated output power from May 24, 2013 to July 12, 2013 (6:00–19:00) in an area of Hefei Anhui. Then, get the sum of one hour rated output power, the sum is named S. Establish model DGM (1,1) with metabolic, among them, n = 5, that is to build mathematical model by using five consecutive sets of data, and according to the mathematical model to predict the sixth total daily output power, the value is named P, and to establish judgment range.

Model DGM (1,1) prediction results are shown in Table 1 (n = 5):

Where the equation for error is:

$$Error = \left| \frac{S - P}{P} \right| * 100\% \quad (15)$$

According to Table 1, error interval distribution charts are shown in Figure 3:

From Figure 3, we can get the following information:

a. Basically the error is between 0% and 30%, accounting for 87% of the total forecast
b. Error range of 10% to 20% accounting 46.7% of the total forecast.

Based on this two information, judged interval is set between 70% and 130% of GM model predicted value. That is, the sum of using BP network predicted power value is between 70% and 130% of GM predicted value at the time, we believe that this prediction is correct. Beyond this range, we believe that neural network prediction is not accurate, needs to be corrected. Correction formula is:

$$P_i = \frac{\sum x_i}{G} * x_i \quad (16)$$

where: P_i represents the one hour rated power output value after correction.

X_i represents the one hour rated power output value after correction.

G represents the total daily output power that forecast by Grey System.

Table 1. Error of the predicted value and the actual total daily power value.

No.	S	P	Error	No.	S	P	Error
1	2595.8			26	2536.1	2703.3	6.19%
2	693.9			27	2848.5	2253.0	26.43%
3	718.8			28	3250.8	2964.3	9.67%
4	1008.8			29	3053.13	3343.5	8.69%
5	1682.6			30	3254.03	3546.5	8.25%
6	2505.1	2239.7	11.8%	31	3083.94	3458.1	10.82%
7	3294.2	3863.3	14.7%	32	3056.1	3241.8	5.73%
8	3307.1	3881.9	14.8%	33	1148.82	2172.2	47.11%
9	1291.8	4293.8	69.9%	34	1979.58	1521.7	30.09%
10	1843.3	1819.8	1.2%	35	1942.17	1529.9	26.94%
11	1387.5	1131.3	22.4%	36	2298.82	1972.0	16.57%
12	2710.4	2364.4	14.4%	37	3116.67	2801.1	11.26%
13	3152.2	2984.6	5.6%	38	2801.48	3306.5	15.28%
14	3340.5	3958.1	15.6%	39	1420.35	1947.4	27.07%
15	2677.6	3008.7	11.0%	40	1478.27	1855.6	20.34%
16	2378.0	2665.7	10.7%	41	1934.48	1614.3	19.83%
17	2677.7	2426.7	10.3%	42	3190.41	2744.4	16.25%
18	1772.9	2189.2	19.0%	43	2925.13	3435.3	14.85%
19	1141.7	1444.7	20.9%	44	3145.24	3697.7	14.94%
20	2913.8	2203.1	32.2%	45	2830.12	3372.	16.08%
21	2858.1	2151.9	32.8%	46	2732.83	2723.5	0.34%
22	2304.9	4513.0	48.9%	47	3109.41	2639.8	17.79%
23	3025.4	2878.8	5.0%	48	2409.96	2897.0	16.81%
24	2541.4	2564.3	0.8%	49	2334.23	2608.0	10.50%
25	2446.6	2522.0	2.9%	50	1527.76	2023.7	24.51%

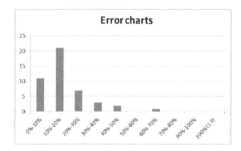

Figure 3. Prediction error chart.

4 BP NEURAL NETWORK

Three-layer feedforward network based on BP algorithm was shown in Figure 1. Next, we determine the neural networks' structure from three aspects.

1. Select the input and output variables.
2. Calculation process.
3. The selection about the number of hidden layer neurons.

4.1 Select the input and output variables

There are a lot of different factors which affects the output power of photovoltaic systems. For example, the installation position and angle of solar panel, weather, temperature, pressure, humidity and so on [13–16]. However, about the neural networks that short-term prediction based on historical data some unnatural factors do not consider.

Choose photovoltaic power stations that is located in Hefei Anhui (31°52′N,117°17′E). Time study is 6:00 ~ 19:00, Study period was 6:00 ~ 19:00, the rest of the time period the output power is almost zero, cannot be considered, the time interval is one hour, corresponding meteorological information from the weather forecast website.

Weather factors on the output power of the PV system is very obvious, season and temperature also have an impact on the photovoltaic power generation system. By analyzing of the data from Photovoltaic power plant in Hefei, we can get the same conclusions.

Therefore, the input variables and output variables are shown in the following Tables 2 and 3 [1].

4.2 Calculation process

Using similar day selection algorithm to train the neural network, specific process is as follows [17]:

Enter the maximum temperature, minimum temperature, date of prediction day.

Table 2. The variables of the input layer.

Input variables	Name of input variables
x_1~x_{14}	Similar day 6: 00–19: 00 per hour output power
x_{15}~x_{16}	Similar daily maximum temperature, daily minimum temperature
x_{17}~x_{18}	Prediction day daily maximum temperature, daily minimum temperature

Table 3. The variables of the output layer.

Output variables	Name of output variables
o_1~o_{14}	Prediction day 6: 00–19: 00 per hour output power

In the historical data, choose the data with same weather type and season type as prediction day, the formation of the training sample A.

According to temperature, calculate the Euclidean distance of training sample A, formula is as follows:

$$a_i = \left[\sum_{j=1}^{2} (y_i - x_{ij})^2 \right]^{\frac{1}{2}} (i = 1, 2, \dots n) \qquad (17)$$

where: y_1, y_2 is the maximum temperature, minimum temperature of prediction day. x_{i1}, x_{i2}, x_{i3} is the highest and lowest temperatures of the NO.i records in sample set A.

1. All data is normalized, T_{max} is the maximum of all the maximum temperature, T_{min} is the minimum of all the minimum temperature, the same with. P_{min}, P_{max} formula is as follows,

$$P_N = \frac{P - P_{min}}{P_{max} - P_{min}} \qquad (18)$$

$$T_N = \frac{T - T_{min}}{T_{max} - T_{min}} \qquad (19)$$

2. Normalized information as an input sample input to the input layer of the neural network, when the neural network train complete. Using it for forecasting the output power.

4.3 Number of hidden layer neurons

In this paper, the number of layers in hidden layer is 1, the number of neurons in input layer is 18, and the number of neurons in the output layer is

14. The number of neurons in the hidden layer neuron number is not only relate to the number of neurons in the input layer and output layer, but also relate to the complexity of the problem. Each application of the basic principles are needed for their own network structure, in order to avoid the phenomenon of over-fitting, ensure a sufficiently high network performance and generalization ability. The basic principles to determine the number of neurons of hidden layer is as low as possible. The following empirical formula about the number of neurons in the hidden layer is given is as follows [18].

$$J = \sqrt{I + L} + a \qquad (20)$$

Among them, I for the number of neurons in the input layer, L is the number of neurons in the output layer, a is a constant between 1 and 10. Therefore, by Eq. (20)

The number of neurons in hidden layer $J \in (7,26)$.

Between (7,26), select 10, 15, 20, and 25 as the number of neurons in the hidden layer neural network, respectively prediction 10 group of photovoltaic power. Four kinds of different number of hidden layer neurons the average error is $\delta_{10} < \delta_{20} < \delta_{15} < \delta_{25}$. Therefore, the number of neurons in hidden layer is 10.

5 EXAMPLE ANALYSIS

Training data and test data of prediction model are from photovoltaic plant in energy limited from December 2011 to August 2013 in the historical data. Select May 14, 2013 as a case analysis.

May 14, 2013 (sunny; maximum temperature: 33; minimum temperature: 21) adjacent the first five days of data in the Table 4:

a. Based on model DGM (1,1), to predict the total daily output power of May 14, 2013 by using the data of 5 days ago adjacent. According to eq. (14), G = $x^{(0)}(6)$ = 3738.64.
   Judged interval is (70%G, 130%G) = (2617.084, 4860.232).

b. According to BP neural networks, to predict one hour rated output. then, $\sum x_i = 2575.54 \notin (70\%G, 130\%G)$.

c. Error analysis, is shown in Table 5: found that the absolute value of the error-corrected is 473.02, far less than the absolute value of the uncorrected errors that is 982.96. And viewed from the graph in Figure 4, the predict data after corrected is closer to actual value than uncorrected one.

Table 4. PV output power from 2013/5/9 to 2013/5/14.

Date	5/9	5/10	5/11	5/12	5/13	5/14
6:00	0	0	0	7.45	19.87	20.48
7:00	0	0	0	77.82	98.52	95.36
8:00	0	45.5	0	137.2	197.93	200.13
9:00	130.07	120.96	280.31	224.32	289.04	295.34
10:00	150.43	166.9	259.58	326.52	256.84	359.2
11:00	236.99	256.39	250.45	406.16	341.67	409.21
12:00	243.33	285.12	255.12	416.31	382.34	467.54
13:00	256.11	297.69	310	314.98	399.65	501.23
14:00	158.55	237.31	219.77	373.61	302.2	490.69
15:00	116.61	179.57	236.12	113.43	319.65	324.6
16:00	63.53	55.08	236.87	200	229.65	161.5
17:00	35.28	55.08	0	131.85	125.29	158.63
18:00	0.23	0	52.62	37.6	38.74	66.75
19:00	0.23	0	7.45	0.37	0.53	7.84
Total	1391.36	1699.6	2108.29	2767.62	3001.92	3558.5

Table 5. Error table before and after correction.

	Actual value	Predict 1	Predict 2	Error 1	Error 2
6:00	20.48	10.59	15.37	9.89	5.11
7:00	95.36	53.43	77.56	41.93	17.8
8:00	200.13	94.83	137.65	105.3	62.48
9:00	295.34	216.88	314.82	78.46	19.48
10:00	359.2	243.26	353.11	115.94	6.09
11:00	409.21	369.84	536.86	39.37	127.65
12:00	467.54	324.59	471.17	142.95	3.63
13:00	501.23	380.88	552.88	120.35	51.65
14:00	490.69	365.43	530.46	125.26	39.77
15:00	324.6	243.22	353.06	81.38	28.46
16:00	161.5	149.77	217.41	11.73	55.91
17:00	158.63	93.73	136.06	64.9	22.57
18:00	66.75	27.31	39.64	39.44	27.11
19:00	7.84	1.78	2.51	6.06	5.33
Total	3558.5	2575.54	3738.57	982.96	473.02

Predict 1: Uncorrected predict
Predict 2: Corrected predict
Error 1: Uncorrected absolute error
Error 2: corrected absolute error.

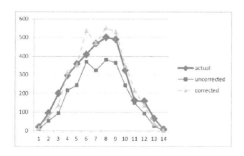

Figure 4. Predict data before and after correction.

6 CONCLUSION

This paper presents a predictive model of photovoltaic power output which combined BP neural network and grey system theory. This method has not only a nonlinear mapping ability from neural network, but also the ability on the evolution of effective monitoring. Meanwhile, the model in the modeling process, was applied similar day selection algorithm and metabolic model DGM(1,1). Not only considered having similarities between similar day, but also consider the continuity between adjacent days. Thus take full advantage of the historical data, improve the prediction accuracy. Finally, a comparative analysis of numerical examples to verify the prediction model indeed can effectively improve the prediction accuracy and feasibility of the algorithm.

REFERENCES

1. Ding Ming, Wang Lei, Bi Rui. et al. A short-term prediction model to forecast output power of photovoltaic system based On improved BP neural network [J]. Power System Protection and Control, 2012, 40(11): 93–99.
2. Pecas Lopes J.A, Moreira C.L, Madureira A.G, et al. Control strategies for micro-grids emergency operation. International Conference on Future Power Systems. Amsterdam, 2005.
3. Md Habibur Rahman, Susumu Yamashim. Novel distributed power generating system of PV-ECaSS using solar energy estimation [J]. 1EEE Transactions on Energy Conversion, 2007, 22(2):358–367.
4. Wang Fei, Mi Zengqiang, Su Shi, et al. A practical model for single-step power prediction of grid-connected PV plant using artificial neural network [C], Proc of IEEE PES Innovative Smart Grid Technologies Asia Conference, Perth, WA, Australia, 13–16 Nov, 2011: 1–4.

5. Moghaddamnia A, Remesan R, Kashani M. Hassanpour, et al. Comparison of LLR, MLP, Elman, NNARX and ANFIS Models-with a case study in solar radiation estimation [J]. Journal of Atmospheric and Solar-Terrestrial Physics, 2009, 71(8–9): 975–982.

6. K. Ishaque, Z. Salam, H. Taheri, et al. Modeling and simulation of photovoltaic (PV) system during partial shading based on a two-diode model [J]. Simulation Modelling Practice and Theory, 2011, 19(7): 1613–1626.

7. Ting-Chung Yu, Hsiao-Tse Chang. The forecast of the electrical energy generated by photovoltaic systems using neural network method [C]. International Conference on Electric Information and Control Engineering (ICEICE).

8. Chen Changsong, Duan Shan-xu, YIN Jin-jun. Design of photovoltaic array power forecasting model based on neutral network [J]. Transactions of China Electrotechnical Society, 2009, 24(9): 153–158.

9. Chang, W.C. (2000) "A Comprehensive Study of Grey Relational Generating," J. Chinese Grey System Association. Vol.3, pp.53–62. (in Chinese).

10. Deng, J.L. (1993) "On Judging the Admissibility of Grey Modelling via Class Ratio," The Journal of Grey System, Vol.5, No.4, pp.249–252.

11. Deng, J.L. (1982) "Control Problems of Grey Systems," Systems and Control Letters, Vol.1, No.5, pp.288–294.

12. Guo, H. (1985) "Identification Coefficient of Relational Grade," Fuzzy mathematics, Vol.5, No.2, pp.55–58. (in Chinese).

13. Liu, S. and Lin, Y. (1998) "An Introduction to Grey Systems: Foundations, Methodology and Applications," IIGSS Academic Publisher, Slippery Rock.

14. Liu, P.L. and Shyr, W.J. (2005) "Another Sufficient Condition for the Stability of Grey Discrete-Time Systems," Journal of the Franklin Institute, Vol.342. pp.15–23.

15. Li Donghui, Wang Hexiong, Zhu Xiao-dan, et al. Research on several critical problems of photovoltaic grid-connected generation system [J]. Power System Protection and Control, 2010, 38(2):208–214.

16. Xu Wei. Research and analysis on the mathematic model of GCPV generation system [J]. Power System Protection and Control, 2010, 38(10):7–21.

17. Davide Caputo, Francesco Grimaccia, Marco Mussetta, et al. Photovoltaic plants predictive model by means of ANN trained by a hybrid evolutionary algorithm [C] The 2010 Imerrational Joint Conference on Neural Networks(IJCNN), Barcelona, Spain, 2010.

18. Tian Yubo. Hybrid neural network technology [M]. Beijing: Science Press, 2009.

231

Power and Energy – Kong (Ed.)
© *2015 Taylor & Francis Group, London, ISBN 978-1-138-02782-4*

A programmed data-processing method for mapping energy allocation Sankey diagram of China

Chinhao Chong, Linwei Ma, Zheng Li, Jia Geng & Tiankan Zhang
State Key Laboratory of Power Systems, Department of Thermal Engineering,
Tsinghua-BP Clean Energy Centre, Tsinghua University, Beijing, China

ABSTRACT: Energy allocation Sankey diagram is popularly used to analyze energy systems of China, and the main data are normally derived from a statistical table named *Energy Balance of China*. To avoid manual calculations required to process the primary data and standardize the data-processing, this paper presented a method that can quickly generate data required for mapping, which is based on a series of standardized steps and rigid equations which can be programmed. Based on this method, it is also presented the energy allocation Sankey diagram of China's energy use and coal use in the year 2012.

1 INTRODUCTION

Energy allocation Sankey diagram has been widely used in previous studies on China's energy system. For example, Ma presented a map of the transformation of energy in China as a Sankey diagram to identify key areas to improve energy efficiency (L. W. Ma, 2012) and mapped a Sankey diagram of China's oil flow to reveal the physical pattern of China's oil supply and consumption (L.W. Ma, 2012), and Pu (H.J. Pu, 2010) and Yu (F.F. Yu, 2013) mapped the energy allocation diagram of coal use in China to illustrate the energy flows from raw coal supply to end-use consumption.

In these studies, the energy data for mapping the Sankey diagram are mainly acquired from the statistical table named *Energy Balance of China* (2014), and researchers need to process the data for mapping by a lot of manual calculations. The manual calculations are still lack of standard methods in different mapping works, and it will cost a lot of time when we need to map a series of Sankey diagrams of different years. In this paper, we attempt to develop a programmed method to process the data from *Energy Balance of China* to generate data required for mapping energy allocation diagram. Based on this method, we can generate energy allocation Sankey diagrams quickly based on a standard data format. For illustration, we present the energy allocation diagram of energy use and coal use in China in the year 2012.

The following contents of this paper are organized as follows: the introduction of methodology in section 2, results in section 3, and conclusions in section 4.

2 METHODOLOGY

The energy data provided in *Energy Balance of China* are expressed in Standard Quantity type (SQ) by heat value, and for secondary energy, the upstream energy losses caused by transformation sector have already been deducted. For mapping energy allocation diagram which does not show any energy losses and only illustrate the energy balance in each stages of energy flows, we need to compensate these energy losses and derive the adjusted data for secondary energy expressed in Primary Energy Quantity type (PEQ) which indicate the total primary energy consumed to produce these secondary energy for mapping energy allocation diagram of China, or expressed in Coal-Based Primary Energy Quantity type (CBPEQ) which indicate the total coal consumed to produce secondary energy for mapping energy allocation diagram of coal. In follows, we first introduce the primary data from *Energy Balance of China*, then re-balanced data and conversion factors for programming, and finally how to generate PEQ and CBPEQ.

2.1 Primary data

Energy Balance of China and *Final Energy Consumption by Industrial Sector* are two tables in China *Energy Statistical Yearbook*[5], which are the main data tables for mapping the Sankey diagram.

The data in *Energy Balance of China* are vertically divided into 6 parts: A) Primary Energy Supply, B) Input (−) and Output (+) of Transformation, C) Loss, D) Final Consumption,

E) Statistical Difference, and F) Total Energy Consumption, and horizontally divided into 30 types of energy, including raw coal, various coal products, crude oil, various oil products, natural gas, electricity, heat and recovery energy. The details of Final Energy Consumption are further presented in another table named *Final Energy Consumption by Industrial Sector* which includes more than 40 subsectors. For the purpose to illustrate the structure of final consumption of industrial sectors in the diagram, we re-categorize the subsectors in industrial sector as shown in Table 1.

2.2 *Re-balanced data*

Before converting the SQ into PEQ or CBPEQ, the Loss and the Statistical Difference in *Energy Balance of China* need to be re-allocated into Total Final Consumption to make a re-balanced data for programming.

In *Energy Balance of China*, the Total of Primary Energy Supply (S), Total of Input (−) and Output (+) of Transformation (T), Loss (L), Total Final Consumption (C) and Statistical Difference (SD) of the fuel j have the relationship:

$$S_j + T_j - L_j - C_j = SD_j \tag{1}$$

Table 1. The re-categorized industrial subsectors.

No	Items	Description
1	Non-energy mining	• Mining and processing of ferrous metal ores • Mining and processing of non-ferrous metal ores • Mining and processing of non-metal ores • Mining of other ores
2	Smelting and pressing of ferrous metals	
3	Smelting and pressing of non-ferrous metals	
4	Manufacture of non-metallic mineral and chemical products	
5	Manufacture of raw chemical materials and chemical products	
6	Manufacture of food and beverage	• Processing of food from agricultural products • Manufacture of food • Manufacture of liquor, beverage and refined tea
7	Manufacture of textile	• Manufacture of textile • Manufacture of textile, wearing apparel and accessories
8	Manufacture of machinery and vehicle	• Manufacture of general purpose machinery • Manufacture of special purpose machinery • Manufacture of automobiles • Manufacture of railway, ship, aerospace and other transport equipment
9	Manufacture of paper and paper products	
10	Other non-energy industrial subsector	• Manufacture of tobacco • Manufacture of leather, fur, feather and related products and footwear • Processing of timber, manufacture of wood, bamboo, rattan, palm and straw products • Manufacture of furniture • Printing and reproduction of recording media • Manufacture of articles for culture, education, arts and crafts, sports and entertainment activities • Manufacture of medicines • Manufacture of chemical fibers • Manufacture of rubber and plastics products • Manufacture of metal products • Manufacture of electrical machinery and apparatus • Manufacture of computers, communication and other electronic equipment • Manufacture of measuring instrument and machinery • Utilization of waste resources • Repair service of metal products, machinery and equipment • Production and supply of water • Others manufacture
11	Energy industry	• Production and supply of electric power and heat power • Production and supply of gas • Mining and preparation of coal • Extraction of petroleum and natural gas • Processing of petroleum, coking and Processing of nuclear fuel

To balance the supply and demand, the Loss and Statistical Difference are shared into Total Final Consumption by using the Loss and Statistical Difference revised factor k_b:

$$S_j + T_j - kb_j \cdot C_j = 0 \tag{2}$$

where

$$k_{bj} = \frac{L_j + SD_j}{C_j} \tag{3}$$

The final consumption of each end-use subsector can be revised as:

$$C'_{ij} = k_{bj} \cdot C_{ij} \tag{4}$$

We notice that, the non-energy use included in industrial sector energy consumption in *Energy Balance of China* does not have a specific declaration in *Final Energy Consumption by Industrial Sector*. Hence, we try to subtract this part of "energy" from each industrial subsector according to the energy consumption share of the subsector in the industrial sector by using (5).

$$I'_{ij} = I_{ij}\left(1 - \frac{N_j}{I_j}\right) \cdot k_{bj} \tag{5}$$

Subscription i: The ith industrial subsector involved
Subscription j: The jth fuel type involved
I'_{ij}: The revised consumption of fuel j in industrial subsector i

Table 2. The sequence for calculation the $k_{PEQ,ij}$ and $k_{coal,ij}$.

	Subsector (i)	Description
1	Coal preparation	(n) Input: raw coal
		(j) Output: cleaned coal and gangue
2	Briquettes	(n) Input: raw coal and cleaned coal
		(j) Output: briquettes
3	Coking	(n) Input: raw coal and cleaned coal
		Output: coke, coking gas, other
		(j) coking products and other gas
4	Gas Work	(n) Input: raw coal and cleaned coal
		(j) Output: coke, coking gas, other coking products and other gas

Integration of Coking and Gas work
a. Coke, coking gas and other coking products are produced both in Coking subsector and Gas work subsector. Hence, we need to integrate these 2 data set.
b. Obtain the k_{PEQ} and k_{coal} of these fuels in both subsectors respectively.
c. We can know the PEQ and CBPEQ were consumed to produce 1 unit of coke/coking gas/other gas in both subsectors respectively.
d. Hence, the k_{PEQ} and k_{coal} of coke/coking gas/other gas can be obtain from the equation (6) and (7):

$$k_{PEQ,j} = \frac{PEQ_{coking,input,j} + PEQ_{gaswork,input,j}}{SQ_{coking,input,j} + SQ_{gaswork,input,j}} \tag{6}$$

$$k_{coal} = \frac{CBPEQ_{coking,input,j} + CBPEQ_{gaswork,input,j}}{PEQ_{coking,input,j} + PEQ_{gaswork,input,j}} \tag{7}$$

	Subsector (i)	Description
5	Natural gas liquefaction	(n) Input: natural gas
		(j) Output: LNG
6	Petroleum refineries	(n) Input: crude oil
		(j) Output: diesel, gasoline, kerosene, fuel oil and other oil products
7	Heat generation	(n) Input: raw coal, cleaned coal, gangue, coking gas, other coking products, other gas, oil products and natural gas.
		(j) Output: heat
8	Electricity Generation	(n) Input: raw coal, cleaned coal, gangue, coking gas, other coking products, other gas, oil products, natural gas and heat.
		(j) Output: electricity

Notice
a. The k_{PEQ} and k_{coal} of fuel j can be obtained when there is a output of fuel j.
b. k_{PEQ} of raw coal, crude oil, natural gas, other energy, blast furnace gas and converted gas are 1.
c. k_{coal} of raw coal is 1. k_{coal} of crude oil, natural gas, other energy, blast furnace gas and converted gas are 0.

I_{ij}: The original consumption of fuel j in industrial subsector i

I_j: The total of original consumption of fuel j in whole industrial sector

N_j: The total of non-energy use of fuel j in whole industrial sector.

2.3 Conversion factor

The PEQ type and CBPEQ type can be expressed as (8) and (9) respectively.

$$E_{PEQ,j} = E_{SQ,j} \cdot k_{PEQ,j} \tag{8}$$

$$E'_{CBPEQ,j} = E_{PEQ,j} \cdot k_{coal,j} \tag{9}$$

where, $k_{PEQ,j}$ represents the total unit of primary energy consumed to produce 1 unit of fuel j, while $k_{coal,j}$ represents the proportion of raw coal in the total unit of primary energy consumed to produce 1 unit of fuel j.

By following the sequence listed in Table 2, we can use (10) and (11) to acquire $k_{PEQ,j}$ and $k_{coal,j}$ step by step.

$$k_{PEQ,ij} = \frac{\sum_n k_{PEQ,in} \cdot SQ_{in,input}}{\sum_j SQ_{ij,output}} \tag{10}$$

$$k_{coal,ij} = \frac{\sum_n k_{coal,in} \cdot PEQ_{in,input}}{PEQ_{ij,output}} \tag{11}$$

Subscript i: ith transformation subsector involved

Subscript n: nth fuel type input in subsector i

Subscript j: jth fuel type output in subsector j

$k_{PEQ,ij}$: PEQ transformation factor of j th fuel output in subsector i

$k_{coal,ij}$: Coal component factor of jth fuel output in subsector i

$k_{PEQ,in}$: PEQ Transformation Factor of n th fuel input in subsector i

$SQ_{in,input}$: Input of nth fuel in subsector i expressed in SQ type

$SQ_{ij,ouput}$: Output of jth fuel in subsector i expressed in SQ type

$k_{coal,in}$: Coal component factor of n th fuel input in subsector i

$PEQ_{in,input}$: Input of nth fuel in subsector i expressed in PEQ type

$PEQ_{in,output}$: Output of jth fuel in subsector i expressed in PEQ type.

The k_{coal} of heat and electricity should be revised by using renewable revised factor $k_{renewable}$ as the calculated processes mentioned above do not consider the renewable energy and nuclear power supply:

$$k_{renewable,j} = \frac{SQ_{transformation,j}}{SQ_{transformation,j} + SQ_{renewable,j}} \tag{12}$$

Subscript j: j th fuel type involved

$SQ_{transformation,j}$: Fuel j produced in transformation sector expressed in SQ type

$SQ_{renewable,j}$: Fuel j produced with renewable energy and nuclear power expressed in SQ type.

Hence, the equation (9) can be expressed as:

$$E_{CBPEQ,j} = E_{PEQ} \cdot k_{coal,j} \cdot k_{renewable,j} \tag{13}$$

2.4 PEQ and CBPEQ

All data provided in *Energy Balance of China* and *Final Energy Consumption by Industrial Sector* can be converted into PEQ type by multiplying the $k_{PEQ,j}$ mentioned in equation (10), except the data in Coking and Gas Works subsectors, those data should be converted by using the $k_{PEQ,ij}$ mentioned in equation (6).

Table 3. The energy types involved in energy allocation diagram of China.

No	Energy flow
1	Raw Coal
2	Cleaned Coal: Cleaned Coal, Other Washed Coal
3	Coke
4	Gas: Coke Oven Gas, Blast Furnace Gas, Converter Gas, Other Gas
5	Other Coal Products: Briquettes, Gangue, Other Coking Products
6	Crude Oil
7	Petrol
8	Diesel
9	Kerosene
10	Others Oil Products: Fuel Oil, Naphtha, Lubricants, Paraffin Waxes, White Spirit, Bitumen Asphalt, Petroleum Coke, LPG, Refinery Gas, Other Petroleum Products
11	Natural Gas: Natural Gas, LNG
12	Heat
13	Electricity

Table 4. The coal-derived energy types involved in energy allocation diagram of coal, China.

No	Coal flow
1	Raw Coal
2	Cleaned Coal: Cleaned Coal, Other Washed Coal
3	Coke
4	Gas: Coke Oven Gas, Other Gas
5	Briquettes
6	Gangue
7	Other Coking Products
8	Crude Oil
9	Heat
10	Electricity

To generate the CBPEQ data, the data provided in Primary Energy Supply Sector, Final Consumption Sector and data in *Final Energy Consumption by Industrial Sector* should be multiplied with $k_{CBPEQ,ij}$ mentioned in equation (13). The data in Transformation Sector should be multiplied with the $k'_{CBPEQ,ij}$ which mentioned in equation (9), except the data in Coking and Gas Works Subsectors, those data should be converted by multiplying the $k_{coal,j}$ which mentioned in equation (7).

The *Energy Balance of China* expressed in PEQ type and CBPEQ type can be acquired by using the methodology mention above. Hence, we can use this data to depict the energy allocation diagram and energy allocation diagram of coal with the e-Sankey Software. For mapping simplification, the 30 types of fuel involved in *Energy Balance*

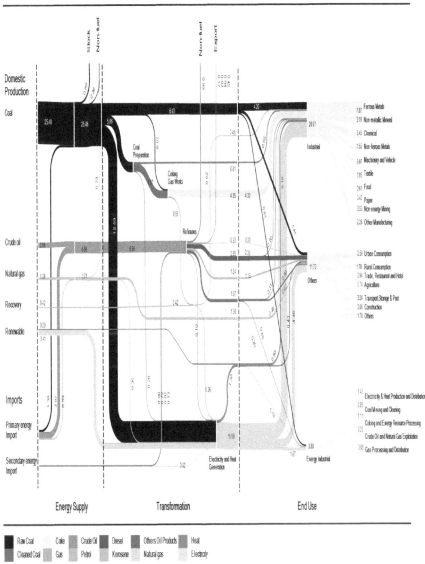

Figure 1.   Energy allocation diagram of China, 2012.

of China are integrated as shown in Table 3 and Table 4.

3 RESULTS

The energy allocation diagram of China and the energy allocation diagram of coal in China are presented in Figure 1 and Figure 2, they can help capturing main features of China's energy use and coal use.

They illustrate that coal is the main energy resource in China, which accounted for 68.7% of the primary energy consumption. The foreign trade dependence of oil products is 60.7%, and crude oil occupied 77.3% of the oil imports. Coal (72.6%) and renewable energy and nuclear power (22.0%) are the main fuels for electricity

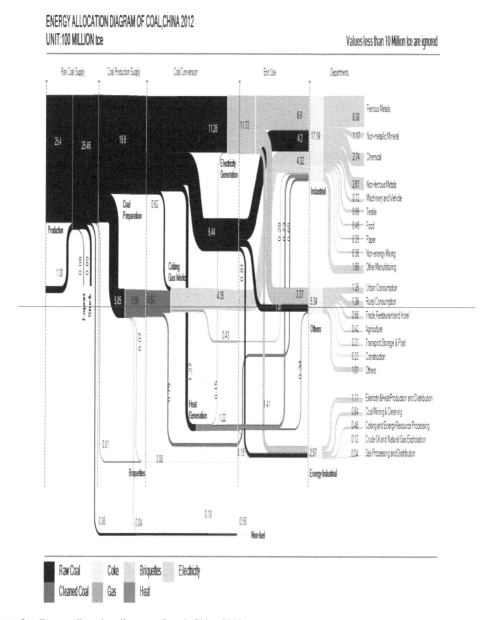

Figure 2.   Energy allocation diagram of coal, China 2012.

generation. In end-use sectors, industrial sector accounted for 57.3% of the primary energy consumption while the non-industrial sectors and energy industrial sectors accounted for 32.1% and 10.6%. The manufacturing of steel & iron, cement and chemicals are the main energy consumers in industrial sector, which in total accounted for 73.0% of the energy consumption of industrial sector.

4 CONCLUSIONS

In this paper, a programmed method for mapping energy allocation diagram and energy allocation diagram of coal is developed and applied to present Sankey diagrams of China's energy use and coal use. This method can quickly generate required data for mapping by a series of standardized steps based on the data from *Energy Balance of China*. For the application of method to other countries, the ideas and factors proposed in this manuscript can also be used but the programming may be different referring to different forms of energy balance table. In next step, it is suggested to further use this method to explore energy use in provinces and cities in China, or applied to research the use of oil and natural gas.

ACKNOWLEDGMENT

The authors gratefully acknowledge the financial support from BP Company in the scope of the Phase II Collaboration between BP and Tsinghua University, and the support to the work of energy allocation diagram of coal use from Tsinghua-Rio Tinto Joint Research Centre for Resources, Energy and Sustainable Development.

REFERENCES

1. China Energy Statistical Yearbook 2013: China Statistics Press, 2014.
2. F.F. Yu, L.W. Ma, L.X. Jiang, Z. li, and W.D. Ni, "Mapping the energy flows of coal utilization in China: the methodology and case studies for the year 2005 and 2010," Advanced Materials Research, vol. 724–725, pp. 1234–1241, 2013.
3. H.J. Pu, The Sustainable Utilization of Coal in China: China University of Mining and Technology Press, 2010.
4. L.W. Ma, F. Fu, Z. Li, and P. Liu, "Oil development in China: Current status and future trends," Energy Policy, vol. 45, pp. 43–53, 6//2012.
5. L.W. Ma, J. M. Allwood, J. M. Cullen, and Z. Li, "The use of energy in China: Tracing the flow of energy from primary source to demand drivers," Energy, vol. 40, pp. 174–188, 2012.

Power and Energy – Kong (Ed.)
© 2015 Taylor & Francis Group, London, ISBN 978-1-138-02782-4

Application and researches of Large Eddy Simulation of the low-level wind over complex terrain

Shuanglong Jin, Feng Shuanglei, Bo Wang, Ju Hu, Zhenqiang Ma & Zongpeng Song
China Electric Power Research Institute, Haidian District, Beijing, China

ABSTRACT: Wind forecast is the basis of wind power prediction and the major determinant of prediction accuracy. Wind speed is affected by wind farm terrain, roughness and other environmental factors. Being restricted by physical parameterization schemes, spatial resolution and other limitations, wind speed can only be described approximately by the meso-scale Weather Research and Forecast (WRF) model, so there are considerable differences between the simulation results and the real situation. With the rapid development of distributed renewable energy, the surrounding environmental factors (such as terrain, roughness, obstacles, and so on) will become more and more complex, and WRF will not be able to meet the application demands. Large Eddy Simulation (LES) is an effective method to finely simulate wind speed under such complex environmental factors. It has rich sub-grid closure schemes, and can provide high accuracy atmospheric motion information with high resolution. To further apply LES method to wind power prediction, we nest LES method into the WRF model, and apply this coupled system to wind forecast in this paper. The results show significant improvement of the prediction accuracy over complex terrain. This work makes contribute to high-precision wind power prediction.

1 INTRODUCTION

Energy crisis, environmental issues and other factors have accelerated the development of distributed generation, especially the rapid development of distributed wind power. "Twelve Five" period, the development of China's wind power has adjusted to "centralized" and "distributed", the distributed wind power is with respect to the concentrated development large-scale wind farms (Wang C, 2005), it is the wind power production and used in the same location or limited to the local area. Distributed wind power generation with flexible, environmentally friendly features is more and more access to the distribution network (Linag Y, 2003), not only to save investment of the high-pressure lines and booster stations, but also to achieve the balance between wind power output and local load (Wang L, 2001; Xu Y, 2011; Dai J, 2011). The installed capacity of distributed wind power in 2015 will reach 500 million kilowatts and in 2020 is expected to reach 15 million kilowatts, accounting for 5% and 7.5% of the target installed capacity, respectively. However, the output with obvious intermittent and random volatility features, will affect the normal operation of the power system (Chen H, 2006; Hu H, 2006; Hadjsaid N, 1999; Puttgen H, 2003), when the large number of distributed wind power access to the distribution network. Studies have shown that accurately forecasting wind of wind farms and predicting the output power of wind turbines can increase the capacity of wind power and will improve power system security and economy (Kariniotakis G, 2003; Giebel G, 2003).

However, the distributed wind power closer to the user or load centers, leading to more complex elements of the surrounding environment. The changes of obstacles, surface roughness and topography have increased the difficulty of numerical weather prediction that used for distributed wind power forecast, which has seriously hampered the development of distributed wind power. Micro-scale weather modeling is an effective way to simulate distributed wind energy prediction. Although encouraging advances in micro-scale flow modeling, the evolution of different technologies and flavors of LES and CFD (Computational Fluid Dynamics) models, have been made in the last decade, the modeling ability for micro-scale flows associated with real weather at distributed wind farm scales is still very limited. In fact, micro-scale weather flow models encounter many challenges. Therefore existing micro-scale models have mostly focused on idealized case study, with idealized initial conditions and/or boundary conditions and/or highly simplified atmospheric physics. The Real-Time Four Dimensional Data Assimilation (RTFDDA) weather forecasting system (Liu Y, 2006), built upon WRF (Liu Y, 2008a. The operational...part 1...; Liu Y, 2008b. The operational...part 2...). It has been downscaled to LES scale modeling grids, through the nested-down grid refinements, this modeling system provides a unique ability for simulating real micro-scale weather processes by incorporating realistic meso-scale weather forcing, high-resolution terrain and land use, and physical processes of solar and long wave radiation and cloud microphysics (Liu Y, 2011). The WRF-LES system is employed to forecast the distribution wind in northeast of China, the results show that the model system can effectively improve the accuracy of wind prediction, and the model system has application potential.

2 DATA AND METHODS

Here we use WRF version 3.4. The model was initialized and forced by meteorological fields from 1° × 1° Global Forecast System (GFS) analysis and forecasts, and the boundary condition data (BC) was updated every 3 hours. The simulation experiment period was for January 2012. The setup consisted of a main domain with two-way nesting three domains with a horizontal resolution of 25 km, 5 km, 1 km and 200 m. The grid numbers for the four domains were 140 × 90, 221 × 176, 351 × 251 and 461 × 251, respectively, the innermost domain covered the area of the distributed wind farms completely (Fig. 1). There were 33 vertical levels with five higher vertical resolution levels below 200 m, which are approximately near to surface 10, 50, 70, 100 and 170 m. The microphysics were determined using a WSM 3-class simple ice scheme (Hong SY, 2004), the longwave radiation option used the RRTM scheme (Mlawer EJ, 1997), the shortwave radiation option adopted the Dudhia

scheme (Dudhia J, 1989), the land-surface option was the thermal diffusion scheme and the cumulus option was the Kain-Fritsch (new Eta) scheme (Kain JS, 1990, 1993, 2004).

WRF–LES makes use of simultaneous nest-down approach with fine-mesh domains running at LES scale. The model simulates meso-scale weather on the coarser grids with a Planetary Boundary Layer (PBL) parameterization along with other model physics parameterizations. In contrast, for the fine-mesh LES modeling, the PBL scheme was replaced by 3D prognostic TKE diffusion scheme (Skamarock W.C., 2008), while all the other atmospheric physics processes, such as radiation, cloud microphysics, and so on are parameterized in the same way as in the coarse meshes. The model system is based on the reference Liu Y (2011).

Domain 1 is large enough to cover the synoptic scale weather of several thousand kilometers, while Domain 4 encloses the small region of the distributed wind farms. The WRF–LES model is employed to simulate the wind speed at the distributed wind farm in northeastern China during the period of 7–10 January 2012. There are 15 meteorological towers in the research region, the observed wind speeds is used to evaluate the WRF-LES simulated wind speed.

To quantify the discrepancies between the simulated and observed data, here we calculate the Mean Absolute Error (MAE), Root Mean Square Error (RMSE), correlation coefficient (R) and index of agreement (IA) to determine the effect of the wind speeds simulation.

The defined formula of MAE is:

$$MAE = \frac{1}{n}\sum_{i=1}^{n}|S_i - O_i| \tag{1}$$

The RMSE definition is:

$$RMSE = \sqrt{\frac{\sum_{i=1}^{n}(S_i - O_i)^2}{n}} \tag{2}$$

The IA is expressed as:

$$IA = 1 - \frac{n \times RMSE^2}{\sum_{i=1}^{n}\left(|S_i - \overline{O}| + |O_i - \overline{O}|\right)^2} \tag{3}$$

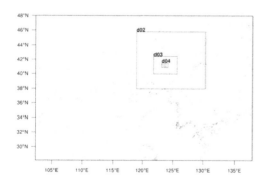

Figure 1. The research domain and WRF configuration setup, the black squares indicate the boundaries of four nested domains. The white pentagle is the position of observed tower in the distributed wind farms.

where n represents the total number of sample size, S_i and O_i are the CFSR reanalysis and observed wind speeds for a given time, respectively. In equation (3) $\overline{O}$ is the mean of observed wind speeds.

3 RESULTS AND DISCUSSION

3.1 *The spatial distribution of different simulate domains*

Local wind flows are complex products of multi-scale weather interactions, and they can be influenced by local and regional scale topography, land–surface and soil heterogeneity, and nonlinear scale interactions. Planetary Boundary Layer parameterizations are applied for the three coarse-mesh domains, whereas the finest meshes are run with the LES settings. The WRF continuously assimilates weather observations and provides continuous boundary forcing to fine-mesh domains for LES modeling. The terrain height in domain 3 and 4 is shown in Figure 2, the same color scales are used for different domains to emphasize the terrain features resolved by each domain. The terrain resolved by fourth domain is finer than the third domain, and the terrain features are more distinct. Domain 4 covers the region where the eastern is higher than the western, and within complex terrain and high slope. Therefore, the local topography and land surface are more considered in fourth domain wind flows.

To describe the multi-scale flow features simulated by high resolution model, a snapshot of the wind speeds at the 10 m layer in domain 3 and domain 4 is given in Figure 3. The area of the LES domain is framed in the meso-scale domain. The LES domain enhanced wind variability compared with the meso-scale domain. The reason could be that the fine-scale variability can be smaller than those that can be captured explicitly by the meso-scale domain, and could therefore excite computing noises in meso-scale domain.

3.2 *Comparison between the observed and simulated values of different domains*

To get an overall evaluation of the capability of the high-resolution model system for improving wind prediction in the distributed wind farms, statistical variables of the observed and the simulated wind speeds at 15 wind towers are computed and compared. Table 1 lists the error statistics of the observed and meso-scale as well as mirco-scale simulated wind speeds for 7–10 January 2012. The MAE, RMSE, correlation coefficient (R) and index of agreement (IA) were calculated at this period.

In Table 1, the black of the tower name means that the wind prediction accurate of high resolution LES is worse than the meso-scale simulation, and the black of statistics signify that the calculate value of the LES simulation is not better than the meso-scale simulation. So the prediction results can divide into three categories, the first is the statistical results of LES is better than the meso-scale simulation, include TA, TE, TF, TJ, TK and TL, the second is the statistical results of fourth domain and the third is only have little different, consist of TD, TG, TH, TM and TN, the last categories is the

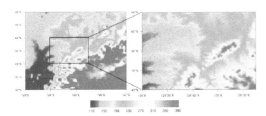

Figure 2. The terrain height of the third and fourth simulation domains.

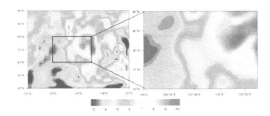

Figure 3. The spatial distribution of wind speed at 10 m above ground level in the third and fourth simulation domains.

Table 1. The statistics of the third and fourth domains simulation results for 15 wind towers.

Tower	MAE (m/s)		RMSE (m/s)		R		IA	
	d04	d03	d04	d03	d04	d03	d04	d03
TA	1.62	1.74	1.95	2.07	0.52*	0.50*	0.71	0.69
TB	**2.16**	**2.09**	**2.68**	**2.53**	**0.29***	**0.27***	**0.56**	**0.56**
TC	**1.57**	**1.42**	**1.98**	**1.80**	**0.51***	**0.60***	**0.69**	**0.76**
TD	1.90	2.03	2.38	2.49	**0.53***	**0.54***	0.71	0.70
TE	3.00	3.24	3.35	3.58	0.57*	0.54*	0.60	0.56
TF	3.06	3.25	3.60	3.84	0.44*	0.42*	0.60	0.59
TG	4.31	4.38	5.29	5.38	**0.59***	**0.60***	**0.61**	**0.61**
TH	2.98	3.10	3.49	3.57	**0.50***	**0.53***	**0.64**	**0.65**
TI	2.32	2.36	**2.93**	**2.91**	**0.08**	**0.09**	**0.46**	**0.47**
TJ	2.05	2.34	2.54	2.81	0.55*	0.45*	0.70	0.62
TK	2.20	2.37	2.64	2.87	0.56*	0.50*	0.69	0.67
TL	2.44	2.58	2.97	3.12	0.54*	0.49*	0.66	0.64
TM	1.38	1.39	1.79	1.83	0.59*	0.58*	**0.74**	**0.76**
TN	2.87	2.91	3.15	3.24	0.57*	**0.59***	0.56	0.58
TO	**1.71**	**1.71**	2.13	2.14	**0.49***	**0.51***	**0.70**	**0.70**

MAE, mean absolute error; RMSE, root mean square error; R, correlation coefficient; IA, index of agreement. *indicates that correlations significant over the 99.95% confidence level.

statistical results of fourth domain is worse than the third, include only TB, TC, TI and TO. It is apparent that the fine grid LES models improved the skills in predicting more accurate winds.

Figure 4 compares the simulated wind speeds of different domains with the observations, and the results are that the LES predicting speeds more accurate than the meso-scale. MAE is in the range of 1.62–3.06, the LES reduced the MAE 0.13–0.29, RMSE between 1.96 and 3.60, and has decreased 0.11–0.27 by LES, R and IA are up to increased by 0.10 and 0.07, respectively.

Figure 5 compares the simulated wind speeds of different domains with the observations, which is only litter different between the meso-scale and LES simulation. Overall speaking, the all statistic results of LES are decreased. MAE and RMSE are reduced more than 0.10, but R and IA are only reduced about 0.01. It is found from the figure that the different domains simulated wind speeds are very similar.

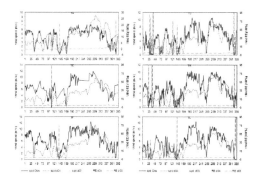

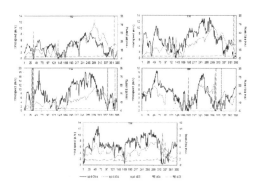

Figure 4. Observed and simulated wind speeds (left axis) and the relative errors of speeds (right axis) of the wind towers that the statistical results of fourth domain is better than the third.

Figure 5. Observed and simulated wind speeds (left axis) and the relative errors of speeds (right axis) of the wind towers that the statistical results of fourth domain and the third have little different.

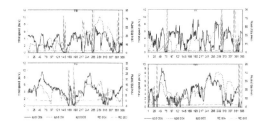

Figure 6. Observed and simulated wind speeds (left axis) and the relative errors of speeds (right axis) of the wind towers that the statistical results of fourth domain is worse than the third.

Figure 6 compares the simulated wind speeds of different domains with the observations, and the results are that the meso-scale predicting speeds more accurate than the LES. MAE and RMSE of LES are increased 0.15 and 0.18 at the most, respectively. R and IA of LES are decreased 0.01–0.09 and 0.01–0.07, respectively.

It is found from the analysis of Figures 4 to 6 that the overall prediction effects of wind speed is improving by LES, but only a small part of the forecast effect is worse than meso-scale. Further analysis of model simulations should be undertaken for better understanding of the factors. However, generally speaking, the high-resolution terrain forcing, explicit PBL mixing that could be the driving force for the better simulation.

4 CONCLUSION AND DISCUSSION

In this paper, the WRF-LES was employed to simulate four days (7–10 January, 2012) of real-time wind speeds at northeastern China distributed wind farms with complex terrain, and four simultaneous nested grid domains at a grid spacing of 25, 5, 1 and 0.2 km, respectively. The three coarse domains were run with meso-scale model setting, while the finest mesh domains were run with LES model configuration. The results show that the overall prediction effects of wind speed is improving by LES model. The nesting LES within WRF model system can improve wind prediction accuracy, it can provide more accurate and reliable support for wind power forecast over complex terrain, and the model system has application potential. The forecast results are also have the room for improvement, it is need to base on this to carry out more in-depth study.

ACKNOWLEDGMENT

Shuanglong Jin appreciate the constructive comments and suggestions of the anonymous reviewers. Many thanks to Dr. Yubao Liu for guiding the configuration of Numerical Weather Prediction system. This work was supported in by the science technology foundation of Beijing (Grant number is D131100002013002) and S&T foundation of State Grid Corporation of China.

REFERENCES

1. Chen Haiyan, Duan Xianzhong, Chen Jinfu. 2006. Impacts of distributed generation on steady state voltage stability of distribution system. Power System Technology. 30(19):27–30.

2. Dai Jiang, Wang Shao, Zhu Jinfeng. 2011. Power flow method for weakly meshed distribution network with distributed generation. Power system protection and control. 39(10):37–41.

3. Dudhia, J. 1989. Numerical study of convection observed during the winter monsoon experiment using a mesoscale two-dimensional model. J. Atmos. Sci. 46:3077–3107.

4. Giebel G, Landberg L, Kariniotakis G. 2003. State-of-the-art on methods and software tools for short-term, prediction of wind energy production. European Wind Energy Conference & Exhibition, Madrid, Spain.

5. Hadjsaid N, Canard J, Dumas F. 1999. Dispersed generation impact on distribution networks. IEEE Computer Applications in Power. 12(2):22–28.

6. Hong SY, Dudhia J, Chen SH. 2004. A revised approach to ice mircophysics process for the bulk parameterization of clouds and precipitation. Mon. Wea. Rev. 132:103–120.

7. Hu Hua, Wu Shan, Xia Xiang, et al. 2006. Computing the maximum penetration level of multiple distributed generators in distribution network taking into account voltage regulation onstraints. Proceedings of the CSEE. 26(19):13–17.

8. Kain JS, Fritsch JM. 1990. A one-dimensional entraining/detraining plume model and its application in convective parameterization. J. Atmos. Sci. 47:2784–2802.

9. Kain JS, Fritsch JM. 1993. Convective parameterization for mesoscale models: The Kain-Fritsch scheme. The representation of cumulus convection in numerical models, K. A. Emanuel and D. J. Raymond, Eds., Amer. Meteor. Soc. 246.

10. Kain JS, John S. 2004. The Kain-Fritsch convective parameterization: an update. J. Appl. Meteor. 43: 170–181.

11. Kariniotakis G, Mayer D, Moussafir J. 2003. ANEMOS: development of a next generation wind power forecasting system for the large-scale integration of onshore & offshore wind farms. European Wind Energy Conference & Exhibition, Madrid, Spain.

12. Liang Youwei, Hu Zhijian, Chen Yunping. 2003. A survey of distributed generation and its application in power system. Power System Technology. 27(12):72–75.

13. Liu, Y., Chen, F., Warner, T., et al. 2006. Verification of a mesoscale data assimilation and forecasting system for the Oklahoma City area during the Joint Urban 2003 Field Project. J. Appl. Meteorol. Clim. 45, 912–929.

14. Liu, Y., Warner, T.T., Bowers, J.F., et al. 2008. The operational mesogamma-scale analysis and forecast system of the U.S. Army Test and Evaluation Command: part 1: overview of the modeling system, the forecast products, and how the products are used. J. Appl. Meteorol. Clim. 47, 1077–1092.

15. Liu, Y., Warner, T.T., Astling, E.G., et al. 2008. The operational mesogamma-scale analysis and forecast system of the U.S. Army Test and Evaluation Command: part 2: inter-range comparison of the accuracy of model analyses and forecasts. J. Appl. Meteorol. Clim. 47, 1093–1104.

16. Liu, Y., et al. 2011. Simultaneous nested modeling from the synoptic scale to the LES scale for wind energy applications. Wind Eng. Ind. Aerodyn. doi:10.1016/j.jweia.2011.01.013.

17. Mlawer EJ, Taubman SJ, Brown PD, Iacono MJ, Clough SA. 1997. Radiation transfer for inhomogeneous atmosphere: RRTM, a validated correlated-k model for the long-wave. J. Geophys. Res. 102(D14): 16663–16682.

18. Puttgen H, MacGregor P, Lambert F. 2003. Distributed generation: semantic hype of the dawn of a new era. IEEE Power and Energy Magazine. 1(1):22–29.

19. Skamarock, W.C., Klemp, J.B., Dudhia J., et al. 2008. A description of the advanced research WRF version 3, NCAR Technical Note.

20. Wang Chengshan, Zheng haifeng, Xie Yinghua, et al. 2005. Probabilistic power flow containing distributed generation in distribution system. Automation of electric power systems. 29(24):39–44.

21. Wang Linchuan, Liang Feng, Li Man, et al. 2011. Research on the distribution network reconfiguration with the distributed generation. Power system protection and control. 39(5):41–45.

22. Xu Yuqin, Li Xuedong, Zhang Jigang, et al. 2011. Research on distribution network planning considering DGs. Power system protection and control. 39(1):87–91.

Power and Energy – Kong (Ed.)
© 2015 Taylor & Francis Group, London, ISBN 978-1-138-02782-4

Quality improvement of hydraulic servo equipment's power supply system

Jianhao Yang, Yanxi Ren, Bo Li & Binxing Wang
Beijing System Equipment Engineering Research Institute, Beijing, China

ABSTRACT: Due to serious power supply problems occurred on a set of servo-hydraulic equipment after installation, it couldn't work properly. The aim of this paper is to investigate the relation of fault phenomenon and ambient operation parameters as well as the test results. It also aims to give a reconstruction scheme in order to improve the operating condition and guarantee the reliable work of the equipment. The whole project includes the employment of harmonic elimination in a higher user power network, SBW voltage stabilizer and so on. The experience from this project indicates that power quality is vital to the equipment's proper work and should be taken into consideration. This work is valuable for the design, planning and installation of large-scale equipment of electric power supply in China rural area.

Keywords: hydraulic servo equipment; power quality; harmonic elimination; performance improvement

1 INTRODUCTION

With the rapid development of economics in China, the consumption of electric power grows up quickly. Various power loads bring about great effects on power quality. Some key equipment might not work regularly if it has power quality problem. This paper illustrates a serious power quality phenomenon, occurred on a set of hydraulic servo equipment, and the improvement steps are given.

2 POWER QUALITY PHENOMENON OVERVIEW

A set of hydraulic servo equipment was established on rural area outside Zhangjiakou, Hebei Province in north China. The equipment used three certificated three-phase asynchronous motors, total power 300 KW, has a 10 KV/420 V transformer at the end of rural electric grid of the State Grid. The motors operate regularly in the factory debugging, whereas the switches of motors turn off unusually after the equipment installed. Simple check up to the equipment found no abnormal from motors, loads and control units. The motors switches have current-, thermo- and balance protection functions. Portable multimeter measurement results show that there is an unstable fluctuation between 410–430 V with the setting voltage of 420 V, so the unbalance of phase voltage may cause switch off and a further comprehensive investigation are taken to solve the phenomenon.

3 TEST AND MEASUREMENT

First, the ground and insulation of the equipment and transformer are tested, founding no out of limits. Then power quality is tested using FLUKE 435 instruments.

Two serious problems have been found after several interval measurements, that is unbalance and harmonics pollution.

Graph 1 on the left is the no running curves of the electric motors three-phase voltage, compared with the right is the running curves of the motors with start and stop. From the graph, we can observe there is only about 5 V voltage change caused by the motors start and stop, but there are more than 10 V voltage fluctuation on the 420 V line voltage unbalance.

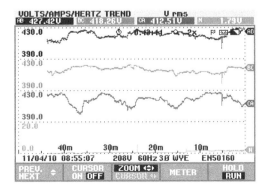

Figure 1a. Curves under motor stop condition. Voltage curves of the equipment.

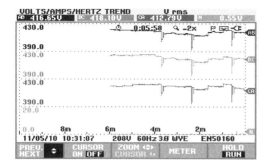

Figure 1b. Curves of motor in start and stop period. Voltage curves of the equipment.

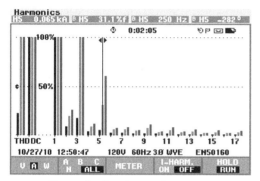

Curve of serious harmonic currents

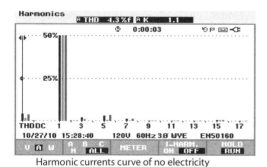

Harmonic currents curve of no electricity

Figure 2. Harmonics curves under test.

Graph 2 is the harmonics charts with and without harmonics in one day monitoring. From the chart we can observe that 3 and 5 orders harmonics are the most serious pollutions, especially the 3-orders can reach to 100%.

According to states power quality standards, three-phase power supply system under 10 KV should meet: voltage deviation range ±7%, voltage unbalance Unb ≤ 2%, voltage frequency range ±0.2 Hz, voltage total harmonics THDU ≤ 4% (95% big probability), voltage flash Pst ≤ 0.9. Further monitoring to the 10 KV line voltage use FLUKE 4370 shows that: the maximum value of the 10 KV line voltage is 11.02 KV, deviation ratio is 10.2%, 3743 groups out of 8644 groups exceed 7% (43.3%), unbalance maximum is 4.11 KV, unbalance ratio is 105.5%, 897 groups out of 8641 groups exceed the states standards (10.38%).

Whether from 420 V or 10 KV monitoring data show voltage deviation, unbalance and harmonics are the major reasons of the breakdown of the equipment. Measurement results also show here must have great DC electric loading on the electric net.

4 SOLUTIONS

A check was performed to the equipments attached to the 10 KV line and the upper level 110 KV line and no large DC equipment was found, so it was concluded that the harmonic was mainly caused by the 220 KV line. Measures to remove harmonic needed to be taken. After the installation of harmonic removal equipment in the 220 KV line, the harmonic met the standard.

The voltage unbalance may be caused by many reasons, such as the unbalance of the upper level line, or the unbalance of the load in 10 KV line. After a thorough of check, it was found that a large number of equipment attached to the 10 KV line. So the problem could be solved by two ways, one is to increase the capacity to 35 KV, the other is to take simple voltage stabilization measures.

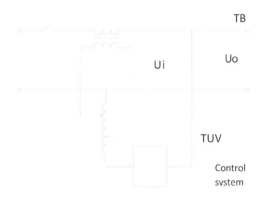

Figure 3. SBW voltage regulating principle.

Table 1. Results of SBW voltage stabilization.

No stabilization	L1	L2	L3
M	247.033	244.837	244.789
Σ	2.093	3.0136	2.672
5–95%	243.2–250	239.7–249.5	240.2–248.8
Fluctuation voltage range	6.8	9.8	8.6
After stabilization	L1	L2	L3
μ	224.11	223.67	225.6
σ	0.85	0.78	0.64
5–95%	221.07–226.78	220.46–227.39	223.–228.36
Fluctuation voltage range	5.71	6.93	5.36

To increase the capacity to 35 KV would take a comparatively long time, so a simple SBW voltage stabilization measure was taken. SBW voltage stabilization uses mechanic-electric feedback control to regulate voltage by phases, it is economical but to a certain extent a response lag exists.

The principle is concisely shown in Figure 3. The system is composed of three-phase compensation transformer TB, three-phase regulation transformer TUV and control system. The control system measures the change of the voltage Uo on the output and controls the servo motor to slide the Carbon Brush, so that the secondary voltage of TUV and then the amplitude and polarity of the voltage on the primary coil of TB are adjusted. As a result, the output voltage can be stabilized automatically and the goal of automatic voltage stabilization can be achieved.

After the installation of SBW voltage stabilization equipment, the line was monitored. It showed that the idle voltage fluctuation within half an hour did not exceed the limit and the unbalance degree decreased obviously. The voltage fluctuation is shown in Table 1. Due to the response lag of the voltage stabilization equipment, when the equipments had been operating for a long time, the voltage unbalance would sometimes exceed the limit and circuit trip would occur.

5 CONCLUSION

Compared with urban grid, the quality of electric power is generally poorer in rural area, especially in the fields of voltage deviation, voltage fluctuation, flickering and three-phase voltage unbalance. So before the installation of large industrial facilities in such areas, the quality of electric power must be monitored to ensure that it meets the standards, or the facilities may be affected seriously to be put into service. Independent power supply is preferred, if possible.

REFERENCES

1. Wang Dai-di. Analysis and research of three-phase unbalance in distribution network, Shenyang University of Technology, 2007.
2. Liu Pei-ling, Huang Xin. The analysis of 10 Kv system voltage unbalance. Peeltroleum Planning & Engineering, 2003, 12(9):63–64.
3. Deng Jia-xiang. Research of power quality problems Beijing gird and its improvement method. North China Electric Power University, 2003.
4. Yu Yong. The research on methods of three phase unbalanced var dynamic compensation. North China Electric Power University, 2008.
5. Zeng Yong-hong, Liu Xiao-liang. The theory and maintenance of SBW voltage regulator. Science & Technology Ecnony Market, 2009, 5:19–20.

Power and Energy – Kong (Ed.)
© 2015 Taylor & Francis Group, London, ISBN 978-1-138-02782-4

Study on the influence of rain flush to insulator surface pollution layer conductivity and dynamic contamination accumulated

Yunhai Song
Maintenance and Test Center, EHV, China Southern Power Grid, China

Xiaoting Huang
Dongguan Power Supply Bureau of Guangdong Power Grid Corporation, Dongguan, China

ABSTRACT: The depravation of atmosphere environment directly aggravates the contamination accumulated of the transmission line insulators; it reduces the insulation level, bringing a series of flashover problems, and serious safety problems to electric power system. So, it is important to study on the influence of rain flush to insulator surface pollution layer conductivity and dynamic contamination accumulated. In this paper, at the condition of lab, flushes the insulator surface contamination in different pollution degree with artificial rainfall, and the procedure of insulator dynamic contamination accumulated, getting the law of insulator surface pollution layer conductivity changing with rain flush, that is the surface conductivity becomes greater, until greatest, and then smaller until smallest in the continue raining rush. It has the guiding meaning to the prevent flashover.

Keywords: transmission line; insulator; dynamic contamination accumulated; pollution flashover; rain; surface conductivity

1   INTRODUCTION

With the rapid development of industrial technology of China, atmospheric environment pollution of China has become increasingly serious, pollutant is more and more complex, and the degree of pollution is high. Power grid pollution flashover accident happens from time to time, and its scope expands unceasingly, has become one of the most serious problems, a threat to the safe operation of power system[1–3]. Under the condition of low humidity, even if the transmission line insulator in heavy volume the dirt status, its insulation performance can also with clean insulator insulation performance, only in damp or under the condition of rainfall, the surface of the insulator produces partial discharge, and develops into arc, finally forms the pollution flashover discharge, and brings serious security hidden danger to the electric power system[4–7]. At home and abroad, the mechanism and the causes of pollution flashover were studied, the development of resistance pollution flashover and pollution flashover of insulator is studied, And also did a lot of work to preventing pollution flashover, online monitoring of transmission line insulator pollution extent, such as on-line monitoring leakage current, optical fiber method to measure salt density, monitoring method of local

electrical conductivity, and so on, these methods and monitoring equipment made an important contribution to preventing pollution flashover work. However, the current research work is rarely on insulator pollution layer conductivity change in the process of rainfall condition and research on the laws of the dynamic product pollution of insulator.

Therefore, it is necessary to rainfall erosion of the changes of the insulator surface fouling layer conductance, as well as the rainfall erosion to study the effect of dynamic product pollution of insulator. The author under the condition of laboratory, and use of insulator surface fouling layer resistance monitoring terminal, research on the change rule of insulator surface fouling layer conductivity in the process of rainfall and the change rule of rainfall erosion effects on dynamic product pollution of insulator.

2   EXPERIMENTAL PRINCIPLE

This principle diagram of the experiment is shown in Figure 1, the main equipments were remote data receiving terminal, insulator surface resistance monitoring terminal, XP-70 type suspension insulator and simulated rainfall watering can.

As shown in Figure 1, this experiment adopts the XP-70 porcelain insulator, the surface of the insulator has gone through special processing, five closed ring metal electrodes was installed on the its surface, annular concave metal electrode embedded in the insulator surface, which is flat and with level surface with the insulator surface, every two ring formed 1 the measurement channel, a total of four, as shown in Figure 2, used as a sensor for measuring insulator surface fouling layer resistance; insulator surface fouling layer resistance monitoring terminal receives signals from the sensors in the insulator surface fouling layer resistance, is used to monitor the insulator surface fouling resistance, then the measured surface of the insulator pollution layer resistance data through the GPRS network to uploaded to the remote data receiving terminal, deposited in the local database; Simulated rainfall watering can used to produce artificial rainfall, to wet wash of polluted insulator surface, the size of the rainfall can be adjusted freely.

The experiment of the insulator adopts the commonly used XP-70 suspension type standard

Table 1. Parameters of XP-70 insulator.

Model	Insulating plate size (mm)	Creepage distance (mm)	Surface area (cm²)	Lower surface area (cm²)
XP-70	255	295	710	803

Table 2. The quantity of XP-70 insulator surface contamination in deference grade contaminated district.

Number	The salt density value (mg/cm²)	NaCl (mg)	Diatomite (mg)
1	0.045	33	198
2	0.08	58	348
3	0.175	126	756
4	0.3	214	1284

insulator of high voltage transmission, its shape parameters are shown in Table 1.

In the process of natural product in the insulator pollution, its impurity composition consists of can be dissolved and cannot be dissolved, Soluble content is mainly composed of monovalent salts and bivalent salts, generally speaking, the content of insoluble substance accounts for most of the filth. In the process of experiment, using a mixture of sodium chloride and diatomite to simulate the soluble content and an insoluble substance of insulator product dirty filth, and the choice of the proportion of sodium chloride and diatomite is 1:6. The experiment selected four different salt density value, specific usage as shown in Table 2.

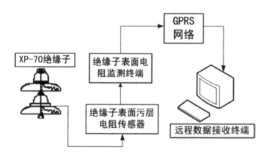

Figure 1. Map of experiment principle.

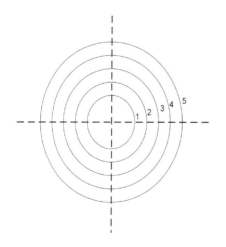

Figure 2. The circular metal electrode insulator surface.

3 EXPERIMENTAL PROCESS

2.1 Rainfall erosion test

In order to guarantee the precision of the experiment, the experimental with precision of 0.1 mg of electronic balance, as shown in Table 2 the dosage of the filth take corresponding sodium chloride and diatomite, measuring 4 of 2 ml deionized water with the measuring cylinder separately, four salt density value corresponding to the filth dissolves in 4 of 2 ml deionized water respectively, configured to dirty solution; take the 2 ml solution of the filth coating on the XP-70 insulator surface uniformly, after drying in the shade naturally, as shown in Figure 1 of the experimental principle diagram connects experiment equipment, makes blur artificial modification of rain with watering can on the surface of the insulator, in the process, the surface of the insulator dirt layer resistance monitoring terminal automatic measuring insulator surface fouling layer resistance, and the measured data through the GPRS network to upload to the remote data receiving terminal.

2.2 Dynamic product experiments

Under the condition of nature, the insulator is in the unclean—rainfall erosion—product again—washed again such a cycle of dynamic process, in order to research the dynamic fouling characteristics of insulator, under the condition of this experiment in the lab, at 0.175 mg/cm² salt density value as a standard, simulating dynamic deposition process of insulator. Make the XP-70 insulator rinsed, and dried completely, and in accordance with the Table 2 standard to configuration filthy solution, then take the filthy solution coating on the insulator surface uniformly, after drying in the shade natural, simulation of artificial rainfall, rainfall erosion of the insulator surface, in the process of erosion, use the insulator surface fouling layer resistance monitoring terminal to measure changes of the insulator surface pollution resistance, when the filth of the insulator surface flush essentially, then stop the rain, the fan was used to simulate the natural wind, accelerate the drying of the rain on the surface of the insulator, in the drying process, using insulator surface fouling layer resistance monitoring terminal to measure the changes of insulator surface fouling layer resistance; According to 0.175 mg/cm² salt density value standard to weigh the corresponding sodium chloride and diatomite, mixing with them, and then make the filth settled on the insulator surface evenly, simulation of artificial rainfall again, rainfall erosion of insulator surface, in the flushing process, using insulator surface fouling layer resistance monitoring terminal to measure the changes of insulator surface pollution resistance, when the filth of the insulator surface flush essentially, then stop the rain, the fan was used to simulate the natural wind, accelerate the drying of the rain on the surface of the insulator, and in the drying process with RTU measuring changes of the insulator surface fouling layer resistance.

4 EXPERIMENTAL RESULTS

In the process of experiment, the insulator surface fouling layer resistance monitoring terminal measured data are insulator surface fouling layer resistance, and the size of the fouling layer resistance will be affected by various factors such as width and length of the fouling layer being measured, inconvenience of text theory research, so bring the insulator surface fouling resistance conversion into the surface of the insulator pollution layer conductivity. The environment temperature of experiment is 25°C, according to the formula (1), (2), (3) the fouling layer resistance conversion into the fouling layer conductivity when in the standard temperature 20°C:

$$\sigma_t = R \cdot f = \frac{1}{G} \cdot f \tag{1}$$

Among them, f for insulator shape factor:

$$f = \int_0^L \frac{dx}{\pi D(x)} \tag{2}$$

Then the conductivity at room temperature converts to conductivity at 20°C standards temperature:

$$\sigma_{20} = \sigma[1 - b(t - 20)] = K_t \cdot \sigma_t \tag{3}$$

Make fouling layer resistance from experiment 1 conversion into insulator surface fouling layer conductivity, drawn into changing curve of electrical conductivity, as shown in Figure 3.

As shown in Figure 3, (a), (b), (c), (d) is when the rains washed out, different salt density value corresponding to the change curve of the conductivity of fouling layer respectively. Then the data from experiment 2 drawn into changing curve of electrical conductivity, as shown in Figure 4.

In Figure 4 (a), (b), (c), (d), is change curve of the fouling layer conductivity when rain washed at the first time, change curve of the fouling layer conductivity when drying at the first time, change curve of the fouling layer conductivity when rain washed at the second time, change curve of the fouling layer conductivity when drying at the second time respectively.

5 RESULTS ANALYSIS

Can know from infiltrating theory[8]: when the water is combined with conductive material, and soluble conductive material, local conductance rise on the surface of the insulator; when the water is combined with non-conductive material, bureau of the ministry of electricity guideline is changeless, the salt and grey guhya affect the performance of the insulation of the insulator are different in nature, insulation performance into a certain relationship with the increase of ash dense, the reason is that the increase of insolubles causes the insulator surface increasing in the number of moisture absorption, formed a more thick water film, thereby unclean in the soluble content can be more dissolved in the water, and reduce the insulation performance

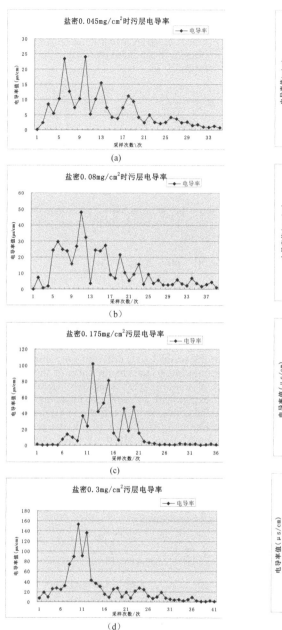

(a)

(b)

(c)

(d)

Figure 3. Conductivity curve of rainfall erosion insulator surface pollution layer.

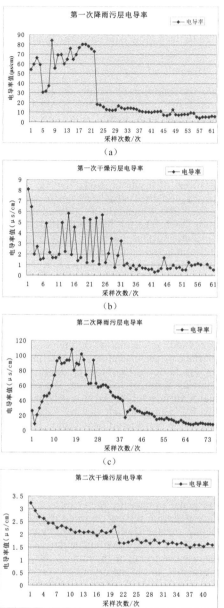

(a)

(b)

(c)

(d)

Figure 4. Conductivity curve of pollution accumulated on the insulator surface pollution layer.

of the insulator indirectly. It can be observed from the Figure 3, when rainfall to scour the filth of the insulator of different salt density value, the rule of insulator surface fouling layer conductivity shows consistent: the early rain, the surface of the insulator is still in a dry relatively, not form a conductive layer, and fouling layer conductivity in 1.0 mu s/cm,

insulator insulation performance and clean insulator insulation performance is the same nearly; As the rain continued, insulator surface fouling layer conductivity dynamic rises, until the fouling layer achieved saturated infiltration status, fouling layer conductivity reach to the maximum, four different salt density value corresponding to the maximum fouling layer conductivity are: 24.0 mu s/cm, 48.0 mu s/cm, 101.9 mu s/cm, 153.6 mu s/cm; Then the rain began to scour of insulator pollution layer, fouling layer conductivity began to decline, until the rain making filth clean, insulator surface fouling layer conductivity and dry insulator is the same nearly, at 1.0 mu s/cm.

In general, in the process of the rain washing the fouling layer, fouling layer conductivity is first increases then decreases; and in the process of drying, insulator surface conductivity is a trend of decrease, but amplitude of decrease is not much, this is because after the rains washed out, insulator surface filth has washed clean, the contribution to the conductivity is mainly comes from the rain, and the conductivity of rain water is far less than the electrical conductivity of wet fouling layer.

6 CONCLUSION

In the laboratory environment, the author simulated artificial rainfall, did the rains washed out simulation experiment for the insulator of artificial smearing, and did rainfall erosion simulation experiment for the impact of insulator dynamic product pollution. In the process of rainfall erosion, insulator pollution layer reached saturated infiltration condition, its surface fouling layer conductivity will reach a maximum value, namely the insulation performance of insulator is the worst, is also the most prone to flashing, but as the rain washed out continued, insulator surface fouling layer conductivity decreased gradually, until the rain making insulator surface washing clean, the insulation of the insulator can restore to the original level. In the process

of dynamic product pollution, after the rain washed clean the insulator filthy, insulator product pollution again, the rain can still washed clean the filthy, and make the insulation of the insulator to restore to the original level.

REFERENCES

1. Guan Zhi-cheng, Wang Shao-wu, Liang Xi-dong,. Application and prospect of polymeric outdoor insulation in China [J]. High Voltage Engineering, 2000, 26(6): 37–39.
2. Yang Yin-hu. Analysis and preventive measure for large area pollution flashover [J]. Electric Power, 1998, 31(4): 74–75.
3. Zhang Hui-yuan, Ding Yang. Research on classification standard for electric porcelain external insulation filthy [J]. Journal of North China Electric Power University, 1997, 24(4): 24–29.
4. He Bo, Lin Hui, Present Status and Future Prospects for Flashover Research of Contaminated Insulators[J]. Insulators And Surge Arresters, 2006,2: 7–9.
5. Wang Tao, Ou Qi-he, Wu Jiang-hong, et al. Test and re-search on cleaning of high voltage transmission line based on salt density [J]. Power System Technology, 2004, 28(4): 22–26.
6. Gao Hai-feng, Fan Ling-Meng, Li Qing-feng, et al. Comparative Analysis on Pollution Deposited Performances of Insulators on the ±500 kV Gao-Zhao DC Transmission Line [J]. High Voltage Engineering, 2010, 36(3): 672–675.
7. Xiao Deng-ming, Pan Long, Li Xiao-dong, et al. On line leakage current monitoring for outdoor insulation in substations [J]. High Voltage Engineering, 1998, 24(1): 28–29.
8. Zhang Yong-ji. Study on AC Flashover Performance of Artificially Polluted XP-160 Insulators at Various ρ_{ESDD} and ρ_{NSDD} [D]. Chongqing University, 2006: 1–5.

Power and Energy – Kong (Ed.)
© 2015 Taylor & Francis Group, London, ISBN 978-1-138-02782-4

Individual blade pitch control design of wind turbine based on load optimization model

Fei Lan, Zhehao Wang, Meng Yang, Xing Jiang, Xiaohua He & Jin Hu
School of Electrical Engineering, Guangxi University, Nanning, Guangxi Autonomous Region, China

ABSTRACT: This paper presents an optimal individual pitch control method for decreasing the unbalanced load caused by wind shear, thus prolonging life of wind turbine. The angles used as reference values for the controller are calculated by a mathematical model of the load optimization, the objective function of which is unbalanced load minimization based on blade element theory and power is the essential restriction as well as the range of pitch angles. The performance of the proposed individual controller is evaluated on the upwind 500 kW wind turbine model and compared with the typical individual pitch control. Simulation results indicate that the controller lessens the unbalanced load significantly, without the loss of power and the frequent pitching.

1 INTRODUCTION

Wind power is developing rapidly and is to grow and mature over last few decades. The variable pitch wind turbine meets the demand of great development of large-scale wind turbine, and the pitch control technology becomes increasingly important under the condition of higher wind to insure wind turbine's safety and efficiency running. The technology contains two methods, Collective Pitch Control (CPC) and Individual Pitch Control (IPC). CPC has been widely applied in actual fields, having one control system and regarding the hub wind as single reference wind of pitch controller, thus adjusting the same angle in the meantime. With the development of wind turbine unit capacity and the increasing of the blade height, wind shear seriously affects every blade's real wind, which changes with the height of the blade. It's inadvisable to regard the hub wind as single reference wind and IPC emerges at the right moment. The controller can adjust every pitch angle according to real wind of blades to lessen the unbalanced load.

Relative to the unbalanced load caused by wind shear, experts at home and abroad have been done full research. The combination of pitch-regulation based on the azimuth angle and advanced controllers was designed in [9] [10] to solve the problem, but ignored the mechanical fatigue. Bossanyi E. *et al.* [11], developed an active control strategy to alleviate the unbalanced load. The controller transformed the singles measured into *d*- and *q*- axes by a special transformation. Each axis controlled by a PI controller working in a feed-forward Filter separately, but the measurement requirement of this method is higher, and there may be a delay.

The paper [13] measured the attack angle and relative velocity and set initial value of pitch angle based on the former inflow parameters, whereas the range of the changed attack angle is smaller and the impact load is bigger, resulting in accurately setting the initial value is difficult.

Based on the above, this paper introduces an Individual Pitch Control based on the Load Optimization (IPCLO) according to the real wind and load, which is capable of reducing the dominant load without much more pitching.

2 INDIVIDUAL PITCH CONTROL

2.1 Pitch control model

Pitch control model includes wind turbine model and pitch control system model. When natural wind inflows the turbine by the way of axis, wind turbine can actually get the useful aerodynamic power and aerodynamic torque as followed:

$$P_r = \frac{1}{2} C_P(\lambda, \beta) \rho \pi R^2 v^3 \tag{1}$$

$$T_r = \frac{1}{2} C_T(\lambda, \beta) \rho \pi R^3 v^2 \tag{2}$$

where ρ is air density; C_P is the turbine power coefficient, C_T is the aerodynamic torque coefficient, $C_p(\lambda, \beta) = \lambda C_T(\lambda, \beta)$. C_p is the pitch angle; R is the turbine radius; λ is the tip speed ratio, that is the ratio of tip linear velocity and the wind speed:

$$\lambda = \frac{\omega R}{v} = \frac{2\pi R n}{v} \tag{3}$$

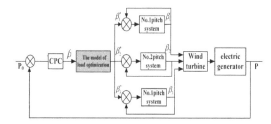

Figure 1. Block diagram of IPCLO.

C_P presents the capacity of absorbing the power from the wind, can be approximated as

$$C_p = (0.44 - 0.0167\beta)\sin\left[\frac{\pi(\lambda - 3)}{15 - 0.3\beta}\right]$$
$$-0.00184(\lambda - 3)\beta \qquad (4)$$

The pitch control system of larger wind turbine generators generally adopts hydraulic drive system. The system has time delay and uses first-order inertia link to simulate with delay, whose transfer function can be expressed as:

$$\frac{\beta(s)}{\beta_r(s)} = \frac{1}{T_\beta s + 1}\exp(-\tau s) \qquad (5)$$

2.2 Individual pitch control

The block of CPC is composed by pitch system, wind turbine and generator, which is simple and can keep the power to a desired level by adopting different controllers. This paper chooses the PID controller generally used on engineering and builds IPCLO of Figure 1 based on the PID controller.

The CPC adjusts the error of actual output power and rated power to get uniform pitch angle β, and then regarding it as initial value solves the load optimization model and gets $\beta_1^*, \beta_2^*, \beta_3^*$, to reduce loads on the blades by varying the pitch of the blades individually.

3 INDIVIDUAL BLADE PITCH CONTROL BASED ON LOAD OPTIMIZATION MODEL

3.1 Calculation model of load

The main components of wind turbine are blades on account of its wind energy capture and wheel force is related to the blades as well as other components [15]. Theoretical load can be calculated based on blade element theory and the formula of load in shimmy direction is below [16].

$$M_{xb} = \frac{1}{2}\int_{r_0}^{R}\rho W^2 c C_n r dr \qquad (6)$$

where R is the turbine radius; c is the chord length; ϕ is the angle of inflow; C_l is the lift coefficient; C_d is the drag coefficient; r is the length of element to blade root.

$$W = \sqrt{V_{x0}^2 + V_{y0}^2} = \sqrt{(1-a)^2 V^2 + (1+b)^2(\Omega r)^2} \qquad (7)$$

where a is the axial; b is the circumferential factors.

$$\frac{a}{(1-a)} = \frac{\sigma C_n}{4\sin^2\phi} \qquad (8)$$

$$\frac{b}{(1+b)} = \frac{\sigma C_t}{4\sin\phi\cos\phi} \qquad (9)$$

where $\sigma = Bc/2\pi r$, B is the number of blades; ϕ is the angle of inflow.

$$\phi = \arctan\frac{(1-a)V_1}{(1+b)\Omega r} \qquad (10)$$

$$C_n = C_l\cos\phi + C_d\sin\phi \qquad (11)$$

$$C_t = C_l\sin\phi - C_d\cos\phi \qquad (12)$$

where C_l and C_d can get from the look-up airfoil profile tables.

3.2 The load optimization model

According to aerodynamic, a correlation between load and pitch exists; hence to solve the load optimization problem is to solve the optimal function and to get optimal pitch. The objective function is to minimize the difference of load from every blade, and pitch and power can be subjected to limits.

The purpose of IPCLO is to reduce the asymmetrical load. According to formula (6), the optimal load function can be designed as:

$$\begin{cases} \min \quad f(x) = \left|\frac{1}{2}\rho W_i^2 c C_n^i - \frac{1}{2}\rho W_j^2 c C_n^j\right|, \\ \quad i, j \in \{1, 2, 3\} \\ \text{Subject} \quad \text{to}: \beta \le \beta^i \le \beta'; P \le P^i \le P'; \end{cases} \qquad (13)$$

where β and β' is uniform pitch angles of minimum and maximal wind speed; P and P' is the output power of minimum and maximal wind speed; $C_n = f(\beta)$ can be obtained by data-fitting. By this way objective function can be changed to research the function of β^i and W_i^2.

3.3 Pitch control based on the load optimization

Because of one-to-one corresponding relations of real wind and azimuth angle, this paper divides azimuth angles into different regions in which uses IPCLO respectively. Basis for the division is azimuth angles whose real wind is very strong and whose difference of wind speeds every blade is larger, to minimize loads.

Figure 1 shows that this method refers to uniform pitch angle and optimizes compute in several areas to realize fewer pitching. Detailed flow diagram of the simulation is showed in Figure 2 [17–18]. When azimuth angles belongs to 0°~30°, the difference of the second and the third blade real wind is the highest, at the same time the unbalanced load reaches the maximum. To realize fewer pitching, the pitch angles of above blades can be calculated to get optimal angles and else blade's pitch angle is to be uniform pitch angle. Three pitch angles of blades respectively are $\beta_1^* = \beta, \beta_2^*$ and β_3^* to apply them to IPCLO.

3.4 Wind speed model considering the wind shear used for simulation

There are many factors to influence the change of wind, in which factor greatly influenced is wind shear [19]. That is the increase of wind speed with height, and the higher the height, the greater the wind speed. Mathematical model of wind speed is as follows [20]:

$$V(r,\theta) = V_h \left(\frac{r\sin\theta + h}{h} \right)^{\partial} \qquad (14)$$

where θ is the azimuth, which is the angle of the blade with horizontal plane. Setting hub as 0° and counter clock-wise direction as positive. V_h is the wind speed of hub up to h. r is the radius of blades, and ∂ is the factor of wind shear is related to the environment of wind turbines, and generally is about 0.3.

When the hub of wind speed is equal to 15 m/s, real time wind speed of the three blades in a wind wheel rotation cycle is showed in Figure 4, with the No. 1 propeller blades, No. 2 and No. 3

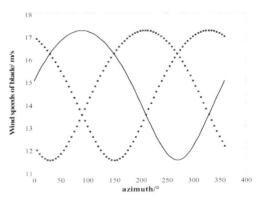

Figure 3. Wind shear effect.

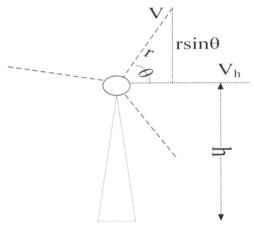

Figure 4. Wind speeds of three blades affected by the wind shear effect.

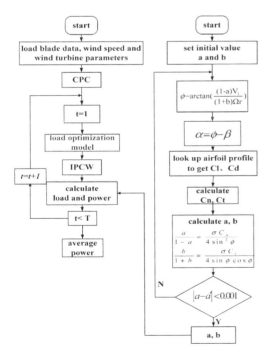

Figure 2. Flow diagram of the simulation.

propeller blades and different shapes to each other to describe the three blades. The horizontal axis is the azimuth angle of the No. 1 blade, the vertical axis is real time wind speed. So the blade azimuth angle difference between each other is 120 degrees in coordinate system. The figure shows the highest wind speeds up to 17.2 m/s, the lowest wind speeds as low as 11.5 m/s.

4 EXAMPLE SIMULATION

NTK 500/41 type produced by Nordtank was used to IPCLO simulation in this paper. The main parameters are in Table 1, paddle specifically described in Table 2.

4.1 The simulation result of CPC

As shown in Figure 5, we adjust the angle of the uniform pitch to 6° and the power can maintain 500 kW. But we can observe from Figure 7, after a rotation period, unbalanced load of CPC up to 5400 Nm, and load amplitude of each blade up to 12,000 Nm.

4.2 The simulation comparison between CPC and IPCW

According to the literature [18], weight distribution was taken for the unified pitch angle and got the simulation results of the 1st paddle IPCW methods and compared with the CPC method. Like Figure 8.

Table 1. Key parameters of wind turbine NTK 500/41.

Motor speed 27.1 r/min	Height of hub 35.0 m
Air density 1.225 kg/m	Input wind speed 4 m/s
Number of blades 3	Output wind speed 25 m/s

Table 2. Detailed descriptions of the blade.

r (m)	Torsional angle	Chord	r (m)	Torsional angle	Chord
4.5	20.00	1.630	12.5	3.15	1.163
5.5	16.3	1.597	13.5	2.60	1.095
6.5	13.00	1.540	14.5	2.02	1.026
7.5	10.05	1.481	15.5	1.36	0.955
8.5	7.45	1.420	16.5	0.77	0.881
9.5	5.85	1.356	17.5	0.33	0.806
10.5	4.85	1.294	18.5	0.14	0.705
11.5	4.00	1.229	19.5	0.05	0.545

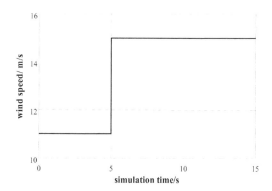

Figure 5. Step wind from 11 m/s to 15 m/s.

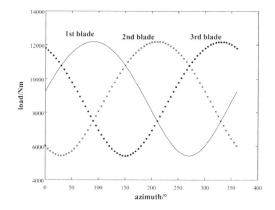

Figure 6. Pitch and power of CPC.

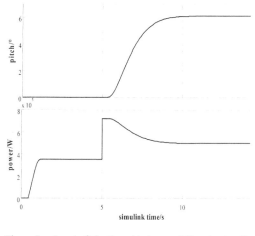

Figure 7. Load of the three blades on different azimuth.

260

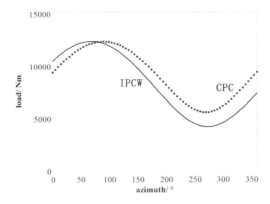

Figure 8. Load and pitch of CPC and IPCW.

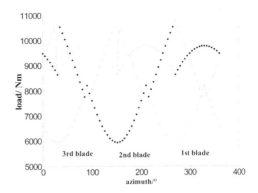

Figure 9. Load by load optimization model.

We can get the impact of the IPCW on the power is not large. The difference of the loads got by the two methods is not large during a rotation cycle. Figure 8 was the curve of the pitch angle corresponding to the two methods. IPCW were repeatedly adjusting the pitch angle compared to CPC. While IPCW method makes multiple paddle angle adjustment, load is not reduced, and there is no good solution to the problem of unbalanced load larger.

4.3 The simulation result of IPCLO

Load and the real-time wind speed of the blade at a given wind speed range with azimuth cycle. We take an example of the 1st blade. To illustrate the changes of the load, the output power and the pitch angle. Within the range from 30° to 90°, the difference of the load between the 1st paddle blade and the 2nd is maximum. So load optimization calculations were taken for the 1st paddle blade and the 2nd and the initial values are set to 6.8 ° and 5.6 °. The Figure 9 shows the pitch angle and the corresponding blade loads calculated by load optimization model. Compared with Figure 8, the load amplitude reduced significantly to less than 11000 Nm. The maximum unbalanced load reduced from 5400 Nm to 4500 Nm, reduced 16.7%. Especially when the azimuth is 90°, the unbalanced load reduced from 4700 Nm to 1900 Nm. In addition, the fluctuation range of the load reduced greatly for each of the blades, output power corresponding to the three blades like Figure 10 is within the allowable range.

In Figure 11, we still use the 1st paddle for example to IPCW and IPCLO simulation. Figure 11 shows that the average power obtained by the two methods are near rating value and have little difference. Figure11 shows that at the load magnitude, the load of IPCLO reduced 19%

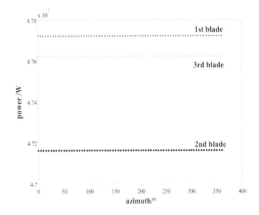

Figure 10. Average power by load optimization model.

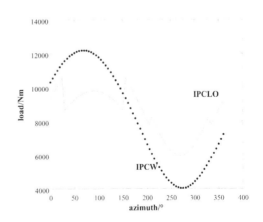

Figure 11. Load by IPCW and IPCLO simulation.

compared with IPCW. The range of the load is 12000 Nm to 4000 Nm for IPCW and 10500 Nm to 6000 Nm for IPCLO. The force of IPCLO is more uniform than IPCW and its range is reduced. It is easy to know that the adjustment to the pitch angle

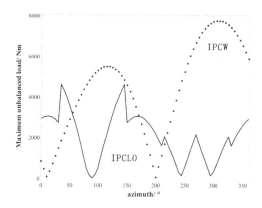

Figure 12. Maximum unbalanced load by IPCW and IPCLO simulation.

of IPCLO reduced, avoiding frequent movement for the variable pitch mechanism. Figure 12 is the maximum value of the unbalanced load for IPCLO and IPCW that indicates the maximum unbalanced load is small and its range of variation is also small. The method proposed in this paper has obvious advantages in reducing the load imbalance and load amplitude. In addition, the output power of the variable pitch mechanism can remain near the ratings without frequent adjustment.

5 CONCLUSION

This paper mainly talk about the unbalanced load of blades caused by shear effect and design load optimization model to carry out independent variable pitch control method. In the premise of ensuring the power, it can not only reduce the unbalanced load between the blades, but also avoid the variable-pitch mechanism adjusted frequently. It is significant to extend the life of the wind turbine.

ACKNOWLEDGMENT

The project is supported by the National Natural Science Foundation of China (51277034, 51377027).

REFERENCES

1. Chen Lei, Xing Zuo-xia, An overview of large grid-connected wind generation system status. renewable energy. 2003, 1.
2. Carlin P W, Laxson A S, Muljadi E B. The history and state of the art of variable speed wind turbine technology [J]. Wind Energy, 2003, 6(2): 129–159.
3. Joselin Herbert G M, Iniyan S, Sreevalsan E, et al. A review of wind energy technologies [J]. Renewable and sustainable energy Reviews, 2007, 11(6): 1117–1145.
4. Yang Xiu-yuan, Liang Gui-shu, Development of Wind Power Generation And Its Market Pro Development of Wind Power Generation and Its Market Prospect, Power System Technology, 2003, 27(7): 78–79.
5. Xing Zuoxia, Chen Lei, Sun Hongli, Wang Zhe, et al. Strategies Study of Individual Variable Pitch Control[J]. Proceedings of the CSEE, 2011, 31(26):131–138.
6. Geng Hua, Yang Geng, Output Power Level Control of Variable-speed Variable-pitch Wind Generators[J]. 2008, 28(25): 130–137.
7. Xu Da-ping, Xiao Yun-qi, Qin Tao, Lü Yue-gang, Cutting-in Control of Variable-Pitch Doubly-Fed Wind Power Generation System and Its Modeling and Simulation[J], Power System Technology, 2008, 32(6): 100–10.
8. Zhang Chunming. Research on Individual Variable Pitch Control Strategies Large-scale Wind Turbine[D]. Shen yang University of Technology. 2011.
9. Yao Xingjia, Ma Xiaoyan, Guo Qingding, et al. Wind turbine individual pitch control basedon single neuron weight coefficient [J]. Renewable Energy Resources, 2010(3):19–23.
10. Yao Xingjia, Liu Yue, Guo Qingding, A Control Method for Split Range Individual Pitch Based on Feed-Forward Azimuth Angle weight Number Assignment[J]. Acta Energiae Solaris Sinica, 2012, 33(004): 532–539.
11. Ma Jia Large-Scale Wind Turbine Individual Pitch Control Technology Research[D]. Shen yang University of Technology, 2012.
12. Bossanyi E. Further load reductions with individual pitch control[J]. Wind Energy, 2005, 8(4): 481–485.
13. Larsen T J, Madsen H A, Thomsen K. Active load reduction using individual pitch, based on local blade flow measurement[J]. Wind Energy, 2005, 8(1): 67–80.
14. Li Jing, Wang Wei-sheng. Modeling And Dynamic Simulation of Variable Speed Wind Turbine[J]. Power System Technology, 2003, 27(9): 14–17.
15. He Dexing. Wind Engineering and Industrial Aerodynamic [M]. BeiJing, Academic Press. 2006.
16. Xiao Jin-song. Aerodynamics of Wind Turbines [M]. BeiJing: China Electric Power Press 2010.
17. Zeng Qingchuan, Liu Hao, Luo Weiqi, et al. Computation of Aerodynamic Performance for Horizontal Axis Wind Turbine Based on Improved Blade Element Momentum Theory [J]. Proceedings of the CSEE, 2011, 31(23): 129–134.
18. David Wood. Small Wind Turbines: Analysis Design and Application[J].2013.
19. Kong Yigang, Gu Hao, Wang Jie, et al. Load analysis and power control of large wind turbine based on wind shear and tower shadow[J] Journal of Southeast University: Natural Science Edition,2010,40(1):228–233.
20. Zhou Wen-ping, Tang Sheng-li, Lü Hong et al. Effect of Transient Wind Shear and Dynamic Inflow on the Wake Structure and Performance of Horizontal Axis Wind Turbine[J]. Proceedings of the CSEE, 2012, 32(14): 122–127.

Power and Energy – Kong (Ed.)
© 2015 Taylor & Francis Group, London, ISBN 978-1-138-02782-4

Research on the set point of feed water system of pressurized water reactor

Hao-Ran Li, Xu-Hong Yang, Peng He & Hai-Zhong Wu

Shanghai Key Laboratory of Power Station Automation Technology, Shanghai University of Electric Power, Shanghai, China

ABSTRACT: The set point of feed water system is an important control system in the pressurized water reactor nuclear power station, which aims to keep target level that steam generator needs. The water flow is generated by steam generator. According to the set point of steam generator, this article controls the main feed water system by fuzzy control. And the experimental results show that the control effect is excellent.

1 INTRODUCTION

The main feed water system is very important in two loop systems of conventional island of pressurized water reactor nuclear power station. It can deliver water to steam generator, which was heated by high pressure heater. The water supply can be adjusted by the feed water system, and then the secondary side water level of evaporator can be maintained in a predetermined reference value with changing turbine loads. In order to eliminate the disturbances of three steam generators controlling the water level, and avoid the frequent actions of feed water regulating valve, as well as improve the circumstances of water level control system, the adjusting speed system of pump is introduced. By doing that, the pressure drop of water supply valve can be maintained as a constant within the normally changing loads, so that the work conditions of water supply valve can be improved.

2 THE STRUCTURE AND THEORY OF THE MAIN WATER FLOW CONTROL SYSTEM

Every steam generator has its own independent control system of water level, which can be controlled through adjusting the valve opening to change the water flow. However, the mother pipe of water supply is common for the three steam generators. If only using the control system of water level, the pressure of mother pipe would change when steam generator's water level deviates from set point, which needs to adjust valve opening to change the water flow. In the meantime, the other two steam generators valve opening are not changed, resulting in the changing of feed water flow owing to the changing pressure of the mother pipe. Therefore, the water level will fluctuate for the imbalance flow of steam and water in the two steam generators. In order to eliminate the disturbances of each other, and avoid the frequent actions of adjusting valve, as well as improve the circumstances of water level control system, the adjusting speed system of pump is introduced. By adjusting the pump speed, the pressure drop of water supply valve can be maintained as a constant within the normally changing loads, so that the work conditions of water supply valve can be improved.

Figure 1 shows the flow diagram of feed water system.

In fact, under the conditions of maintaining pressure drop of valve as constant, the differential pressure, between the mother pipe of feed water and of the evaporator, varies as the patterns of parabola according to the changing loads.

As an approximation, a crease line can be indicated, showing in Figure 2.

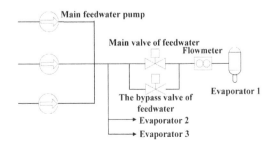

Figure 1. The flow diagram of feed water system.

The total pressure drop of the mother pipe of feed water and of the evaporator consists of four parts:

$$\Delta P = \Delta P_1 + \Delta P_2 + \Delta P_3 + \Delta P_4 \qquad (1)$$

In the formula, ΔP_1—the potential difference between the export of feed pump and the feed water import of evaporator, which is a constant;

ΔP_2—the pressure drop of adjusting valve, which needs to keep unchanging;

ΔP_3—the pressure drop of secondary side of evaporator, changing with the loads;

ΔP_4—the pressure drop of the steam pipe line and water-supply line, changing with the loads.

We can guarantee the outlet pressure and the flow of pump will change accordingly with the changing loads by adjusting the speed of feed pump. This can not only maintain the pressure drop of the feed water valve unchanging, but also help the pressure fit well with the total pressure drop curve shown in Figure 2. Finally, it can eliminate the defective coupling of separate flow adjustment among the three evaporators.

The primary purpose of speed control of main feed water pump is to ensure the pressure drop, between the mother pipe of feed water and of the evaporator, as a setting value with changing loads, and also to keep pressure drop of the regulating valve of feed water control system approximately constant. Then the coupling effect within the three evaporators can be eliminated and the require flow of evaporator contented.

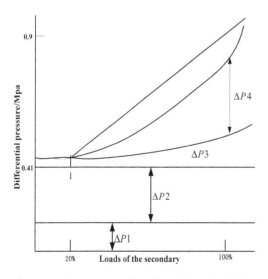

Figure 2. The pressure drop of mother pipe of feed water and of the evaporator.

3 THE LIQUID LEVEL CONTROL SYSTEM OF STEAM GENERATOR

The function of liquid level control system of steam generator can be achieved by the main feed water control system. The feed water of the two loops, which has been heated and deoxidized by the deaerator and boosted by three parallel main feed water pumps, is collected to the mother pipe of feed water through high pressure feed water heating system. Finally, it will be delivered to the three steam generators.

Each liquid level control system of steam generator configures a regulating station of feed water flow, consisting of a main feed water regulating valve and a bypass valve, which can control feed water flow automatically under any power levels. The design capacity of bypass valve, applying to start-up and adjusting feed water flow in low loads and keeping all open in high loads, is at least 20% of nominal flow. The design capacity of main feed water valve is at least 90% of nominal flow, adjusting the feed water flow in high loads.

3.1 Control strategy

3.1.1 The set point of liquid level of steam generator

The level settings of CPR1000 evaporator is an increasing function of loads when loads change from 0% to 20%;when the loads are bigger than 20%, the level settings is as a constant as 50% of NR level.

3.1.2 The control of feed water flow in high loads

The control of feed water flow in high loads is realized by the regulating valve of main feed water flow. The control system is a three pulse cascade of closed loop system, which consists of two controllers named liquid level regulator and flow regulator. The liquid level regulator is PID and made up of two parts: a component of PI and a differential component in series with PI. The deviation of the liquid level of settings and measured value is delivered to liquid level regulator. Its output signal is added to measured signal of evaporator flow which has been filtered. Then the sum is provided to set value of feed water flow. Compared with the signal of feed water set value with measured feed water flow of evaporator, the signal difference is transported to flow regulator in PI patterns, which outputs the signal controlling the valve opening of main feed water.

3.1.3 The control of feed water flow in low loads

The control of feed water flow in low loads is realized by the bypass flow valves. Considering that the measured precision is insufficient and the liquid level regulator's performance is not good in low loads, as a result, we take the total steam flow (the turbine loads + ADG flow + GCT flow) as feed-forward and add to the output of liquid level regulator and through the function generator. In the end, the bypass valve takes it as opening signals. The precision of measuring total steam flow directly does not suffice, therefore, we use a narrow range of turbine pressure measured signal plus steam discharge flow signal in total as a substitute.

The bypass valve of feed water configures manual/automatic control stations, as well as signal tracking system to ensure non disturbance of manual and automatic control.

The liquid level control system of evaporator is realized by feed water control system. Under the conditions of high and low loads, the feed water's changing scope is very large and the frequent actions of main valve can not only wear the valve excessively, but also have a bad effect of adjusting performance.

In high loads, the set point of feed water flow is generated by steam generator: the deviation of the liquid level of settings and measured value is delivered to liquid level regulator. Its output signal is added to measured signal of evaporator flow which has been filtered. The sum is provided to set value of feed water flow. For the steam flow is determined by loads, therefore, the set point of feed water flow would not be affected too much by the disturbance if the liquid level of evaporator was controlled well. The liquid level of evaporator is controlled by three pulses, which help the set value of main feed water flow not to change too much.

4 FUZZY CONTROL OF MAIN FEED WATER FLOW

According to the characteristic experiment in nuclear power plant simulator, we can gain the characteristic curve of main feed water flow. By using the technique of identification system, the transfer function of main feed water flow is as follows:

$$G(s) = \frac{0.0167}{4880s^2 + 140s + 0.935} \qquad (2)$$

To discuss the system of main feed water flow, a two-dimensional fuzzy controller is chosen. We take the system deviation e and deviation rate ec as input signals, and control variable u as output signal. According to the experience, the ranges of deviation, deviation rate and control variable

are [−6,6]. In the fuzzy control rule, the linguistic variables of e, ec and u are taken as 'Negative Big', 'Negative Middle', 'Negative Small', 'ZERO', 'Positive Small', 'Positive Middle', 'Positive Big', simplified as NB, NM, NS, ZO, PS, PM, PB. The output of fuzzy control is a fuzzy subset, which needs to be operated by solution of fuzzy taken gravity model approach to obtain corresponding control variable. Their membership functions are showed in Figure 3.

The design principles of fuzzy control system are: while the deviations are greater, the change of control variable should decrease the deviations quickly; while the deviations are smaller, except eliminating the deviations, the stability of system should be considered to prevent unnecessary overshoot or shock. The control rules are shown in Table 1.

After repeated experiments, the final image compared with the common is shown in Figure 4.

The step responses of two control methods come to stability quickly; however, there are some differences in comparing. At the process of response curve rising to the wave peak, the fuzzy PID control takes 270 s, while the traditional PID 320 s, so the former is faster. When reaching the stable values, the fuzzy PID takes 220 s, while traditional

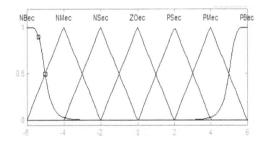

Figure 3.   The membership functions of e, ec and u in fuzzy controller.

Table 1.   Rule base of fuzzy control.

e u ec	NB	NM	NS	ZO	PS	PM	PB
NB	NB	NB	NM	NM	NM	NS	PS
NM	NB	NM	NM	NM	NS	ZO	PS
NS	NM	NM	NS	NS	ZO	PS	PM
ZO	NM	NS	NS	ZO	PS	PS	PM
PS	NM	NS	ZO	PS	PS	PM	PM
PM	NS	ZO	PS	PM	PM	PM	PB
PB	NS	PS	PM	PM	PM	PM	PB

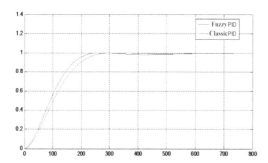

Figure 4. The response curve compared with fuzzy PID control and traditional PID control.

PID 270 s. Obviously, the response speed of fuzzy PID is very fast. And both the two responses have no overshoots. As we can see from the picture, the main feed water flow is controlled very well.

5 CONCLUSION

The set point of Feed water System is an important control system in the pressurized water reactor nuclear power station, which aims to keep target level steam generator needs. The article proposes a method to control the level of evaporator by using three pulses control at first, which can help the set value of main feed water flow not change too big, then to control the main feed water flow over fuzzy control. Eventually, the simulation shows that the control effect of main feed water flow is better.

ACKNOWLEDGEMENTS

This paper is supported by National Natural Science Foundation of China (Project Number: 61040013), National Natural Science Foundation of China (Project Number: 61203224), The Foundation of Shanghai Key Laboratory of Power Station Automation Technology (No.13DZ2273800), project of Science and Technology Commission of Shanghai (project number: 12510500800), Shanghai Natural Science Foundation (13ZR1417800), and Innovation Program of Shanghai Municipal Education Commission (Project Number: 09YZ347).

REFERENCES

1. Hao Y., Qiu L., Wang L. Thermo-economic performance comparison between two steam source schemes in the auxiliary steam system in nuclear power plants[C]//Mechanic Automation and Control Engineering (MACE), 2010 International Conference on. IEEE, 2010: 5136–5139.
2. Huang W., Huang Q., He Q., et al. Decoupling control for steam and water system[C]//Systems and Control in Aerospace and Astronautics, 2008. ISSCAA 2008. 2nd International Symposium on. IEEE, 2008: 1–5.
3. Roldan-Villasana E.J., Vazquez A.K. Model of the Feed Water System Including a Generic Model of the Deaerator for a Full Scope Combined Cycle Power Plant Simulator[C]//Computer Modeling and Simulation (EMS), 2010 Fourth UKSim European Symposium on. IEEE, 2010: 271–276.
4. Wang J., Wang Z. Boiler feed water control system based on improved MFAC[C]//Industrial Engineering and Engineering Management (IE & EM), 2011 IEEE 18th International Conference on. IEEE, 2011: 907–910.
5. Nilsson J., Wojciechowski A., Stromberg A.B., et al. An opportunistic maintenance optimization model for shaft seals in feed-water pump systems in nuclear power plants[C]//PowerTech, 2009 IEEE Bucharest. IEEE, 2009: 1–8.

Power and Energy – Kong (Ed.)
© *2015 Taylor & Francis Group, London, ISBN 978-1-138-02782-4*

The present situation of solar thermal power in China and its policy-thinking

Wei Wang
School of Humanity and Social Science, Research Center for Beijing Energy Development,
North China Electric Power University, Beijing, China

Anhui Li
School of Humanity and Social Science, North China Electric Power University, Baoding, China

ABSTRACT: Solar thermal power is attracting more attention from many countries because of energy shortages. In China, solar thermal power is in a preliminary stage, its development still face many difficulties. This paper introduces the present situation of solar thermal power and makes an analysis of its problems, and then provides some preferential policies.

Keywords: solar thermal power; present situation; policy thinking

1 INTRODUCTION

Energy problem has always been the focus of all countries. With conventional energy shortages and environmental degradation, states have stepped up the development of renewable energy sources. In 2010, The International Energy Agency published "solar thermal power technology roadmap" and mentioned that the global cumulative installed capacity of solar thermal power will reach to 10 GW, accounting for 11.3% of global electricity production by 2050, and China solar thermal power production will account for 4% of global electricity production. In some areas solar energy resource is rich, solar thermal power is expected to become a competitive bulk power and serve as peak regulating power load and intermediate power load until 2020 and based power load from 2025 to 2030. European countries attached great importance to the solar thermal power and got great achievements. The U.S. and Spain are the main market of solar thermal power. In Spain, there are 45 solar thermal power stations in commercial operation, with a total installed capacity of 525 MWe. In the United States, there are 5 solar thermal power stations under construction, with a total installed capacity of 1312 MWe. China will pay more attention on solar thermal power and support it by investing more capital and providing advanced technology in the future, thus promoting the industrialization of the operation of solar thermal power generation, which will have a great influence on China's economic and energy safety.

2 DEVELOPMENT OF CHINA'S SOLAR THERMAL POWER

Currently, the development of solar thermal power is still in a preliminary stage in China. China has abundant solar energy resources, so the development of solar thermal power has great feasibility and potential.

2.1 *Feasibility and potential of China's solar thermal power*

China is rich in solar energy resources. According to the amount of accepting solar radiation, the whole country can be divided into five regions. The first region is the most abundant area; the total annual solar radiation is 6680 ~ 8400 KWh/m², just like the western part of Tibet. The total annual solar radiation of the second region is 5850~6680 KWh/m², which includes the southern part of Xinjiang. The third region is 5000~5850 KWh/m², the fourth is 4200~5000 KWh/m². The poorest region includes Sichuan and Guizhou, of which the total annual solar radiation is 3350~4200 KWh/m².

Solar thermal power generation is such a technology that gathers together solar energy to produce a high temperature and then heat the working medium to drive generator and generate electric power (Wang, 2006). China has abundant solar energy resources, which provides a solid foundation for the development of solar thermal power. It also has obvious advantages compared with other power generation. Thermal power generation is the traditional way which used mineral resources such as coal and oil to generate electricity, thus consuming a large amount of primary energy and causing serious environmental pollution. By contrast, solar thermal power generation is clean and it generates solid waste little. As for wind power, wind power plant must be built in the outlet, it applies to a small geographical. Compared with wind power station, solar thermal power is suitable for a wide area, which can be built in the northwest region. What's more, nuclear power is questioned by the public after nuclear accident happened at Japan's Fukushima nuclear plant. While solar thermal power plants convert solar radiation energy into electrical energy without the use of high-risk chemicals. As for Hydropower, a large amount of fixed assets must be invested in Hydropower, while the initial investment cost of solar thermal power station is far lower. Comparison with photovoltaic generation, solar thermal power storage system can also maintain 2 or 3 days to keep the power plant running when the sunlight is short. In general, with the background of the growing energy shortage and the increasing demand for electricity, solar thermal power will have broad market prospects and play an increasingly large proportion in the energy structure and then change into imperative energy source.

2.2 The development status of China's solar thermal power

In Inner Mongolia, Erdos 50 MW solar parabolic trough type thermal power project was carried franchise tender, this project was considered the China's solar thermal power industry awakening. This paper will analyze the development status of China solar thermal power from three aspects: status of industrialization, status of policy, status of technology.

Industrial chain is formed but not complete. Figure 1 shows that solar thermal power industry chain includes five links: raw materials production, parts production, power plant systems integration, industry service system and production equipment. In raw materials production link: China is one of the suppliers of raw materials and even exports aluminium brackets to other countries. In parts production link: some companies have got

small achievements in the research on key components and could produce some critical components independently, such as trough type vacuum tubes and mirrors. The amount of domestic manufacturers of trough type vacuum tubes have reached to 17, mirror manufacturers also reached to 7. In power plant system integration link: electricity design firms and power development corporation are making efforts to the study of solar thermal power plants. In service system link: China is still in the blank stage. In production equipment link: China's production capacity is weak, there is a big gap between Chinese level and international advanced level.

Demonstration projects and commercial projects are at a started but preliminary stage. In recent years, China attaches great importance to the development of solar thermal power projects, promoting a number of demonstration projects, such as CAS electrician 1 MW tower type power station, Chinese Academy of Sciences electrician trough type collecting system, Shanghai Yi Cobo Company's 1 MW power plant. Commercial projects also entered the preparatory stages, such as SGCC Delhi 50 MW project, SGCC Golmud 50 MW project, Datang Group's Erdos 50 MW trough type power plant project.

In 2006 China issued "Renewable Energy Related Regulations", indicating that the state encourages and supports renewable sources connect into power grid. In 2007, China issued "Notice on the renewable energy tariff subsidies and quota trading scheme", marking that the renewable energy price subsidies were formally implemented. Solar thermal power development is still in its infancy in China, the state supports it mainly through informed development plan. In 2010, The national Ministry of science and technology launched the "Twelve Five-Year" national basic research project. That is "Efficient Large-scale Basic Research Program About Solar Thermal Power"; in 2011, the National Development and Reform Commission launched "Guiding Catalogue of Industrial Structure Adjustment (2011)", solar thermal power was in the first place of the guiding Catalogue; in 2013, the National Solar Thermal Industry Technology Innovation Strategic Alliance and the National Renewable Energy Center issued "China Solar Thermal Power Industry Policy Research". But specialized incentive policies for solar thermal power generation have not yet been introduced.

Our country started doing research on solar thermal power technology in 1979. After 30 years' development, we have got great achievements on many aspects, such as solar heat absorption, thermoelectric conversion and thermal materials storage. Although our country has made great progress on the study of solar thermal power technology, we

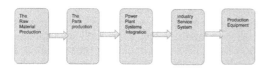

Figure 1. Solar thermal power industry chain.

still face major technical bottlenecks. This paper will illustrate bottlenecks from three aspects as follows, including: solar thermal power plant technology, production technology of key components, technical support service system.

3 PROBLEMS OF CHINA'S SOLAR THERMAL POWER

The development of solar thermal power in China faces many problems. This paper will analyze three main problems: the cost, the technology and the price.

3.1 The cost problem

Cost is a key factor in the promotion of solar thermal power. Take solar thermal power plants in foreign countries for example, the 50 MW trough type power plant of 7.5 hours' molten salt thermal storage has a total investment of 310 million euros, the cost has become an obstacle of solar thermal power technology promotion. Figure 2 shows that the cost of solar thermal power mainly includes three aspects: early research cost, the initial investment cost, operating and maintenance cost. The construction of solar thermal power plants completely relies on advanced technology and requires sufficient funds, thus leading to the high cost of early research. The initial investment cost of solar thermal power stations includes many aspects: building solar farm, installing thermal oil systems, installation of electrical systems and heating systems, maintenance of the control system, the cost of land. The worse is, our country does not have the production capacity for some needed key components, so we have to buy some needed key components from abroad, thus increasing the initial investment cost. Meanwhile, a good run on solar thermal power plants requires good maintenance: examining the equipment, replacing failed parts regularly and keeping up with international trends, which causes high cost of operation and maintenance.

3.2 The technology problem

Solar thermal power plant technology includes: tower type technology, trough type technology, dish Stirling type technology, linear Fresnel type technology (Li et al.2004). These power plant technologies have varying degrees of technical bottlenecks. Technical bottlenecks of tower solar thermal power stations: firstly, photo-thermal conversion efficiency is only about 60%. Secondly, different heliostats have different radius of curvature, the complexity of the optical design cause the manufacturing cost increased, and China cannot produce cheap heliostat. Thirdly, numerous heliostats were built around the central tower, each heliostat needs two-dimensional control, so the control system is extremely complex. Technical bottlenecks of trough type solar thermal power stations: firstly, heliostats are prone to appear the phenomenon of two-phase laminar flow inside the collector pipes, so trough type vacuum tubes are easy to be damaged. Secondly, the price of used oil is high. Thirdly, the design of control system and components is relatively complex. Technical bottlenecks of dish Stirling type solar thermal power stations: on the one hand, thermal storage is difficult; on the other hand, the risk of hot molten salt heat storage technology is high. Technical bottlenecks of linear Fresnel type solar thermal power stations: firstly, concentrating solar collecting mirrors were arranged in a relatively dense way, blocking each other seriously, causing huge heat loss and low system efficiency. Secondly, there is a long distance between mirrors and heat absorbing tube, condenser is needed for condensing the second time, thus increasing the cost.

Technology of trough type solar concentrator and high-temperature solar collecting vacuum tubes is still in the experimental stage. Control techniques of tower heliostat, precision manufacturing technology of high performance concentrating collectors and production technology of dish Stirling engine are not mastered by China. There are still many technical bottlenecks in the key components technology.

Overall operation of domestic solar thermal power plant is still in the experimental stage, lacking experience on overall design and operation of power plants and a complete set of standards and testing system to evaluate the system health. Modeling system and simulation technology is in blank.

3.3 The price problem

On the one hand, the total cost of solar thermal power is too high, resulting in the unit price to

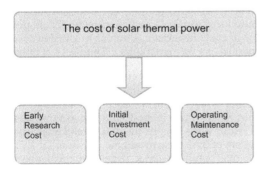

Figure 2.   The cost of solar thermal power.

too high. Take first domestic solar thermal power demonstration projects—Inner Mongolia 50 MW trough type solar thermal power projects for example, tax tariff is 2.26 Yuan/kWh, the cost of power plant is 2.6 million per kilowatt, suppressing purchasing willingness of power grid enterprises. On the other hand, solar thermal power does not enjoy various preferential policies and price subsidies.

4 POLICY THINKING OF SOLAR THERMAL POWER INDUSTRY

To promote development of solar thermal power industry, the government must establish and implement a system of preferential policies.

4.1 Tax preferential policy

The development of solar thermal power is in a difficult stage, it is supposed to receive policy support. A series of tax subsidies for solar thermal power can not only give the solar thermal power companies financial support, but also give investors confidence. Terms of tax base preferential policy, the state should allow the amount of gain or loss after netting as tax basis in order to reduce the overall tax burden. Terms of tax liability preferential policy, the state is supposed to provide Income Tax preferential policy by reducing reasonable proportion and implement VAT preferential policies; as for imported parts used in solar thermal power generation, the state could implement Tariff preferential policy of exempting or reducing. Terms of tax time preferential policy, the state had better allow solar thermal power enterprises delay paying tax time and give them tax holiday. What is more, public infrastructure projects enjoy tax preferential policy in China, the state should introduce provision to add solar thermal power generation into public infrastructure projects, making solar thermal power generation enjoy tax preferential policy.

4.2 Tariff incentive policy

Solar thermal power is in the preliminary stage: technology is not yet mature, cost of power station is high, and these disadvantages will hinder its development. The government could learn advanced experience from foreign countries to introduce specific price subsidy policy for solar thermal power (Xie et al.2013). Learning advanced experience from Germany, the state should establish a linkage mechanism of price subsidies: adopt a fixed price policy to achieve capital subsidies, introduce the provisions of the standard solar thermal power subsidy, establish automatic price adjustment mechanism and government price adjustment mechanisms to ensure subsidy price can be adjusted according to market supply and demand. Learning advanced experience from Denmark: subsidy price of solar thermal should be adjusted according to the floating market trading price, and the subsidy is set up not only for price but also for technique.

Reasonable tariff policy could guide enterprises control costs of solar thermal power to achieve the incentive effect. The government should introduce tariff policy for solar thermal power generation, using way of combining agreeing price and government price (Zhang, 2008). Firstly, the power grid enterprises and solar thermal power companies initially achieved agreement price, then the government finally determine the final price after considering interests of many factors (grid power companies, solar thermal power enterprises, public acceptance of electricity, the national energy development strategy). Terms of government pricing, the state could learn from wind power tariff policy and then divide solar energy resources region into different parts and make different parts perform different price (Wu et al.). After the promulgation of solar thermal tariff policy, the government is supposed to pay close attention to the implementation and enforcement of policies to make sure that each power grid enterprise and solar thermal power company keep accurate and complete records on the solar thermal power projects' online trading volume and accept government departments' supervision.

4.3 Credit incentive policy

Lacking of funds has become an obstacle to the development of solar thermal power generation. So the special credit preferential policy will make a big difference. Firstly, DOE should introduce loan guarantee program to provide loan guarantee for the solar—fired power plants and establish guarantee fund. Guarantee fund is supposed to pay off loan losses caused by the solar thermal power project. Secondly, the state could make efforts to introduce a series of orientation guiding credit policy to guide financial institutions to invest in solar thermal power generation. Thirdly, increasing disposable loan amount for solar thermal power companies will be a good measure. Fourthly, the state should introduce provision of solar thermal special power company benchmark lending rate, thus reducing the interest burden and repayment pressure. What is more, the state should establish a system of loan repayment time extension for solar thermal power business, if solar thermal power companies could not pay money back on time, they could apply for deferment.

4.4 *Land incentive policy*

Land policy has a significant impact on the industry. Therefore, the state should introduce incentive policy for solar thermal power land: reduce the price of land used in solar thermal power project; relief the land use tax, reducing the financial burden of solar thermal power generation enterprises; the state should give priority to solar thermal power plants if the solar thermal power generation companies compete with other types of enterprises for one land; provide good infrastructure facilities for the land, such as roads and water supply.

5 CONCLUSION

Solar thermal power generation occupies an important position in the national energy strategy, so it is imperative to develop it. This paper recommends that government should strengthen the research of solar thermal power and provide a system of preferential policies in order to accelerate the market acceptance of that.

ACKNOWLEDGEMENTS

This article is supported by the Beijing municipal education commission special funding and supported by the Fundamental Research Funds for the Central Universities.

REFERENCES

1. BinLi, Anding Li. 2004. Solar thermal power technology. *Electrical equipment* (04):80–82.
2. Xuxuan Xie, Zhongying Wang, Hu Gao. 2013. Advanced national renewable energy development subsidy policy trends and Enlightenment to China. *China energy* (08):15–19.
3. Yinan Wang. 2006. A point view of development of solar thermal power in China. *Chinese energy* (08):5–10.
4. Yu Zhang. 2008. Feed-in tariff mechanism in theory and practice. *NCEP technology* (07):28–31.
5. Yu Wu,Yongmin Bian. 2013. China's wind power feed-in tariff subsidy policy research under the WTO vision. *Macroeconomic Research* (10):40–46.

Power and Energy – Kong (Ed.)
© 2015 Taylor & Francis Group, London, ISBN 978-1-138-02782-4

Application of a novel TD-LTE power private network at discrete narrow band

Xiaobin Wei & Jianming Zhang
School of Electronic Engineering, Beijing University of Posts and Telecommunications, Beijing, China

Zhan Shi
Power Dispatch and Control Center of Guangdong Power Grid, Guangzhou, China

ABSTRACT: TD-LTE, the 4th generation mobile communication technology, with the advantages of spectrum utilization and asymmetric transmission, now is popular with the power private network builders. However, it is difficult to construct TD-LTE private network because of the scarce spectrum resource and the high loan cost. In this paper, we will take China power industry for example to explore the construction of a TD-LTE power private network by using the VHF resource, and also will carry on a simulation to analyze the performance of the Distribution and Utilization Information Acquisition Private Network, in order to provide references for the construction of TD-LTE power private network.

1 INTRODUCTION

For public security, army, as well as oil fields, electricity, rail transit, inland waterway navigation, airports and other key industries, the public communication network, which focus on civilian communication, can hardly meet their unique demands: 1st, stability, when a sudden emergency happens at the public places, it may issue in the communication cut, and also may lead the information platform to paralysis, which will cause a secondary hazard with much more damage; 2nd, safe differentiated design for operation authority, different permission should be set in the industry manufacturing platform according to the jobs, and cross-tier communication is forbidden, it is hard to be realized by the public communication technology; 3rd, cluster scheduling, which is essential for the multimedia communication. The public communication also can hardly provide. Therefore, to construct a private network, linking closely to the safety in production, management administration and emergency response, has been a key project for those fields like power, transportation and public security.

In this paper, we will analyze the construction of the private network for power industry, where Electric Power Monitoring and Management System, Electric Power Telecommunication and Data Network must be realized by the private network. And a huge amount of terminals are distributed in the network discretely. Distribution and Utilization

Communication, the main business of the power private network, is primarily responsible for data transfer of the distribution products and marketing operations.

In 2009, China State Grid Corp put forward the goal of "Strong Smart Grid". And so far, a phase objective has been achieved: the 110 kV and above voltage level power transmission network has been 100% optical fiber, and the 35 kV and 10 kV mid-voltage distribution network has been part-fiber. However, at some places of the city, to build a fiber network needs too much resource, which is a cost way. In this case, to build a wireless private network to extend the wired fiber network would be a good option.

In 1995, China Radio Regulatory Commission allocated 40 frequency points during 223.025~235.000 MHz to the power industry. Wireless digital broadcasting station used to be used for the power wireless communication. However with the rapid development of mobile multimedia services, now it can hardly provide broadband services to meet these needs.

In this paper, we propose a novel construction project for the power wireless private network. It applies the technologies of TD-LTE to realize the information acquisition and remote control. TD-LTE is based on Orthogonal Frequency Division Multiplexing (OFDM), makes use of Time Division Duplex (TDD) to arrange the transmission slots, and provides low transmission delay through a flat network structure. By using carrier aggregation, TD-LTE also can effectively make use of the power industry's discrete spectrum resource, which will substantially reduce the cost.

The rest of the paper is organized as follows: In Section 2, we present the usage of carrier aggregation. Section 3 analyzes the capacity of 230 MHz TD-LTE private network. In Section 4, we introduce the business requirements of Distribution and Utilization Communication System. Section 5 analyzes the performance through a simulation. Finally in Section 6, we conclude our paper.

2 CARRIER AGGREGATION

According to the regulations of China Radio Regulatory Commission, there are 40 frequency points during 223.025~235.000 MHz, 1 MHz in total, allocated to the power industry in China. The spectral distribution is shown in Figure 1.

The wireless digital broadcasting station technology has been used for power industry. However as wireless digital broadcasting station can only use one frequency point for data transmission, it can hardly meet the need of the power wireless broadband communication. By using carrier aggregation, TD-LTE can effectively make use of those discrete frequency bands as one continuous broad band, and get a higher data transmission speed. All of these operations could be realized by a base band processing module.

Figure 2 shows Discrete Carrier Aggregation, which can effectively make use of those discrete narrow frequency bands. However, in order to reduce the interference caused by the leakage at the adjacent frequency points, there is a need to reverse 10% of the frequency for the protection band, so after carrier aggregation, we can get 0.9 MHz band for the private network.

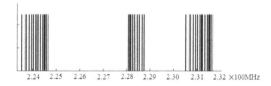

Figure 1.   The spectral distribution.

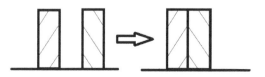

Figure 2.   Discrete carrier aggregation.

3   THE CAPACITY OF 230MHZ TD-LTE PRIVATE NETWORK

The capacity of TD-LTE system is decided by many factors: system bandwidth, scheduling algorithm, transmitter power, length of the cyclic prefix, resource allocation, sub-carrier spacing, time slot schedule, link overhead, MIMO, interference cancellation and so on. But in this paper, we only focus on the effects of the system bandwidth, resource allocation and link overhead.

3.1   *System bandwidth*

TD-LTE assigns no special filter in time domain, but sets a transitional band to eliminate the impact of "waveform broadening" and "waveform oscillation" in the time domain. So not only a guard band should be set for carrier aggregation, but also some bands should be reversed for the protecting bandwidth in TD-LTE. Even with more guard bands, the leakage of energy out of the system bandwidth will be smaller. But considering a balance between the frequency spectrum efficiency and the system protection, we assign 10 percent of the frequency resource for the guard band. As a result, we would get 20 kHz per discrete frequency point at 230 MHz for the power wireless communication.

In TD-LTE, the sub-carrier spacing Δf is 15 kHz. So the total number of subcarriers in 230 MHz TD-LTE system would be:

$$N_{sc} = \left[\frac{20000}{\Delta f} \times 40 \right] = 53 \qquad (1)$$

The number of symbols per time slot would be:

$$N_{symb} = \left[\frac{0.5}{T_{symbol}} \right] \times 2 = 14 \qquad (2)$$

The symbol period is: $T_{symbol} \approx 66.67 \ \mu s$.
The number of the Resource Blocks per time slot would be:

$$N_{RB} = \left[\frac{N_{sc}}{12} \right] = 4 \qquad (3)$$

3.2   *The configuration of uplink-and-downlink time slot and special sub-frame*

TD-LTE system supports two switching period: 5 ms and 10 ms, in total 7 kinds time slot configuration. Meanwhile, in order to save the cost of the network, special time slot such as DwPTS and UpPTS can be used to transfer data and the system control information.

3.3 Resource grid

In TD-LTE system, whether uplink or downlink, the minimum resource unit is Resource Element (RE), the physical channels correspond to a range of REs to carry the high-level information. The signals, transferred in a slot, can be described as a resource grid.

3.4 Cyclic Prefix

In order to avoid serious Inter-Symbol Interference (ISI) and Inter-Carrier Interference (ICI), the length of CP should be much larger than the maximum allowable delay spread of the channel. In TD-LTE, when it is normal CP, the number of symbols in uplink each slot N_{symb}^{UL} would be 7, when it is Extended CP, N_{symb}^{UL} would be 6.

3.5 Downlink overhead

In TD-LTE system, the downlink overhead includes Reference Signal (RS), Synchronization Signal, Physical Broadcast Channel (PBCH), Physical Control Format Indicator Channel (PCFICH), Physical HARQ Indicator Channel (PHICH) and Physical Downlink Control Channel (PDCCH).

RS: RS is used for channel estimation and demodulation. There are 3 kinds of RS: Cell Specific Reference Signals, MBSFN Reference Signal and UE Reference Signal.

P-SCH/S-SCH: Primary Synchronization Channel (P-SCH) is mapped to the 3rd OFDM symbol of the DwPTS. Secondary Synchronization Channel (S-SCH) is mapped to the last OFDM symbol of sub-frame-NUM0. 72 REs will be assigned for P-SCH, as well as S-SCH.

PBCH: PBCH is mapped to the front 4 OFDM symbols in the 2nd slot of sub-frame 0, in total 288 REs. Since there are 24 REs would be assigned for RS, so 264 REs are assigned for PBCH.

PCFICH: PCFICH is mapped to the first OFDM symbol of the DwPTS, and is used to transmit the information of PDCCH. 16 REs are assigned for PCFICH.

PHICH: PHICH is mapped to the first one OFDM symbol or two OFDM symbols of the DwPTS, and is used for carrying HARQ ACK/NAK information. 12 REs are assigned for PHICH.

PDCCH: PDCCH is mainly responsible for the uplink and downlink resources allocation, MCS allocation, the transmission of ACK/NAK, power control and the transmission of most of the L1/L2 signaling.

This article assumes that RS, PCFICH, PHICH and PDCCH completely occupies the first one or two OFDM symbols of the DwPTS, so PDSCH data will not be transmitted at the first one and two OFDM symbols of DwPTS.

3.6 Uplink overhead

In TD-LTE, uplink overhead includes DMRS (Demodulation Reference Signal), PUCCH (Physical Uplink Control Channel) and SRS (Sounding Reference Signal).

DMRS: DMRS is used for the base station to estimate the channel at the time of coherent detecting, and schedule the uplink users. It is transmitted with the uplink traffic data, which applies for PUSCH, or control signaling which applies for PUCCH. When RS and uplink service data are transmitted together, 12 REs per RB would be used for DMRS.

PUCCH: PUCCH is used for carrying uplink control information, like ACK/NACK, CQI, MIMO feedback information, SR and so on. The RBs at the edge of the working frequency band are assigned for PUCCH, in total 288 REs.

UpPTS: When the length of UpPTS is 2 OFDM symbols, the UpPTS can be used for PRACH or SRS; when the length of UpPTS is 1 OFDM symbol, it can be used for SRS, and UpPTS won't carry uplink data.

3.7 Transmission rate of 230 MHz TD-LTE power private network

In this paper, the system configuration is: 1 MHz system bandwidth, 2×2 antenna, normal CP, the uplink and downlink configuration-NUM0, the uplink and downlink sub-frame ratio of 3: 1, the special sub-frame configuration of 10: 2: 2, 2 OFDM symbols in DwPTS for the transmission of control information.

3.7.1 Downlink transmission rate

Downlink transmission speed is related to the number of REs for PDSCH in downlink sub-frame and the number of RE for PDCCH in DwPTS.

Each frame has two sub-frames for downlink, in total 1344 REs. Except the first two OFDM symbol in DwPTS for PDCCH, there are 768 REs left. And, 744 REs is assigned for the link spending, so there are 1080 REs left for PDSCH. With different channel environments, the downlink transmission rate of 230 MHz TD-LTE private networks would be:

$$Rate_{DL} = N_{RE} \times ModType \times \frac{CodeRate}{1024} \qquad (4)$$

In the formula, N_{RE} is the number of REs for PDSCH per frame, ModType is the modulation scheme, CodeRate/1024 is the validity of the modulation coding.

3.7.2 Uplink transmission rate

Uplink transmission rate is related to the number of REs for PUSCH in uplink sub-frame. Each

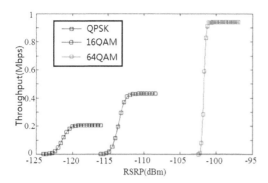

Figure 3. Transmission rate of uplink with different MCS.

frame has six sub-frames for uplink. At the edge of the frequency band, 2 RBs are assigned to PUCCH per time slot, so 2 RBs are left for PUSCH. Besides, 1 OFDM symbol would be assigned for transmitting the uplink RS per time slot, so in total 1728 REs can be used for PUSCH. With different channel environments, TD-LTE uplink transmission rate would be:

$$Rate_{UL} = N_{RE} \times ModType \times \frac{CodeRate}{1024} \qquad (5)$$

In this formula, N_{RE} is the number of REs for PUSCH per frame, ModType is the modulation scheme, CodeRate/1024 is the validity of the modulation coding. Figure 3 shows the uplink maximum transmission rate through AWGN channel with different modulation and coding schemes, and we can see that: with the QPSK modulation mode, the uplink maximum transmission rate can reach 198.4 kbps; 16QAM, the uplink maximum transmission rate can reach 406.05kbps; 64QAM, the uplink maximum transmission rate reaches 959.9kbps.

4 THE BUSINESS OF POWER WIRELESS NETWORK

The power wireless network is at the end of the power communication network, and contacts with the users directly, which has many characteristics, such as: a large number of terminal nodes, widespread distribution, unbalance density of the nodes and low communication traffic per node. Distribution and utilization communication, as the primary application, is mainly responsible for the transmission of the power distribution production business data and marketing business data.

4.1 Distribution business

The main devices of the distribution network are pole top breaker, ring main unit, switching station, pole transformer and box transformer. The data volumes of these devices are shown in the Table 1.

4.2 Utilization business

According to the property of the power users and the requirement of marketing business, the power users can be divided into six types: class A is large-scale transformer user, class B is Small and medium-scale transformer user, class C is three-phase commercial user, Class D is single-phase commercial user, class E is residential users, class F is public assessment measurement points.

The data acquisition of the power user power-expenditure information mainly includes the functions of automatic information acquisition, interrogation, information report, tele-control and parameter setting. The data volumes of the power users are shown in Table 2.

4.3 Business requirements

In this paper, we assume that each distribution area has K output lines, and each distribution line has 15 pole top breakers, 2 switching stations, 8 ring main units, 30 box type substations, and 50 pole transformers. And there are N electrical terminals are distributed in the distribution area, including 10 terminals of Class A, 90 terminals of Class B, 9 terminals of Class C, 81 terminals of Class D, 730 terminals of Class E, 60 terminals of Class F per 1000 utilization terminals. In order to simplify the calculation, K and N satisfy the following relationship:

$$K = N/1000 \qquad (6)$$

In the acquisition system, the uplink terminal number N_{Event}, who will transfer event information to the station per minute, submits to Poisson Distribution:

$$P(N_{Event} = k) = \frac{e^{-\lambda}\lambda^k}{k!}, \lambda = 0.001 \times N \qquad (7)$$

Figure 4 shows the uplink traffic distribution of the distribution and utilization information acquisition system with different numbers of the distribution and utilization terminals, it can be calculated as follows:

$$T_{rUL} = Per_{Dis} + Per_{Uti} + E \times Event \qquad (8)$$

In the formula, Per_{Dis} is the uplink periodic service of the distribution terminals, including

Table 1. Distribution service.

Terminal	Business data (byte)			
	Tele-command	Tele-metering	Tele-control	Degrees of ammeters
Pole Top Breaker	2	22	10	0
Switching Station	10	80	75	112
RMU	3	32	45	0
Box Transformer	2	28	30	0
Pole Transformer	1	26	0	0

Table 2. Utilization service.

Terminal	Periodical service (bps)	Events upload	Download control	Parameter setting
Class A	145.92	240 bit	87 bit	400 bit
Class B	86.19	220 bit		
Class C	8.53	227 bit		
Class D	8.53	193 bit		
Class E	14.93	218 bit		
Class F	21.1	210 bit		

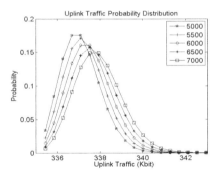

Figure 4. Uplink traffic probability distribution.

tele-command, tele-metering and degrees of ammeters. And, Per_{Uu} is the uplink periodic service of the utilization terminals, E is the number of the utilization terminals who upload event information, which submits to Poisson distribution, Event is the data size of the event information of one utilization terminal.

5 THE PERFORMANCE ANALYSIS OF 230MHZ TD-LTE POWER PRIVATE NETWORK

5.1 Maximum urgency first

In this simulation, we use Maximum Urgency First (MUF) algorithm to schedule the users. MUF

Figure 5. Task model.

takes the static priority scheduling algorithm and dynamic priority scheduling algorithm into account, and makes use of the dynamic priority and static priority to ensure the effective transmission of the burst traffic.

The priority of each task is called as the urgency, which consists of a static priority, a dynamic sub-priority and a static sub-priority.

And the dynamic priority is inversely proportional to the task-laxity.

$$Task\,Laxity = (d - a) - (e - b) \qquad (9)$$

The symbol 'a' is the arrival time of the task, the symbol 'd' is the deadline of the task, the symbol 'b' is the time when the task is executed, and the symbol 'e' is the time when the task is finished. And, static priority is decided by the type of the terminal, static sub-priority is decided by the traffic type.

5.2 Simulation environment

The simulation is based on TD-LTE dynamic simulation. And the sampling interval is a TTI (1 ms), and the simulation time is 600 s. Each simulation has a different number of the terminals. Then we will get the success rate of Distribution and Utilization information acquisition with different conditions. The system parameters have been shown in Table 3.

5.3 Analysis of the simulation result

In the Distribution and Utilization Information Acquisition System, the uplink capacity is limited. Therefore, the simulation focuses on the uplink channel.

When there are 6000 utilization terminals and 6 distribution outlines, we can see that 230 MHz TD-LTE power wireless private network can effectively

Table 3. System parameters.

Parameter types	Setting	Parameter types	Setting
Simulation scene	Dense urban	Channel type	AWGN
Path loss model	Loss = 139.3 + 20lg(f) −E(q,t,f,h_{Tri},d) dB	Uplink and downlink time slot allocation	Sub 0 and Sub 5 used for downlink, Sub 1 and Sub 6 used for special slot, the other sub slots used for uplink
OFDM symbol number per sub-frame	14	MCS level	29 levels according to 3GPP TR36.913
Sub wave interval	15 kHz	Scheduling algorithm	MUF
Resource block size	180 kHz (12 subcarriers)	CQI Feedback delay	5 ms
eNB Transmitting power	46 dBm	UE Transmitting power	17 dBm
The height of eNB antenna	30 m	The height of UE antenna	1.5 m

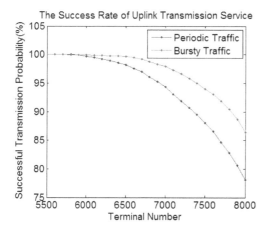

Figure 6. The success rate of uplink transmission service.

meet the needs of the distribution and utilization information acquisition. And it is in accordance with Power User Electric Energy Data Acquire System Functional Specification, that the success rate of the periodic information acquisition should be more than 99.5%.

6 CONCLUSION

This paper offers a novel network, 230 MHz TD-LTE power private network, to build the wireless private network, which can provide a guarantee for the Strong Smart Grid. By using carrier aggregation, we can make effective use of those discrete frequencies at 230 MHz frequency band. And through the analysis of the system simulation, we get the performance of the private network, which can provide references for the following construction of the private network.

REFERENCES

1. Cao Jinping, Liu Jianming, Li Xiangzhen. Carrier Aggregation Technology on 230 MHz Dedicated Spectrum of Power Systems [J]. Automation of Electric Power Systems, 2013, 37(12):63–68.
2. China Academy of Telecommunication Research of MIIT. Research on China private network technology and development stratagem [R]. Beijing: 2009.
3. Deng Xu, Zhang Jianguo. Analysis of TD-LTE Capacity [J]. Mobile Communications. 2011, 35: 49–52.
4. Gong Gangjun, Xiong Chen, Xu Gang. Research of communication business model of power distribution and utilization [J]. Power System Protection and Control. 2013, 22, 19–24.
5. Kang Zhiyi. Telecom service and communication equipment [R]. Beijing: 2010.
6. Q/GDW 373–2009, Power user electric energy data acquire system function specification [S].
7. Wang Xibo, Tong Xin. A Modified Maximum Urgency First Algorithm and Its Implement [J]. Control and Automation.2010, 26:40, 41, 88.
8. Xiao Qinghua, Mao Zhuohua, Ling Wenjie, Zhang jianguo. Comprehensive Analysis of TD-LTE Capacity Cpability [J]. Designing Techniques of Posts and Telecommunications, 2012, 4: 36–40

Power and Energy – Kong (Ed.)
© *2015 Taylor & Francis Group, London, ISBN 978-1-138-02782-4*

Effect of interfacial tension on viscous fingering patterns by DLA modelling

Qin Shenggao
China Center for Industrial Security Research, Beijing Jiaotong University, Beijing, China
Sinopec International Petroleum E&P Corp., Beijing, China

Jingchun Wu
EOR Key Laboratory of Ministry of Education of China, Northeast Petroleum University, Daqing, Heilongjiang, China

Jianwen Yan
Research Institute of Petroleum Exploration and Development, CNPC, Beijing, China

ABSTRACT: Based on analysis of DLA random walk of particles, the kinetic mechanism of viscous fingering process was explained comparatively with its counterpart of DLA in the same Laplace Field. Several quantitative variables of DLA model were changed to produce different clusters at different interfacial tensions. Their corresponding fractal dimensions were also calculated. Elaboration was made about the correlation between DLA model and viscous fingering process. The results show that during the multiphase fluid flow period, the displacing fluid with low interfacial tension has the same effects to reduce viscous fingering extent for oil recovery enhancement as the multi-directionally connected pores. This method overcomes the limitation of growth scale in the physical experiment and provides a convenient tool for further analysis of the effect of different parameters on viscous fingering patterns.

1 INTRODUCTION

Viscous fingering is one of the important phenomena during multi-phase flow (Jia et al. 2001a,b), especially in the environment where low viscosity fluid displaces high viscosity fluid. It exists commonly in the processes like oil displacement and the non-equilibrium growth. On this point, the pore structure complexity (Qin 2012, Yao 2012, Qin 2013, Wu 2014a, b) is the internal factor to cause such viscous fingering while the differences of physical properties among multi-phase fluids act as the external factors. To analyze the viscous fingering mechanism in porous media, some researchers tried to change the properties of injected fluids to produce viscous fingering patterns which have been restricted by the experiment conditions and scale.

The experimental pattern of viscous fingering process shares the same characteristics like self-similarity with DLA (Diffusion Limited Aggregation) clusters in the modeling process. They follow the same form of kinetic mechanism of the Laplace Field in the self-growth process. Hence DLA can be adopted to study the evolution features of viscous fingering process. This method overcomes the limitation of growth scale in the physical experiment

and provides a convenient tool for further analysis of the effect of different parameters on viscous fingering patterns.

2 DLA MECHANISMS

Figure 1 shows how DLA grows. The black particle in the center is the nucleation core here called seed I. A dozen of particles has attached to it to form a cluster with a radius of r_{max}. At a random location on the circle with radius r_{max+5}, a new particle S1 appears and then goes randomly in the horizontal matrix in the direction decided by a random number the computer produce between 0 and 1. For example the number in 0~0.25 decides a step up and the number in 0.25~0.5 decides a step down etc.. After several steps in t1 it would walk in the vicinity of the central cluster (see the dashed grid) until it stops walking and becomes a part of the cluster. To save time, any particle K which goes outside of $3r_{max}$ in t2 will be neglected because it needs to walked many steps to attach the central cluster. Afterwards the particles born in the distance will repeat the aforesaid steps until a big enough DLA cluster forms. So the growth-up of the DLA cluster is a result of surface nucleation

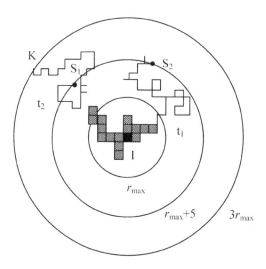

Figure 1. Mechanism of DLA modelling.

process of the diffusion particles. Fractal feature can be observed in the process.

If the cluster of particles was taken as the displacing fluid while the remaining space that hadn't been occupied by particles was taken as the displaced fluid, then the seed could make the role of an injector. Because any particle is assumed no intrusion inside the existing cluster it is not possible to find a free particle inside. So the displacement front between displacing phase and displaced phase moves at a rate that is decided by the appearance possibility of a particle near the front.

3 EFFECT OF INTERFACIAL TENSION IN DLA MODEL

The fractal growth of DLA follows the same form of mathematical model with the viscous fingering process. Its particle density U meets the below Laplace equations (Sun 2006, Liu 2009):

$$\nabla^2 U(\vec{r},t) = 0 \tag{1}$$

$$V_n = -a\vec{n} \cdot \nabla U(\vec{r},t) \tag{2}$$

$$U_\infty = c \tag{3}$$

where $\vec{r}$ is the radial vector; t denotes time; $\vec{n}$ is the unit vector perpendicular to the interface; V_n is the growth rate in the direction $\vec{n}$; a and c are constants. Although the non-linear differential equation (1) is very difficult to resolve, the DLA model can generate its corresponding diagram of the fractal Laplace growth cluster conveniently.

According to the discussion above, in the evolution process, the viscous fingering during displacement shares the same forms of kinetics equations group with DLA model. Both belong to the self-growth process in the Laplace Field. If the particles in DLA field were replaced by the molecules of displacing or displaced fluids, the function of particle density can be treated as the pressures between phase interfaces. The fractal growth theory can be introduced to the study of viscous fingering phenomenon with quantitative description by the fractal dimension. It provides a new method to analyze this physical process in the fluid flow through porous media.

However, in the flow process of multi-phase fluids the interfacial tension exists and acts as a very important factor to stabilize the phase interface. In the aforesaid DLA model (1)–(3) no variable describes the role of interfacial tension, so the obtained cluster has an open tree structure. To reflect the actual displacement process with interfacial tension, one of the effective methods is to introduce a new physical parameter called adsorption possibility p_s to describe whether the particle will go randomly around the cluster boundary and attach it or continue movement far from the cluster with a possibility $1-p_s$. p_s can be represented by the local curvature of the cluster κ as follows.

$$p_s(\kappa) = A\kappa + B \tag{4}$$

$$\kappa = \frac{N(r)}{\pi r^2} - n_0 = \frac{N(r)}{\pi r^2} - \frac{\pi}{4\sqrt{3}} \tag{5}$$

where A and B are constants; $N(r)$ is the particle number in a circle area with radius r; $n_0 = \pi/4\sqrt{3}$ is the area density of close surface stack.

The adsorption possibility in DLA model functions to change the interface shape, not the surrounded area. Hence the fractal growth of DLA is also a self-organization process. It is a good way to model the kinetic process of fluids.

4 EXPERIMENTAL RESULT

In this study, 20 cases were simulated to consider factors like interfacial tension (with and without interfacial tension), number of particles (5 groups: 500, 1000, 2000, 3000 and 4000), 2 walk types (4 directions: 0°, 90°, 180° and 270°; 8 directions: 0°, 45°, 90°, 135°, 180°, 225°, 270° and 315°). The fractal dimension of each obtained cluster was calculated by Sandbox method; see Tables 1 and 2 and Figures 2 and 3.

According to the discussion above, A is proportional to the interfacial tension. That is the big value of A will increase the possibility of a particle's transverse diffusion to reduce the

Table 1. Fractal dimensions vs. particles without considering the interfacial tension of DLA clusters.

Particle number	500	1000	2000	3000	4000
4 Direction	1.5751	1.7025	1.7181	1.7008	1.7221
8 Directoin	1.5923	1.6270	1.7755	1.7954	1.7303

Table 2. Fractal dimensions vs. the interfacial tensions of the DLA clusters (A).

Interfacial tension (A)	No	2	10	20
4 Direction	1.7181	1.7203	1.7368	1.7775
8 Directoin	1.7755	1.7600	1.7602	1.7855

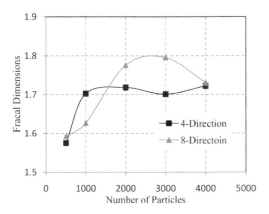

Figure 2. Correlation of fractal dimension vs. particles numbers.

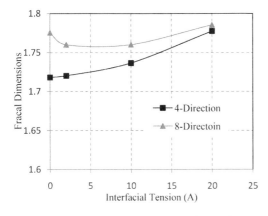

Figure 3. Correlation of fractal dimension vs. interfacial tensions (values of A).

growth rate of cluster's tips while to expedite the fractal growth rate at the hollow of clusters, then to make the DLA cluster seem compact and fat; on the contrary the obtained DLA cluster looks slim. Thus the interfacial tension acts to change the particles' movement direction. The clusters including 2000 particles in both 4 and 8 walk directions demonstrate the same characteristics and their fractal dimensions increase with the value of A. In Figure 2, it can be observed that the movement mode in 8 directions will increase its fractal dimensions quickly to a bigger peak value than the 4 walk direction, but will approach the same level. Figure 3 shows that with the increase of A value (opposite change of interfacial tensions), the cluster will have higher values of fractal dimensions, of which the incremental rate will become bigger and bigger. But actually, it is harder and harder to reduce the interfacial tension of a chemical fluid when the interfacial tension of this chemical agent has already reached its very low level. In the development process of low permeability reservoir, the formation pressure conductivity is always low due to poor physical properties, strong heterogeneity and big pore-throat ratio. Well stimulation fluids with low interfacial tension are very necessary to build an effective displacement pressure difference between oil and water wells to book the remaining reserves.

There is a great difference between the clusters from 4 walk directions and 8 walk directions. The first walk way produced a tree structure with thin and long trucks while the second one produced a tree structure with short and fat trunks. Quantitatively, the cluster obtained from first walk way has smaller fractal dimension than the second walk way. It indicates that the multi-directional connection functions in the same way as the displacing fluid with low interfacial tension to improve the sweep efficiency. The low permeability reservoir always has a great portion of pore space connected in one or two direction which is very bad for oil displacement.

5 CONCLUSION

Based on the same form of kinetic mechanism, quantitative variables were introduced into the general DLA model to analyze the effect of interfacial tension on the clusters' pattern. Its correlation with viscous fingering process was also presented.

1. The interfacial tension affects the particles' diffusion way: low interfacial tension will increase the possibility of a particle's transverse diffusion to reduce the growth rate of cluster's tips while to expedite the fractal growth rate at the hollow

of clusters, then to make the DLA cluster seem compact and fat to get big fractal dimension.

2. The 8 walk direction mode will produce thicker DLA clusters than 4 walk direction model at the same number of particles. Quantitatively, the cluster obtained from first walk way has bigger fractal dimension than the second walk way.

3. The multi-directionally connected pore space functions in the same way as the displacing fluid with low interfacial tension to improve the sweep efficiency.

ACKNOWLEDGEMENT

Here I must give thanks to the National Natural Science Foundation of China for its sponsorship under Project No. 51374075. In this research team, I have learned a lot of knowledge of not only physical modeling techniques but also simulation methods by a computer with the helps from all other members. Brainstorm are always been adopted by us to make discovery in the field of oil and gas development. It works very well and prepared us not to missing the important inspirations and physical phenomena in the study.

Thanks to J.C. Wu for his generous, patient and professional guide when I worked in his team. His encouragement pushed us to overcome difficulties in the process of researches.

I want to appreciate J.W. Yan's support here. He made many corrections to perfect this paper. It's very important to have his help to finish this paper, especially in my life outside of work.

I would like to thank all the professors, colleagues and friends who have share their ideas and friendship during research process.

REFERENCES

1. J.C. Wu, W.Z. Tuo and S.G. Qin, "Multi-Loops of Mercury Intrusion-withdrawal Processes and the Fractal Analyses," Proceedings of 2014 International Conference on Geological and Civil Engineering -ICGCE 2014 at Macau, 2014.1: 85–89.
2. J.C. Wu, Z.L. Zhang and S.G. Qin, "Fractal Description of Pores at Low Permeability Sandstone and the Inside Nonlinear Fluid Flowing," Proceedings of 2014 International Conference on Geological and Civil Engineering -ICGCE 2014 at Macau, 2014.1: 79–84.
3. J.H. Liu, J.H. Zhang, "DLA Simulation and Fractal Dimensions of Viscous Fingering for Various Experiments," Bulletin of Science and Technology, 2009.3: 131–135. (in Chinese).
4. L.S. Yao, S.G. Qin and G.H. Wang, et al. "Multifractal Characteristics of the Heterogeneous Pore Distribution of Cores," Proceedings of Conference of CDCIEM2012 at Zhangjiajie, 2012.3: 424–427.
5. S.G. Qin, H.L. Wu and M.B. Tian, et al. "Fractal Characteristics of the Pore Structure of Low Permeability Sandstone". Digital Manufacturing & Automation III, 2012: 482–486.
6. S.G. Qin, "Fractal Characteristics of One loop of Mercury intrusion-withdrawal Curves," Inner Mongolia Petrochemical Industry, Vol. 39, No. 12, 2013: 54–56. (in Chinese).
7. X. Sun, Z.Q. Wu and Y. Huang, Fractal Fundamental and its Application. Press of University of Science and Technology of China, 2006.3: 23–142. (in Chinese).
8. Z.Q. Jia, "Complexity of dynamic reservoir system and new theory of reservoir dynamics,". Journal of Daqing Petroleum Institute, Vol. 25, No.1, 2001: 14~17. (in Chinese).
9. Z.Q. Jia, Y.F. Wang and J.L. Fu, et al. "Characteristics of non-Darcy percolation and under the condition of low-permeability and low-velocity," Journal of Daqing Petroleum Institute, Vol. 25, No. 3, 2001: 73~76. (in Chinese).

Power and Energy – Kong (Ed.)
© *2015 Taylor & Francis Group, London, ISBN 978-1-138-02782-4*

Intaglio printing press intelligent drying system design based on WinCE

Zhong-ming Luo, Peng Cao & Bin Zhang
Beijing Institute of Graphic Communication, Beijing, China

ABSTRACT: Intaglio printing press drying control system has a great influence on the quality of printed matter and energy consumption, the general drying system of Intaglio printing press use PLC or industrial computer, its cost perfor mance is usually lower. In this paper, design a drying system based on WinCE control, its use Visual Studio 2005 developed a fuzzy adaptive PID intelligent control procedures, and designed a user-friendly interface, also put forward good advice to the stability of the system. This not only reduces the cost of production system and design difficulty, also solves the problem such as the energy waste, drying effect instability.

1 INTRODUCTION

Intaglio printing press occupies an important position in the whole packaging production, it impacts the drying system and the quality of product factory environmental is significant. Intaglio printing press printing production costs a large part of energy for the drying system, optimizing the control part of the drying system can significantly reduce energy consumption. General drying system of Intaglio printing press uses PLC or industrial computer to control, the control system is high cost, and not commonality. So to improve the control part of Intaglio printing press's drying system, it has very important significance.

2 THE SYSTEM DESIGN

In order to decrease the cost of control system, improve its versatility, the system uses WinCE touch one machine to be host computer to improve the system availability, timeliness, and provide a convenient touch operation for the drying system and let the system run more intelligent. Drying process uses air heat pump and electric heating devices. So let the system drying temperature more stable, reducing energy consumption significantly.

2.1 System architecture

This article describes the dryer used in unit-type Intaglio printing press, each machine are equipped with an Energy saving device (Fig. 1). Each unit and the host computer communicate via RS-485 bus. System optimizes the drying process, using heat pump technology for heat exchange, which produces hot air into the oven to drying prints,

cold air from heat exchange to governance workshop environment. In addition to air heat pump, but also equipped with electric heating devices. Only when heat pump working, the oven temperature has reached the target temperature, the electric heating does not work. Only when air heat pump reaches full capacity, the oven temperature cannot reach the target temperature, electric heating start. Fan promote the heat circulation in the system. WinCE touch one coordinated system components responsible for the whole system running, setting parameters and troubleshooting. In order to timely feedback to adjust the current oven temperature, the system uses thermocouple to measure oven temperature real-time. Temperature data transfer via RS-485 bus to the host computer, and then adjust the machine running parameters via fuzzy Adaptive PID algorithm.

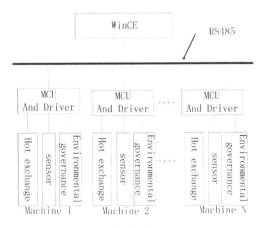

Figure 1. The system block diagram.

2.2 WinCE program development

The development of WinCE software is use Visual Studio 2005, with using the MFC class library, MFC is Microsoft Foundation Class Library, it is a C++ library developed by Microsoft, mainly encapsulates most of the windows API functions. MFC also provides many common frameworks, to reduce the workload and accelerate development progress. The main part of the software includes serial communication, intelligent control algorithms and software UI design, software flow chart shown in Figure 2.

The system for communication between PC and the drying machine via RS-485 bus. RS-485 use balanced drive and differential receiver combination, anti-common-mode interference, interface signal level than the RS-232-C reduced, difficult to damage the chip interface, and the level compatible with TTL level, the maximum data transfer rate to 10 mbps, the maximum transmission distance is 1219 m approximately, support many nodes. The main function of the serial port operations are:

CreateFile() open serial port.
SetCommState() According to the structure configuration communication equipment.
ReadFile() Read the data from serial port.
WriteFile() Write data from serial port.
CloseHandle() Close the handle of opened serial port.

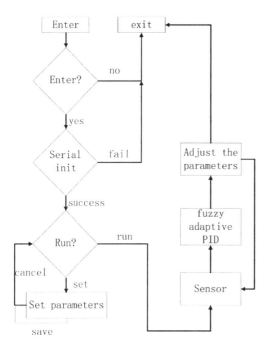

Figure 2. The flow chart of WinCE program.

In the program the serial read and write in a new thread, to ensure serial communication real-time. Receiving and sending process are based on an event-driven approach. OnComm events and CommEvent attribute to capture and examine communications event. When Communication event or an error occurs, it will trigger OnComm event, CommEvent property value will be changed, the application checks CommEvent property value and respond accordingly, and the final completion of the communication process.

System during operation using the fuzzy adaptive PID control algorithm. Intaglio printing press drying system is temperature controlled in open environment, so the system's temperature change with inertia, regulatory response delay characteristics. Conventional PID control algorithm has small amount of calculation, real-time, programming easy, widely used in the control field, correctly set the parameters Kp, Ki and Kd, and to establish a precise mathematical model of the control object can be achieved control action, but for the time-varying and nonlinear uncertainties, conventional PID control is not ideal. To make PID controller with parameter-adaptive function, you can put the parameters of conventional PID controller Kp, Ki and Kd in different conditions to be adjusted. The system uses fuzzy control rules on PID parameters automatically corrected, fuzzy adaptive strategy formation. Input error e (difference between the current measured temperature and the set temperature) and changes in the error rate ec, using the fuzzy control rules to correct the three parameters of PID controller in real time.

Software's interactive interface (referred to as UI) design is very important, UI design of the system with simplicity, easy to understand language, memory burden minimal, consistent principles, human and computer interaction logic is simple, interface generous. Figure 3 is screenshot

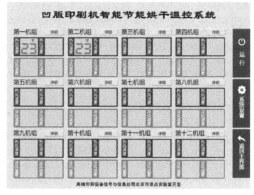

Figure 3. Screenshot of the drying system.

of the system's program. It allows the operator to operate the machine real-time and know the status of the machine; the operator can start up or stop the machine via the button on the right side of the screen. In the screen, show oven temperature by simulate digital tube, in this way the operator can observe the running system with a convenient distance.

3 IMPROVE THE SYSTEM'S RELIABILITY

Due to space and design reasons, the factory's electrical equipment is very complicated; there is much interference, so the system put forward a very high demand in stability. This system from several aspects to the reduce interference and improve the stability.

Ideal quality of power supply voltage should be stable, pure sine wave, frequency accuracy, uninterrupted power supply, but due to the factory has high-power machine, load changes, relay protection system switch-off, which will inevitably lead to the power voltage ups and downs. WinCE touch one machine control system has a large part of interference from the power load change. The system uses a shielded transformer to isolate most part of the electrical pulse interference. Also add the voltage sensitive resistance in the power cord to absorb electrical interference, voltage sensitive resistance can fixed voltage value to specific values, so realize the circuit protection.

In the factory, so a lot of harmful electromagnetic interference, it has great influence on the electronic devices, even when it so serious jeopardize to the device cannot work properly. The system uses shielding technology for WinCE touch one machine, isolates the WinCE touch one machine from outside, and prevents electric interference and magnetic interference.

The system uses the RS-485 bus to transfer data, in order to ensure the reliability of WinCE transfer instructions, so take some measures on the RS-485 bus to improve its anti-jamming capability. In its terminal add matching resistor, the resistance equal to the characteristic impedance of the transmission cable. Use shielded twisted-pair cable, shielded cable as the ground, and at one end of the cable (such as end of the main station) reliable access to ground.

4 THE EXPERIMENT RESULTS

In the tests, the system improves drying effect, energy efficiency and reduces labor. Ambient temperature is 30 degrees, set the printing drying temperature to 56 degrees, oven's temperature rapidly

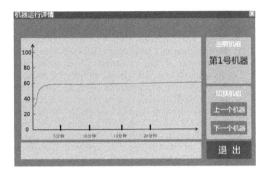

Figure 4. The software records oven temperature change.

increases when the system just started, when close to the target temperature the rising speed decreased gradually and reaches the target temperature within three minutes eventually. Dithering amplitude near the target temperature within plus or minus 2 degrees (Fig. 4). After entering the steady state, the system temperature fluctuations did not appear. In the usually a Intaglio printing press consumption 12–18 kwh power in one hour, the power consumption of the system only 6–8 12–18 kwh, more than 40% energy savings. At the same time, ambient temperature dropped about 3–5 °C near the printing units. And print quality has been significantly improved, reducing ink consumption. Experimental results show that the system has a good prospect and economic benefits.

5 CONCLUSION

Overall, Intaglio printing press reduces power consumption significantly after using the techniques described above, and increases system stability. Easy operation, but the drying system energy consumption still occupies a considerable part. In the future, drying system can also designed to heat recovery, governance workshop environment and other functions, to further improve the efficiency, make contribution to energy conservation and emissions reduction.

REFERENCES

1. Yujie Zhu, Mian Yang. Study on the Application of Activity Based Costing in Packaging and Printing Enterprise. Chinese Industrial Engineering Institution, CMES, Institute of Electrical and Electronic Engineers, Beijing Section. Proceedings of 2012 IEEE 19th International Conference on Industrial Engineering and Engineering Management (IE&EM 2012) [C]. Chinese Industrial Engineering Institution, CMES, Institute of Electrical and Electronic Engineers, Beijing Section, 2012:4.

2. J. Emmanouil D. Fylladitakis, Antonios X. Moronis, Konstantinos Kiousis. Design of a Prototype EHD Air Pump for Electronic Chip Cooling Applications. Plasma Science and Technology, 2014, 05:491–501.

3. J. Performance analysis of air-water dual source heat pump water heater with heat recovery [J]. Science China (Technological Sciences), 2012, 08:2148–2156.

4. J. Mo Yuqing (Hunan College of Information, Changsha, Hunan 410200). Research on Data Transmission Based on VC++MFC[A]. IEEE, Zhengzhou Institute of Aeronautical Industry Management, Henan University of Technology, University of Electronics Science and Technology of China, Sichuan Institute of Electronics. Proceedings of 2011 3rd IEEE International Conference on Information Management and Engineering (ICIME 2011) Vol. 03 [C]. IEEE, Zhengzhou Institute of Aeronautical Industry Management, Henan University of Technology, University of Electronics Science and Technology of China, Sichuan Institute of Electronics, 2011:6.

5. J. Huang Qing-ming, Xu Peng, Bao Neng-sheng, Chen Fang-yuan, Bai Wen-hua. Energy-saving & Emission Reduction Technological Control Research of Drying System to Gravre Press. Light industry machinery, 2009, 02:97–100.

Power and Energy – Kong (Ed.)
© *2015 Taylor & Francis Group, London, ISBN 978-1-138-02782-4*

Impacts of shale gas on the energy system in the United States

Qiaozi Zhao & Huikai Ding
College of Economic and Management, Shanghai Institute of Electric Power, Shanghai, China

Bin Chen
School of Economic and Trade, Hunan University, Changsha, Hunan, China

ABSTRACT: Shale gas has been found as a large area of continuous distribution of unconventional gas reservoirs in the United States. The application of horizontal well technology enables the exploitation of shale gas to have exponential growth and will completely change the supply and demand of U.S. energy structure. Based on the statistical data and a comprehensive analysis, this paper discussed the developing prospects of shale gas as well as its impacts on the energy system in the United States, and therefore pointed that the increased production of shale gas would lower the energy prices, remodel the distribution of energy consumptions and investment, and possibly become a magic factor in the revitalization of U.S. manufacturing industry.

1 INTRODUCTION

As a kind of very important unconventional gas resources, shale gas has large thickness, wide distribution and long production cycle, and causes the new attention in the world energy system. Although it is the regulatory issues affecting unconventional gas development in Europe (Peter 2012), it is illustrated that shale gas developments may have substantial implications for regional gas balances, gas flows, and infrastructure requirements throughout Europe in the next decades (Jeroen et al. 2012). They examine best practices in nanotechnology and financial industries (Gentzoglanis 2012) and conclude that non-conventional gas resources at competitive prices may have a big impact in global gas markets. The extent to which the boom in U.S. shale gas production can spread to other countries (Alejando and Marta, 2010).

The application of horizontal well technology enables the development of U.S. shale gas to enter a new historical stage since 2003. It was only 19.6 billion cubic meters of shale gas production in 2005 and it showed the exponential growth in consequential six years. In the year of 2011, an annual average was increased up to 43.1% and the production was 172 billion cubic meters. The successful exploitation of shale gas has revolutionized the natural gas supply and demand structure. As a result of that, the exploitation has caused the subsequent change of energy prices and energy structure recombination.

2 THE SUPPORT FOR SHALE GAS BY THE U.S. GOVERNMENT

The U.S. government administrations in the past as well as the present have always pursued "energy independence" as the ideals and objectives for the national energy policy. In general, the U.S. government's energy policy objective is to enhance the security of the energy supply which includes two aspects. One aspect is to reduce the dependence on the imported oil and to increase domestic oil and gas production, strategic reserves and the efforts to expand energy self-sufficiency. Second aspect is to reduce the dependence on oil by searching for the alternative energy sources (for example, nuclear, wind, solar, etc.) production to promote diversification of energy consumption structure.

In particular, the U.S. government also has supported shale gas development through various national initiatives and tax policies. For instance, during the late 1970s, U.S. government provides tax subsidies on the unconventional energy development such as "Energy Windfall Law". In Texas, since the early 1990s, production tax associated with the development of shale gas was also exempted. In addition, United States government set up a special research fund just to promote the development of the unconventional oil and gas resources. The importance of the shale gas development by the U.S. government cannot be underemphasized.

3 THE DEVELOPMENT OF SHALE GAS IN THE UNITED STATES

U.S. shale gas early development began in 1821. In 2000, United States began to accelerate the pace of shale gas development and achieved the astonishing speed in 2009 with shale gas production of more than 87.8 billion cubic meters, which accounts for the share of 13% in overall natural gas production of the nation, The percent share of above 13% of production in 2009 has reached to 17% in 2010. As the result of that, with the increase percentage in shale gas production, the overall U.S. natural gas imports has gradually decreased as demonstrated in Figure 1.

It is in recent years that shale gas exploration and exploitation have been developed rapidly. Apart from the breakthroughs of the horizontal well technology and fracturing technology, the U.S. government incentives of fiscal policy is the major force that sharp warming the shale gas exploration! Under the jointly promote by the mature technology, flexible market operation and policy incentives, American's enthusiasm in shale gas investment has been surging because of high freedom of energy exploration and development market that enjoys too many qualified energy market players and variety wide source of funding.

Substantially increased investment in exploration and development, the United States accelerated the pace of shale gas industrialization. At the same time, it has been a strong support that the United States has developed the natural gas pipeline network for the shale gas exploration and development. According to the Energy Information Administration (EIA) data show that the rapid growth of the United States shale gas production was an annual growth rate of 9.9% since 2000.

And it was 2006 a turning point that shale gas production was from the Antrim shale and basic Barnett Shale before 2006 and after that, Fayetteville, Woodford, Haynesville, Marcellus, Eagle Ford and other shale have been developed, particularly rapid developed of the Barnett Shale.

4 IMPACTS OF SHALE GAS ON THE ENERGY SYSTEM

There is a huge amount of shale gas resources worldwide and the amount is predicted 456 trillion cubic meters, more than 2 times the amount of conventional natural gas resources. The worldwide shale oil reserves are about 11 trillion tons to 13 trillion tons, far more than 4,000 tons of the world's conventional oil reserves. Unconventional gas reserves in North America and the Asia-Pacific region exceeds the conventional gas reserves with a huge potential for development. The success factors for the U.S. achievements in the shale gas expropriation are due to its dedicated decades-long period continuous technology development, government support policy and the maturity of the technology associated with the drilling.

China energy service data show that U.S. shale gas production is less than 1% of the natural gas supply in 2000. Today shale gas has accounted for 30% of the natural gas supply and the number is rising. U.S. natural gas production capacity was to reach to 684 billion cubic meters in 2012. Compared to 2006 data, it is a 31% increase as shown in Figure 2. According to global energy statistical yearbook 2013, the weight of North America in the world production of gas is 24% in 2012.

4.1 Impact on natural gas price

Due to the rich reserves of U.S. shale gas, future shale gas production will gradually increase.

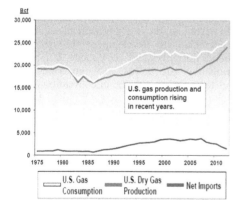

Figure 1.   U.S. gas consumption and source. Source: SWN(2014).

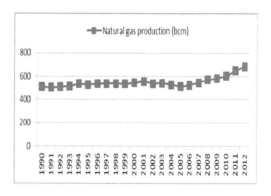

Figure 2.   Natural gas production—United States. Source: http://yearbook.enerdata.net.

That will become a major contributor to the disappearance of gas gap and have an impact on the gas prices. The successful exploitation of U.S. shale gas has revolutionized structure of the natural gas supply and demand chain. U.S. Energy Information Administration (EIA) expected in 2005 that future natural gas supply gap would continue to expand. With the shale gas exploitation technology fully developed in 2011, U.S. Energy Information Administration expected that natural gas will become a net exporter in 2022.

Furthermore, shale gas is expected to be accounted for 49% of total production of natural gas in 2035, while it only accounted for 23% in 2010 (Ting Gong. 2013). The data showed the percentage has more than doubled in 25 years. In addition, shale gas is the only one in the family of various types of natural gas that will increase in the percentage. Since the 2008 world financial crisis, oil prices increased up to 200%, but natural gas prices have remained low, (see Fig. 3). This is due to the development of the shale gas, which has become the supply end of the natural gas.

Southwestern Energy Company has thrived because of a deep commitment to providing the energy whose work is principally focused on the development of natural gas in the Fayetteville Shale. According to the data of SWN Corporation, shown in Figure 4, completed lateral length has more than doubled since 2007 while total well costs have decreased 17% due to the continuous and ongoing improvements in their Fayetteville Shale operations.

4.2 *Impact on coal demand*

The decline in natural gas prices will encourage the consumption of natural gas, replacing the

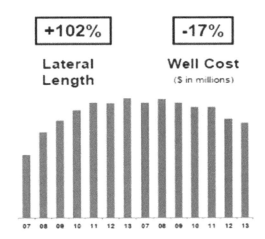
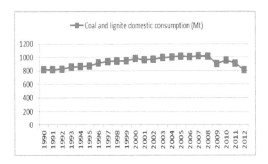

Figure 4. SWN improvement in shale gas.
Source: SWN (2014).

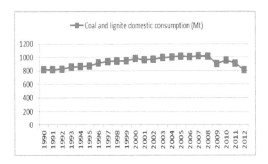

Figure 5. Coal and lignite domestic consumption—United States.
Source: http://yearbook.enerdata.net.

traditional coal consumption. Because of that, U.S. coal consumption and demand will be decreased and exports of coal will be increased.

The United States is the second largest coal producer, output of about 10 million tons per year. The electric power plants replacing coal with natural gas led to decreasing demand for the coal. The sharp decline of coal consumption in the United States results in coal's stocks value reached eight-year high since 2012. According to U.S. Energy Information Administration (EIA), the power sector of coal consumption in 2012 would decline 133 million tons compared with 2011, corresponding to a 14.3% drop as shown in Figure 5. Coal enterprises became the exporters of coal and exports coal of 107 million tons to foreign country and share the global coal's trade of approximately 8.8% in 2011.

In fact, from the beginning of 2009, the cost of coal relative to natural gas began to decrease.

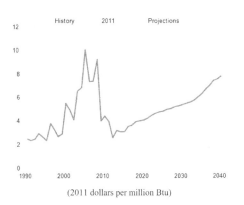

Figure 3. Annual average Henry Hub spot natural gas price, 1990–2040.
Source: U.S. Energy Information Administration.

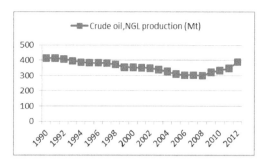

Figure 6. Crude oil, NGL production—United States. Source: http://yearbook.enerdata.net.

The establishment of natural gas power plants will also solve the problem of inadequate transport of infrastructure pipeline through grid generation. The U.S. Department of Energy stated that the United States will continue to eliminate a large number of coal-fired units in the next four years which summing up to 27 million kilowatts of coal-fired power generation plants. All these benefits were due to shale gas revolution. The contributing factors of these successes are due to the United States government's commitment to substantially eliminate old fashion coal generation units.

4.3 Impact on oil consumption

Based on a huge difference in pricing of natural gas and oil on the market, many developers of shale gas exploitation gradually move to higher commercial value of shale oil and gas liquids. This shift will further promote the rapid growth of oil production in the Unites States.

The U.S. energy self-sufficiency rate is also increased because of shale gas production. In 2011, U.S. daily crude oil imports were 891 million barrels, which was the lowest amount in nearly a decade. Due to the huge difference between natural gas and oil prices in the market, many developers of shale gas exploitation gradually move to the higher commercial value of shale oil and gas condensate to seek better profits. In 2012, U.S. shale drilling development plan had turned from the dry gas to petroleum products and natural gas liquids-rich products. As shown in Figure 6, U.S. Department of Energy data demonstrate that U.S. daily crude oil production capacity was 500 million barrels in 2008. The capacity has risen up to about 550 million barrels in 2010. The increased daily crude oil production capacity was 76 million barrels in 2012, marking the largest increase since the beginning of commercial production of crude oil in 1859.

5 CONCLUSION

Due to the government support policy and it's the dedicated decades-long period of continuous R&D in technology, the rapid development of U.S. shale gas has not only reduced the proportion of consumption of coal and other energy sources, but also decreased the dependence on foreign oil imported. The increased production of shale gas will lower the energy prices and possibly become a magic factor in the revitalization of U.S. manufacturing industry. Meanwhile, the funds that were used in the past for the U.S. to import energy will now be used for domestic investments instead, and therefore help United States to enhance its ability to resist the world's economic fluctuations.

ACKNOWLEDGMENTS

This research was supported by the Scholarship for Overseas Study from Shanghai Municipal Education Commission. Professor Qiaozi Zhao and Dr. Bin Chen are the joint first authors of this paper.

REFERENCES

1. Alejandro, A., Mingo, M. 2010. The expansion of "Non-conventional" production of natural gas (Tight gas, gas shale and coal bed methane). A silent revolution. 2010 7th International Conference on the European Energy Market (EEM); 2010 June 23–25.
2. Coulson, C., He, X., Placona, A. 2011. Investment in a natural gas economy: The future of the Marcellus Shale and Southwestern Pennsylvania. Systems and Information Engineering Design Symposium (SIEDS); 2011 April 29; University of Virginia Charlottesville, VA.
3. Gentzoglanis, A. 2012. Emerging technologies and regulatory hold-up: the case of shale gas in Europe and North America. 2012 9th International Conference on the European Energy Market (EEM); 2012 May 10–12.
4. Jeroen de Joode, Arjan Plomp and Ozge Ozdemir. 2012. A model-based analysis of the implications of shale gas developments for the European gas market. 2012 9th International Conference on the European Energy Market (EEM); 2012 May 10–12.
5. Ting Gong. 2013. Status and prospects for the U.S. shale gas revolution. China Institute of International Researches <http://www.ciis.org.cn/chinese/2013–12/11/content_6529465.htm.> 11 Dec 2013.
6. Zeniewski, P. 2012. Surface level challenges for shale gas development in Europe; a regulatory perspective; 2012 9th International Conference on the European Energy Market (EEM); 2012 May 10–12.

Power and Energy – Kong (Ed.)
© 2015 Taylor & Francis Group, London, ISBN 978-1-138-02782-4

Design of three-phase APFC circuit based on the average current mode

Xi-Ping Huang, Jia Wang, Qiang Sun & Gui-Tao Chen
Xian University of Technology, Xian, Shaanxi, China

ABSTRACT: Active Power Factor Correction technology is an effective method of reducing harmonic pollution of power electronic devices. This article used a circuit of three-phase single-switch single inductor to be the main circuit topology of APFC, and designed the double closed-loop based on the average current mode control circuit. Then the relevant parameters of the three phase APFC with UC3854 were designed. Finally the simulation and experiment were executed. The results show that the studied three-phase high-power factor correction circuit has achieved the design requirements, and the power factor above 0.938.

1 INTRODUCTION

Usually power electronic devices are interfaced with the grid by rectifier, the traditional rectifier circuit often uses the diode or thyristor phased rectifier, the power factor of the network side is low and the harmonic content is high, then improving the power factor and suppressing the harmonic have became major issue in research of power electronics. Active Power Factor Correction (APFC) technology has been widely used and has the small volume and lightweight. Single phase APFC technology is relatively mature while the three-phase APFC is still in the stage of development.

2 APFC ANALYSIS OF THREE-PHASE SINGLE SWITCH SINGLE-INDUCTOR CIRCUIT

In this paper, we studied the three-phase APFC control system of 18 kW high power air plasma cutting inverter power supply, the total power supply block diagram [4] has been shown in Figure 1.

Among these, the APFC part finished the booster function of AC/DC circuit and correction the PF value of the whole cutting power system at the same time. Indicators of the APFC circuit are shown in Table 1. According to the design index of the circuit, APFC main circuit is shown in Figure 2 which used the three-phase single switch single inductance structure [2][3], the inductor worked in CCM mode and the program has the advantages of simple structure and low cost. Following is the theory analysis of correction effect to the main circuit.

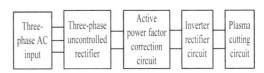

Figure 1. High-power air plasma cutting inverter power system.

Table 1. Three-phase APFC circuit design index.

Parameter	Numerical
Three-phase input voltage	$u_a = u_b = u_c = 220$ V
The output voltage	$V_0 = 750$ V
Switching frequency	$f_s = 18$ kHz
The grid frequency	$f = 50$ Hz
The output power	$P_0 = 18$ kW
PF	≥ 0.9

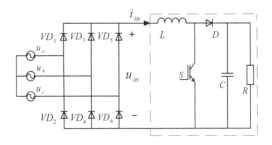

Figure 2. Boost type of three-phase single switch single inductor.

In Figure 2, the correction objects respectively are voltage u_{in} and inductance current i_{in} after the rectifier bridge, after the correction, i_{in} can be completely follow u_{in}, then the circuit part after the rectification bridge can be equivalent to a pure resistance R_{in}, At this $u_{in=} i_{in}*R_{in}$. Assuming the three-phase input voltage: A phase can be positive conducted between $\omega t \in [(1/6 + 2k_1)\pi, (5/6 + 2k_1)\pi]$ by the symmetry, to the negative between $\omega t \in [(7/6 + 2k_1)\pi_1, (11/6 + 2k_1)\pi]$ and cut off the rest of the time. When A phase to the positive conduction is $i_a = i_{in}$, when A phase to the negative is $i_a = -i_{in}$, when A phase is cut off $i_a = 0$.

The u_{in} and i_{in} can be obtained by the formula (1):

$$k \in [6k_1, 6k_1 + 5] \qquad (k \in N^+)$$

$$\begin{cases} u_a = \sqrt{2}U \sin(\omega t) \\ u_b = \sqrt{2}U \sin\left(\omega t - \dfrac{2\pi}{3}\right) \\ u_c = \sqrt{2}U \sin\left(\omega t + \dfrac{2\pi}{3}\right) \\ u_{in} = \sqrt{6}U \cos\left(\omega t - \dfrac{k+1}{3}\pi\right) \\ i_{in} = \sqrt{6}\dfrac{U}{R_{in}} \cos\left(\omega t - \dfrac{k+1}{3}\pi\right) \\ \omega t \in \left(\dfrac{2k+1}{6}\pi, \dfrac{2k+3}{6}\pi\right) \end{cases} \qquad (1)$$

A phase PF value can be calculated by the formula (2) according to the definition of PF.

$$PF = \frac{\dfrac{1}{T}\int_0^T u_a * i_a dt}{\sqrt{\dfrac{1}{T}\int_0^T u_a^2 dt}\sqrt{\dfrac{1}{T}\int_0^T i_a^2 dt}} = 0.956 \qquad (2)$$

It can be learned that the chosen main circuit topology can meet the requirements of calibration from the Table 1. As long as the input voltage wave form is followed by the input current waveform after the rectifier bridge, it can make the three-phase system PF value to meet the design requirements when using the shown main circuit topology.

3 DOUBLE CLOSED LOOP CONTROL SYSTEM BASED ON AVERAGE CURRENT MODE

Because the rectifier bridge circuit of the three-phase APFC circuit is consistent with the single

phase, the average current mode was chosen in this study to achieve the PF correction function. It has the advantages that THD and EMI are relatively small, is not sensitive to noise and can work under a constant frequency. The principle diagram of the correction [5] is shown in Figure 3.

Refer to Figure 3, it can be concluded the small signal model structure diagram of the double closed-loop control system based on average current mode shown in Figure 4.

When it comes to the design of the double closed loop, including the design of the inner current regulator ACR and the outer voltage regulator AVR, inner ring needs to emphasize the follow performance of current while the outer ring needs to emphasize on anti-jamming performance of the output voltage.

A) The design of the current inner ring
The current inner loop transfer function block diagram can got by Figure 4, as shown in Figure 5.

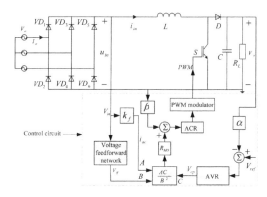

Figure 3. Three-phase APFC based on average current mode control principle diagram.

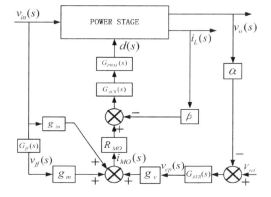

Figure 4. APFC small signal model based on UC3854 control block diagram.

Figure 5. Current inner loop transfer function block diagram.

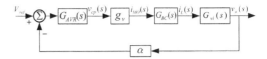

Figure 6. Outer voltage loop transfer function block diagram.

The open-loop transfer function of current closed loop can be concluded by the small signal model that in the case of not join current regulator ACR is (3):

$$G_{ca}(s) = \frac{\beta V_o}{V_s L s} \qquad (3)$$

It needs a low frequency $s_{z1} = j\omega_{cz}$ to compensate the original open-loop transfer function, to provide enough gain to make the circuit work stable. To ensure that the transfer function of the current loop can across zero by the -20 dB/dec, slope the same time, joined a pole $s_{p1} = 0$, in addition to restraining the noise of the interference of current loop, a switching frequency $s_{p2} = j\omega_{cp}$, $\omega_{cz} \ll \omega_{cp}$ near the poles was joined. So the transfer function of current regulator ACR is as (4):

$$G_{ACR}(s) = \frac{k_{ACR}\left(\dfrac{s}{\omega_{cz}} + 1\right)}{s\left(\dfrac{s}{\omega_{cp}} + 1\right)} \qquad (4)$$

B) The design of the voltage loop
The block diagram of voltage loop transfer function of the double closed-loop control system can got by Figure 4, as shown in Figure 6.

The open-loop transfer function was concluded through the small signal model in the case of not join voltage regulator AVR (5):

$$G_{va}(s) = g_v \frac{\alpha R_{MO}}{\beta} \frac{V_{rms}}{V_o} \frac{R_L}{1 + \dfrac{R_L C}{2}s} \qquad (5)$$

Because the voltage disturbance is contained in outer ring, before the disturbance need to join an integral part in order to guarantee the feedback voltage astatic, so to add a pole $s_1 = 0$ in the voltage regulator. At the same time due to the existence of the output capacitance, frequency $fr = 6$ f of the voltage of the output contains six harmonics, the pole $s_2 = j\omega_{vp} = (1/5)2\pi f_r j$ need to be added to reduce the harmonic interference on the system of the voltage regulators. In addition, a zero $s_3 = j\omega_{vz}$ was added to offset the poles of the open-loop transfer function of the original voltage loop.

So the transfer function of the identified voltage is (6).

$$G_{AVR}(s) = \frac{k_{AVR}\left(\dfrac{s}{\omega_{vz}} + 1\right)}{s\left(\dfrac{s}{\omega_{vp}} + 1\right)} \qquad (6)$$

4 THE CALCULATION OF THREE-PHASE APFC PARAMETERS BASED ON THE AVERAGE CURRENT CONTROL

Combining the adopted circuit topology with the circuit design index, the key parameters can be calculated: $L = 1.4$ mH, $C_o = 2057 \mu F$. UC3854 chip was chosen to complete the implement according to the average current mode when design the hardware circuit, then determine the key parameters of peripheral chip according to the double closed-loop structure.

A) Current inner loop
Set the $G_{KC}(s)$ through the frequency in the zero transition frequency, is also the zero point of current regulator $G_{ACR}(s)$. In order to make the correction for the entire current loop with 45° phase margin. And through the frequency is set commonly in one 6 of the switching frequency, the pole current regulator is set in the switching frequency. From which the transfer function of current regulator ACR as follows (7):

$$G_{ACR}(s) = \frac{2.58s + 4.8 \times 10^3}{0.88 \times 10^{-5} s^2 + s} \qquad (7)$$

Take $R_{ci} = R_{MO} = 1.5$ K, correspondence $G_{ACR}(s)$, the current regulator parameters can be calculated: current: actually $R_{cz} = 37$ K, $C_{cz} = 2.3$ nF, actually take 2.2 nF. $C_{cp} = 11.5$ nF, actually take 10 nF.

B) Voltage outer loop
Set the $G_{kv}(s)$ through frequency on $s = j\omega_{vz}$ to adjust the voltage loop with 45° phase margin. Thus the voltage regulator transfer function $G_{ACR}(s)$ can be calculated:

$$G_{AVR}(s) = \frac{6.32s + 179.51}{0.88 \times 10^{-3} s^2 + s}$$

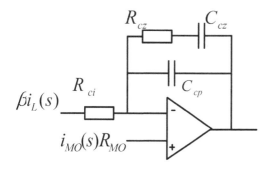

Figure 7. Current inner loop transfer function block diagram.

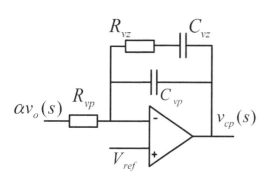

Figure 8. Voltage outer loop transfer function block diagram.

Combination the internal structure of UC3854, use the circuit shown in Figure 8 to realize the implementation of regulator AVR.

Correspondence with $G_{ACR}(s)$, take $R_{vp} = 1$ K. The voltage regulator parameters can be concluded as follows: $R_{vz} = 15$ K, $C_{vz} = 0.33$ uF, $C_{vp} = 2.2$ uF.

5 SIMULATION AND EXPERIMENTAL RESULTS

5.1 Simulation

Build the three-phase APFC simulation circuit based on the average current mode control method according to the designed parameters for the simulation research. The simulation conditions are shown in Table 1, for load conditions $R_L = 32 \ \Omega$, the input voltage and inductance current waveform are shown in Figure 9.

The output voltage waveform and the measured a phase PF value are shown in Figure 10.

It can be seen from the above Figures 9 and 10 waveform that the inductor current waveform

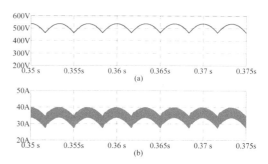

Figure 9. (a) Input voltage after the rectifier bridge. (b) The induct-or current waveform.

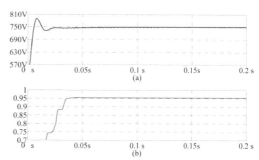

Figure 10. (a), (b) A phase output voltage values of PF.

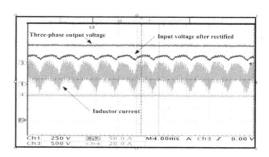

Figure 11. Three-phase APFC correction effect.

basic can completely follow the three-phase rectifier bridge after the input voltage waveform, and the three-phase output voltage can achieve the desired output value, and the phase power factor after correction can reach above 0.95.

5.2 Research of experimental

The input voltage, inductance current and three-phase output voltage after the rectifier bridge of the experimental platform are shown in Figure 11.

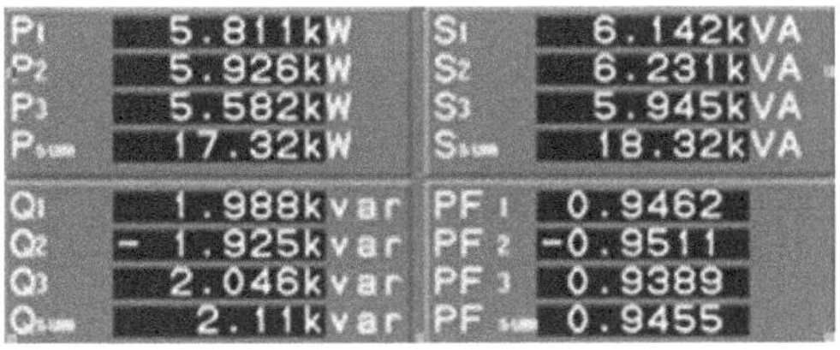

P_1	5.811kW	S_1	6.142kVA
P_2	5.926kW	S_2	6.231kVA
P_3	5.582kW	S_3	5.945kVA
P_{sum}	17.32kW	S_{sum}	18.32kVA
Q_1	1.988kvar	PF_1	0.9462
Q_2	– 1.925kvar	PF_2	–0.9511
Q_3	2.046kvar	PF_3	0.9389
Q_{sum}	2.11kvar	PF_{sum}	0.9455

Figure 12.   Power output and PF value of system.

Then analysis the power quality of the system by the analyzer for power quality, the PF value was measured as shown in Figure 11.

The rationality of the design of three-phase APFC circuit has proved by the Figures 11 and 12. Finally reached the design target by all the phase power factor above 0.938.

6   CONCLUSION

This paper studied the single switch single inductance of three-phase APFC circuit, completed the design of average current mode of the double closed-loop control system, then selected UC3854 as the realization of the control circuit chips and designed peripheral circuit parameters. Finally the simulation and experiment were executed. The results show that the studied three-phase high-power factor correction circuit has achieved the design requirements, and reduced the cost.

REFERENCES

1. Ding Qiang, Miao Ze-ceng and Bao Yun-jie, "Research on the inverter air plasma cutting machine of non-HF contact pilot arc," Electric welding machine, February 2009, pp. 56–58.
2. Nan Yurong, Zhong Degang, Wu Zhigang, MA Dezhong and Zhang Zhihui, "Design of Switching Power Supply for Large Capacity," Modern electronic technology, June 2003, pp. 52–54.
3. Philip C. Todd, Controlled Power Factor Correction Circuit Design, 1999.
4. Yao Kai1, Ruan Xinbo1, Zou Chiland Ye Zhihong2, "Three-phase Single-switch Boost PFC Converters With High Input Power Factor," Chinese Society for Electrical Engineeing, February 2012.
5. Zhu Yong-hua, Tou Yong-jie and Ru Chao, "Simulation and Application of the Boost Circuit Based on UC3854," Technology of West China, June 2011, pp. 32–34.

Power and Energy – Kong (Ed.)
© *2015 Taylor & Francis Group, London, ISBN 978-1-138-02782-4*

Application of Monte Carlo method on heat transfer properties of gas

Y.B. Xie, J.W. Liu & J.Y. Chen
Department of Power Engineering, North China Electric Power University, Baoding, Hebei, China

ABSTRACT: Monte Carlo method is a numerical method based on random probability, and it can be used for a large number of molecules of macroscopic physicals to simulate. In this paper, take nitrogen, hydrogen and helium as examples, make model, and then calculate the thermal conductivity of these three gases. The results show the dates are similar to NIST standards.

1 INTRODUCTION

There are lots of experimental researches and calculation methods at home and abroad on material thermophysical properties determination, which mainly included experimental methods such as transient hot-wire method [18], hot disk thermal constant analyzer measurement [6], comparison method [2], and Russell [14] model at early stage, Maxwell-Eucken [20] model, Y. Agari [4] model, calculation method such as the minimum thermal resistance method [19], percolation theory method [5], fractal theory method [7]. Recently, methods by combination of experimental study and simulation to calculate material thermophysical properties especially the thermal conductivity become more and more, for instance in porous media [16], ceramic powder [12], glass fiber [10], gas thermodynamic properties [3] etc.

Determination of physical parameter with high precision, which unfortunately fitted based on the experimental data at current, is the basis for engineering calculation and scientific research, such as M-H equation. Interestingly, because of the high input on experimental process, the effect of measurement accuracy on experimental data, and the limitation of measuring range which cause a difficult extrapolation, numerical study is being considered as a more suitable method.

Physical properties of material is a macro behavior of a multitude of molecular irregular movements, and the movements decide the microcosm of material full of different accidental and random phenomenon. Therefore, in essence, all physical problems can be solved as random problems. Random sampling theory in Monte Carlo method is just a way that cannot be replaced by other methods in material's microscopic calculation.

Nitrogen, hydrogen, and helium are common and conventional gases; researches about them are wide and mature. Their thermal conductivities were simulated by Monte Carlo method and compared with the existing database in this paper, it showed that Monte Carlo method can be well used in molecular motion.

2 BASIC PRINCIPLE

Monte Carlo, which is also known as the random sampling method and statistical experiment method, is different with other general numerical methods. Based on the approximation probability, a random number is extracted for a large number of statistical tests. Its eigenvalue is seen as a numerical solution of problems.

The origin of Monte Carlo is that the famous French scholar Buffon used needle problem to solve circumference ratio in 18th century [21]. During world war II, Von Neumann and Uiam made the secret work of atomic bomb development to be named Monte Carlo, which is the world famous gambling capital [23]. From then on, the random simulation method of using the computer is called as Monte Carlo.

The theoretical basis of Monte Carlo, is the law of large numbers and central limit theorem [11], which both have asymptotic property to make multiple sampling for the better results. Monte Carlo method is divided into two categories in physics, which respectively is stochastic and certain problem. The problems, which are solved by Monte Carlo method, are seen as a random event. That is to using random variables for hypothetical test in the computer. When the test time is enough, the probability of event or arithmetic average value can be approximate solution of problems.

Monte Carlo, which could reduce the algorithm complexity, is a usual probability algorithm. With the development of computer technology, Monte Carlo has been the statistical model of the random event and high dimensional problems. And it has been extended to the complex physical computer system of atomic bomb project [13], gas dynamics

[24], physic engineering [1], computational biology [8] etc.

3 CALCULATION

Nomenclature

b constant
c constant volume specific heat (JK^{-1})
d molecular effective diameter (m)
k Boltzmann constant (JK^{-1})
l mean free path (m)
m molecular mass (kg)
n molecular number density
q heat flux ($Jm^{-2}\,s^{-1}$)
r rotational degrees of freedom
s vibrational degrees of freedom
t time (s)
w translational degrees of freedom
v the rate of a single molecule (ms^{-1})
x displacement (m)
A cross-sectional area (m^2)
M relative molecular mass ($kgmol^{-1}$)
N_A avogadro's constant (mol^{-1})
P pressure (Pa)
Q quantity of heat (J)
R the universal gas constant ($Jmol^{-1}\,K^{-1}$)
T temperature (K)
V volume (m^3)
α approximately one
β constant
ε energy of a single molecule (J)
ρ density (kgm^{-3})
λ thermal conductivity ($Wm^{-1}\,K^{-1}$)

Subscripts

h calculation of M-H equation
m calculation of Monte Carlo
n calculation of NIST
x x-axis direction
y y-axis direction
z z-axis direction

3.1 *Martin-Hou equation of state*

Martin-Hou equation of state (referred to as M-H equation) is a more accurate multi-parameter equation of state, and has been widely used at home and abroad. M-H equation is proposed by Martin and Yujun Hou in 1955, and further improved in 1981 [15].

The general form [26] of M-H equation is

$$P = \sum_{i=1}^{5} \frac{f_i(T)}{(V-b)^i} = \frac{f_1(T)}{(V-b)} + \frac{f_2(T)}{(V-b)^2}$$
$$+ \frac{f_3(T)}{(V-b)^3} + \frac{f_4(T)}{(V-b)^4} + \frac{f_5(T)}{(V-b)^5} \quad (1)$$

Where the coefficients are a function of temperature,

$$f_i(T) = A_i + B_i T + C_i \exp\left(-KT/Tc\right) \quad (2)$$

M-H equation has been successfully applied to calculate the various thermodynamic properties, but because of its many parameters, and more complex form, thus making the process of solving the physical properties becomes more difficult. Information about its solution process is rarely mentioned, and its application range is limited.

3.2 *Velocity distribution*

A large number of molecules are composed of the gas. The individual molecules have different rates for the random motion in different directions, and it will frequently collide to change the speed and direction of the molecules, so that, for each molecule, its rate is random. But in the equilibrium state, for a large number of molecules, they have a certain statistical law of velocity distribution. According to Maxwell velocity distribution rate [25],

$$\iiint_{-\infty}^{+\infty} f\left(v_x, v_y, v_z\right) dv_x dv_y dv_z$$
$$= n\left(\frac{m}{2\pi kT}\right)^{\frac{3}{2}} e^{-\frac{m}{2kT}\left(v_x^2+v_y^2+v_z^2\right)} dv_x dv_y dv_z = n \quad (3)$$

To introduce spherical coordinates, velocity distribution function can be obtained

$$f(v) = 4\pi\left(\frac{m}{2\pi kT}\right)^{\frac{3}{2}} e^{-\frac{mv^2}{2kT}} v^2 \quad (4)$$

Then the average rate of molecule can be obtained

$$\bar{v} = \int v f(v) dv = \int 4\pi\left(\frac{m}{2\pi kT}\right)^{\frac{3}{2}} e^{-\frac{mv^2}{2kT}} v^3 dv \quad (5)$$

The numerical solution of the integral $\int_a^b f(x)dx$ is the average rate of molecules by Monte Carlo method, which can be divided into three steps totally:

1. Rewrite the integral to the form of $\int_a^b f(x)\psi(x)dx$, then the random number x is generated continuously on basis of probability distribution, and then calculate the value of $f(x)$;

2. Accumulate the values, and calculate the approximation of the average at the interval of [a, b]

$$\overline{f} = \frac{1}{N}\sum_{i=1}^{N} f(x_i)$$

3. N is set to generate random numbers x, when arrive at a stop condition, exit operations.

The greater the N is, the closer the approximation to the true value. We calculated the average rate of the molecule by the above method. When N is large enough, the result is very close to the calculation of the formula (6).

$$v = \sqrt{\frac{8kT}{\pi m}} = \sqrt{\frac{8RT}{\pi M}} \qquad (6)$$

3.3 Determine the appropriate number of particles

When the particle number N is large enough, the approximation obtained by the Monte Carlo method is very close to the true value. At room temperature, take the calculation of the nitrogen molecules' rate as an example, and then to determine the appropriate number of particles N.

According to Figure 1, when the number of particles is become greater, the value of randomized trials is small floating around the true value. However, when N is increased to 10^8, while the floating in the vicinity of the true value is very small, but the calculation is slow and time consuming. Therefore, the number of particles determined as 10^6 or 10^7 is appropriate, and it is also very close to the true value, and the calculation is simple and rapid.

3.4 Calculation of the thermal conductivity

Thermal conductivity is a sign of heat capacity. It is defined as the heat which the heat transfer area

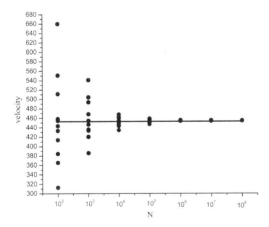

Figure 1. Relationship between the number of particles N and molecular velocity.

is 1 m², the surface temperature difference between both sides is 1 °C, the thick of material is 1 m, under the stable heat transfer conditions within one hour. In value, it is equal to the heat flux at the role of the unit temperature gradient [22]. According to Fourier's law, one-dimensional heat flux q_y is [9],

$$q_y = -\lambda \frac{\partial T}{\partial y}\bigg|_{y_r} \qquad (7)$$

So that obtained the heat dQ from the surface dA at place of y in the time of dt is,

$$dQ = q_y dA dt = -\left(\frac{dx}{dT}\right)\lambda\left(\frac{\partial T}{\partial y}\right)_{y_r} dA dt \qquad (8)$$

From the microscopic, we used molecular dynamic view of the theory to discuss gas heat conduction phenomenon. Based on the average energy and the number density of molecules, using molecular mean free path and collision frequency can be obtained the heat dQ along the y direction form surface dA within the same time of dt is [17],

$$dQ = -\frac{1}{2}\beta n \overline{v} l\left(\frac{\partial \overline{\varepsilon}}{\partial y}\right)_{y_r} \cdot dA \cdot dt \qquad (9)$$

The constant volume specific heat c_v is,

$$mc_v = \left(\frac{\partial \overline{\varepsilon}}{\partial x}\right)_{y_r} = \left(\frac{\partial \overline{\varepsilon}}{\partial x}\right)_{y_r}\left(\frac{dx}{dT}\right)_{y_r} \qquad (10)$$

Combined formula (9) and (10) can be obtained,

$$dQ = -\frac{1}{2}\beta \rho \overline{v} l c_v\left(\frac{dx}{dT}\right)_{y_r} dA dt \qquad (11)$$

Compared with macroscopic heat conduction equation, thermal conductivity λ is,

$$\lambda = \frac{1}{2}\beta \rho \overline{v} l c_v \qquad (12)$$

where,

$$l = \frac{1}{\sqrt{2}\pi n d^2}$$

The above equations show, the gas thermal conductivity λ is relate to the average rate of molecule, the density ρ, the gas mean free path l, and the constant volume specific heat c_v, therefore λ depends on the nature and state of the gas.

At standard atmospheric pressure, when the number of particles taken is 10^6 and 10^7, the thermal conductivity of three gases varies with temperature as shown in Figures 2, 3, and 4.

3.5 *Error analysis*

According to Table 1, when the number of particles N is 10^7, the calculated value is more accurate. When in normal temperature conditions (T in the order of 10^2 K), the results are similar the standard rigid molecular model values. But in the low temperature (less than 10^1 K middleweight) and high temperature (greater than 10^3 K middleweight) case, the results are quite different with the standard values. This is because as the temperature increases, the molecular degrees of freedom are gradually excited. Take hydrogen for example, at low temperatures, only three translational degrees of freedom; at room temperature, there are three

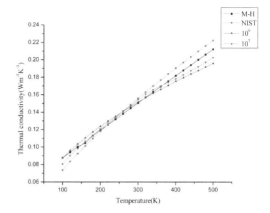

Figure 4.   Thermal conductivity curve of helium.

Table 1.   Error analysis.

		M-H	10^6 (N)	10^7 (N)
N_2	Maximum error	3.23%	22.28%	11.58%
	Minimum error	−0.77%	−13.07%	−7.36%
	Average error	1.75%	10.34%	6.16%
H_2	Maximum error	2.50%	18.73%	14.23%
	Minimum error	−10.62%	−13.69%	−12.16%
	Average error	7.91%	13.03%	7.68%
He	Maximum error	18.89%	18.61%	8.15%
	Minimum error	−4.63%	−11.92%	−9.53%
	Average error	8.66%	8.79%	5.81%

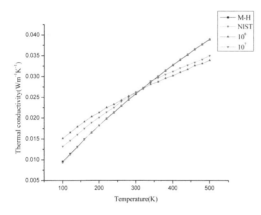

Figure 2.   Thermal conductivity curve of nitrogen.

translational degrees of freedom and two rotational degrees of freedom; but at high temperatures, in addition to translational and rotational degrees of freedom, there is a degree of freedom vibration.

4   CONCLUSIONS

Monte Carlo is a Stochastic modeling which is efficiency, economy, convenience, high accuracy and easy. Compared with other numerical methods, Monte Carlo method has its own advantages: (1) the time of calculated results has nothing to do with the dimensions of the unanswered questions. (2) A little was affected by the problem conditions. (3) Procedure is simple, the structure is clear and easy to understand, Monte Carlo method is easy to implement on the computer, easy preparation and inspection.(4) Especially for a physical problem such as neutron transport has an irreplaceable role.

In this paper, calculating gas thermal physical properties from micro, illustrate the Monte Carlo method on physical parameters calculation is reasonable and feasible.

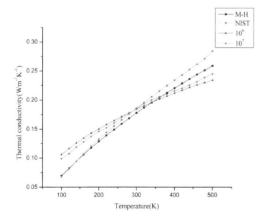

Figure 3.   Thermal conductivity curve of hydrogen.

Today quite a few problems in the learning object has beyond the existing test conditions; using Monte Carlo method has great advantages in numerical calculation and computer simulation. With the continuous development of computer technology, Monte Carlo method will make a more and more important role on various aspects in the future.

ACKNOWLEDGMENTS

The Project is supported by Natural Science Foundation of Hebei province. (No. E2014502085).

REFERENCES

1. B. Eliasson et al. Monte Carlo Simulation of Runaway Electrons in O2/N2Mixture [J]. Herbsltagung der SPG/SSP, 1987.
2. Carson J.K. Lovatt S.J. Experimental measurements of the effective thermal conductivity of a peudoporous food analogue over a range of porosities and mean pore sizes [J]. Journal of Food Engineering 2004, 63(1):87–95.
3. Chang Yong-qiang, Cao Zi-dong, Zhao Zhen-xing, Liu Hong. Calculation Method for Thermal Properties of Mult-i component Gas [J]. Journal of Power Engineering. 2010, 30(10):772–776.
4. Deng Xiaoyan, Rao Baolin. Heat-Conduction Theories of Powder-Filled Polymer Composites [J]. Aerospace Materials & Technology. 2008, (2):
5. Dong Qi-wu, Liu Lin-lin, Liu Min-shan. Prediction Methods of Thermal Conductivity of Polymer-Based Composites [J]. Chemical Propellants & Polymeric Materials. 2007, 5(6):36.
6. He Yi. Rapid thermal conductivity measurement with a hot disk sensor Part 1. Theoretical considerations [J]. Themochimica A cta, 2005, 436(1):122–129.
7. Huai Xiulan, Wang Weiwei, Li Zhigang. Analysis of the effective thermal conductivity of fractal porous media [J]. Applied Thermal Enginneering, 2007, 27(17):2815–1821.
8. Lei Gui yuan. Studies of monte carlo and quasi monte carlo method [D]. Zhe jiang: Zhejiang University, 2003
9. Li Hong fang. calorifics [M]. Bei jing, Higher Education Press, 2001:283–305.
10. Li Qian, Cao Xuan, LI Chen-yu, Peng Yue-nuan, Liu Jun-jie. Theoretical Calculation and Experimental Study of Thermal Conductivity Coefficiency of Glass Fiber Heat Insulation Paper [J]. China Pulp & Paper. 2013, 32(1):35–41.
11. Liang Fei bao. probability and mathematical statistics [M]. Fu jian, Fujian Science and Technology Press, 2012:128–134.
12. Lu Lin Jiang Li Feng Qing Wang Heping. A Method for Calculating Thermal Conductivity of the Solid Phase in Ceramic Powder [J]. Journal of Ceramics. 2012, 33(3):361–164.
13. Pei Lu cheng, Zhang Xiao ze. Monte carlo method and its application in particle transport problem [M]. Bei jing. Science Press. 1980.
14. Russell H.W. Principles of Heat Flow in Porous Insulators [J]. Am. Ceram. Soc. 1935, 18(1):1.
15. Shen Wei dao, Jiang Zhi min, Tong Jun gen. engineering thermodynamics [M]. Bei jing, Higher Education Press, 2000.
16. Shi Yu-feng, Liu Hong, Sun Wen-ce. Experim ent and Numerical Simulation of Effective Thermal Conductivity of PorousMedia [J].* Journal of Sichuan University(Engineering Science Edition). 2011, 43(3):198–203.
17. Tian Chang lin, statistical thermodynamics [M]. Bei jing, Tsinghua University Press. 1987:342–344.
18. Wang Buxuan. Yu Weiping. The measuring echniques for the simultaneous determination of thermal conductivity And diffusivity of moist prous media by at ransient hot-wire method [J]. Journal of Engineering Thermophysics. 1986, 7(4):381–386.
19. Wang Lei. ToPology Optimization for Materials and Structures Based onThermal Conductivity ProPerty [D]. Northwestern Polyteehnieal Universiyt. 2006.
20. Wang Puyu, Hu Xux iao, Zhou Jie, Yang Keji. Research Progress and Application of Thermal Conductivity Models for Polymer Matrix Composite. Materials Review. 010, 24(5):108.
21. Xie Guo-rui. probability and mathematical statistics [M]. Beijing: Higher Education Press, 2002.
22. Yang Shi ming, Tao Wen quan. heat transfer theory [M]. Bei jing, Higher Education Press, 2006:33–41.
23. Yin Zengqian, Guan Jingfeng, Zhang Xiaohong, Cao Chunmei. The Monte carlo Method and Its Application [J]. Physics and Engineering. 2002, 12(3):47.
24. Zhang Ling-fen. The APPlieiaton of Monet Carlo Mehtod in Kinetic Theory of Gases [J]. Journal of Hunan Institute of Science and Technology(Natural Sciences) 2003, 16(1):72–74.
25. Zhou Wei, Li Deng hua. Thermal physics tutorial [M]. Shan dong: Shandong university press. 2010:156–158.
26. Zhu Zhao you, Zhang Fang kun, Xu Chao. Development of the Martin-Hou(MH) Equations of State Application [J]. Shanghai Chemical Industry. 2011, 36(7):12–15.

Power and Energy – Kong (Ed.)
© 2015 Taylor & Francis Group, London, ISBN 978-1-138-02782-4

Optimization of fracturing damage mitigation agents for low permeability reservoir

Shanglin Yao
China Center for Industrial Security Research, Beijing Jiaotong University, Beijing, China
Research Institute of Petroleum Exploration and Development, CNPC, Beijing, China

Dong Lin
Engineering and Technical Brigade, No. 1 Oil Production Company of Daqing Oilfield, Daqing, China

ABSTRACT: A mathematical model was derived to analyze the pressure distribution inside porous media with fractures. Then physical experiments were carried out to evaluate the mitigation effects of different chemical agents to release the formation damage caused by fracturing fluid in the process of hydraulic fracturing. The derived model demonstrated that the cross-flow pressure platform from fractures to matrix can be adopted as an important measure of formation damage by injected fluids. The cross-flow pressure together with the flow conductivity was used as the index to optimize the four fracturing damage mitigation agents. The results show that the recipe of A1 can improve the fluid flow capacity in the porous media to the best effect.

1 INTRODUCTION

In the development process, many issues like low water injectivity and poor oil flowing capacity etc. due to big starting flowing pressure always exist at low permeability reservoirs (Van 1989). Stimulation measures such as hydraulic fracturing are required to improve the fluid flow capacity (Shen 1995). However, the external fluid which doesn't match the in site formation fluid often renders formation damage on matrix as the major oil storage space due to the infiltration of a great amount of fracturing fluid cross fracture plane (Ely 1990). So many wells demonstrate little or negative effects. In this paper physical experiments were used to optimize several chemical agents to mitigate formation damage on fracture plane to get the best one for extension application.

2 EXPERIMENT PRINCIPLES

Inside the low permeability reservoir, fracture is the primary flowing path while matrix is the storage space, of which the recovery factor contributes greatly to the oil productivity. Figure 1 shows the physical model of fractured media. Assuming the same displacing and displaced fluids, no effects of gravity and capillary pressure, and slightly compressible fluid, then the fluid flow through porous media blow can be described by the following mathematical model as:

$$k_f \frac{\partial^2 p_f}{\partial x^2} + \alpha k_m (P_m - P_f) = (\Phi \mu C_t)_f \frac{\partial p_f}{\partial t} \qquad (1)$$

$$\alpha k_m (P_f - P_m) = (\Phi \mu C_t)_f \frac{\partial p_m}{\partial t} \qquad (2)$$

$$P_m(t=0) = P_f(t=0) = P_0 \qquad (3)$$

$$P_m(x=0) = P_f(x=L) = P_0 \qquad (4)$$

$$\lim_{x \to 0} \frac{\partial p_f}{\partial x} = -\frac{q_1 \mu}{\pi \gamma_w^2 k_f} \qquad (5)$$

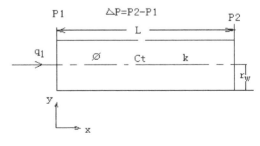

Figure 1. Physical model of fractured media.

Corresponding dimensionless expressions are used to transform the above model into dimensionless form:

$$P_{jD} = \frac{\Delta P_j k_j \sqrt{A}}{q_l \mu}, \; j = m, f; \; \mu_m = \mu_f = \mu;$$

$$x_D = \frac{x}{\sqrt{A}}, \; k_D = \frac{L}{\sqrt{A}}$$

$$t_D = \frac{k_f t}{(\Phi \mu C_t)_{mtf} A}, \; \omega_j = \frac{(\Phi \mu C_t)_j}{(\Phi \mu C_t)_{mtf}},$$

$$\lambda = \frac{\alpha k_m A}{K_f}; \; \omega_f + \omega_m = 1$$

The dimensionless counterpart can be written as;

$$\frac{\partial^2 p_{fD}}{\partial x_D} + \lambda(P_{mD} - P_{jD}) = \omega_f \frac{\partial p_{fD}}{\partial t_D} \quad (6)$$

$$\lambda(P_{jD} - P_{mD}) = \omega_m \frac{\partial p_{mD}}{\partial t_D} \quad (7)$$

$$p_{mD}(t_D = 0) = P_{fD}(t_D = 0) = 0 \quad (8)$$

$$p_{mD}(x_D = L_D) = P_{fD}(x_D = L_D) = 0 \quad (9)$$

$$\lim_{x \to 0} \frac{\partial P_{fD}}{\partial x} = -1 \quad (10)$$

Laplace transformation is done to get the model below:

$$\frac{d\bar{P}_{fD}}{dx_D} = sf(s)\bar{P}_{fD}, \; f(s) = \frac{\omega(1-\omega)s + \lambda}{(1-\omega)s + \lambda} \quad (11)$$

$$\bar{P}_{fD}(x_D = L_D) = 0 \quad (12)$$

$$\lim \frac{\partial \bar{P}_D}{\partial x_D} = -\frac{1}{s} \quad (13)$$

If s is replaced by $f(s)$, the solution can be obtained as:

$$\frac{1}{s\sqrt{sf(s)}} \frac{\sinh\left[(L_D - x_D)\sqrt{sf(s)}\right]}{\cosh(L_D\sqrt{sf(s)})} \quad (14)$$

Stehfest numerical inverse transformation (Stefest 1970) is used to get the results in Figure 2.

The above theoretically derived curves show that in the water injection process there is a cross-flow platform from fracture to matrix in fractured formation. Its pressure marks the resistance amount

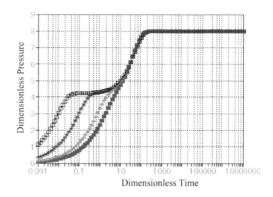

Figure 2. Correlation of dimensionless pressure vs. time for different energy storage ratios.

of the fluid flow from fracture to matrix. So the bigger cross-flow platform pressure, the more difficult fluid flowing into matrix, then the lower recovery factor. Therefore this pressure test can be used to evaluate the damage extent of fracture plane of an oil reservoir. That is the heavier damage the bigger cross-flow platform pressure; and vice versa. This conclusion is adopted to appraise several recipes of damage mitigation agents.

3 PHYSICAL EXPERIMENTS

Here physical experiments were designed to analyze the reason for the low oil flowing capacity through matrix due to fracture plane damage in the process of stimulation at low permeability reservoir. The mixed oil by a neutral white oil and a crude oil from D oilfield has a viscosity of 9.85 mPa·S under 60°C while the formation water is the standard saline water of KCl type for clay antiswelling purpose. The fracturing fluid is mixed by a hydroxypropyl guanidine gum of 1% and a sodium borate of 0.05%. After enough agitation, sodium carbonate or sodium bicarbonate can be used to change the PH value of the fracturing fluid in 6–11. The 4 fracturing fluids A1, A2, S3 and S4 are developed by an Oilfield Institute for evaluation. A1 and A2 are mixtures while S3 and S4 are composed liquids.

Two types of core samples were prepared, the matrix core and fractured core. The matrix core samples will be scanned by X-CT after oil flushing to make sure no fractures and micro fractures inside cores. Enough matrix core samples should be drilled from the same big size sample. The parameters of the matrix core samples see Table 1.

The selected matrix core samples were divided into two groups by experimental requirements,

Table 1. Basic core parameters.

No.	Formation	Porosity (%)	Permeability (10^{-3} μm^2)	Diameter (cm)	Length (cm)	Irreducible water saturation (%)	Note
1	G2	12.2	11.5	2.50	6.8	32.8	Matrix
2	G2	12.6	911.8	2.51	6.9	33.5	Fracture
3	G1	11.2	1212.3	2.51	5.8	40.2	Fracture
4	G1	11.0	12.5	2.50	5.8	44.5	Matrix
5	Y1	11.3	2532.8	2.50	4.5	48.2	Fracture
6	Y1	9.8	2.9	2.50	4.3	51.3	Matrix
7	S1	16.8	30.9	2.49	8.8	29.5	Matrix
8	S1	16.9	2035.2	2.49	8.3	28.6	Fracture
9	S3	21.5	128.5	2.49	7.8	20.8	Matrix
10	S3	21.6	1331.8	2.48	7.2	25.6	Fracture
11	F1	9.6	680.3	2.51	6.9	49.2	Fracture
12	F1	8.7	0.21	2.51	6.8	44.8	Matrix

one for fluid flowing test and the other for fracture stimulation test.

According to the experimental requirements, all core samples were measured in size, dried at 105°C and weighted. Their gas permeability and porosity were also measured. Then all core samples and fluids were vacuumed under a vacuum degree −760 mmHg for 72 hours to remove the dissolved air. Subsequently the core samples were saturated by the standard saline water and weighted for porosity calculation. The experiments were carried out in the constant environment temperature 60°C.

4 EXPERIMENT PROCEDURES AND RESULTS

1. Water displacement process by oil to make irreducible water: after measurement of oil phase permeability the pressure was built inside cores to perform the cross-flow capability test to get the cross-flow pressure of the mixed oil.
2. Fracturing fluid experiment: some amount of sodium persulfate of 0.02 mg/L was added into the fracturing fluid to measure the fracture's conductivity and cross-flow capacity (see Table 2 and Fig. 3).
3. Effect of several damage mitigation agents for fracture plane: 4 types of recipes were injected into the core samples after fracturing fluid injection to measure the conductivity and cross-flow pressure.

The results show that all damage mitigation agents except S4 have improved the filtration characteristics of the fracture plane. A1 has the best

Table 2. Experiment results of core sample 2.

Fluid	Crude	Fracture fluid	A1
Effective permeability (10^{-3} μm^2)	421.8	2.15	382.5
Fluid conductivity (10^{-3} $\mu m \cdot mm$)	8454.1	3.1	7300.5
Cross-flow pressure (MPa)	13.6	27.6	14.1

Table 3. Comparison of experiment results of different core samples.

Parameters	Crude	Fracture fluid	Mitigation agent	No. of cores samples
Effective permeability (10^{-3} μm^2)	785.2	3.8	9.8	2
	2438.2	–	–	3
	1289.7	6.2	95.3	8
	892.3	4.8	7.5	10
	305.6	–	227.6	12
Fluid conductivity (10^{-3} $\mu m \cdot mm$)	22774.7	8.3	31.9	2
	124066.7	–	–	3
	39028.4	14.2	913.0	8
	19866.9	7.2	15.3	10
	5972.9	–	5157.9	12
Cross-flow pressure (Mpa)	12.3	26.0	17.6	2
	7.1	20.7	17.9	3
	3.1	12.7	12.0	8
	22.6	–	24.4	10

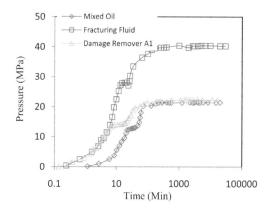

Figure 3. Cross-flow pressure curves for 3 chemical agent.

performance and A2 and S3 follow successively. So A1 is recommended to mitigate the formation damage on fracture plane by fracture fluid after systematic evaluation.

5 CONCLUSION

Here some conclusions are summarized as follows:

1. Formation hydraulic fracturing process improves the fluid flow capacity while causes damage on the fracture plane to different extents at the same time, sometimes to a great amount. All mitigation agents except S4 that are evaluated in the experiment are able to release the infiltration effects in the following sequence, A1 > A2 > S3.
2. A1 is recommended to mitigate the formation damage on fracture plane by fracture fluid after systematic evaluation.
3. It is feasible to use the cross-flow pressure experiment to optimize the best formation damage mitigation agent for field extension application.

REFERENCES

1. H. Stehfest, Numerical Inversion of Laplace Transforms, Communications of ACM, 1970, 13: 47–49.
2. J.W. Ely, W.T. Arnold, and S.A. Holditch, "New Techniques and Quality Control Find Success in Enhancing Productivity and Minimizing Proppant Flowback", SPE20708, 1990.
3. P.P. Shen, et al. Petrophysical Experiment Technologies, Petroleum Industry Press, Sep. 1995.
4. T.D. Van Golf-Racht, Fractured Reservoir Engineering Fundamental, Petroleum Science Progress 12 and translated by Chen Zhongxiang, Jin Linnian and Qin Tongluo, Petroleum Industry Press, May 1989.

Power and Energy – Kong (Ed.)
© *2015 Taylor & Francis Group, London, ISBN 978-1-138-02782-4*

Parameters optimization for Wu'an sintering waste heat power generation unit

Zhi Li
Hebei Electric Power Design and Research Institute, Shijiazhuang, China

Weigang Shi
Hebei Xibaipo Generation Co. Ltd., Shijiazhuan, China

ABSTRACT: An optimization was performed for a sintering waste heat power unit with all data obtained in the site and under the unit normal operating conditions. The physical and mathematical model for the process of cooling and generation is established, which makes the net power generation as an objective function of the cooling machine imported ventilation, the thickness of sinter and the main steam pressure. Optimizing for single parameter, we found that each parameter had an optimal value for the system. In order to further optimize the system's operating parameters, genetic algorithm was used to make the combinatorial optimization of the three parameters. Optimization results show that power generation capacity per ton is increased by 13.10%, and net power generation is increased by 16.17%. The optimization is instructive to the operation of sintering waste heat power unit.

1 INTRODUCTION

As one of the main methods of waste heat utilization in the steel industry, sintering waste heat power generation plays an increasingly important role in energy conservation, environmental protection and improving energy efficiency, etc., which is paid more and more attention and favor by the steel companies. To achieve better efficiency, optimizing the unit operation is a very effective measure. Under the premise of the sintering waste heat power generation system's safety, improving the net power generation and waste heat utilization efficiency is the goal to optimize the operation of heat recovery unit. Therefore, researching the dynamic characteristics of waste heat power unit, analyzing the influencing factors and variation laws of the evaluation index system, then determining the optimum range of operation parameters, have significant practical meaning to the high efficiency operation of waste heat power unit [5].

The sintering waste heat power generation project in Wu'an Iron & Steel Co., Ltd. Hebei province, China, has three sinter production lines (total 320 m2), with 1×85 t/h of superheated steam boiler and 1×12 MW condensing turbine. In this paper, according to the cogeneration projects in Wu'an Iron & Steel Co., Ltd., by establishing the physical and mathematical models for the process of cooling and generation, then programming and simulation, we can obtain the impact of the cooling machine imported ventilation, the thickness of sinter and the main steam pressure on the system net power generation, finally using genetic algorithms to optimize the overall system.

2 SINGLE PARAMETER OPTIMIZATION

In order to study the effect of various parameters on the sintering waste heat power generation system, it is necessary to analyze single parameter optimization. Firstly, establishing an optimal mathematical model [1] that makes the net power generation as an objective function of the cooling machine imported ventilation, the thickness of sinter and the main steam pressure, it will provide the basic analysis for the dynamic characteristics of the unit [2].

Eq. (1) shows the calculation of Net power generation:

$$N_{net} = P_i \eta_m \eta_e - Q_{fan} - Q_{pump} \tag{1}$$

where is turbine power; is turbine mechanical efficiency; is generator efficiency; is pump power consumption; is fan power consumption.

2.1 Optimization of the cooling machine imported ventilation

While the sinter production is constant, namely the heat resource is constant, with the increase of the cooling machine imported ventilation, outlet air temperature is gradually reduced, therefore the heat quantity carried by the outlet hot air has a peak value.

However the power consumption of booster fan increases as the ventilation increases, there is certainly an optimal cooling air imported ventilation value that makes the net power generation reach maximum [6]. As other variables are constant, net power generation is calculated while the cooling air imported ventilation is changed. The optimization results are shown in Figure 1. From Figure 1(c) we

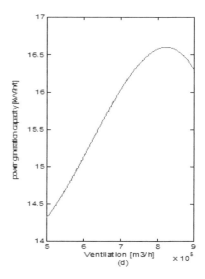

Figure 1. (c) Net power generation as a function of thickness.

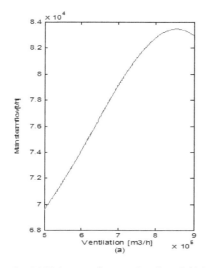

Figure 1. (a) Main steam flow as a function of thickness.

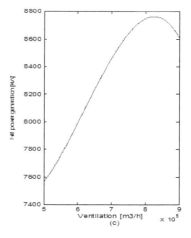

Figure 1. (d) Power generation capacity as a function of thickness.

can find that while the cooling air imported ventilation is 851696 m3/h, the net power generation reaches maximum of 8869 kW.

2.2 Optimization of the thickness of sinter

In the sinter sensible heat recovery system, with the increase of the thickness of sinter, the contact time between air and sinter becomes longer and heat exchange proceeds more sufficiently, however, it also makes the resistance increase when cooling air goes through the material layers. Furthermore,

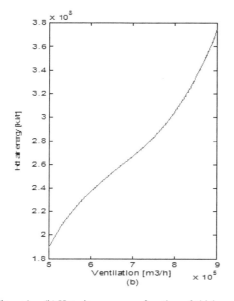

Figure 1. (b) Hot air energy as a function of thickness.

the energy loss will rapidly increase. There is certainly an optimal thickness value of sinter that makes the net power generation reach maximum [7]. As other variables are constant, net power generation is calculated while the thickness of sinter is changed. The optimization results are shown in Figure 2. From Figure 2(c) we can find that while the thickness of sinter is 1.036 m, the net power generation reaches maximum of 8589 kW.

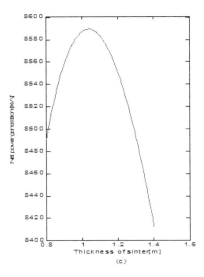

Figure 2. (c) Net power generation as a function of thickness.

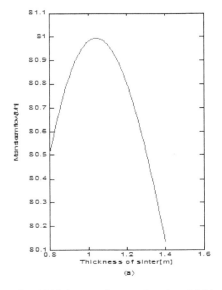

Figure 2. (a) Main steam flow as a function of thickness.

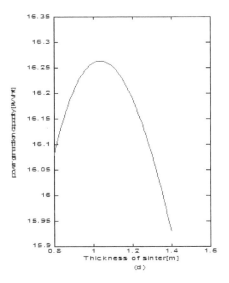

Figure 2. (d) Power generation capacity as a function of thickness.

2.3 *Optimization of the main steam pressure*

The main steam pressure and net power generation is not a linear relationship in sintering waste heat power generation system. The increase of main steam pressure makes the main steam power capability increase, and so does the temperature and pressure of working medium. However, the increase of main steam pressure also makes the system water pump power consumption increase.

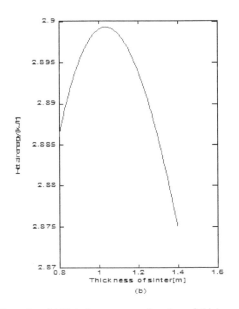

Figure 2. (b) Hot air energy as a function of thickness.

Therefore, it exists an optimal main steam pressure value in the system to make the net power generation reach maximum [4]. As other variables are constant, net power generation is calculated respectively while the main steam pressure is changed. The optimization results are shown in Figure 3. From Figure 3(c) we can find that while the main steam pressure is 1.876 MPa, the net power generation reach maximum power of 9040 kW.

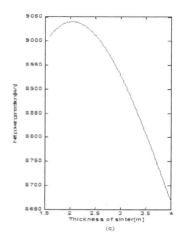

Figure 3. (c) Net power generation as a function of main steam pressure.

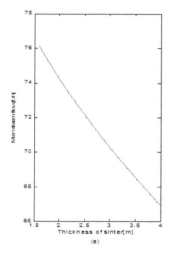

Figure 3. (a) Main steam flow as a function of main steam pressure.

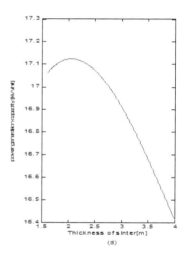

Figure 3. (d) Power generation capacity as a function of main steam pressure.

3 MULTI-PARAMETER COMBINATORIAL OPTIMIZATION

In the former section, the optimization is just for the single parameter. However, in actual operation, the performance of the unit is affected by various parameters rather than a single parameter. Therefore, combinatorial optimization is needed to obtain a more comprehensive analysis results, and provide a more effective basis for unit performance optimization. On the basis of single parameter optimization, combinatorial optimization of three parameters is performed by the means of genetic algorithm [3].

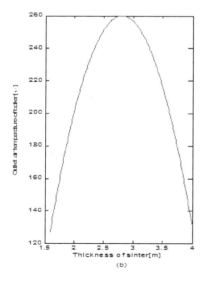

Figure 3. (b) Boiler outlet air temperature of main steam pressure.

Table 1. Combinatorial optimization results.

Optimization methods	$\bar{X} = \left(P_{sat}, L_1, G_g \right)$	Power generation [kW·h/t] / Increasing rate [%]	Net power generation [kW] / Increasing rate [%]
Original parameters	(0.8, 1.1, 750000)	16.26 / —	8353 / —
Optimization for G_g	(0.8, 1.1, 851696)	16.8 / 3.32	8869 / 6.18
Optimization for L_1	(0.8, 1.036, 750000)	16.37 / 0.61	8589 / 2.83
Optimization for P_{sat}	(1.876, 1.1, 750000)	17.12 / 5.29	9040 / 8.22
Combinatorial optimization	(1.765, 1.278, 1045000)	18.39 / 13.10	9704 / 16.17

Establish an optimal model for the parameters of the sintering waste heat power generation system, making the net power generation as an objective function, as shown in Eq.(2):

$$N_{net.\max} = f_{\max}(\bar{X}) \qquad (2)$$

Optimization variables are shown in Eq.(3):

$$\bar{X} = (P_{sat}, L_1, G_g) \qquad (3)$$

where G_g is the cooling machine imported ventilation; L_1 is the thickness of sinter; P_{sat} is the main steam pressure.

Constraints are shown in Eq.(4):

$$\begin{cases} P_{sat.min} < P_{sat} < P_{sat.max}, L_{1.min} < L_1 < L_{1.max}, \\ G_{g.min} < G_g < G_{g.max} \\ d \leq d_{max}, T_{g.in} < T_{g.m.out} < T_{g.m.in} < T_{g.out} < T_s \\ 8°C \leq \Delta T_{ap} \leq 20°C, 8°C \leq \Delta T_{pp} \leq 20°C \\ T_{g.in} = 150°C, T_s = 380°C \end{cases} \qquad (4)$$

4 RESULTS AND ANALYSIS

While using the genetic algorithm to optimize, individuals is set to 32, generations is set to 128, mutation rate is set to 0.35. It cost 620.7 s, after iterations for 2147 times, it reaches convergence when the generations is 64. Optimization results are shown in Table 1.

From Table 1, after the combinatorial optimization of the three parameters by genetic algorithm, power generation capacity per ton is increased by 13.1% while the net power generation is 16.17% compared with the original parameters, and the optimized values satisfy the optimization constraints and the stable operation of unit, which shows that the optimization purpose is reached.

5 CONCLUSIONS

In this paper, the model of waste heat recovery system and generation system of the sintering waste heat power plant is established, making the maximum net power generation as the objective function. Taking the cooling machine imported ventilation, the thickness of sinter and the main steam pressure as the controllable operation parameters, genetic algorithm is used to optimize the operation parameters. In the case of all the parameters satisfy the optimization constraints and the stable operation of unit, we can find that power generation capacity per ton is increased by 13.10%, and net power generation is increased by 16.17%, which means it is succeeded to reach the purpose of optimization.

REFERENCES

1. Alessandro Franco, Nicola giannini. A general method for the optimum design of heat recovery steam generators [J]. Energy, 2006, 31(15):3342–3361.
2. C. Casarosaa, F. Donatinib, A. Franco. Thermo economic optimization of heat recovery steam generators operating parameters for combined plants [J]. Energy, 2004, 29:389–414.
3. Genku Kayo, Ryozo Ooka. Building energy system optimizations with utilization of waste heat from cogenerations by means of genetic algorithm [J]. Energy and Buildings, 2010, 42(7):985–991.
4. Huoji Huang. Gas Turbine Technology, 2006, 19(4):60–63,70. (In Chinese).
5. Shujian Cui. Gas Turbine Technology, 2001, 14(2):14–23. (In Chinese).
6. Shumei Yu, Song Fu, Haiping Chen. Journal of Engineering for Thermal Energy and Power, 2002, 17(99):285–287. (In Chinese).
7. Xiaowen Lu. Exergy Analysis and Steam Parameters Optimization on Sintering Waste Heat Power Generation System [D]. Tangshan:Hebei United University, 2010. (In Chinese).

Study on micro-grid inverters parallel control based on Virtual Synchronous Generator

Yu-Qin Xu & Huan-Jun Ma
School of Electrical and Electronic Engineering, North China Electric Power University, Baoding, Hebei, China

ABSTRACT: Micro-grid can operate in grid-connected or islanding mode. In grid-connected mode, the system voltage and frequency are supported by the distribution network. While in islanding mode, droop control strategy is used to maintain the system voltage and frequency stability. However, when the line impedance is inconsistent, droop control is difficult to achieve a reasonable distribution of load power. To solve this problem, an inverter based on Virtual Synchronous Generator (VSG) is presented. It exhibits characteristics of Synchronous Generator (SG), and has double functions of controlling power and regulating frequency and voltage. Therefore, the control theory of conventional power system can be conveniently applied in micro-grid, maintaining the micro-grid voltage and frequency stability. Two inverter units with the same capacity and different line parameters were carried out in the matlab/simulink platform, the simulation results showed that the inverters shared the load equally, and electrical fluctuation was small.

1 INTRODUCTION

Micro-grid is a recent concept developed by grouping loads and Distributed Generation (DG) units in a local area. It can operate in grid-connected or islanding mode. In grid-connected mode, energy management is the main control objective. While in islanding mode, maintaining the system voltage and frequency stability is the key. Since the interfaces of DG units to micro-grid are usually inverters, parallel inverters control is critical. Droop control is generally adopted and the performance is satisfying.

However, due to the dispersion of DG units, line impedances are different, resulting in unequal load sharing among the parallel operated DG units. To solve this problem, an improved droop control based on virtual impedance was proposed, in which virtual impedance was introduced to eliminate reactive power sharing errors caused by imbalanced line impedance. However, with the introduction of virtual impedance, the inverter performance was affected and the system instability factors increased. A control method of combining the Q–ΔU droop control, in which Q was the reactive power output and ΔU was the variation of voltage of inverter, with restoration mechanism of ΔU was used to control both reactive power output and voltage of inverter. This method improved the reactive power output of inverter while the voltage decreased. Another control method was adjusting the droop coefficient in voltage control equation according to the connection impedance (which

equaled output impedance plus line impedance). This program improved the poor current sharing caused by the connection impedance difference, but the output impedance of inverter was required to be matched with line impedance and droop coefficient denominator existed sine function, the constraints were many and difficult to achieve.

SG has self-balancing capacity, sagging features and large inertia, which are conducive to stable operation and load power distribution. This provides a new idea for micro-grid inverter control. The new inverter exhibits SG characteristics, and has double functions of power control and frequency and voltage regulation. Therefore, the control theory of conventional power system, such as primary and secondary regulation, can be applied in micro-grid conveniently, maintaining the micro-grid voltage and frequency stability. When the line parameters are different, each inverter can achieve accurate proportional load sharing, and electrical fluctuation in the dynamic adjustment process is small.

2 TRADITIONAL DROOP CONTROL THEORY

Traditional droop control simulates operating characteristics of SG. Relationships of frequency and active power, voltage amplitude and reactive power are used to regulate the frequency and voltage.

Active power-frequency droop characteristic curve was shown in Figure 1. When the load active

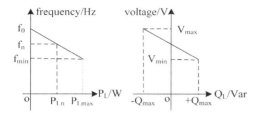

Figure 1. Droop control characteristics: primary regulation.

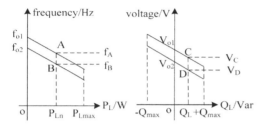

Figure 2. Droop control characteristics: secondary regulation.

power P_L changed, to meet the power balance, active power output of the SG changed, which caused the generator rotation speed (frequency) changed. Then the output of prime mover was changed by the governor to adjust the generator active power output, the grid frequency was regulated ultimately. These changes were known as "primary frequency regulation". Namely generator changed its active power to regulate the frequency along with the static frequency characteristic curve of prime mover, and it was a difference regulation. Similarly, change along with the reactive power-voltage characteristic curve was called "primary voltage regulation".

As illustrated in Figure 2, when the deviation from the frequency or voltage ratings was too large, the system could be adjusted by translating the droop characteristic curve. This was known as "secondary regulation", including "secondary frequency regulation" and "secondary voltage regulation".

Taking secondary frequency regulation for example, when the change of load active power was large, the prime mover operated tuner manually or automatically, so the active power-frequency droop characteristic curve was panned. Assuming load was constant in the adjustment process, the system would enter a new steady state, and the frequency regulating function was achieved. Similarly, the adjustment of reactive power-voltage characteristic curve was known as "secondary voltage regulation".

It was this self-balancing capacity that offered the possibility for synchronous generators operating in parallel. Currently, it has also been applied in parallel inverter control in micro-grid.

3 DROOP CONTROL THEORY OF MICRO-GRID INVERTER

The key difference between micro-grid and conventional power system is that it's based on power electronic devices, without inertia, and more flexible power management. Taking two parallel inverters for example, the structure was shown in Figure 3. Each inverter DC side was equipped with energy storage system, which maintained the DC voltage constant basically, so an ideal DC power supply was used to simplify the analysis.

The Point of Common Coupling (PCC) voltage was set to $U_{PCC} \angle 0°$ and taken as a reference. Output voltage of inverter n ($n = 1,2$) via LC filter was $U_{on} \angle \varphi_n$, the line impedance between inverter n and load was $Z_{ln} \angle \theta_n = R_{ln} + jX_{ln}$, Powers transmitted to load, P_n and Q_n, were available and shown in formula (1):

$$\begin{cases} P_n = \dfrac{U_{PCC}}{Z_{ln}} \Big[(U_{on} \cos\varphi_n - U_{pcc})\cos\theta_n \\ \qquad\qquad + U_{on} \sin\varphi_n \sin\theta_n \Big] \\ Q_n = \dfrac{U_{PCC}}{Z_{ln}} \Big[(U_{on} \cos\varphi_n - U_{pcc})\sin\theta_n \\ \qquad\qquad - U_{on} \sin\varphi_n \cos\theta_n \Big] \end{cases} \quad (1)$$

Different from conventional power system, micro-grid transmission lines are mainly resistance. To decouple the frequency and voltage control, coordinate rotation transformation was used to the output powers. Assuming that the phase difference φ_n was very small, because the inverters were first synchronized by the phase-locked loop modules. Then decoupling expressions were obtained and shown in formula (2):

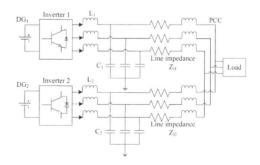

Figure 3. Parallel operation system structure.

$$\begin{cases} P_{tn} = \dfrac{U_{pcc}}{Z_{ln}} U_{on} \sin\varphi_n \approx \dfrac{U_{pcc}}{Z_{ln}} U_{on}\varphi_n \\ Q_{tn} = \dfrac{U_{pcc}}{Z_{ln}} (U_{on}\cos\varphi_n - U_{pcc}) \approx \dfrac{U_{pcc}}{Z_{ln}} (U_{on} - U_{pcc}) \end{cases} \tag{2}$$

It can be seen from (2) that Q_{tn} was positively associated with the voltage amplitude U_{on} and P_{tn} with phase difference φ_n. Because of the calculus relationship between phase and frequency, the frequency could be adjusted by P_{tn}, then the voltage phase was regulated. The voltage amplitude could be adjusted by Q_{tn}.

Therefore, the droop control equations in microgrid inverters were expressed in formula (3):

$$\begin{cases} f_n = f_o - mP_{tn} = f_o - m\sin\theta_n P_n + m\cos\theta_n Q_n \\ U_n = U_{on} - nQ_{tn} = U_{on} - n\cos\theta_n P_n - n\sin\theta_n Q_n \end{cases} \tag{3}$$

where f_n and U_n, frequency and output voltage amplitude of the inverter n, respectively; f_o and U_{on}, corresponding values under no-load condition; m and n, droop coefficients, which were determined by the allowable deviations of voltage and frequency under rated load and no-load conditions.

Micro-grid inverters with droop control, the voltage and frequency adjusting process were similar to SG in conventional power system:

1. Adjusting micro-grid frequency according to the virtual active power, and adjusting the voltage amplitude according to the virtual reactive power. Imbalanced power distribution could also be achieved by adjusting the droop characteristic slope. This method was similar to primary regulation in conventional power system. When micro-grid operated in islanding mode, for small load fluctuation, frequency and voltage amplitude were adjusted based on the current droop characteristics.

2. Adjusting the position of droop characteristic curve depending on parameters in droop control equation: no-load operating frequency and voltage. This method was similar to secondary regulation in conventional power system, which achieved frequency and voltage regulation without difference. For large load fluctuation, primary regulation may not guarantee the frequency and voltage offsets within the allowable range, secondary regulation was required in this case.

In conventional power system, the SG internal impedance is usually large enough to prevent the circulating current. However, in the radioactive micro-grid, the inverter output impedance and line impedance are usually small, so controlling circulating current is a key factor determining whether the micro-grid can operate normally.

4 REALIZATION OF MICROGRID INVERTER DROOP CONTROL

In steady state, frequency of parallel inverter units was equal. However, if the line impedance was different, output voltage amplitude would no longer be equal. It can be seen from (3) that in this case, $P_{t1} = P_{t2}$, $Q_{t1} \neq Q_{t2}$. Since K was a nonsingular matrix, then $P_1 \neq P_2$, $Q_1 \neq Q_2$. That was to say, active and reactive powers could not be distributed equally. In this paper, primary regulation and secondary voltage regulation were used to eliminate the impact of imbalanced line impedance, and to achieve the rational distribution of load power.

4.1 Achievement of primary regulation

It can be seen from (2) that the higher output voltage amplitude of an inverter, the more virtual reactive power, so virtual reactive power could be used as a negative feedback to voltage amplitude droop coefficient. Therefore, the virtual reactive power-voltage amplitude droop coefficient was adjusted dynamically. The primary regulation control equation was shown in formula (4):

$$\begin{cases} f_n = f_o - m\sin\theta_n P_n + m\cos\theta_n Q_n \\ U_n = U_{on} - n(1 + KQ_{tn})Q_{tn} \\ \quad = U_{on} - n\cos\theta_n[1 + K(\cos\theta_n P_n + \sin\theta_n Q_n)]P_n \\ \quad - n\sin\theta_n[1 + K(\cos\theta_n P_n + \sin\theta_n Q_n)]Q_n \end{cases} \tag{4}$$

Taking increase in load power for example, primary regulating process of two parallel inverters was as follows: Droop characteristic curve intersection with the vertical axis was constant, when the load increased, the inverter output power increased, then the output frequency and voltage fell, regulating with difference is an inherent defect of primary regulation.

Suppose inverter 1 was closer to PCC, the line impedance was smaller, and its output power at PCC was larger. It could be seen from formula (4) that its droop characteristic slope was larger, which leaded to lower voltage in steady state, and the output power was reduced. Similarly, the output power of inverters 2 was smaller, the output voltage was higher via the primary regulation, and the output power was increased. In this way, the output power sharing error and circulating current between two parallel inverters were reduced.

4.2 Achievement of secondary voltage regulation

When the load power fluctuation exceeded the threshold, primary regulation may not meet the requirements of the micro-grid operation, so secondary regulation was needed. In this paper, secondary voltage regulation referred to calculating the voltage drop according to the line impedance and the power flow through it, which was compensated to the reference voltage amplitude, therefore, the virtual reactive power-voltage amplitude droop characteristic curve was translated. To accurately determine the position of the curve, the adopted formula was shown in (5):

$$U_{on} = \sqrt{2}\left[\left(220 + \frac{X_{ln}Q_n + R_{ln}P_n}{3\times220}\right)^2 + \left(\frac{X_{ln}P_n - R_{ln}Q_n}{3\times220}\right)^2\right] \quad (5)$$

where U_{on}, the compensated voltage amplitude, which was adjusted automatically with P_n and Q_n.

The regulating process was as follows: When the load power fluctuation exceeded the threshold, to meet the power balance, output power of inverters changed correspondingly. To avoid a significant deviation from the rated output voltage, U_{on} changed according to equation (5), to shift the virtual reactive power-voltage amplitude droop characteristic curve up and down.

The secondary regulation could achieve regulation without error, and the position of the curve ensured the rated voltage and desired output power of load. This regulation could not only guarantee accurate proportional load sharing, but also ensure the quality of power supply at PCC.

However, since secondary regulation was to large load variation, and the inverter units had no inertia, so large amount of electrical fluctuation may occur.

4.3 Present of improved droop control method

Considering the primary and secondary regulation comprehensively, the improved droop control algorithm was put forward and shown in (6):

$$\begin{cases} f_n = f_o - m\sin\theta_n P_n + m\cos\theta_n Q_n \\ U_n = U_{on} - n\cos\theta_n[1 + K(\cos\theta_n P_n + \sin\theta_n Q_n)]P_n \\ \quad - n\sin\theta_n[1 + K(\cos\theta_n P_n + \sin\theta_n Q_n)]Q_n \\ U_{on} = \sqrt{2}\left[\left(220 + \frac{X_{ln}Q_n + R_{ln}P_n}{3\times220}\right)^2 \\ \quad + \left(\frac{X_{ln}P_n - R_{ln}Q_n}{3\times220}\right)^2\right] \end{cases} \quad (6)$$

In the above algorithm, secondary voltage regulation (U_{on} adjustment) was enabled only when the output power fluctuation exceeded the threshold. When the fluctuation was less than the threshold, secondary regulation was exited, and primary regulation was solely relied on regulating the micro-grid frequency and voltage amplitude dynamically.

5 SIMULATION RESULTS AND ANALYSIS

According to Figure 3, two parallel inverters model was built in the Matlab/Simulink platform, inverters power control used the control algorithm in section 4.3, and voltage control used the typical double loop controller of output voltage and inductor current.

Simulation parameters were: Simulation step: 5×10^{-5} s, simulation time: 0.3 s, filter parameters: $R_{f1} = R_{f2} = 0.01\Omega$, $L_{f1} = L_{f2} = 0.6$ mH, $C_{f1} = C_{f2} = 1500\mu F$, line parameters: $Z_{l1} = 3.21\Omega + j0.415\Omega$, $Z_{l2} = 6.42\Omega + j0.83\Omega$, rated capacity of each inverter: $(5 + j2)kVA$, rated voltage of three-phase load: 380 V. At 0.1 s, load power increased $(6 + j4)$ kVA from $(10 + j4)$ kVA, and at 0.2 s, load power reduced $(4 + j2)$ kVA. The waveforms of inverters output voltage amplitude were shown in Figures 4 and 5.

As can be seen, secondary voltage regulation was functioned (substantial changes in the curves

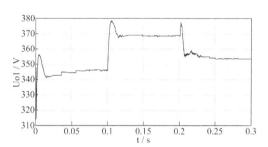

Figure 4. Output voltage amplitude of the inverter 1.

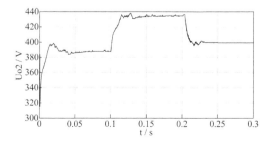

Figure 5. Output voltage amplitude of the inverter 2.

occurred) only when a substantial change in load was detected, and the virtual reactive power-voltage amplitude droop characteristic curve was panned up and down to adapt to changes in load power. The rest of time, secondary voltage regulation was exited, primary regulation was relied on regulating the system dynamically, and the output voltage fluctuation was flat.

Waveforms of inverters active and reactive power output were shown in Figures 6 and 7. As can be seen, when the transmission line impedances were different, parallel inverters shared equal load power dynamically through the coordination of primary and secondary regulation. It was worth stressing that in load-changed time, fluctuation

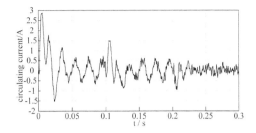

Figure 9. Circulating current waveform of system.

of inverters output was small, which was a great improvement of the control strategy.

In Figure 8, output current waveform of phase A at PCC was marked with solid line and voltage waveform with dotted line. Both waveforms had no obvious distortion, and the output voltage was almost unaffected by load change. This illustrated that the improved control algorithm guaranteed good power quality of load.

Waveform of the circulating current between two inverters was shown in Figure 9. As can be seen, the circulating current was small, which was conductive to inverters safety and system stability.

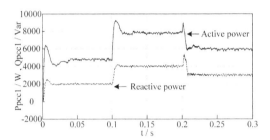

Figure 6. Power waveforms of the inverter 1.

6 CONCLUSIONS

Micro-grid inverters can operate in parallel without communication by droop control method. However, when the line impedance is inconsistent, droop control is difficult to achieve a rational distribution of load power.

To solve this problem, an inverter based on virtual SG was presented. It behaved like SG from the external characteristic. Therefore, the control theories of power system can be applied to micro-grid inverter conveniently. To maintain the micro-grid voltage and frequency stability, primary and secondary regulation applicable to the micro-grid inverters were put forward. Two parallel inverter units were simulated under different line impedances, the simulation results showed that the proposed control strategy not only achieved accurate proportional load sharing, but also ensured good power quality of load. Meanwhile, the circulating current and system electrical fluctuation were small. Both steady performance and dynamic performance of the system have been improved.

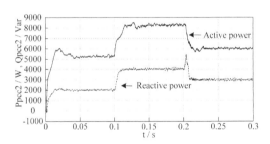

Figure 7. Power waveforms of the inverter 2.

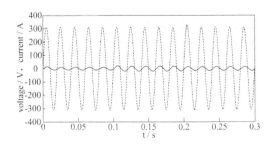

Figure 8. Voltage and current waveforms of phase A at PCC.

REFERENCES

1. Bao, W. Hu, X.H. Li, G.H. & Bao, W.Y. 2013. An improved droop control strategy based on virtual impedance in island micro-grid. *Power System Protection and Control* 41(16):7–13.

2. Ding, M. Yang, X.Z. & Su, J.H. 2009. Control strategies of inverters based on virtual synchronous generator in a microgrid. *Automation of Electric Power Systems* 33(8): 89–93.

3. Fan, Y.L. & Miao, Y.Q. 2012. Small signal stability analysis of microgrid droop controlled power allocation loop. *Power System Protection and Control* 40(4):1–13.

4. Huang, X. & Jin, X.M. 2012. A voltage and frequency droop control method for microsources. *Transactions on China Electrotechnical Society* 27(8):93–100.

5. Kan, J.R. Xie, S.J. & Wu, Y.Y. 2008. Research on decoupling droop characteristic for parallel inverters without control interconnection. *Proceedings of the CSEE* 28(21):40–45.

6. Li, T. Ma, H. Liu, Q.W. & Yin, Y.D. 2013. A wireless parallel control method of inverters with connection impedance differences being taken into account. *Power Electronics* 47(4):30–38.

7. Li, Y. & Li, Y.W. 2001. Power management of inverter interfaced autonomous micro-grid based on virtual frequency-voltage frame. *IEEE Transactions on Smart Grid* 2(1):30–40.

8. Sun, X.F. & Lv, Q.Q. 2012. Improved PV control of grid-connected inverter in low voltage microgrid. *Transactions on China Electrotechnical Society* 27(8):77–84.

9. Xie, L.L. Shi, B. Hua, G.Y. Wen, F. Yang, L.L. & Dong, Q.L. 2013. Parallel operation technology of distributed generations based on improved droop control. *Power System Technology* 37(4):992–998.

10. Yao, W. Chen, M. Mou, S.K. Gao, M.Z. & Qian, Z.M. 2009. Paralleling control technique of microgrid inverters based on improved droop method. *Automation of Electric Power Systems* 33(6):77–94.

11. Zhang, C. Chen, M.Y. & Wang, Z.C. 2011. Study on control scheme for smooth transition of microgrid operation modes. *Power System Protection and Control* 39(20):1–10.

12. Zhang, L. Xu, Y.Q. Wang, Z.P. Li, X.D. & Li, P. 2011. Control scheme of microgrid fed by synchronous generator and voltage source inverter. *Power System Technology* 35(3):170–176.

13. Zhou, X.Z. Rong, F. Lv, Z.P. Luo, A. & Peng, S.J. 2012. A coordinate rotational transformation based virtual power *V/f* droop control method for low voltage microgrid. *Automation of Electric Power Systems* 36(2):47–63.

14. Zhu, D. Su, J.H. & Wu, B.B. 2010. Research on control methods of microgrid based on the virtual synchronous generators. *Electrical Automation* 32(4):59–62.

Power and Energy – Kong (Ed.)
© 2015 Taylor & Francis Group, London, ISBN 978-1-138-02782-4

Numerical simulation of aerothermodynamics performance deterioration of compressor with fouling

Xing He & You-Hong Yu
College of Power Engineering, Naval University of Engineering, Wuhan, Hubei, China

ABSTRACT: The purpose of this research is to explore the mechanism of aerodynamic characteristic and the degradation trend and quantify of thermodynamics for the compressor with various fouling. In order to solve the problem described above, a numerical analysis is carried out by employing Spalart-Allmaras turbulence modelling which with Extended Wall Function. The result shows that the compressor performance deterioration is attributable to the flow of boundary layer, which is under the influence of the roughness and the blade profile of wall in conjunction with the shock-wave/boundary-layer interactions. The aerodynamics of the root and tip region is worsened than the mid-span of blade. The thermodynamics parameters is non-linear decreased with the fouling worsen.

1 INTRODUCTION

The performance of compressor, which is the major component of gas turbine, has an obvious effect on the function of the machine. But at the moment when compressor was used, the trend of performance degradation was on the way. Many factors could lead to degradation, such as the environment, the design and the use et al. The degradation pattern could be concluded as fouling, tip clearance increase, abrasion and cauterization et al results show that the 70%–85% performance degradation is caused by fouling. Fortunately the fouling could be washed to recover the performance of compressor in some degree (X. He, 2010, Z. Li, 2011).

Ref (J. P. Stalder, 2001, F.C. Mund, 2006, M.P. Boyce, 2007) analyzed the fouling mechanism and kinds of washing methods detailed. The effects of environment on the fouling were analysed. E. Syverud et al (2007) introduced the experiment method by injecting thick salt water at the inlet to research on the accelerated compressor blades fouling based on the GE J85-13. M. Naeem (2008) researches the transformation rule of the thrust of turbine engine under the condition of different altitude, Mach number, fouling degree and kinds of controlling modes. B.-W. Li et al (2010) simulated the aerodynamic performance of compressor fouling. Y.-H. Yu & X. he (2012) simulated the dynamic characteristic of gas turbine after the compressor fouling performance degradation.

In the open literature, the special linear transformation method was adopted to simulate the variable of compressor aero-thermal characteristic after fouling. The linear method has obvious limitation on the high non-linear compressor aerodynamic performance. In the research of aerodynamic performance, the effect of fouling on the variable of total aero-thermal performance and the local variety and the mechanism of the variable has not been researched deep.

Computational Fluid Dynamic (CFD) method was used to explore the deterioration of compressor performance after different fouling degree based on a certain compressor pattern. The mechanism and rule of the deterioration were analyzed. Results could apply theoretic for the washing of compressor and the gas path fault diagnosis.

2 FOULING MECHANISM

Fouling pollution problem always happened at the gas turbine compressor. The main reason is the stack of impurity at the surface of flow component and blade. The impurity contains smoke, oil, salt and carbon compound et al, which can be seen from Figure 1. The particle was adsorbed to the flow component and blade surface by several kinds of power. Electrostatic effect and hit affect et al for instance. The component of impurity is an important factor of determining the contamination level. The pollution degree will be different under different working conditions. In the ocean environment condition, the dust particle always covers a salt layer. The natural viscosity enforced the pollution effect. The layer will change the surface of the flow component and blade, which will decrease the mass flow and pressure ratio of the

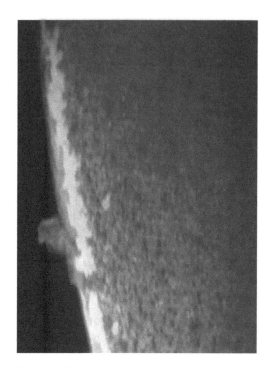

Figure 1. The fouling at the leading edge of the compressor blade.

gas turbine. When the output power inclined, the dissipative of the oil will increase and the efficiency will decrease.

3 NUMERICAL METHOD

The Navier-Stokes (RANS) equation was used as the control equation. The performance parameter of the compressor working at the steady condition was simulated. In the Cartesian coordinate, the consecutive and momentum equation are:

$$\frac{\partial \rho}{\partial t} + \frac{\partial \rho}{\partial x_i}(\rho u_i) = 0 \qquad (1)$$

$$\frac{\partial(\rho u_i)}{\partial t} + \nabla \cdot (\rho u_i \boldsymbol{u})$$
$$= -\frac{\partial p}{\partial x_i} + \nabla \cdot (\mu \mathrm{grad} \boldsymbol{u}) + \frac{\partial}{\partial x_j}(-\rho \overline{u_i' u_j'}) \qquad (2)$$

For the Spalart-Allmaras (S-A) equation is simple and has a steady computation identity which is easy to convergence. For the reasons above, the S-A model was acquired as the turbulence model in this paper. The S-A model is based on the Boussinesq assumption, which considers that the Reynolds

stress and the averaged velocity grad have a direct proportion:

$$-\rho \overline{u_i' u_j'} = \mu_i \left(\frac{\partial u_i}{\partial x_j} + \frac{\partial u_j}{\partial x_i} \right) - \frac{2}{3} \left(\rho k + \mu_i \frac{\partial u_i}{\partial x_i} \right) \delta_{ij} \qquad (3)$$

The viscidity parameter can be calculated by equation (4):

$$\mu_t = \rho \tilde{v} f_{v1} \qquad (4)$$

where: $f_{v1} = \chi^3 / \chi^3 - C_{v1}^3$ is the viscidity damp function, $\chi = \tilde{v}/v$, v is the gas molecule viscidity parameter, $\tilde{v}$ is the turbulent viscidity parameter.

For the near-wall area, the non-dimension velocity parameter is:

$$u^+ = \frac{u}{u_\tau} = \frac{1}{\kappa} \ln y^+ + B + \Delta B \qquad (5)$$

where: $\kappa = 0.41$ is the Von Karman constant, For the slip wall, $B = 5.5$; ΔB is a minus value and will decrease with the increase of the rough degree.

4 THE MODEL CONFIGURATION AND THE FOULING SIMULATION

4.1 The compressor model configuration

The 3D configuration data of the compressor contains the blade number, the coordinate of the pressure side and suction side, the hub and shroud, blade tip clearance information which can be seen at Figure 2.

In the grid plot, the single passage was plotted as HOH type. Total grid number was 509776. The blade to blade mesh of the 50% span can be seen at Figure 3.

4.2 Fouling simulation

In this paper, the steady aerodynamic performance after fouling was simulated based on the

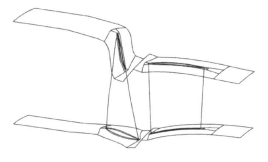

Figure 2. The 3D model of the compressor blade.

320

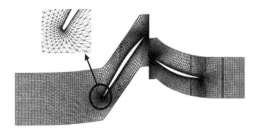

Figure 3. The blade to blade mesh of 50% span.

Table 1. Characteristic parameter.

Reynolds number		
Characteristic length (m)	Characteristic velocity (m/s)	Density (kg/m³)
0.5	180	1.2

Table 2. Working environment.

Environment	
Temperature (K)	Pressure (Pa)
293	101325

NUMECA FINE/Turbo platform. The whole passage has the same condition so a single passage was simulated in this paper. The process of the simulation can be concluded as follow:

① the medium type: real gas;
② the flow type: steady, Turbulent Navier-Stokes control equation, Spalart-Allmaras (Extended Wall Function) turbulent model, the working environment and parameters can be seen at Table 1 and Table 2;
③ the rotation component: −17188.7r/min rotating speed;
④ the rotor-stator type: Conservative Coupling by Pitch-wise Row;
⑤ boundary condition: flow axial inlet, 101 325 Pa inlet total pressure, 293 K inlet total temperature, 0.000 05 m²/s inlet turbulent viscosity, 130 000 Pa outlet averaged pressure; "Smooth" or "rough" wall type, if the "rough" type, $(B + \Delta B)$ value must be set;
⑥ CFL = 3 and multigrain parameter was set;
⑦ initialization field: turbomachinery type, the inlet pressure of vane and blade were set;
⑧ case saved;
⑨ case replicated, modify the wall condition for different fouling degree simulation;
⑩ run the computation batch and get results.

5 RESULTS AND ANALYSIS

5.1 Validation

The CFD result and the experiment result were compared for validation of the clean model of the compressor blade in this paper. The experiment data of the design condition were: 20.20 kg/s mass flow, pressure ratio was 1.773. The relative errors were 2.58% and −1.15%. The simulation model was valid in the allowed error bound.

When the blade surface was clean, the limit streamline of the suction side of the rotating blade and the relative Mach number of the blade to blade section at the 50% span can be seen at Figure 4 and Figure 5 respectively.

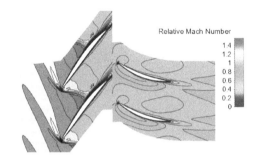

Figure 4. The blade to blade section relative Mach number of the 50% span (smooth surface).

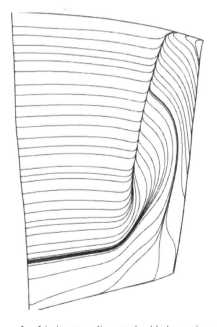

Figure 5. Limit streamline at the blade suction side (smooth surface).

5.2 Aerodynamic performance analyses

The loading analysis distribution can be concluded from Figure 6 to Figure 10. The fouling has a tiny effect on the vane and has an obvious effect on the pressure side of the rotating blade. In the pressure side of the rotating blade, the main affect area was the 20% to 35% axial chord length near the leading edge which has the largest variation near the blade hub. In the suction side of the blade, the main affect area was the 60% to 95% axial chord area near the trailing edge which has the largest variation near the blade hub. General speaking, the fouling has the most obvious effect on the blade hub.

The variation of static pressure and total temperature at the outlet of the vane were plotted as Figure 9 and Figure 10. Results can be concluded that the fouling has little effect on the total temperature and has an obvious effect on the distribution

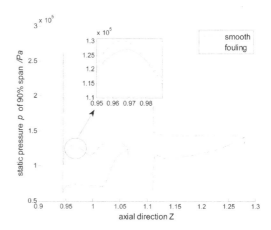

Figure 8. Loading curve of the blades at 90% span.

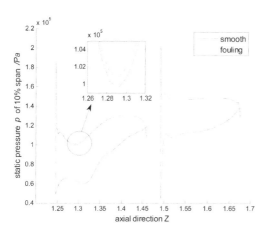

Figure 6. Loading curve of the blades at 10% span.

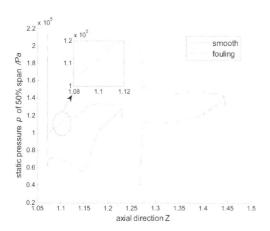

Figure 7. Loading curve of the blades at 50% span.

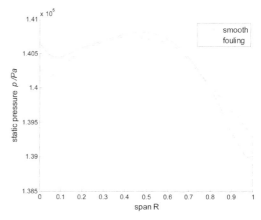

Figure 9. Static pressure distribution at outlet of the vane.

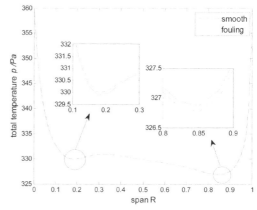

Figure 10. Total temperature distribution at outlet of the vane.

of the static pressure, especially for the blade tip and hub. The reason can be concluded that the shock-wave and the boundary layer have an enhanced interactional effect which enlarges the static pressure. The efficiency of the compressor decreased when the gas entropy increased.

5.3 *Analysis of aerothermodynamics performance*

In order to research the relationship between the aerothermodynamics performance and the fouling degree, numerical research was made for different fouling conditions. The variety data of the mass flow, efficient coefficient and pressure ratio were acquired. Polynomial was used to acquire the equation between the mass flow, coefficient, pressure ratio and the variable of deterioration parameter ΔB, which can be seen from equation (6)–(8).

$$\delta G_{in} = 25.4807 \times (\Delta B)^3 + 6.1742 \times (\Delta B)^2 + 0.7154 \times (\Delta B) \tag{6}$$

$$\delta \eta_C = -31.0874 \times (\Delta B)^3 - 25.5154 \times (\Delta B)^2 - 1.6360 \times (\Delta B) \tag{7}$$

$$\delta \pi_C = 20.8133 \times (\Delta B)^3 - 20.1702 \times (\Delta B)^2 - 1.8404 \times (\Delta B) \tag{8}$$

The trend of the effect of blade fouling on the performance of compressor can be seen from Figures 11–13. The linear deterioration trend was also given for comparison, which can be seen as the dashed line.

As can be seen from the figure, with the variety of fouling performance degeneration parameter ΔB, the mass flow, efficient and pressure ratio reduce in a non-linear way. When $\Delta B = -0.35$, the mass flow will reduce 0.59%, the efficient will

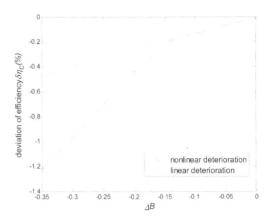

Figure 12. The variation trend between the efficiency and compressor fouling deterioration performance.

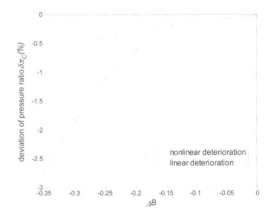

Figure 13. The variation trend between the pressure ratio and compressor fouling deterioration performance.

reduce 1.22%, pressure ratio will reduce 2.71%. Results show that the pressure ration plays the most important role while the efficient less and the mass flow is the least.

6 CONCLUSION

CFD method was used in this paper. The fouling of compressor was simulated in the NUMECA FINE/ Turbo platform. The aero-thermal parameters was acquired which developed with the compressor performance degradation. Results show that.

When the compressor was fouling, the blade configuration was changed and the blade surface turned to be roughness. The end wall condition changed and the disturb of coupled shock-wave will affect the boundary layer flow and then the compressor performance will be influenced.

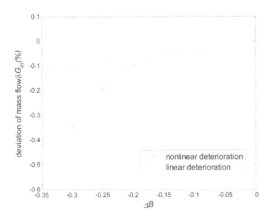

Figure 11. The variation trend between the mass-flow and compressor fouling deterioration performance.

Parameter varies at different span. Compared to the aerodynamic performance of the averaged radius, the performance degeneration influenced much at the blade root and blade tip. The obtained results could provide advice for the washing of compressor fouling.

Results show that, with the development of the compressor fouling, the linear thermodynamic parameter degeneration error is much larger than the non-linear degeneration. The changes of thermodynamic performance rely on the change of aerodynamic performance for the compressor which has a high non-linear characteristic. So the linear method is not recommended to get the degeneration data.

The controlling method of the compressor is the fixed rotational speed at design point. The modelling technique and method could be extended to non-design proceeding point of other kinds of compressors. The variation rule of the aerothermal performance of the compressor could be acquired in different control method.

REFERENCES

1. B.-W. Li, D. Li, J.-H. Li, et al. Quantitative Research on Performance Degradation of Single Stage Compressor [J]. Journal of Aerospace Power, 2010, 25(7): 1588–1594. (in Chinese)

2. E. Syverud, O. Brekke, L.E. Bakken. Axial compressor deterioration caused by saltwater ingestion [J]. Journal of Turbomachinery, 2007, 129(1): 119–126.

3. F.C. Mund, P. Pilidis. Gas turbine compressor washing, historical development, trends and main design parameters for online systems [J]. Transactions of ASME, 2006, 128(2): 344–353.

4. J.P. Stalder. Gas turbine compressor washing state of the art: field experiences [J]. Journal of Engineering for Gas Turbine and Power, 2001, 123(2): 363–370.

5. M. Naeem. Impacts of low-pressure (LP) compressors' fouling of a turbofan upon operational-effectiveness of a military aircraft [J]. Applied energy, 2008, 85(4): 243–270.

6. M.P. Boyce, F. Gonzalez. A study of on-line and off-line turbine washing to optimize the operation of a gas turbine [J]. Transactions of ASME, 2007, 129(1): 114–122.

7. X. He. Study on gas turbine performance deterioration based on the thermodynamic work potential [D]. Wuhan: Naval University of Engineering, 2010. (in Chinese)

8. Y.-H. Yu, X. He. Transient state characteristics of performance deterioration in gas turbine [J]. Journal of Naval University of Engineering, 2012, 24(5): 39–42. (in Chinese)

9. Z. Li, B.-W. Li, D.-Y. Wang, et al. Analysis of Sensitivity of Compressor Performance Parameters to Fouling [J]. Aeronautical Computing Technique, 2011, 41(6): 41–44. (in Chinese)

Power and Energy – Kong (Ed.)
© 2015 Taylor & Francis Group, London, ISBN 978-1-138-02782-4

Technical review on protection issues of the matrix converter

Jain Taruna
Barkatullah University Institute of Technology, Barkatullah University, Bhopal, India

Upadhyaya Nikhil
Accenture UKI, London, UK

Dubey Sakshi
Barkatullah University Institute of Technology, Barkatullah University, Bhopal, India

ABSTRACT: Protection issues are perhaps seen to be the most essential aspect of research associated with the Matrix converter. It is critically important to resolve these issues to allow the Matrix converter to make inroads and present itself as a superior replacement in areas where other conventional converters have dominated the market. Strong competitors are the back-to-back VSI or the CSI. One of the potential applications of the Matrix converter is integrating the wind turbine system with the grid in power generation. The purpose of this paper is to carry out an investigation on the various faults that can possibly occur in a Matrix converter system in power generation. The hardware and control for a typical wind turbine Matrix converter system is probed to identify potential modes of failures and various fault protection strategies have been examined. A technical review of different protection methods adopted for both over-voltage and over-current situations are performed. Methods such as the clamp circuit, varistor protection, suppressor diode (or active voltage clamping), active voltage control, relays & circuit breakers, input filter design and FPGA control have been discussed.

1 INTRODUCTION

The converter has been a fast emerging technology and much research has been done on its prospects of being used in motor drive applications. The Matrix converter has a number of attractive features like sinusoidal input and output currents, unity power factor operation on any load, ability to function in the regenerative mode etc. The overall design is straightforward and compact. Despite all the advantages the Matrix converter possesses in 3-phase AC-to-AC conversion, it still faces severe technical difficulties in its induction into the mainframe applications for motor drives. It has failed to replace the widely accepted and conventionally used converters such as the back-to-back VSI and the CSI. In the past, there have been issues of safe current commutation and complexity of control algorithms which now have been resolved to quite an extent.

However, there are fundamental practical issues of protection that need to be addressed before industry can contemplate using Matrix converters. This implies that an in depth investigation of all possible faults likely to affect a Matrix converter used in a grid application must be looked into. A typical application of the Matrix converter system used to integrate the grid supply with the wind turbine power generation has been selected in this paper for a detailed study. Various failure modes in the power electronics associated with the control system have been considered. Mechanical faults linked with the wind-turbine system are also outlined. A detailed review of various protection strategies for over-voltage and fault current is presented. Numerous widely published methods such as the clamp circuit, varistor protection, suppressor diode, active voltage control, relays & Circuit Breakers (CBs), input filter design and FPGA control have been critically analyzed in this paper.

2 WIND TURBINE POWER GENERATION SYSTEM DESIGN

The Matrix Converter-Induction Generator AC-AC power system architecture is represented as a simplified block diagram in Figure 1. This is an integrated wind-turbine system depicting the power, control and interface sections [1]. In the power area, the supply lines from the grid incorporate an LC—filter to suppress high current spikes on the input side [2]. The Switch Matrix (converter) is connected to the load side, which is an induction

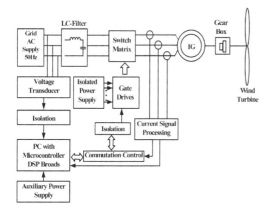

Figure 1. Block diagram showing the Matrix converter-Induction Generator AC-AC power system architecture.

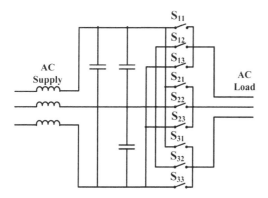

Figure 2. A typical 3-phase to 3-phase Matrix converter.

generator here. The generator is linked to the wind-turbine via the gear box and braking systems. The gear box provides the desired optimum speed regulation. The IGBTs in the Switch Matrix have independent gate drive circuits and isolated power supplies. The Control section constitutes of a processor, A/D converters, Micro-controller/DSP Boards, means of signal data storage and communication, commutation control and gate drive circuits. A standard PC is used which serves the operator to monitor the system functioning. This control system generates and monitors the PWM signals to the gate drive circuits.

The interfacing between the power and control circuits is provided by voltage transducers and current signal processing. This incorporates the current transducers, sign detection (of the output current), etc. There are various kinds of control system configuration used to control the switching in the Matrix converter for motor drive applications which have been researched extensively and

widely published [3], [6], [8]–[9], [11]–[12]. The Switch Matrix is the Matrix converter configured by nine bi-directional switches distributed in groups of three, each forming a "leg" for the AC-to-AC conversion as shown in Figure 2. The bi-directional switch used in the Matrix converter is in the common emitter IGBT mode configuration.

3 MODES OF FAILURE IN WIND TURBINE GENERATION USING MATRIX CONVERTER SYSTEM

As the system hardware is complex, there are many areas where a fault could occur and disrupt the normal operation of power generation from the wind turbine. Thus to ensure that protective measures are installed to prevent such a crisis, it becomes essential to probe into the system for all possibilities of a failure. This has been investigated in the sections that follow:

3.1 *Gate drive power supply*

This is a major issue concerning the Matrix converters, as they lack a natural free-wheeling path as compared to the VSI. If the gate drive supply is lost (possibly due to a shutdown of the grid or any other emergency), the IGBTs turn off rapidly. However, the stored stray circuit inductive energy and energy stored in the generator has to dissipate through the switches. This leads to a surge of voltage across the switches to dangerously high levels.

3.2 *Input LC-filter*

The Matrix converter is expected to provide a smooth sinusoidal transition of 3-phase frequency from the input side to a desired frequency on the output side. Due to presence of noise and switching frequency harmonics, the input currents drawn into the Matrix converter from the grid are not perfectly sinusoidal. This can affect the synchronization of the wind turbine system with the grid and potentially generate undesirable high harmonics in the input waveform. The LC-configuration can also affect the 'power-up' process of the Matrix converter and cause over-voltages during transient operation [12].

3.3 *IGBT*

The IGBTs used in the Matrix converter are subjected to high-stress switching situations and expected to be very rugged in their operation. Since the load is inductive in nature here, they can cause an inrush of current through the switch even after it is turned off or till this current is re-directed to other circuit elements. For high power

IGBT modules, large currents at full voltage outputs are switched in a very short duration of time. Each switch shares the switching transient of the load current at a di/dt determined by its gate drive circuit. Now the current on the output side is the summation of the currents and the di/dts passing through each IGBT chip. The instant when the device turns off to protect the IGBT during short circuit is the most dangerous moment. As at this time, di/dt could exceed several thousand A/μs. To shut down the device and to protect it from a fault condition, care must be taken to minimize the accompanying over-voltage, or the device could be destroyed [4], [13]. Over-voltages can also occur under the following conditions resulting in the failure of the device [8]:

a. Input side, due to lightning or line disturbances.
b. Output side, if the converter is shutdown due to an over current fault.

The cause of IGBT destruction is attributed to the tremendous thermal stress in the device and generation of internal heat due to package design constraints. Other inherent failures in the power IGBT are:

1. Gate Stress: Long-term gate stress can cause rupture of the gate oxide implying a resistive short either between gate and emitter or gate and collector, which may appear like a low breakdown diode between the gate and source.
2. Thermal Fatigue: Irregular heating and cooling of the silicon/metal interfaces can cause electrical/thermal performances to degrade. This can lead to the wire bond to 'detach' from the metal surface resulting in a loss of connectivity and malfunctioning of the device.
3. Mechanical Stress: Excessive mechanical stress can be a direct implication of drastic temperature fluctuations causing possible fractures in the internal silicon structure. This may cause shorted gates or affect the breakdown characteristics in an adverse manner.
4. Structural Design: Due to the structural design and casing of capsule IGBTs, they are more likely to fail short-circuits.

3.4 Mechanical faults

a. Induction generator: Two primary causes of fault occurrences in the induction generator used in the wind turbine could be either due to insulation degradation in the windings or related to environmental factors. External exposure could cause moisture or oil in combination with dirt to settle on the coil surfaces outside the stator slots. These can lead to electrical tracking discharges in the winding which can eventually puncture the ground wall and causes loss of conductivity in the cables.

b. Gear box: Normally the gear box used in the wind turbine system is ensured to be of very high standards. However, there are specific problems associated with the operation of the gear box which are localized to the type of application it is used in. Possible causes may be due to:

1. Misalignment: Although rare, but if there is an alignment error in the parts of the gear, it could lead to unnecessary fatigue of the mechanical parts and increase friction. This may have a detrimental affect the gear box system, thereby affecting its overall efficiency.
2. Unexpected loads and thermal instability: The wind turbines are subjected to very high loads (due to gale winds) which can be quite irregular at times. This can lead to more mechanical stress on the shaft and is first reflected as a temperature rise of the oil used inside the gear box system. Subsequently, this could damage the other parts associated with the gear box and lead to thermal instability.
3. Torsion and lateral vibration: As the gear box has moving parts, their rotational movement may not be expected to be smooth at all times. This may be caused due to jerks which may lead to torsion and unwanted vibrations of the parts.
4. Electric discharge: The use of high switching frequency IGBTs has seen to cause severe electrical discharge in the bearings and thus reducing their useful life.

c. AC drive: Faults likely to occur are:

1. Line-to-Line short due to faulty wiring, short circuiting of the motor leads, phase-to-phase insulation failure.
2. Ground fault may be caused by break down of the insulation between the phases of the motor.
3. Shoot through occurring due to false turn-on of the IGBT.

d. Rotor brake: The rotor brakes are mechanical disc brakes made of iron which can rust and lead to corrosion (as they are not used that often) and thus it can be difficult to operate them at times. As a result, the rotor brakes could fail causing dangerously high speeds of the wind turbine. This can make the wind turbine system unstable and potentially damage the generator and the transformer. Usually, the wind farm operators pre-empt bad weather situations and shut down the wind turbines as a precautionary measure avoiding an emergency.

3.5 Safe current commutation

Extensive research has been done and many papers have been published: [3]–[5], [14]–[15] addressing the problem of safe current commutation. This had been a major hurdle in the acceptance of the Matrix converter in the industrial market. With nine Bi-Directional Switches (BDSs), there are numerous current paths that can be realized in a 3-phase Matrix converter.

Careful investigation on considering a single output phase of a two phase (Vr & Vy) Matrix converter shows two possible situations that can arise [4]–[5] –

a. Short-circuit condition: If the two Bi-Directional Switches (BDS1 & BDS2) of the input phase are switched "on" simultaneously, it results in a line-to-line short-circuit condition as indicated in Figure 3. Currents now flowing through the circuit would be much higher than the rated current of the IGBT. This will cause permanent damage to the bi-directional switches and eventually lead to destruction of the converter.

b. Open-circuit conditions: It is also observed that the bi-directional switches pertaining to each output phase should not all be in the "off" state at the same time. If this occurs, the inductive load current has no path to flow, thus causing an abrupt open-circuit condition between the input and the output phase as shown in Figure 4. Consequently, large over-voltages appear across the switches that could exceed their maximum voltage handling capacity on an open circuit and eventually destroy them.

3.6 *Matrix converter control system hardware*

The critical areas susceptible to faults associated with the control aspects of the Matrix converter are:

a. PC with μ-controller/DSP boards: The PC and the micro-controller/DSP boards are powered by

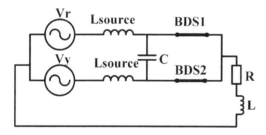

Figure 3.   Line-to-Line short-circuit conditions.

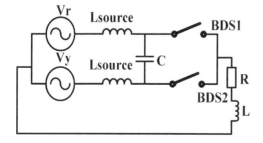

Figure 4.   Open-circuit conditions.

an auxiliary power supply. Loss of this supply would result in loss of signals necessary for the commutation control and firing of the gate drives. There are high possibilities that the protective circuit may eventually become unstable if other faults such as, over-voltages & over-currents take place. The μ-Controller/DSP boards can also malfunction if proper isolation techniques from the power side are not implemented.

b. The gate drive circuit: Appropriate triggering of the gate drives is extremely critical in the smooth commutation and functioning of the Matrix converter. As each of the gate drive circuits have isolated power supplies, a fault in the supply implies loss of gate signals and flawed firing of the IGBTs. This will set off unsafe commutation and give rise to open-circuit or short-circuit situations during the switching transients.

c. Voltage transducers: One of the voltage transducer type used is the Capacitive Voltage Transformer (CVT). The difficulty here is that since the CVT is used to measure the voltages directly from the power lines, they can be affected by harmonics and electromagnetic interference [32]–[33]. In the event of a voltage surge occurring either due to a lightning strike or any unbalance in the power system, it is likely that the voltage transformation ratio may not be adequate and the range may exceed the specified value of the transducer used.

d. Current signal processing: This module contains the current transducers and the current detection circuit. The current transducer is quite sensitive and can be affected by short-circuit currents, in the event of a line-to-line short [3], [8]. This causes disruption and generates faulty command signals to the commutation control leading to malfunctioning in the control of the switching of the devices. If this were to happen, no feedback signal would be available for the current directions from the output side under an over current situation. Thus, the faults occurring in the control system are usually interconnected in nature. In general, internal failures within the control system (DSP, μ-controller, logic circuits, decision units, etc.) are not that likely to occur so easily, as they are usually isolated from the power side.

4   DETECTION AND PROTECTION TECHNIQUES USED WITH A MATRIX CONVERTER SYSTEM

4.1 *Detection*

Both input and output voltage/current sensors are used for regulation and analysis of a Matrix converter system. If there is a fault or surge in the current or voltages occurring as mentioned in the previous section, most of the possible failures can be identified as explained in [16]. As regards to the mechanical faults of the over-all system such as that described for a wind-turbine system, condition monitoring can be used for the generator (Generator Condition Monitor) and gear box.

For control circuits—Based on the output currents, current sign detectors are deployed to ensure a semi soft-commutation and also to identify an over-current situation [8].

4.2 Protection

1. Over-voltage

a. Clamp circuit: The clamp circuit shown in Figure 5 is a protective circuit connected across the Matrix converter for a typical 3-phase AC grid supply system used to feed a 3-phase AC load (represented by a motor load here). It consists of a bridge rectifier using twelve fast recovery diodes linked through a capacitor and a discharge resistor. Whenever the switches are abruptly turned off, the load current gets interrupted all of a sudden and the energy stored in the motor inductance has to be discharged safely without causing any surge in the voltage. The diodes in the clamp circuit on the output side provide a path for the inductive currents to flow which then charges the capacitor. If there is an over-voltage situation occurring on the input side (i.e. the grid), similar action is taken by the clamping diodes in the bridge circuit [4]. In both these cases, the resistor, connected parallel to the capacitor discharges the stored energy of the capacitor. This way the system is protected from damage arising due to over-voltages in case of a fault situation. This circuit gives a common solution for over-voltage protection resulting from line disturbances appearing on the input side and also due to over-current faults affecting the output side. A clamp circuit configuration has been patented in [12] and published widely in [4], [8], [17].

The size of the diodes is dependent on the over-current levels and is mounted on a Power Stage Board, thus minimizing the stray inductance. The electrolytic capacitor used in the clamp circuit is mounted externally along with the input L-C filter, which allows a degree of freedom to change the parameters if required [7]. The size of the clamp capacitor is a critical issue. The clamp circuit design would normally be to provide for the worst-case condition for an over-voltage occurring due to a fault.

Comparative studies did in [16] assess the overall semiconductor losses in the VSI against the Matrix converter including the input filter. Findings reveal that for a 460 V 7.5 kW induction motor drive, the DC-link capacitor value of the VSI is 750 μF/900 V and that for the clamp capacitor is 135 μF/900 V. This shows that the DC-link capacitor used for the VSI is approximately six times larger than the clamp capacitor value of the Matrix converter.

This study draws out the fact that, the clamp capacitor used in the protective bridge circuit in the Matrix converter is not in operation at all times and operates only in the time of a fault situation. On the other hand, the DC-link capacitor is charged and discharged on each instant of operation of the VSI. The size of the clamp capacitor thus, need not be very large. A lower rating capacitor may be sufficient to cater for the maximum over-voltage resulting from the inductive load current and also to handle the kinetic energy regenerated from the induction motor on the output side.

The clamp capacitor used thus, does not have to be a "high performance" capacitor. This theory does not apply to the DC-link capacitor as it is inherently activated in the normal operation of the VSI. The selection of this capacitor has to be such that, it has a longer life time and a higher capacity. This proves that, essentially, the DC-link capacitor has to be of a larger size as compared to the clamp circuit capacitor used for the Matrix converter.

These findings are very encouraging as they address the issues of size & cost on using an additional clamping capacitor in the Matrix converter. The Matrix converter topology was initially proposed to present an "all silicon" solution in providing direct AC-to-AC conversion but on stability grounds, the clamp configuration has been incorporated and widely used.

Despite its benefits, the inherent shortcomings in this clamp circuit design such as conduction loss, turn-off & turn-on losses and electrolytic capacity losses cannot be ignored. Turn-on losses may be disregarded for their negligible ratio however the other mentioned losses cannot be overlooked on account of their significant effect on the performance of the Matrix converter in terms of reliability & efficiency which are both critical for industrial applications.

b. Dynamic clamp circuit: The dynamic clamp circuit, shown in Figure 6 [18] is an improvement over the standard clamp circuit described earlier.

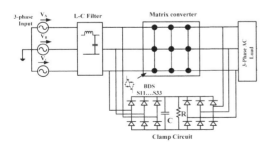

Figure 5.   Matrix converter using a basic clamp circuit.

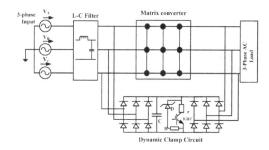

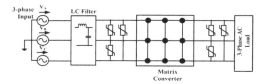

Figure 6. Matrix converter with dynamic clamp circuit.

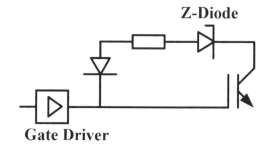

Figure 7. Matrix converter with varistor protection.

The dynamic clamp constitutes of an IGBT that operates in active region during a protection operation and dissipates reactive load energy, whilst all the other devices are turned off. The heating of the PN-junction is avoided here by connecting a resistance in series with IGBT. The resistance value is selected such that the voltage drop due to the output current from the load should be smaller than the clamp voltage. The clamping voltage of the dynamic clamp circuit is decided by the voltage across the zener diode between the gate and collector terminals. The dynamic clamp circuit demonstrates a much superior performance over the conventional clamp circuit when compared on protection aspects of the Matrix converter as:

1. Due to fast operation of the dynamic clamp circuit, sizing of the required capacitor is reduced further. This leads to increased life span of the circuit and thus, avoids using an electrolytic capacitor.

2. Additional control or drive circuit for the IGBT is not necessary.

3. As current flows for a very short time on the input and output side of the diode bridge; the bridge diodes used in the dynamic clamp circuit have a 10% to 20% nominal current rating range compared to the nominal current of the main IGBT.

For the above mentioned benefits, the dynamic clamp circuit shown in Figure 7 is preferred over the standard design as it is simpler to control, costs less due to smaller size of capacitor and is more reliable.

However, both protection schemes implementing the clamp circuit illustrated in a) & b) above

are unable to diagnose the cause of fault. Due to absence of selectivity, malfunctioning may happen during load transients or external switching where complete shutdown of the Matrix converter is not required in a grid operation.

c. Varistor protection: This technique avoids the use of the diode clamped circuit for over-voltage protection. This method can be implemented provided that, the converter does not undergo a hard pulse-off shut down. This may appear under the circumstances of an emergency off or a converter breakdown. There should also be no grid shutdown and the signal processing is ensured to be fail-safe.

As published in [12], [17] if the induction motor is shut down due to the pulses going off, the recovered energy is expected to get stored in the motor leakage inductance. This amount of energy is small and that a device such as a varistor can be used to absorb this quantity of energy. This was concluded by taking measurements on the varistor operating on a 30 kW, 4 pole induction motor. The studies reveal that during pulse-off, the varistor voltage was capable to handle a voltage surge up to 1.25 kV and that the ratings of the varistor must have been quite high. The energy absorbed by the varistor mentioned is 3 Ws.

The varistors are installed in a triangular configuration both on the input and output lines of the converter as shown in Figure 7. The varistors on the input side provide protection of the converter from voltage transients from the grid and that on the output side absorbs the energy of the induction motor under a pulse-off condition.

2. Fault current protection

a. Suppressor diode or active voltage clamping: This protection circuit shown in Figure 8 consists of an avalanche diode Z in series with a blocking diode. The selection of the avalanche diode should be such that, its breakdown voltage is lesser than the maximum blocking or allowable voltage at the IGBT module terminals. Now if this voltage limit is exceeded during a turn-off and the collector-emitter voltage of the IGBT becomes higher than the breakdown voltage of the suppressor diode,

Z-Diode

Gate Driver

Figure 8. Gate Driver with suppressor circuit protection.

then it starts to conduct and re-energizes the gate. In this process an avalanche current flowing in the gate resistor, would raise the gate-emitter voltage above its threshold level and keep the IGBT in the "conducting state". This feedback mechanism, thus clamps V_{CE} to a safe value.

When a fault occurs, the circuit under consideration will react only at turn-off. If there is even a slightest of a delay in the re-energizing of the IGBT, it would reflect as a potentially dangerous voltage spike [13]. Thus, appropriate selection of the Z-diode and the blocking diode is extremely crucial. The value of the gate resistance R_G should also not be too low as, it would make the operation of this protection circuit less effective.

There will be an additional loss in the transistor with this method of protection, but this will be there for a very short time until all the transistors are eventually turned-off. Thus, it is going to be only a one-time operation. This method is more effective when coupled with the varistor protection and studies done in [12] for a 7.5 kW induction motor show promising results as a means of protection of the Matrix converter. The usefulness of this circuit is therefore limited to fault current protection.

The high input impedance of the IGBT at the gate allows more control over the switching of gate drive. Another feature of the modern IGBTs makes it possible to stretch its active region to the extremes of the safe operating area typically for 10 μs as reported in [21]. This has been exploited here in the active voltage clamping circuit. The advantages of using the IGBT under this mode for active snubbing, short-circuit protection and for series operations have been established [19], [22]–[25], [32]–[33].

b. Active Voltage Control: The Active Voltage Control (AVC) scheme is a method used to provide direct control of the collector-emitter voltage of the IGBT using a feedback loop. In this method, V_{CE} is the feedback in the control loop that gets compared to with a reference voltage, V_{REF}. These two voltages i.e., V_{CE} & V_{REF} are inputs to a low gain op-amp and the resulting error signal is applied to the gate drive circuit. Thus, the gate resistance is enabled and in this way, the IGBT is controlled [21]. AVC has found applications in many areas where IGBT devices have been used as seen in [20], [27], [30]–[31].

The feedback signal from the IGBT as depicted in Figure 9, will give an indication whether it has failed short-circuit or not. This method effectively controls the gate drive of the IGBT and provides an immediate shutdown of the bi-directional switch undergone failure. It is assumed that as soon as this is initiated from the control circuit; other protection mechanisms are working together

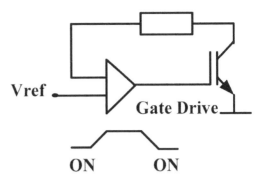

Figure 9. Active Voltage Control (AVC).

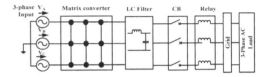

Figure 10. Output protection using relays & circuit breakers.

so that the over-all system can be shut down safely. It is imperative that the control-system is activated within a safe period of 10 microseconds, as this is the period under which the power device can handle a short circuit under full voltage.

c. Output protection scheme using relays & circuit breakers: The protection schemes reviewed earlier perform exceptionally well for non-linear load applications such as induction motor drives. However, non-selectivity and inability to diagnose the cause of protection operation, limits them from being applied to wind energy applications. As for grid connected wind turbine systems, the nature of the input parameters require a more secure and reliable protection of Matrix converter. A more comprehensive protection system that addresses the selectivity and reliability issues in the Matrix converter operation for a grid connected application has been published in [28].

As shown in Figure 10, this protection scheme consists of an LC ripple filter, standard protective relays and a CB applied at the output terminals of the Matrix converter. This design also provides protection against line-to-line faults occurring on the grid.

When a fault occurs, a command to reduce the output voltage on the faulted phases to zero is given to the Matrix Converter. As the instantaneous parameters within the converter are set at higher levels, this command would only be given for dangerously high levels of fault currents. The zero voltage output limits the fault current and allows the relay to send a trip command to open the CBs. However, the performance of the relays is influenced by the selection of switching frequency in this method. For lower switching frequency, the range of fault selection is increased, but the currents need to be sampled at a faster rate than the switching frequency. This is important for detecting abnormal current levels that may result in damage of switches as command signal cycle may become too long. Thus, higher switching frequency is recommended for better fault protection near the converter.

This output protection technique avoids the use of the bridge diode clamped circuit for over-voltage protection. This method can be implemented provided the converter does not undergo a hard pulse-off shut down due to drive circuit failure, which is a rare case otherwise. This may occur under circumstances of an emergency off or a converter breakdown. Also, a grid shutdown event or erroneous signal processing event should not occur.

d. Input filter design: As means of tackling the problems associated with the simple design of the LC-filter as investigated in [2], connection of damping resistors has been proposed in [8], [26]. These damping resistors are to reduce the over-voltages. The obvious question arising is the placement of these resistors on the input line. They can be placed before the LC-filter configuration or connected in parallel to the input reactors. This method of parallel connection which introduces a "damping" in the filter circumvents the Matrix converter input currents containing a component close to that of the resonant frequency of the LC-filter [2]. Once the converter is running, the damping resistors are "short-circuited". Thus, the selection of the input filter becomes crucial and a perfect tuning of the LC—filter is absolutely essential to attenuate these harmonics.

3. *Protection of Matrix converter through series of hardware blocks implemented in FPGA*
In [29], Field-Programmable Gate Array (FPGA) is shown to control the Matrix converter alone, replacing traditional control using DSP. FPGA control system comprise of fundamental elements such as the basic cells, input & output blocks, interconnection matrix, RAM and clock synchronization interconnected through standard buses. The FPGA control involves Double-Sided Space Vector Modulation (DS SVM) and safe commutation converter with a series of hardware modules like 7432 Flip-Flops (FFs), 10125 Look-Up Tables (LUTs), 23 signal-processors (SPs) and 2 Block RAM modules (BRAMs). This published FPGA control as shown in Figure 11, is built using System on Chips (SoCs) that enable the entire complex blocks (hardware or software) to be integrated and designed in a single device. The SoCs used thus,

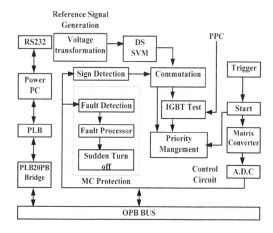

Figure 11. Control bus diagram for FPGA based Matrix converter.

offers cost advantages due to lesser consumption, reduced chip size and increased simulation handling for simultaneous series and parallel execution capacity.

FPGA protected Matrix converter system exhibits more robust performance with least complex and faster control options as it operates as a single unit compared to a DSP controlled Matrix converter, wherein execution speed is low. Thus, FPGA controlled Matrix converter may be sighted as a future alternative over DSP as it provides safe and compact hardware for practical implementation.

5 DISCUSSION

Revolutionary changes in the energy scenario in last few decades have motivated researchers towards providing more efficient and reliable renewable energy systems. Many grid operators across the world are thinking about implementing Smart Grid applications wherein protective devices and meters would be able to pre-empt fault occurrences thereby minimizing outages and providing greater grid security. There is a global effort to inject more and more renewable power onto the grid with minimum grid disturbances. In such a scenario, power converters are bound to make a significant impact especially in the area of grid connected wind-turbine generation. Currently this critical power electronic piece in the system is catered by the traditional interacting converters such as the VSI which have dominated the market for many decades and accepted as an industry standard. But with increasing costs of energy storage size, there is a pressing need for systems to economize on space usage of converters, growing need for dynamic AC n-phase to n-phase conversion with faster switching capabilities, etc. Research on the Matrix converter over the past few years has seen tremendous progress in finding a more efficient & compact converter to meet this gap. Matrix converter is anticipated to supersede the traditional interfacing devices, if the protection issues related to their isolation fault protection are completely resolved. This paper has reviewed in detail the various protection issues related to the Matrix converter and assessed the merits & de-merits of its remedies to potentially pose as a serious contender in providing a complete AC to AC power conversion solution to the industry.

Despite all the benefits that the Matrix converter brings, there is still a lot of scope for further research to make it a converter that can handle multiple fault scenarios occurring in a grid connected system. The converter needs to develop robust inherent protection & control mechanisms so that it can be implemented in various field applications such as that used in the renewable energy, marine, aviation industries to name a few.

Authors have planned to investigate further the safe and secure measures for protection of matrix converter integrated to renewable energy applications, through real time implementation of techniques mentioned in 2) c and 3) in hybrid manner.

REFERENCES

1. Zhang L., Watthanasarn C. & Shepherd W., 9–14 Nov. 1997, Application of a Matrix Converter for the Power Control of a Variable-Speed Wind-Turbine Driving a Doubly-Fed Induction Generator, *IEEE IECON International Conference on Industrial Electronics, Control and Instrumentation,* New Orleans, USA, pp. 906–911.
2. Wheeler P., Grant D., Jan. 1997, Optimised Input Filter Design and Low-Loss Switching Techniques for a Practical Matrix Converter, IEE *Proceeding on Electrical Power Applications,* vol. 144, pp. 53–60.
3. Wheeler P., Clare J.C., Empringham L., Bland M. & Apap M., April 2002, Gate Drive Level Intelligence and Current Sensing for Matrix Converter Current Commutation, *IEEE Transaction on Industrial Electronics,* vol. 49, no. 2, pp. 382–389.
4. Wheeler P., Rodriguez J., Clare J.C., Empringham L., Weinstein A., April 2002, Matrix Converters: A Technology Review, *IEEE Transaction on Industrial Electronics,* vol. 49, no. 2, pp. 276–287.
5. Wheeler P., Clare J., Empringham L., Apap M. & Bland M., Dec. 2002, Matrix Converters, *Power Engineering Journal.*
6. Klumpner C., Nielsen P., Boldea I. & Blaabjerg F., Oct. 2000, New Steps Towards a Low-Cost Power Electronic Building Block for Matrix Converters, *IEEE Industry Applications Conference,* vol. 3, pp. 1964–1971.
7. Klupner C., Blaabjerg F., Neilsen P., 2001, Speeding-up the Maturation Process of the Matrix Converter technology, *IEEE PESC.*
8. Klumpner C., Blaabjerg F., April 2002, Experimental Evaluation of Ride-Through Capabilities for a Matrix Converter under Short Power Interruptions, *IEEE Transaction on Industrial Electronics,* vol. 49, no. 2, pp. 315–324.
9. Neilsen P., Blaabjerg F. & Pederson J.K., Sept/Oct 1999, New Protection Issues of a Matrix Converter: Design Considerations for Adjustable-Speed Drives, *IEEE Transaction on Industry Applications,* vol. 35, no. 5, pp. 1150–1161.
10. Blaabjerg F., Casadei D., Klumpner C. & Matteini M., April 2002, Comparison of Two Current Modulation Strategies for Matrix Converters under Unbalanced Input Voltage Conditions, *IEEE Transaction on Industrial Electronics,* vol. 49, no. 2, pp. 276–287.
11. Chang J., Sun T., Wang A., April 2002, Highly Compact AC-AC Converter Achieving a High Voltage Transfer Ratio, *IEEE Transaction on Industrial Electronics,* vol. 49, no. 2, pp. 345–352.
12. Mahlein J., Igney J., Weigold J., Braun M. & Simon O., April 2002, Matrix Converter Commutation Strategies with and without Explicit Input Voltage Sign Measurement, *IEEE Transaction on Industrial Electronics,* vol. 49, no. 2, pp. 276–287.
13. Chokhawala R., Sobhani S., 13–17 Feb. 1994, Switching Voltage Transient Protection Schemes For High Current IGBT Modules, *APEC Conference Proceeding,* vol.1, pp. 459–468.
14. Kwon H., Min D., Kim H., July 1998, Novel Commutation Technique of AC-AC Converters, *IEE Proceeding for Electric Power Applications,* vol. 145, no. 4, pp. 295–300.
15. Burany N., Oct. 1989, Safe Control of Four-Quadrant Switches, *IEEE Conference on Industrial Application Society,* vol.1, pp. 1190–1194.
16. Bernet S., Ponnaluri S., Teichmann R., April 2002, Design and Loss Comparison of Matrix Converters and Voltage-Source Converters for Modern AC Drives, *IEEE Transaction on Industrial Electronics,* vol. 49, no. 2, pp. 304–314.
17. Mahlein J., Bruckmann M., Braun M., April 2002, Passive Protection Strategy for a Drive System with a Matrix Converter and an Induction Machine, *IEEE Transaction on Industrial Electronics,* vol. 49, no. 2, pp. 297–303.
18. Mahlein J., Braun M., Aug 2002, A Matrix Converter without Diode Clamped Over-voltage Protection, *The 3rd International Power Electronics and Motion Control Conference,* vol. 2, pp. 817–822.
19. Palmer P.R., Githiari A.N., 1995, The Series Connection of IGBTs with Optimized Voltage Sharing in the Switching Transient, *Proceeding of IEEE PESC'95 Conference,* Atlanta, GA, pp. 44–50.
20. Palmer P.R., Githiari A.N., Leedham R.J., 1997, Some Scaling Issues in the Active Control of IGBT Modules for High Power Applications, *Proceeding IEEE PESC'97 Conference,* St. Louis, MO, pp. 854–860.
21. Palmer P.R., Rajamani H.S., July 2004, Active Voltage Control of IGBTs for High Power Applications, *IEEE Transaction on Power Electronics,* vol. 19, no. 4, pp. 894–901.

22. Hefner A., March 1991, An Investigation of the Drive Circuit Requirements for the Power Insulated Gate Bipolar Transistor, *IEEE Transaction Power Electronics*, vol. 6, pp. 208–219.
23. Castino G., Dubashi A., Clements S. & B. Pelly, 1991, Protecting IGBTs Against short circuit, *IR IGBT Designer's Manual*. El Segundo, CA: International Rectifier.
24. Saiz J., Mermet M., Frey D., Jeannin P.O., Schanen J.L. & P. Muszicki, 2001, Optimization and Integration of an Active Clamping Circuit for IGBT Series Association, *Proceeding of IEEE Industry Applications Soc. Meeting (IAS'01)*, Chicago, IL.
25. Gediga S., Marquardt R., Sommer R., 1995, High Power IGBT Converters with New Gate Drive and Protection Circuit, *Proceeding of 5th European Power Electronics Conference (EPE'95)*, vol. 1, Seville, Spain, pp. 66–70.
26. Neft C.L., Schauder C.D., Oct. 1988, Theory and Design of a 30-hp Matrix Converter, *IEEE Transaction on IAS*, vol. 1, pp. 934–939.
27. Itoh J., Sato I., Odaka A., Ohguchi H., Kodachi & Eguchi N., Nov. 2005, A Novel Approach to Practical Matrix Converter Motor Drive System with Reverse Blocking IGBT, *IEEE Transaction on Power Electronics*, vol. 20, no. 6, pp. 1356–1363.
28. Augdahl B.W., Hess H.L., Johnson B.K., Katsis D.C., October 2005, Output Protection Strategies for Battlefield Power Supplied by Matrix Converter, *Proceeding of the 37th Annual North American* Power Symposium, pp. 151–158.
29. Ormaetxea E., Andreu J., Kortabarria I., Bidrate U., Martinez de Alegria I., Ibarra E. & Olaguenaga E., January 2011, Matrix Converter Protection and Computational Capabilities Based on a System on Chip Design with an FPGA, *IEEE Transaction on Power Electronics*, vol. 26, no.1, pp. 272–287.
30. Andreu, J., De Diego J.M., de Alegria I.M., Kortabarria I., Martin J.L., Ceballos S., 2008, New Protection Circuit for High-Speed Switching and Start-Up of a Practical Matrix Converter, *IEEE Transactions on Industrial Electronics*, vol. 55, issue 8, pp. 3100–3114.
31. Zhou D., Sun K., Liu Z., Huang L., Matsuse K., Sasagawa K., 2007, A Novel Driving and Protection Circuit for Reverse-Blocking IGBT Used in Matrix Converter, *IEEE Transactions on Industry applications*, vol. 43, issue 1, pp. 3–13.
32. Shi M., Zhou B., Wei Z., Zhang Z., Mao Y., Han C., 2011, Design and Practical Implementation of a Novel Variable-Speed Generation System, IEEE Transactions on Industrial Electronics, vol. 58, issue 11, pp. 5032–5040.
33. Pfeifer M., Schroder G., 2010, Matrix converter with overvoltage protection circuit, MELECON 2010–2010 15th IEEE Mediterranean Electrotechnical Conference, pp. 1293–1296.

Power and Energy – Kong (Ed.)
© 2015 Taylor & Francis Group, London, ISBN 978-1-138-02782-4

The excitation control strategy of Doubly-Fed Wind Generator for smart grid infrastructure

H.H. Song & Y.B. Qu
School of Information and Electrical Engineering, Harbin Institute of Technology, Weihai, China

D.H. Chu
School of Computer Science and Technology, Harbin Institute of Technology, Weihai, China

S.M. Xu
NARI Technology Co. Ltd., Nanjing, China

H.P. Zhao
Huntech Technology Co. Ltd., Beijing, China

ABSTRACT: As the penetration of wind energy increases, the intermittent nature may cause oscillations in the grid which, in consequence, degrade the excitation controllability of Doubly Fed Wind Generators (DFWGs). The rotor overcurrent of the generator under grid voltage dips can cause damage to rotor-side converter. Therefore, strong grid is needed; meanwhile, robust excitation control strategy for small voltage sag ride-through is also necessary. An improved control strategy of Low Voltage Ride-Through (LVRT) by inserting compensated terms considered the variation of magnetizing current was proposed to restrict the DFWG's rotor overcurrent, protect the rotor side converter, and implement the LVRT. The mathematical model for the DFWG was established, and the improved control strategy was analyzed. Finally, the validity of the proposed control strategy was verified by MATLAB/Simulink.

1 INTRODUCTION

Smart grid will be strong, flexible, reliable, self-healing and fully controllable grid with great amounts of distributed generations and large-scale centralized power plants. Figure 1 illustrates an example of a common smart grid with different power levels (G. Benysek et al. 2011).

As observed in Figure 1, heavy-duty power electronics converters are employed in the smart grid, maintaining and improving the state of the power supply security and quality. Equally important is the wind generators need fast and flexible control strategies to damp out transient power oscillations and to contribute to system stability after grid disturbance clearance as the penetration of wind power continues to increase (D. Xiang et al. 2006). This compulsive requirement is referred to as the LVRT standard. Figure 2 illustrated the Low Voltage Ride-Through (LVRT) standard made by the USA Federal Energy Regulatory Commission.

The DFIG rotor side converter usually employs the IGBT as the semiconductor-switching device, and the back-to-back converter can be rated at only 30% of the overall generator power. The rotor overcurrent of the generator under voltage dips can cause damage to the rotor-side converter (F. Hachicha & L. Krichen. 2012, M. Mohseni et al. 2011). Therefore, some LVRT strategies should be adopted. Today, to achieve the LVRT ability for the DFIG, the following three technologies are mainly used: (i) the active Crowbar

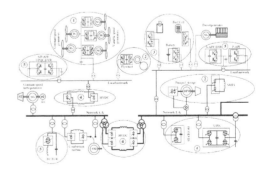

Figure 1. A smart grid infrastructure: 1 wind generators, 2 energy storage, 3 power supply systems from low-voltage sources, 4 network couplers, 5 devices for improvement of energy quality.

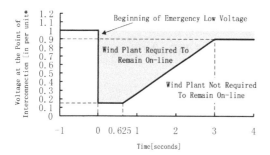

Figure 2. Low voltage ride-through standard (USA).

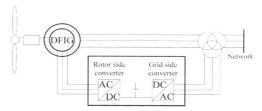

Figure 3. Configuration of the DFIG wind turbines system.

circuit protection; (ii) the DC bus energy storage circuit; (iii) the rotor current control (C. Niu & G. Liu 2012, A. D. Hansen & G. Michalke 2007).

To increase efficiency and save cost, the rotor current control is adopted during the grid slight sag. By analyzing the mathematical model, a new excitation control strategy is presented in this paper. The new strategy can reduce the rotor current oscillation and protect the back-to-back converter.

The remaining of the paper is organized as follows. Section 2 provides an improved mathematic model of the DFIG. Based on the model, a new control strategy for LVRT realization is performed in Section 3. Simulation results via MATLAB/Simulink are illustrated in Section 4. Conclusions are drawn in Section 5.

2 IMPROVE MATHEMATICAL MODEL OF DFIG

As illustrated in Figure 3, the rotor side converter connects to the rotor of DFIG directly, so the DFIG mathematical model is one of the basic theories for the design of rotor side converter control.

The precise DFIG mathematical model considering the variation of exciting current is established under synchronous rotating coordinate system firstly.

The DFIG voltage equations can be represented as follows (F. Hachicha & L. Krichen, 2012):

$$
\begin{cases}
u_{ds} = R_s i_{ds} + p\psi_{ds} - \omega_s \psi_{qs} \\
u_{qs} = R_s i_{qs} + p\psi_{qs} + \omega_s \psi_{ds} \\
u_{dr} = R_r i_{dr} + p\psi_{dr} - s\omega_s \psi_{qr} \\
u_{qr} = R_r i_{qr} + p\psi_{qr} + s\omega_s \psi_{dr}
\end{cases}
\tag{1}
$$

And Equation 1 can be simplified as:

$$
\begin{cases}
\mathbf{U_s} = R_s \mathbf{I_s} + p\mathbf{\Psi_s} + j\omega_s \mathbf{\Psi_s} \\
\mathbf{U_r} = R_r \mathbf{I_r} + p\mathbf{\Psi_r} + js\omega_s \mathbf{\Psi_r}
\end{cases}
\tag{2}
$$

The DFIG magnetic equations give the flux expressions as follows:

$$
\begin{cases}
\psi_{ds} = L_s i_{ds} + L_m i_{dr} \\
\psi_{qs} = L_s i_{qs} + L_m i_{qr} \\
\psi_{dr} = L_m i_{ds} + L_r i_{dr} \\
\psi_{qr} = L_m i_{qs} + L_r i_{qr}
\end{cases}
\tag{3}
$$

And Equation 3 can be simplified as:

$$
\begin{cases}
\mathbf{\Psi_s} = L_s \mathbf{I_s} + L_m \mathbf{I_r} \\
\mathbf{\Psi_r} = L_r \mathbf{I_r} + L_m \mathbf{I_s}
\end{cases}
\tag{4}
$$

where the subscript 's' and 'r' distinguish quantities or parameters on the stator and rotor windings. p represents differential of the variables. $\mathbf{\Psi}$ represents vector of the flux. $\mathbf{U}$ represents vector of the voltage. $\mathbf{I}$ represents vector of the current. ω_s is the angular speed of the generator. s is the rotor slip (L. Piegari et al. 2009).

The equivalent circuit of the DFIG according to Equations 2 and 4 is shown in Figure 4.

The state vector $\mathbf{I_m}$ which is defined as the stator exciting current can be represented as:

$$
\mathbf{I_m} = \frac{\mathbf{\Psi_s}}{L_m} = \frac{L_s}{L_m}\mathbf{I_s} + \mathbf{I_r}
\tag{5}
$$

From the above equation, stator current $\mathbf{I_s}$ is represented as

$$
\mathbf{I_s} = \frac{L_m}{L_s}(\mathbf{I_m} - \mathbf{I_r})
\tag{6}
$$

The rotor flux vector $\mathbf{\Psi_r}$ can be expressed according to Equation 7 as

$$
\mathbf{\Psi_r} = L_r \mathbf{I_r} + L_m \frac{\mathbf{\Psi_s} - L_m \mathbf{I_r}}{L_s} = \frac{L_m^2}{L_s}\mathbf{I_m} + \sigma L_r \mathbf{I_r}
\tag{7}
$$

The constant term σ equals to:

$$
\sigma = 1 - \frac{L_m^2}{L_s L_r}
\tag{8}
$$

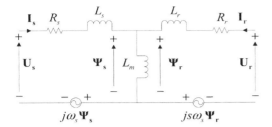

Figure 4. Equivalent circuit of DFIG.

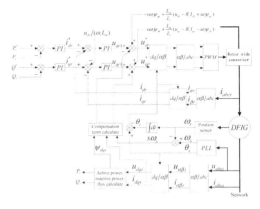

Figure 5. Schematic diagram of the excitation control strategy.

The DFIG voltage equation can be expressed according to Equations 6–8 as

$$
\begin{cases}
\mathbf{U_s} = R_s \mathbf{I_s} + L_m p \mathbf{I_m} + j\omega_s \boldsymbol{\Psi_s} \\[4pt]
\mathbf{U_r} = R_r \mathbf{I_r} + \sigma L_r p \mathbf{I_r} + \dfrac{L_m^2}{L_s} p \mathbf{I_m} \\[8pt]
\quad + js\omega_s \left(\dfrac{L_m}{L_s} \boldsymbol{\Psi_s} + \sigma L_r \mathbf{I_r} \right)
\end{cases}
\tag{9}
$$

Equation 9 is the improved mathematical model of DFIG considering the variation of exciting current. The state vector $\mathbf{I_m}$ is changing under the voltage dips. The new excitation control strategy will be designed according to this equation.

3 THE EXCITATION CONTROL STRATEGY FOR LVRT

The traditional control strategy performs well during normal operation. As it is designed according to a lower order mathematical model of the DFIG without considering the stator exciting current fluctuation under the grid voltage dips, the traditional one can't guarantee the rotor overcurrent remains within the safe range of power electronic components in the rotor side converter (J. Hu et al. 2006). To solve this problem, a new excitation control strategy is designed as follows.

According to Equation 10, we have

$$
p\mathbf{I_m} = \frac{\mathbf{U_s} - R_s \mathbf{I_s} - j\omega_s \boldsymbol{\Psi_s}}{L_m}
\tag{10}
$$

Then, eliminate the $p\mathbf{I_m}$ in $\mathbf{U_r}$ of Equation 10,

$$
\mathbf{U_r} = (R_r + js\omega_s \sigma L_r)\mathbf{I_r} + \sigma L_r p\mathbf{I_r}
$$

$$
+ js\omega_s \frac{L_m}{L_s} \boldsymbol{\Psi_s} + \frac{L_m}{L_s}(\mathbf{U_s} - R_s \mathbf{I_s} - j\omega_s \boldsymbol{\Psi_s}) \quad (11)
$$

Therefore, the DFIG voltage equation considering the variation of exciting current can be represented as:

$$
\begin{cases}
u_{dr} = R_r i_{dr} + \sigma L_r p i_{dr} - s\omega_s \sigma L_r i_{qr} \\[4pt]
\quad - s\omega_s \dfrac{L_m}{L_s} \psi_{qs} + \dfrac{L_m}{L_s}(u_{ds} - R_s i_{ds} + \omega_s \psi_{qs}) \\[8pt]
u_{qr} = R_r i_{qr} + \sigma L_r p i_{qr} + s\omega_s \sigma L_r i_{dr} \\[4pt]
\quad + s\omega_s \dfrac{L_m}{L_s} \psi_{ds} + \dfrac{L_m}{L_s}(u_{qs} - R_s i_{qs} - \omega_s \psi_{ds})
\end{cases}
\tag{12}
$$

Equation 10 is the theoretical foundation of the excitation control strategy. The improvement is inserting a compensated term can improve the voltage sensitivity to the current variation $L_m/L_s (\mathbf{U_s} - R_s \mathbf{I_s} - j\omega_s \boldsymbol{\Psi_s})$. Figure 5 shows the schematic diagram of the excitation control strategy.

4 SIMULATIONS

To prove the validation of the excitation control strategy for the LVRT of the DFIG and to compare the excitation control strategy with the traditional one, a simulation test is performed in MATLAB/ Simulink environment. The model is established based on the reference (G. Richard et al. 2010). The simulation model is composed of 6 wind turbine generators, and the rated active power is 9MW. The wind speed is retained at 15 m/s. The grid voltage drops to 0.9pu at 0.2 s and is recovered at 0.4 s. The simulation results are shown in Figure 6 to Figure 11. In these figures, the black waveforms represent the traditional control strategy results, and the red ones are obtained by the new control strategy.

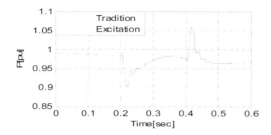

Figure 6. Active power.

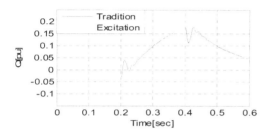

Figure 7. Reactive power.

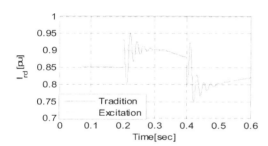

Figure 8. d-axis current of the rotor.

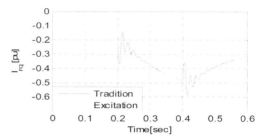

Figure 9. q-axis current of the rotor.

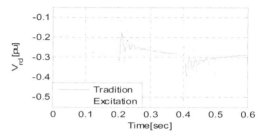

Figure 10. d-axis voltage of the rotor.

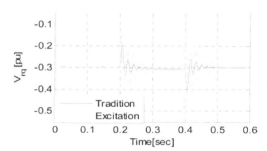

Figure 11. q-axis voltage of the rotor.

During the grid slight sag, the DFIG can't supply energy to the network normally. Hence, the change of the control strategy has little influence on the active and the reactive power. It can be seen that the behaviors of the active power and reactive power in Figure 6 and Figure 7, are shown to be very similar. Figure 8 and Figure 9 show the oscillations of the dq-axis rotor current under the two control strategies. It shows that the oscillations reduce obviously when the compensated term is inserted in the control process. Therefore, the new control strategy is more effective than the traditional one on the prevention of the overcurrent.

On the contrary, it can be found in Figure 10 and Figure 11 that the rotor voltage oscillations increase when the new control strategy is utilized.

Thus, the excitation control strategy is to reduce current oscillation at the expense of increasing voltage oscillation. The excessive energy is consumed internally by the DFIG. This means this kind of LVRT strategy is only suitable for the grid slight sag.

5 CONCLUSION

To realize the LVRT of the DFIG under smart grid slight sag, an excitation control strategy for LVRT is proposed based on the studying of the mathematical model of DFIG. By inserting a compensated term to the traditional control strategy, the new control strategy is formed which can improve the voltage sensitivity to the current variation. Different dynamic properties of the DFIG under the grid small voltage sag between the new strategy and the traditional one are illustrated in the simulation. The result shows the oscillations of the dq-axis rotor current can be reduced obviously by the new excitation control strategy, and consequently confirms its validity.

ACKNOWLEDGMENT

This work was supported by the National Science Technology Support Plan Project (2012BAA13B01), the National Development and Reform Commission in 2012, the Internet of Things Technology Research and Development and Industrialization Projects, and the Natural Scientific Research Foundation in Harbin Institute of Technology. (HIT.NSRIF.2014138 & HIT.IBRSEM.A.2014014).

REFERENCES

1. D. Hansen and G. Michalke. "Fault ride-through capability of DFIG wind turbines," Renewable Energy. Vol. 32, 2007, pp. 1594–1610.
2. G. Benysek, M.P. Kazmierkowski, J. Popczyk, and R. Strzelecki. "Power electronic systems as a crucial part of smart grid infrastructure-a survey," Bulletin of the polish academy of sciences technical sciences. Vol. 59, No. 4, 2011, pp. 455–473.
3. Niu and G. Liu. "The Requirements and Technical Analysis of Low Voltage Ride Through for the Doubly-Fed Induction Wind Turbines," Energy Procedia. Vol. 12, 2012, pp. 799–807.
4. Xiang, R. Li, P.J. Tavner, and S. Yang. "Control of a Doubly Fed Induction Generator in a Wind Turbine During Grid Fault Ride-Through," IEEE Transactions on Energy Conversion, Vol. 21, No. 3, 2006, pp. 652–662.
5. F. Hachicha, and L. Krichen. "Rotor power control in doubly fed induction generator wind turbine under grid faults," Energy. Vol. 44, 2012, pp. 853–861.
6. J. Hu, D. Sun, Y. He, and R. Zhao. "Modeling and Control of DFIG Wind Energy Generation System Under Grid Voltage Dip," Automation of Electric Power Systems. Vol.30, No.8, 2006, pp. 21–26.
7. L. Piegari, R. Rizzo, and P. Tricoli. "Optimized Design of a Back-to-Back Rotor-side Converter for Doubly Fed Induction Generator equipped Wind Turbines," IEEE International Conference on Clean Electrical Power, June 9–11, 2009, pp. 679–684.
8. M. Mohseni, M.A.S. Masoum, and S.M. Islam. "Low and high voltage ride-through of DFIG wind turbines using hybrid current controlled converters," Electric Power Systems Research. Vol.81, 2011, pp. 1456–1465.
9. G. Richard, T. Gilbert, L. Christian, B. Jacques, S. Gilbert, and F. Martin. Large-Scale Real-Time Simulation of Wind Power Plants into Hydro-Québec Power System. IEEE 9th International Workshop on Large-Scale Integration of Wind Power into Power Systems as well as on Transmission Networks for Offshore Wind Power Plants, October 18–19, 2010, pp. 1–8.

Power and Energy – Kong (Ed.)
© 2015 Taylor & Francis Group, London, ISBN 978-1-138-02782-4

Desalination engineering ship based on tidal current energy generation device

Yong Ma, Xiaojun Tan, Yang Cao, Xiang Xu & Li An
Harbin Engineering University, Harbin, Heilongjiang Province, China

ABSTRACT: As a new special boat, desalination engineering ship based on tidal current energy generation device is applied to the tidal current energy of ocean to carry out electric power development and fresh water production, transporting electric power resource and fresh water resource. The engineering ship is applicable the wave-piercing catamaran to use vertical shaft turbine as the tidal current energy conversion device, which has solved the adaptability problem of flow direction in tidal current energy generating set; the angle of attack for hydraulic turbine blade by magnetic control is introduced to improve the energy conversion efficiency and working stability of water turbine. The application of engineering ship is built the method of development and utilization in new tidal current energy to offer the engineering experience to diversified development of tidal current energy; the supply mode for electric power resource and fresh water resource of offshore islands has been changed to supply electric power and fresh water for remote island, facilitating the economic development on the offshore islands to some extent.

Keywords: engineering ship; tidal current energy; desalination; power transmission; a water turbine with magnetic control in angle of attack

1   THE OVERALL DESIGN OF ENGINEERING SHIP

The engineering ship is consisted of hull, tidal current energy generating set, sea water desalting plant, casting anchor system and control section. Hull is the wave-piercing catamaran, which is fixed by mooring system in its load when operating. As a tidal current energy conversion device,

Figure 1.   System total schematic diagram.

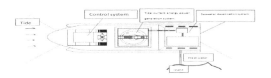

Figure 2.   The working total schematic diagram of desalination engineering ship.

the electric generator of vertical axis water turbine is viewed as the permanent magnetic generator directly driven by wind turbine. The water turbine of tidal current energy generating set is the vertical axis straight blade lift resistance tidal current energy water turbine, which is adopted the new-type control device for magnetic variable angle of attack; the sea water desalting plant is applied to JHH-SW200 sea water desalting equipment. The system total schematic diagram is shown as below (Fig. 1).

The engineering ship is firstly come to an anchor and fixed on the suitable sea area of tidal current energy; then the tidal current energy device is laid down for power generation to extract seawater for desalination. Finally, the fresh water and electric power is transported to the island through water duct and cable. The working total schematic diagram is shown as Figure 2.

2   CAPACITATION DEVICE DESIGN OF ENGINEERING SHIP

In this project, a type of non-contact control device—magnetic control angle of attack device is applied to the vertical axis water turbine. Through a test, it is superior to use repulsive force between magnets to control blade angle of attack subject to attraction, which can manage the blade by fair adjustment of the distance between magnet.

The structure of water turbine is shown in Figure 3 and the magnetic control principle is shown in Figure 4.

1. Blade; 2. Blade axle; 3. Turnplate; 4. Blade bearing; 5. Rocking beam; 6. Oscillating bar magnet; 7. Fixed magnet; 8. Mainshaft bearing; 9. Turbine main shaft; 10. Engine base of water turbine.

Through a preliminary trial, the magnet variable angle of attack water turbine is excellent in performance.

Energy on the tide flowing in the water turbine in unit time

$$W = 0.5\rho AU^3 \qquad (1)$$

ρ is current's density, A is incident flow area and U is flow velocity.

Thus, the energy utilization coefficient is as follows:

$$C_p = w/W \qquad (2)$$

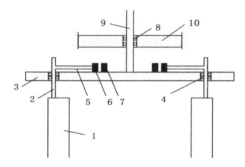

Figure 3. A structural map of water turbine.

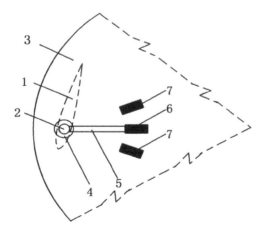

Figure 4.   Schematic diagram of magnet control angle of attack.

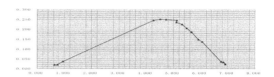

Figure 5.   A curve for energy utilization coefficient.

In this form, w' is the shaft power of water turbine.

C_P, the energy utilization coefficient of water turbine under different speed ratio is shown in Figure 5.

In operation, the engineering ship is needed to put tidal current energy water turbine into the water for energy collection; when sailing, the hydrodynamic performance of engineering ship will be affected by the tidal current energy water turbine. To avoid it, a jacking system is proposed to control the location of the relative water surface of water turbine. The upgrade device is used the four-pile stationary type slide through pile to fix the location of water turbine. Therefore, the water turbine can begin to slide up and down in different circumstances, fulfilling the up and-down movement of water turbine, to avoid the influence of hydrodynamic performance from the water turbine when the engineering ship is sailing.

3 HULL DESIGN OF ENGINEERING SHIP

3.1 *The basic requirements of landing design*

The landing is the carrying landing of tidal current energy conversion device, sea water desalting plant and its subsystem and attachment. Its design should be met the following requirements:

1. The shape design of the landing should have stream guidance, which is able to improve the hydrodynamic environment of water turbine.
2. The integrated design of the landing should be strong and stiff to ensure the safety of tidal current energy under the extreme adverse sea condition, focusing on fatigue strength of floor space.
3. There should be big enough floor space in the landing, which is convenient to install the water turbine, sea water desalting plant and its auxiliary system.
4. With regarding to the sea severe weather and other environmental factors and the needs of installation and overhaul, the carrier should be able to have self-propulsion or towing capability.
5. Its appearance is attractive and generous, with traveling ornamental value in its shape and color.

6. In design, we should also consider the construction difficulty, hydrodynamic force and tide weather.

3.2 Unit type

As for the essential requirements of (1)–(6) in 2.2.1 and the working environment of the future power station, wave-piercing bow and catamaran body is applied.

Wave-piercing bow and catamaran body is good in seaworthiness, with the expansion of wide deck in the catamaran. The characteristics of this structure are as follows:

1. The front part of these two hulls is deep V-shaped and becomes trapezoid backwards. Thus, this shape seems a fairing to an incoming flow seawater, which is beneficial to improve the dynamic environment of water turbine.
2. The wave piercing catamaran has the advantages like the catamaran and wide deck, which is good to arrangement of the water turbine, sea water desalting plant and other accessories.
3. The self-power wave piercing catamaran can protect itself from the severe weathers to ensure safety due to typhoon in the coastal regions like Zhejiang and Guangdong with rich tidal current energy.
4. The floating unit is more advantageous to Zhoushan sea of Zhejiang where much in depth and more in rock is.
5. The normal operation of power station can be ensured by good stability and seakeeping in the wave-piercing catamaran.

3.3 Measurement of principal dimension of unit

1. Estimation of space weight

Approximate estimation is conducted to the light weight of catamaran in different methods. It is assumed that the equilibrium condition for ship floating on the quiet water is the displacement of main body being equal to the sum between the light weight Lw and load capacity Dw, from the equation the expression:

$$\Delta = Lw + Dw \quad (3)$$

Light weight is included the weight of hull structure Wh, fitting-out weight Wo and the weight of electromechanical device Wm, from the equation the expression:

$$Lw = Wh + Wo + Wm \quad (4)$$

The weight of hull structure Wh: due to lack of the reliable statistical information about the structural weight of catamaran hull, the cubical

module calculation method is applied to suppose that the part weight and its volume of hull structure is in direct proportion, from the equation the expression:

$$Wh = dBV_b + dcBV_{cb} \quad (5)$$

In this formula:

V_b and V_{cb} is respectively the volume of main body and the transverse connection, with the unit of m^3;

dB and dcB are respectively the related coefficients of structure density, with unit of f/m^3.

The structure density coefficient value should be calculated as below: we can assume the designing scheme of hull structure in different tonnages, and its stress value of each part under the effect of applied load and local load shouldn't exceed than the allowable stress. Then, we can calculate the weight of the main sheet materials and components, such as the planking, longitudinal bulkhead, transverse bulkhead, intermediate deck and stiffener to estimate the weight of the general construction, count d, the coefficient of structure density according to volume of each part. In the stage of preliminary estimate, the formula is $dB = dcB = 0.1 t/m^3$.

The fitting-out weight Wo: this weight includes the system, carpentry, coating and fitting-out of equipment systems.

In the initial design, the following formula is calculated, with an unit of t:

$$Wo = 0.55 + 0.1Vcb \quad (6)$$

Furthermore, when we estimate the light weight, we should usually think about the redundancy of certain displacement. In the initial stage, the redundancy of displacement is 4% of the light weight.

The volume and weight for hull and cross structure of 8 ship units are shown in Table 1.

3.4 Check for unit stability

We can suppose that the ship is in the process of equivoluminal and small-angle inclination. The movement curve of buoyant centre is viewed stability, M as the center of a circle and BM as the arc of radius (B is the location of buoyant centre when it is floating on even keel). The location of M dot is unchanged, and the buoyancy action line passes M.

In accordance with assumption, when the small angle heels under the exogenous process, the catamaran inclines to waterline $W\phi L\phi$ from waterline $WoLo$ by equal volume. The volume of displacement is still the same before and after the incline. The intersecting line axle of the front and

343

Table 1. Table for volume and weight of hull and connecting bridge.

Serial number	Volume of hull (m^3)	Hull quality (t)	Volume of lateral connection bridge (m^3)	Quality of lateral connection bridge (t)	Fitting-out (t)	Light weight (t)
1	769	84.6	435.6	42.59	4.85	132.02
2	712.3	78.4	591.5	65.07	6.56	149.98
3	858.3	94.4	605.0	66.55	6.71	167.67
4	801.7	88.3	665.4	65.07	7.37	160.63
5	966.1	106.3	680.6	66.55	7.54	180.37

Table 2. Crosswise stability vertical elevation value.

| | Crosswise stability vertical elevation value | | | | | | | |
| | L = 30 | | | | L = 34 | | | |
Draught (m)	D = 4, B = 2.5	D = 4, B = 3	D = 5, B = 2.5	D = 5, B = 3	D = 4, B = 2.5	D = 4, B = 3	D = 5, B = 2.5	D = 5, B = 3
2.1	24.0	25.8	24.2	25.8	23.9	25.5	24.1	25.7
2.2	22.9	24.6	23.1	24.6	22.8	24.4	23.0	24.6
2.3	21.9	23.5	22.1	23.5	21.8	23.3	22.0	23.5
2.4	21.0	22.5	21.1	22.6	20.9	22.4	21.1	22.5
2.5	20.2	21.7	20.3	21.7	20.1	21.5	20.3	21.7

Table 3. Table of hull principal dimension.

Molded length (m)	Molded breadth (m)	Molded depth (m)	Draught (m)	Hull length (m)	Hull width (m)	Total weight (kg)
36	14	7	2.4	30	3	4400

Figure 6. The modeling diagram.

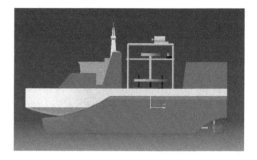

Figure 7. Diagram of engineering model.

backwater planes, $O - O$ gets through the centre of flotation of water plane before incline. $WoOL\phi$ and $LoOL\phi$ are regarded as a wedge whose wedge is emptied, from the equation the expression:

$$\overline{BM} = \frac{I_1 + A_{WS1}b_1^2 + I_2 + A_{WS2}b_2^2}{\nabla} \qquad (7)$$

In this formula, I_1 and I_2 is respectively the inertia moment of the lengthways longitudinal central axis to the positive floating water plane of 2 hulls.

A_{WS1} and A_{WS2} are the areas of even keel water plane;

b_1 and b_2 are hull span;

∇_1 and ∇_2 is the volume of displacement when the hull is floating on even keel, $\nabla = \nabla_1 + \nabla_2$.

According to the above-mentioned principle, the data of all models and the comparison in the stable vertical elevation of all schemes are calculated as shown in Table 2.

To sum up, based on the former data comparison and certain proportion setting, we choose the principal dimension of hull, as shown in Table 3.

The modeling diagram of engineering ship is shown in (Fig. 6) and the hull material object is shown in (Fig. 7).

4 DESIGN OF MOORING SYSTEM

Design for mooring system: (1) Catenary mooring; (2) In order to support the large load produced by the normal operation from water turbine and high strength, according to the relative stability and economic efficiency of tide power station, the multi-component mooring line integrated with nonlinearity high elastic cable and anchor chain is applied, that is anchor chain-high elastic cable-anchor chain;(3)The six-link symmetric expression is adopted as the arrangement form of mooring line because the power station is located near the complicated underground sea area and the incident flow direction's mooring line is bigger and the rupture will happen in the mooring line, as shown in Figure 8.

The catenary form is shown under the pre-stressed calculation in Figure 9.

To sum up, mooring system is applied in this ship to reduce the oscillating of ship to a large extent and ensure the smooth operation of water turbine be on charge.

Figure 8. Six-link symmetric anchoring.

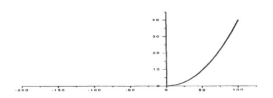

Figure 9. Catenary outline drawing.

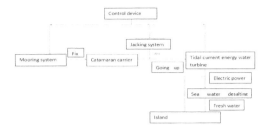

Figure 10. Desalination flow diagram.

5 DESIGN FOR SEA WATER DESALTING PLANT

Engineering ship embarkation JHH-SW200 sea water desalting equipment is used the reverse osmosis process, with 10 m³/h of water yield and high efficiency. The fresh water can be supplied to the island constantly, and the power consumption per ton is only 4 kilowatts, which can greatly save the electricity consumption. Furthermore, the boundary dimension of such sea water desalting equipment is $3400 \times 2000 \times 1800$ (mm), being convenient to install to the ship, with a little space in volume.

Desalination flow diagram is shown in Figure 10.

6 CONCLUSION

A new concept is proposed to the application of tidal current energy based on the desalination engineering ship of tidal current energy generating set. The main characteristics are as follows:

1. By using the renewable energy sources—tidal current energy production electric energy being clean and free of contamination, some electric energy is sent to island and the other is consumed on the spot, which is applicable to sea water desalination, produce and transport fresh water. In generation of electrical energy, sea water desalination and transportation of fresh water, the consumption of conventional source of energy is reduced to protect the environment. In addition, the following problems are solved by the site generation produced by the power generation, such as difficulty in storage and transportation of electric energy and high cost; the shortage of fresh water resource can be relieved on the earth to a certain extent.

2. The structure of wave piercing catamaran is beneficial to the equipment layout and improve the electricity generation of tide water turbine and ship navigation. Therefore, the engineering ship can get back to the island or inland on the

adverse sea conditions rapidly and safely, and transport fresh water safely and rapidly under emergency situation.

3. The engineering ship can transport the fresh water to island by water duct and can get through the water line to island.
4. By using vertical axis water turbine as the tidal current energy conversion device, the engineering ship is adaptable in the flow direction through generate electricity of tidal current energy.
5. The angle of attack of blade of hydraulic turbine is managed by magnet. The angle of attack of the blade of hydraulic turbine by magnetic control can be replaced the pitman-type structure, with the stable magnet performance, to enhance the efficiency of energy conversion and working stability of the water turbine, being low in processing and manufacturing cost.

The desalination engineering ship based on the tidal current energy generating set is a successful example through combined utilization between the engineering ship and tidal current energy generating set to solve the problems in transportation of power generation in tidal current energy and power supply of the island, being accelerative to the economic development of the offshore islands in China.

ACKNOWLEDGEMENTS

This paper was financially supported by the National Natural Science Foundation of China (Grant No. 51309069), the Special Funded of Innovational Talents of Science and Technology in Harbin (Grant No. RC2014QN001008), the Open Fund of the Laboratory of Multihull Technology National Defense Key Discipline (Grant No. HEUDTC1407).

REFERENCES

1. Gu Chunjiang. Model selection of tide power station and design research of mooring system[D]. A Master's Thesis on Harbin Engineering University, 2009.
2. Sheng Qihu, Luo Qingjie and Zhang Liang. Design in Carrier of 40kw Tide Power station[C]//Collected Works on the First Session Colloquium in the Ocean Energy Specialized Committee of Chinese Renewable Energy Society. Hangzhou, Zhejiang, 2008.
3. Luo Xianwu. A Guiding Device of Vertical Axis Tide Electric Generator. China: CN 101319648A, 10th Dec., 2008.
4. Zhang Liang. Design in Hydrodynamic Force of Fairing in the Tide Water Turbine[J]. Journal of Harbin Engineering University, July 2007.
5. Sun Baichao and Zhu Dianming. Working Principle of Straight Blade Water Turbine for Angle of Attack by Spring Control[C]// The Essays for the 2nd China's Utilization of Ocean Energy Colloquium (1983). Beijing, 1984.
6. Ma Qingwei. Performance study in the Adjustable Angle Straight-Blade Water Turbine[D]. A Master's Thesis on College of Ship building Engineering of Harbin Engineering University, 1984.
7. Guo Xiaotian. Design Research in Elasticity Mooring System of Floating Tide Power Station [D]. A Master's Thesis on College of Ship building Engineering of Harbin Engineering University, 2013.

Power and Energy – Kong (Ed.)
© 2015 Taylor & Francis Group, London, ISBN 978-1-138-02782-4

Power and RF GaN transistors for flexible antennas

A. Timoshenko, A. Bakhtin & K. Tsarik
National Research University of Electronic Technology, Moscow, Russia

ABSTRACT: This article presents considerations about using GaN transistors for implementation of flexible antennas compatible with other flexible electronic components and systems. The patch antenna layouts were chosen in order to obtain most significant degradation in performance characteristics which can be handled with using compensation methods during antenna and transceiver design. As the compensation methods, switched capacitors matrix and switched patch-pattern matrix are proposed.

1 INTRODUCTION

Growing interest in "flexible" electronics both among producers and among users [1] has spurred new researches. The aim is reducing weight, cost, size, ensure high-energy efficiency as well as their accessibility to electronic devices of new generation [2]. This suggests the need for integration of devices with flexible antenna operating in the frequency range of modern wireless communication. Antennas need to be mechanically strong, efficient in a wide range of frequencies with a predetermined radiation pattern, and its effectiveness will depend on the characteristics of the integrated circuits with two signal processing.

The material of the antenna on a flexible substrate must meet the basic requirements: high mechanical strength, low loss, high reliability and allows the integration of electronic components. Thus the base can serve as a polyamide film with a multilayer coating of copper or other conductive material [3–4]. Microstrip antennas are not compatible with flexible solutions in connection with a narrow band of frequencies which is dependent on the thickness of the substrate. In [5], we have proposed a variant of the use of patch antennas, which can use a coplanar waveguide as a power feeding method, since it does not require the creation of holes and provides low radiation losses, wide range, the best impedance matching and forms in one and the as the film layer of the radiation element.

The main field of flexible planar antenna application assumes the occurrence of degradation in antenna characteristics curved states (any changes in the geometric configuration of the antenna due to bending, buckling, twisting, etc.) and configuration changes in the states of the environment (the change in material characteristics immediately adjacent to the antenna).

The goal of this work is to analyze the possibilities of using power microwave GaN transistors for controlling the characteristics of flexible antennas.

2 MICROWAVE TRANSISTORS

2.1 *Technological facilities*

GaN due to its unique properties are promising for the various electronic devices. High thermal and radiation resistance, high values of the breakdown field, pronounced polarization effects make these materials attractive in the power electronics and the creation of powerful microwave transistors.

For our purpose we conducted our growing multilayer nitride heterostructures and transistors on equipment STE3 N3 on Al_2O_3 (0001) and 6H-SiC substrate. For GaN has a number of problems, one of which is the expensive metal nitride substrates of the third group [1]. Therefore, we grow a layer on the substrates that have mismatch in the parameters of the crystal lattice and thermal expansion coefficients. For this, we use the metal organic chemical vapor deposition and molecular beam epitaxy using ammonia as the nitrogen source.

We get good electron mobility in the GaN layer (Fig. 1) comprised with other works [7–8].

2.2 *Transistor creation*

Electrophysical parameters of a two-dimensional electron gas AlN/AlGaN/GaN/AlGaN, as well as the electrical parameters of GaN, are on par with the best values achieved in Russia on a similar technology equipment (electron mobility about 1500–1650 $cm^2/V \cdot s$, layer concentration $1,24 \cdot 10^{13} - 1,5 \cdot 10^{13}$ cm^{-2}, sheet resistance of ~ 300 $\Omega/$).

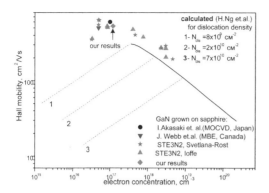

Figure 1. Electron mobility for the GaN layer.

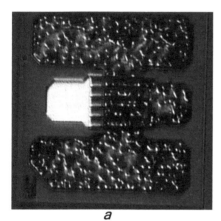

a

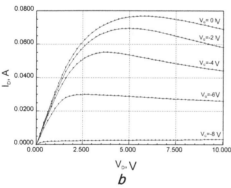

b

Figure 2. GaN transistor (a) and current-voltage characteristic (b).

On the designs, we have implemented test transistors with a gate length of 0.6 m (Fig. 2a). Ohmic contacts Ti\Al\Ni\Au were created by heating at a temperature of 850 °C for 30 seconds and had a resistance of $5 \cdot 10^{-6} - 7 \cdot 10^{-6} \ \Omega \cdot cm^2$. Drain-source current of 0.8–1 A/mm (Fig. 2b), transconductance of 200 mS/mm, the breakdown voltages over 100 V.

Small signal s-parameters f_T about 20 GHz, f_{max} up to 55 GHz. These transistors could be used in flexible antenna design as a power and RF switch.

3 PROPOSED APPROACH

For the electromagnetic characteristics forming of the antenna are in a degraded condition in real time have been identified sources of distortion and distortion compensation methods of analyzes in the design of a transceiver with an antenna composed of a flexible electronic system.

To compensate for the distortion characteristics of the antenna are proposed for non-linear component of the coefficient of external influence when setting up the radio. Payment can be made via the control S-parameters using the matching circuits. Although the integrated devices for such compensation schemes can be used gyrators and varactors, their limitation in the frequency range of parasitic circuit elements and low stability leads to the desired result. Another way is by using matrices matching capacitance connected or connectable elements metallization (Fig. 3), which can lead to an increase in the application and of the radio communication system with antennas on a flexible basis as a whole.

Also, gyrator circuits might have certain stability issues, needed to be addressed by thorough design of transconductances, and higher Q values can be obtained only in expenses of decreased dumping factor [9]. Then only capacitors are yet to be matched which can be performed either by using varicaps or capacitance matrixes.

The processed signal should be limited by frequency due to symmetry of measured dependence of reflection coefficient from measuring

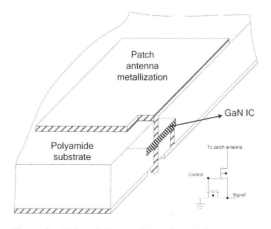

Figure 3. Using GaN transistors in switch controlled matrix.

signal frequency. Otherwise, changes can be interpreted wrong which can cause occurrence of positive feedback.

All before mentioned considerations allow concluding that there could be two options for layout of planar flexible measurement antenna: patch antenna with great capacitance component and planar long antenna with big inductive component. Then for designing such antennas the same principles as for capacitate or inductive MEMS devices can be used, when the feedback only affects the input cascades of transceiver without changing it. Other option is when antenna itself changes impedance significantly when being bent, for instance, like Split Ring printed monopole antenna. Though while measuring, it is necessary to consider all possible harmonic signals which excite the resonator.

The calculation shows that the probe feed location should be at 5.06 mm. But according to technology limitations [10] the grid pitch should be 25 um (Table 1). It causes the correction of antenna input impedance just to 0.328 Ω both for 1 and 2 mil polyamide.

According to parallel and serial commutation capacitors unit capacitance was taken at 250 fF (about 300 um^2 for MIM capacitance in 180 nm CMOS). The array of the commutation capacitors contain 1024 unit capacitors that could be switched off or switched serial/parallel. The lowest capacitance is about 244 aF and the highest is 256 pF. But due to non-uniform changing of load capacitance, we lose about 20 dB of the dynamic range.

To address the issue of changeability of flexible antenna parameters due to degradation in bent conditions the following concept was proposed (Fig. 2). Flexible film with target flexible antenna also contains test pattern with high Q value which can be reconfigured by 'Influence cancelation control' circuit. Evaluated S-parameters of target antenna measured by means of test inductive structure allow forming a control signal to adjust parameters of Power Amplifier (PA) and

low signal to adjust parameters of Power Amplifier (PA) and Low Noise Amplifier (LNA) of antenna-feeder tract of radio transceiver. Radio frontend adjustment allows expecting a partial compensation of parameters degradation of target antenna due to bent conditions. Exact layout of test pattern can be used to maximize the efficiency of the retrieved correction coefficients for target flexible antenna.

4 CONCLUSIONS

After measurement, considerations for implementation together with transceiver the planar antenna or matrix patch antenna is proposed. Antennas will be manufactured using one-sided copper-clad polyamide which allows embedment of antenna with electronic system and hence allows being less careful with input impedance mismatch. GaN transistor IC will be implemented between copper layers.

ACKNOWLEDGMENTS

This work was supported in part by the Council for Grants of President of Russia (Grants No. 14.Y30.14.5369-MK and MK-1922.2014.8) and Ministry of Science and Education of Russia (Grants No. 14.575.21.0097 and 14.575.21.0019).

REFERENCES

1. Hu, J. (2010). Overview of flexible electronics from ITRI's viewpoint. VLSI Test Symposium (VTS), 2010 28th, 19–22 April, 84.
2. Yongan, Huang, Chen, Jiankui, Yin, Zhouping, & Xiong, Youlun. (2011). Roll-to-Roll Processing of Flexible Heterogeneous Electronics With Low Interfacial Residual Stress. Components, Packaging and Manufacturing Technology IEEE Trans. on Sept., 1(9), 1368–1377.
3. Hitachi Chemical: MCF-500ID, http://www.hitachi-chem.co.jp/english/report/054/54.pdf.
4. Hertleer, C., Tronquo, A., Rogier, H., Vallozzi, L., & Van Langenhove, L. (2007). Aperture-Coupled Patch Antenna for Integration Into Wearable Textile Systems. Antennas and Wireless Propagation Letters, IEEE, 6, 392–395.
5. Timoshenko, A., Lomovskaya, K., On Possible Application Areas and Layout Configurations of Flexible Antennas//Proceedings of ICUMT—2014—Oct. 6–8., pp. 735–737.
6. Kukushkin, S.A., Osipov, A.V., Bessolov, V.N., et al., Substrates for Epitaxy of Gallium Nitride: New Materials and Techniques, Rev. Adv. Mater. Sci., 2008, vol. 17, pp. 1–32.
7. Ng, H.M., Doppalapudi, D., Moustakas, T.D., et al., The Role of Dislocation Scattering in n-Type GaN Films, Appl. Phys. Lett., 1998, vol. 73, no. 6, pp. 821–823.

Table 1. Summary of values chosen for patch antenna array (3×3) and capacitance array.

Parameter	Dimension
Patch width	19 mm
Patch length	15.6 mm
Element spacing $\lambda/2$	27 mm
E-plane separation distance	11.4 mm
H-plane separation distance	8 mm
Probe feed location	5.05 mm
Unit commutated capacitance	250 fF
Effective capacitance dynamic range	100 dB

8. Alekseev A.N., Petrov S.I., Nevolin V.K., Tsarik K.A., Krasovitskii D.M., Chalyi V.P., MBE Grown Nano-heterostructures with Increased Electron Mobility// Russian Microelectronics, 2012, Vol. 41, No. 7, pp. 400–404.

9. "Advancement in Microstrip Antennas with Recent Applications", book edited by Ahmed Kishk, ISBN 978-953-51-1019-4, Published: March 6, 2013 under CC BY 3.0 license ch.

10. Vertyanov, D.V., Tikhonov, K.S., Timoshenkov, S.P., Petrov, V.S., Blinov, G.A., "Peculiarities of multichip micro module frameless design with ball contacts on the flexible board," Electronics and Nanotechnology (ELNANO), 16–19 April 2013 IEEE XXXIII International Scientific Conference, vol., no., pp. 417–419.

Conglamorate

Power and Energy – Kong (Ed.)
© *2015 Taylor & Francis Group, London, ISBN 978-1-138-02782-4*

Field charging test on arresters applied in 1000 kV ultra-high voltage substation

Changcheng Zhu & ZhiWei Lin
State Grid Hubei Electric Power Research Institute, Wuhan, Hubei Province, China

ABSTRACT: This paper introduces the purposes, contents and principles of 1000 kV arrester field charging test at Jingmen substation. During large load power supply, electrified test of three groups of 1000 kV arrester at substation side, side and line side is completed in Jingmen substation smoothly, the test results are the same with the 1000 kV Wuhan UHV AC base surge arrester electrified test results. This testing accumulates rich experience for 1000 kV equipments on site commissioning test and measurement.

1 INTRODUCTION

The 1000 kV Jindongnan-Nanyang-Jingmen UHV AC Demonstration Project, which is an initial step project to develop ultra-high voltage transmission and distribution technology in China. As the pioneer there are neither standards available to apply immediately, nor comprehensive and mature technology and experience to draw lessons from. So installations used in the 1000 kV UHV transforming project field are demanded for high voltage and large capacity, more than these, the field commissioning test are much more difficult to take and requires technology much higher than routine high voltage project.

The request for trail installation used in online test on arrester of 1000 kV is easy to accomplish, compared to that on transformer, GIS and other facilities, which only needs a LCD-4 leakage current tester and a current divider with ratio 4:1. Combining with the arrester online test course and experience in 1000 kV UHV AC Test Base, Wuhan, we accomplished online test on three groups of 1000 kV arresters applied to main transformer, busbar and line successfully during the period of transmitting heavy load, with the same result as that tested on arresters from Fushun electric porcelain Manufacturing Co., Ltd in 1000 kV UHV AC Test Base.

2 THE PURPOSE, CONTENT AND PRINCIPLE OF CHARGING TEST

As the same as the 500 kV arrester, the 1000 kV arrester also have inner capacitance stem, which means that commissioning test needs to employ AC test items including capacitance stem. For the large

difficulty in routine offline AC commissioning test on 1000 kV arresters, the purpose to online testing adds the meaning of commissioning test comparing with that on common high voltage arresters.

The structures of 1000 kV arrester include two kinds, one have four parts while the other is five parts. According to the demand from SGCC the 1000 kV Jindongnan-Nanyang-Jingmen UHV AC Demonstration Project Electric Installation Commissioning Test Standard, the continuous operating current of 1000 kV arrester is no more than 20 mA, and the continuous operating voltage can reach to 638 kV, and all phases test requires much to step-up installations. Even test part by part,

Figure 1. Arrester.

the step-up installation also needs 200 kV/50 mA. More than this the height of the peak of 1000 kV is already up to 12.76+8 m, and the diameter of shield ring exceeds 3.5 m, and the shield depth is the same as that of the peak part, as shown in Figure 1. After installation, the error of AC test on the peak part with shield ring leans to large. Therefore, it's necessary to test the full current and resistive current of 1000 kV arrester under operating voltage.

As well as the theory applied on routine arrester online test items, the principle is to utilize the operating voltage signal to separate resistance current from full current of arresters, then check and judge degrees of arrester inner affected with damp, disc deterioration and capacitance stem damage.

3 THE PROCESS AND CONSEQUENCE OF CHARGING TEST

There are three groups of 1000 kV arresters, made by Fushun electric porcelain Manufacturing Co., Ltd., being put into operation in substation of Jingmen; while the installation height of counter is 2.2 meters; the distance from the CVT on transmission line, as well as on the main transformer, of the second of the main transformer terminal box to the furthest arrester of A phase is about 100 meters.

We used LCD-4 leakage current tester for the measurement of the voltage of CVT on 1000 kV line, of arrester on transmission line, on busbar, and on main transformer respectively at 9:30 am,

Dec. 29th, 2008. When it was 10:30 am at the same day, we tested the voltage of CVT of the main transformer for the same parameters. When it was 11:30 am that day, we recorded the waveforms on steady state of the voltage, full current, resistive current and the power consumption. The result is shown in Table 1.

4 COMPARISON ANALYSIS OF TEST DATA

Testing that's morning's maximum load transmission, with voltage as 1067 kV, loads increasing from 800 MW to 2800 MW gradually, the 1000 kV system's working voltage vibrates as load changes, and corresponding data are shown in Table 2. As the LCD-4 isn't a digital device, and the resolution of voltage meter is low, which is less than the current meter. As a result, the voltage value in Table 1 varied little, and the amplitude of full current changes apparently.

Different from UHV AC Test Base, Wuhan, there are three groups of 1000 kV arresters in operation in Jingmen substation now, these arresters are the products of Fushun electric porcelain Manufacturing Co., Ltd., and the amplitude range of full current is 10.4–13.2 mA, consistent with the 9–11.2 mA of three phase arresters in UHV AC Test Base.

Same as the 500 kV arresters, the Phase characteristics of the 1000 kV arresters' resistive current and power consumption are obvious, A phase high and C phase low, but the difference is small. With no other same voltage grade electrical devices

Table 1. The first online testing data of 1000 kV arresters in substation, Jingmen.

Operation number	Phase	U (kV)	Ix (mA)	Irp– (mA)	Irp+ (mA)	Px (W)	Monitor (mA)
Nanyan-Jingmen	A	620	10.8	2.04	2.00	800	12.0
Line I	B	620	10.4	1.24	1.20	440	11.5
	C	620	11.6	0.84	0.80	280	12.5
Busbar	A	620	11.6	3.52	3.52	1440	12.5
	B	620	10.8	2.12	2.08	840	11.5
	C	620	10.8	2.08	2.12	840	12.0
Main transformer	A	620	13.2	3.52	3.48	1400	13.0
	B	620	11.6	2.24	2.24	920	11.5
	C	620	11.6	2.44	2.40	960	13.0
Nanyan-Jingmen	A	620	10.8	2.16	2.12	840	11.5
Line I	B	620	10.8	1.28	1.24	440	11.5
	C	620	11.6	1.00	0.96	320	12.5
Busbar	A	620	12.0	3.52	3.52	1440	12.5
	B	620	10.8	2.56	2.52	1040	11.5
	C	620	10.8	1.72	1.68	640	11.5
Main transformer	A	620	11.6	3.16	3.08	1240	12.5
	B	620	10.4	2.60	2.56	960	11.5
	C	620	11.2	2.56	2.52	1000	13.0
Base	C	630	11.2	2.20	2.20	815	11.0

Table 2. Changes of voltage and load during test of heavy load transmission.

Time	P/MW	Ua/kV	Ub/kV	Uc/kV	Uab/kV
9:30	800	617.28	615.38	616.85	
10:00	1100	617.13	615.00	616.02	1067
10:30	2800	615.86	615.28	616.56	

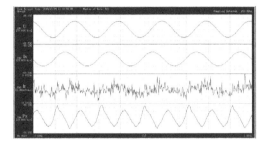

Figure 2. The typical signal waveform of arresters recorded by LCD-4.

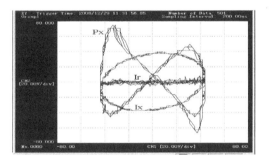

Figure 3. The Lissajous figure of Ix, Ir and Px with a reference of U.

around, the line arresters have no negative power consumption phenomenon, which is related to the four shunt posts of the 1000 kV arresters. Although the 1000 kV arresters of Jingmen substation have two more sections than the 500 kV arresters, its whole capacitance are about three times larger the 500 kV arresters'. As distances between phases two times larger, the influence of phase to phase stray capacitance decreases. As a result, for the 1000 kV arresters, the influence of phase to phase stray capacitance is little and no negative power consumption will appear. The signal waveform recorded by LCD-4 is shown in Figure 2, and the Lissajous figure of the full current Ix, resistive current Ir and power Px with a reference of Voltage U is given in Figure 3.

In Figure 2, the second voltage signal of C phase CVT in 1000 kV lines is obviously affected by harmonic. The full current is almost capacitive alone and the peak value of resistive current is small, about 1/10 of the full current. The fundamental component of resistive current is smaller, with the same phase of the voltage, and the power frequency vibration is visible. The power is consistent with the multiplication of voltage and full current, and frequency is two times of the working frequency. The full current waveform in Figure 3 is not right a round, the long axis is 45° to the horizontal line. Frequency characteristic of the resistive current is not obvious, and the multi-frequency of power and the magnitude characteristics are very notable, as an obvious inverse 8 graph.

The data in Table 1 show that the testing place has little influence on resistive current measurement for CVT of the same manufacturer, and the difference is less than 20%, do little influence on judgment of CVT feature.

5 CONCLUSION

The features of three groups of 1000 kV arresters under operating voltage in Jingmen substation are normal.

Under the operating voltage, no negative power consumption appears in the three groups of 1000 kV arresters of Jingmen substation.

The two groups of CVT, 100 meters apart from each other, do no influence on the feature tests of the arresters under operating voltage.

REFERENCES

1. Andoh H, Nishiwaki S, Suzuki H. Failure mechanisms and re-cent improvements in ZnO arrester elements [J]. IEEE Electrical Insulation Magazine, 2000,16(1):25–31.
2. Hu Dao-ming, Pan Wen-xia. Commenting on MOA leakage current online monitoring [J]. Jiangxi Electric Power, 2004,(4):26–28.
3. Wang Xing-gui, Li Qing-ling, et al. Case study of MOA [J]. High Voltage Apparatus, 2008,44(2):175–177.
4. Xie Peng, Zhang Guo-dong. The development and application of MOA test methods [J]. Insulators and Surge Arresters, 2006,(5):36–38.
5. Yan Yu-ting, Jiang Jian-wu, et al. MOA accident analysis and comparative study of testing methods [J]. Insulators and Surge.
6. Arresters, 2011,(5):63–69.
7. Yan Xiong-kai, Shao Tao, Gao Xiang, et al. Present state and future of detecting methods of MOA [J]. High Voltage Engineering, 2002,8(6):34–36.
8. Yan Zhang. Electrical insulation online testing technology [M]. Beijing: China Electric Power Press, 1995.

Power and Energy – Kong (Ed.)

A voltage stability index based on local branch measurement with good characteristics

Sanen Du, Yifeng Dong & Shiying Ma
China Electric Power Research Institute, Beijing, China

ABSTRACT: The online voltage stability monitoring based on local phasor measurement is a research focus in voltage stability field recently. A good online voltage stability assessment index, should meet the accuracy, linearity, fast calculation of performance requirements. A voltage stability index based on local branch phasor measurement is proposed, which only need the branch power and two nodes voltage phasor, with simple method of calculation and clear physical meaning. A comparative analysis between this index and other several branch indices shows that this index has better accuracy and linearity. The good characteristics of this index are proved by two test systems.

1 INTRODUCTION

Between the late 70's and end of the last century, several large blackout associated with voltage collapse has undergone in the world [1–7]. The common feature is that the accident occurred suddenly and concealment, the operating personnel is very difficult to detect during voltage instability accidents, not timely take effective control measures. Once the voltage collapse accident occurred, recovery of power supply will take several hours, or even more than ten hours. Therefore, the real-time continuous monitoring of power system voltage stability, and finding and locating weakness of voltage stability timely, have very important realistic meanings. In Wide Area Measurement System (WAMS), voltage phasor, current phasor, internal potential and excitation current of generator can be measured in real time and synchronously, providing new physical conditions for voltage stability monitoring. Power system voltage stability monitoring and analysis of power system voltage stability by using the local amount of Power Measurement Unite (PMU), because it does not depend on the state estimation results, global power flow calculation, few the data quantity is few without global trend, therefore become a research hotspot.

On-line monitoring voltage stability index based on local measurement can be divided into two categories: node-based index and branch-based index. The theoretical basis of branch-based index is the presence of branch impedance. When transmission power through the branch exceeds the limit, the power flow equation of this branch will have no real solution, and will not be able to meet load demand further. Many scholars have done a lot of research in this direction, made a variety of indices [8–11]. But their effectiveness and working

condition are different, even they are queried by some researchers [12–14]. Reference [8] and [9] proposed indices respectively based on the real solution condition of real power flow equation and the real solution condition of reactive power flow equation, but ignored the fact that voltage instability is associated with both real power and reactive power. Reference [8–10] directly applied the indices derived from two buses system in the complex system, lack of rigorous theoretical basis. In the load increasing process, one line's power exceeding the limit cannot draw the conclusion of system voltage collapse directly. Because if load demand increase further, there will be a number of lines reaching the limits. When this occurs in large range and loads continuously attempt to increase current to obtain more power (real power and reactive power), voltage collapse will be happened [14–15].

A new branch-based voltage stability index is proposed, which is derived from the real solution condition of the quadratic equation about sending voltage, and two test systems is designed for comparative analysis with the indices in Reference [8–11]. Point out the limitations of indices in Reference [8–9], validates the good characteristics of the index proposed in this paper.

2 DEDUCTION OF THE NEW BRANCH-BASED VOLTAGE STABILITY INDEX

For the simple power system with two buses in Figure 1, the real power and reactive power can be expressed as equation (1) and (2),

$$P_j = \frac{V_i V_j}{Z} \cos(\theta - \delta) - \frac{V_j^2}{Z} \cos\theta \qquad (1)$$

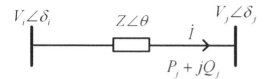

Figure 1. Two buses system.

$$Q_j = \frac{V_i V_j}{Z} \sin(\theta - \delta) - \frac{V_j^2}{Z} \sin\theta \qquad (2)$$

where $\delta = \delta_i - \delta_j$. Equation (1) and (2) can be rewritten as equation (3) and (4).

$$V_j^2 \cos\theta - V_i \cos(\theta - \delta)V_j + P_j Z = 0 \qquad (3)$$

$$V_j^2 \sin\theta - V_i \sin(\theta - \delta)V_j + Q_j Z = 0 \qquad (4)$$

Adding equation (3) and (4), we can get equation (5),

$$V_j^2(\cos\theta + \sin\theta) - V_i[\cos(\theta - \delta) \\ + \sin(\theta - \delta)]V_j + (P_j + Q_j)Z = 0 \qquad (5)$$

which is a quadratic equation on V_j. According to the discriminant of quadratic equation which has real root, the following conditions need to be satisfied,

$$V_i^2[\cos(\theta - \delta) + \sin(\theta - \delta)]^2 \\ - 4(\cos\theta + \sin\theta)(P_j + Q_j)Z \geq 0 \qquad (6)$$

or

$$V_i^2[1 + \sin 2(\theta - \delta)] \geq 4(\cos\theta + \sin\theta)(P_j + Q_j)Z \qquad (7)$$

rearranging (7) and simplifying

$$0 \leq \frac{4(\cos\theta + \sin\theta)(P_j + Q_j)Z}{V_i^2[1 + \sin 2(\theta - \delta)]} \leq 1 \qquad (8)$$

a voltage stability index is defined as follows:

$$L_{PQ} = \frac{4(\cos\theta + \sin\theta)(P_j + Q_j)Z}{V_i^2[1 + \sin 2(\theta - \delta)]} \qquad (9)$$

L_{PQ} is termed the voltage stability index of the branch whose value range is [0, 1]. As long as the index remains less than 1, the branch is stable. When the index is 1, the branch reaches critical point. The value is closer to 1, the branch is less stable.

3 OTHER SEVERAL INDICES FOR COMPARISON

From equation (1), reference [8] derived a voltage stability index as follows, called on-line voltage stability index (L_{VSI}).

$$L_{VSI} = \frac{4P_j Z \cos\theta}{[V_i \cos(\theta - \delta)]^2} \qquad (10)$$

The index's value range is [0, 1]. The value is closer to 1, the branch is less stable. When the index is 1, the branch reaches critical point.

From equation (2), reference [8] defined a line stability index as follows:

$$L_{mn} = \frac{4Q_j Z \sin\theta}{[V_i \sin(\theta - \delta)]^2} \qquad (11)$$

The index's value range is [0, 1]. The value is closer to 1, the branch is less stable. When the index is 1, the branch reaches critical point.

Reference [8] defined a voltage amplitude discriminant index (A_{VM}) as (12) from a two bus system as shown in Figure 1.

$$A_{VM} = [2(P_j R + Q_j X) - V_i^2]^2 \\ - 4[(P_j R + Q_j X)^2 + (P_j X - Q_j R)^2] \qquad (12)$$

The index's value range is [0, V_i^4]. The value is closer to 0, the branch is less stable.

Starting with a two bus system, reference [8] derived a Transmission Path Stability Index (*TPSI*).

$$TPSI = 2V_m \prod_{i=0}^{m-2} \cos\delta_{m-i-1, m-i} - V_1 \qquad (13)$$

The index's value range is [0, V_1]. The value is closer to 0, the branch is less stable.

4 TEST AND ANALYSIS

A good voltage stability index should have the following characteristics: accuracy, linearity, fast calculation, and other various information [16]. Five indices, mentioned in this paper, are comparative analyzed from these aspects, through two test system.

Two test systems are designed in this paper. In test system one, the voltage of sending bus remain constant while load is increased gradually. In testing system two, the voltage of the sending bus will drop down while load is increased gradually. In these two test system, load model type is constant power, and load is increased respectively by constant power factor, constant real power, and constant reactive power. The testing tool is PSD-BPA power flow program v4.0 by China Electric Power Research Institute. By increasing load gradually and calculating power flow repeatedly, until power flow does not converge, the process of the voltage of receiving bus trending to critical point can be simulated. Although more factors than load increasing affect power flow convergence, but because the other experimental conditions are the same, the results are comparable.

4.1 Sending bus of branch with constant voltage

A two buses test system is shown in Figure 2, where the voltage of sending bus BUSA remains constant, and the basic load on BUSB is 50 + j20 MVA, other detailed parameters in Table 1. When the load is gradually increased in steps, respectively by constant power factor ($\cos\varphi = 0.9285$), constant real power ($P_j = 0.5$ p.u.), constant reactive power ($Q_j = 0.2$ p.u.), as a result, five voltage stability indices and PV curves are shown in Figures 3–5, where V_j is the p.u. voltage of BUSB and Sj is the p.u. apparent power of load on BUSB.

The comparing results of indices characteristics are shown in Table 2, where voltage of sending

Figure 2.   Two buses test system.

Table 1.   The parameters of two buses system.

Line parameters				
BUSI	BUSJ	R (p.u.)	X (p.u.)	B/2 (p.u.)
BUSA	BUSB	0.02	0.05	0.0
Bus voltage setting				
Name	Amplitude (p.u.)		Angle (degree)	
BUSA	1.0		0.0	
Base information				
Power (MVA)			Voltage (kV)	
100.0			130.0	

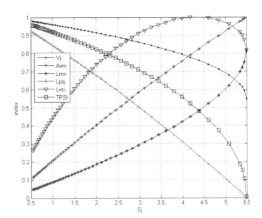

Figure 3.   Indices curves while loading increases by constant power factor ($\cos\varphi = 0.9285$).

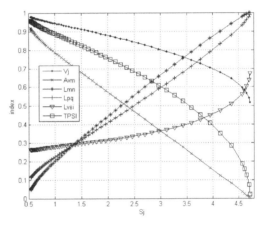

Figure 4.   Indices curves while loading increases by constant real power ($P_j = 0.5$ p.u.).

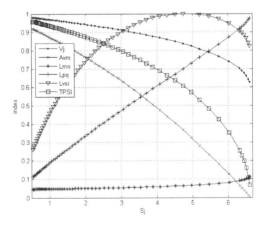

Figure 5.   Indices curves while loading increases by constant reactive power ($Q_j = 0.2$ p.u.).

Table 2. Characteristics comparison of indices with constant sending voltage.

	Constant power factor			Constant real power			Constant reactive power		
	Speed	Accuracy	Linearity	Speed	Accuracy	Linearity	Speed	Accuracy	Linearity
A_{VM}	★★★	★★★	★★★	★★★	★★★	★★★	★★★	★★★	★★★
L_{mn}	★★★	★★	★★	★★★	★★★	★★★	★★★	★	★★★
L_{PQ}	★★★	★★★	★★★	★★★	★★★	★★★	★★★	★★★	★★★
L_{VSI}	★★★	★	★	★★★	★★	★★	★★★	★	★
$TPSI$	★★	★★★	★★	★★	★★★	★★	★★	★★	★★

Figure 6. Four buses test system.

Table 3. The parameters of four buses system.

Line parameters

BUSI	BUSJ	R (p.u.)	X (p.u.)	B/2 (p.u.)
BUSB	BUSC(1)	0.04	0.10	0.0
BUSB	BUSC(2)	0.04	0.10	0.0
BUSC	BUSD	0.01	0.08	0.0

Transformer parameters

BUSI	BUSJ	R (p.u.)	X (p.u.)
BUSA	BUSB	0.0	0.002

Bus voltage setting

Name	Amplitude (p.u.)	Angle (degree)
BUSA	1.01	0.0

Base information

Power (MVA)	Voltage 1 (kV)	Voltage 2 (kV)
100.0	230.0	16.5

bus keeps constant. It can be observed, in the case of constant voltage of sending bus, L_{VSI} and L_{mn} cannot adapt to different load increasing modes. The main problem of L_{VSI} is highly nonlinear and inaccurate, not monotonic in the case of increasing load by constant power factor and by constant reactive power, and not accurate also in the case of increasing load by constant real power. L_{mn} is nonlinear also, and inaccurate in the case of increasing load by constant reactive power. A_{VM}, L_{PQ} and $TPSI$ can make an accurate assessment of voltage stability, although they have different linear.

4.2 Sending bus of branch with variable voltage

In actual network, along with the load demand increasing, the both buses voltage of the transmission line will change. The characteristics of these indices are studied in a test system where sending bus voltage is variable.

A four buses test system is shown in Figure 6, where the voltage of BUSA with generator remains constant, and the basic load on BUSD is 50+j20 MVA, other detailed parameters in Table 3. When the load is gradually increased in steps, respectively by constant power factor ($\cos\varphi = 0.9285$), constant real power ($P_j = 0.5$ p.u.), constant reactive power ($Q_j = 0.2$ p.u.), as a result, five voltage stability indices and PV curves are shown in Figures 7–9., where V_j is the p.u. voltage of BUSD and Sj is the p.u. apparent power of load on BUSD.

The comparing results of indices characteristics are shown in Table 4, where voltage of sending bus is variable. It can be observed, in the case of variable voltage of sending bus, L_{VSI} and L_{mn} cannot adapt to different load increasing modes. The main problem of L_{VSI} is highly nonlinear and inaccurate,

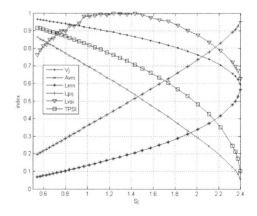

Figure 7. Indices curves of branch BUSC-BUSD while loading increases by constant power factor ($\cos\varphi = 0.9285$).

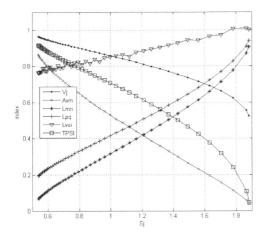

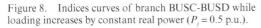

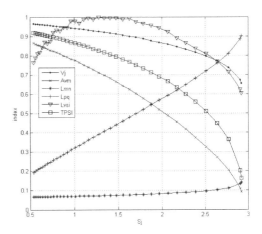

Figure 8. Indices curves of branch BUSC-BUSD while loading increases by constant real power ($P_j = 0.5$ p.u.).

Figure 9. Indices curves of branch BUSC-BUSD while loading increases by constant reactive power ($Q_j = 0.2$ p.u.).

Table 4. Characteristics comparison of indices with variable sending voltage.

	Constant power factor			Constant real power			Constant reactive power		
	Speed	Accuracy	Linearity	Speed	Accuracy	Linearity	Speed	Accuracy	Linearity
A_{VM}	★★★	★★★	★★★	★★★	★★★	★★★	★★★	★★★	★★★
L_{mn}	★★★	★	★★	★★★	★★	★★★	★★★	★	★★★
L_{PQ}	★★★	★★★	★★★	★★★	★★★	★★★	★★★	★★★	★★★
L_{VSI}	★★★	★	★	★★★	★	★★★	★★★	★	★
$TPSI$	★★	★★	★★	★★	★★★	★★	★★	★★★	★★

not monotonic in the case of increasing load by constant power factor and by constant reactive power, and not accurate also in the case of increasing load by constant real power. L_{mn} is nonlinear also, and inaccurate in the case of increasing load by constant reactive power. A_{VM}, L_{PQ} and $TPSI$ can make an accurate assessment of voltage stability, although they have different linear.

5 CONCLUSIONS

1. After comparative analysis on the five line-based voltage stability indices, points out that L_{VSI} and L_{mn} can make correct branch voltage stability assessment only under very specific conditions. $TPSI$ shows highly nonlinear, when power flow tends to voltage stability critical point, and only when approaching the critical point the index obvious changes. As a result, the operators have not sufficient time for reaction, difficultly in engineering applications. Because the value range of A_{VM} is associated with the sending

voltage, it is not conducive to assess voltage stability margin from current state.
2. The index L_{PQ} presented in this paper has good characteristics of calculation speed, accuracy, linearity, and has clear value range, is easy to use in engineering applications.
3. The index L_{PQ} presented in this paper is derived from two buses system, how to be used in complex system has to be studied further.

REFERENCES

1. Y.M. Wang, J.C. Wu, D.Z. Meng, "A large power system technology," Beijing: China power press, 1991.
2. H.Z. Cheng, "Study on static voltage stability of power system," Shanghai Jiao Tong University, 1998.
3. D.Y. He, "A preliminary understanding of WSCC disturbance and separation on July 2, 1996 in the United States," Power System Technology, Vol. 20, No. 9, 1996, pp. 35–39.
4. Y. Mansour, et al. Voltage stability of power systems, Concepts, analytical tools and industry experience[C]. IEEE Publication 90TH0358-2-PWR, 1990.

5. Z.H. Feng, "Study on power system voltage stability," Tsinghua University, Ph. D. thesis, 1990.

6. R.P. Guo, "Study on power system voltage stability," Zhejiang University, Ph. D. thesis, 1999.

7. X. Zhang, "Investigation of power system voltage stability contingency ranking," China Electric Power Research Institute, Master degree thesis, 2007.

8. D.W. Liu, X.R. Xie, G. Mu, P. Li, "An on-line voltage stability index of power system based on synchronized phasor measurement," Proceedings of the CSEE, Vol. 20, No. 1, 2005, pp. 13–17.

9. M. Moghavvemi, F.M. Omar. Technique for contingency monitoring and voltage collapse prediction[J]. IEE Proc.–Generation, Transmission and Distribution, Vol. 145, No. 6, 1998, pp. 634–640.

10. X.Z. Sun, X.Z. Duan, Y.Z. He, "Local voltage stability security index research on load nodes in power system," Automation of Electric Power Systems, Vol. 22, No. 9, 1998, pp. 61–64.

11. N.C. Zhou, M.X. Zhong, G.Y. Xu, Y.Z. Ren, "Voltage stability index in power system based on voltage phasors," Proceedings of the CSEE, Vol. 17, No. 6, 1997, pp. 425–428.

12. B.Z. Liu, Z. Qi, B.L. Li, "On-line real-time voltage stability analysis based on feasible equilibrium solution region of branch power flow," Proceedings of the CSEE, Vol. 28, No. 10, 2008, pp. 63–68.

13. J. Yu, W.Y. Li, W. Yan, "Querying effectiveness of several existing line-based voltage stability indices," Proceedings of the CSEE, Vol. 29, No. 19, 2009, pp. 27–35.

14. J.Q. Zhao, Y.D. Yang, Z.H. Gao, "A review on-line voltage stability monitoring indices and methods based on local phasor measurement," Automation of Electric Power Systems, Vol. 34, No. 20, 2010, pp. 1–6.

15. Y. Tang, "Voltage stability analysis of power system," Beijing, Science Press, 2011.

16. S.X. Zhou, Y. Jiang, L.Z. Zhu, "Review on steady state voltage stability indices of power systems," Power System Technology, Vol. 25, No. 1, 2000, pp. 1–7.

Power and Energy – Kong (Ed.)
© 2015 Taylor & Francis Group, London, ISBN 978-1-138-02782-4

Using an open source platform to manage EHV substation protective relays

Ming-Yuan Cho
Professor, National Kaohsiung University of Applied Sciences, Taiwan

Kunta Hsieh & KunChan Wang
National Kaohsiung University of Applied Sciences, Taiwan

ABSTRACT: The protective relays in the Taipower Extra High Voltage (EHV) substations have different brand models because they were procured during different time periods. There are differential current protective relays and directional distance protective relays from SEL, differential current protective relays from GE, and protective relays from Toshiba. The different relays each have different settings and maintenance modes, including those with the RS232 and RS485 serial ports and ASCII text command mode maintenance operations, those using the factory GUI utility to connect to the remote relay, to login, and to change the flag settings, those that use MODBUS protocol to communicate with remote intelligent devices, and those that exchange information with other substations using DNP3 protocol. A small portion of the old SEL devices include an FTP protocol mode to download flag situations. As learned from actual operating experience, as all protective relay flags are manually modified, it is impossible to judge whether they were modified properly until the time trip results are available when an accident occurs. Therefore, if the electrical protective relays at a substation were to be altered by unauthorized persons, supervisors have no method of detection. Thus, this paper proposes an open source hardware and software architecture with regular data inquiries, storage, and comparisons to determine whether online relay settings have been altered.

1 INTRODUCTION

The Taipower company manages all the EHV substation protective relays, in a direct manner or through the converted manner via Ethernet connection, passing information easily and conducting maintenance remotely. This is because maintenance personnel do not need to conduct their work on site, as they can change the settings of the device flags remotely. The process looks very easy, but its ease allows for many potential crises due to security vulnerabilities or issues related to the validation of flag settings. Due to construction, maintenance and other reasons, occasions arise when the feeder line must undergo a temporary blackout; managers will temporarily modify the flag setting, and then restore the settings while awaiting completion. Before a power outage, a variety of simulations are done in advance and finally decided on to choose a program to modify the settings of the feeder protective relays. Therefore, these protective relay settings are not set up as permanently unchanged; rather, with power feed line status changes, the relationships between the various protective relays become complex, and the current settings cannot be directly and properly observed by an operator's eyes. As such, the maintenance staff will, before a

flag value change, try to simulate a variety of state failures, and then find the best settings, which are maintenance personnel to manually way to modify remotely via network logon. However, this procedure does not allow a third party or the director to confirm the action, and so cases of mistaken entries still occur. In order to reduce such problems, and to increase the level of intelligence and automation of substations, embedded systems have been installed in the field to attempt regular, automatic checks of each relay flag's settings, with the previous settings values stored for comparison to see if there are any changes, and also in order to make a report of the device number and status differences. In this way, relay settings can determine the status of all substations. Of course these can be done as early as possible to make any special responses or repairs.

Figure 1 shows the substation management platform infrastructure. For certain EHV substations, there are several kinds of protective relays, classified as follows:

1.1 *SEL only legacy serial port connection*

Such protection relays have to use converters to convert the serial port to connect to Ethernet

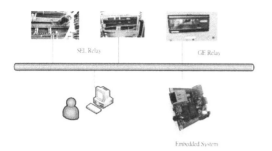

Figure 1. The substation management platform infrastructure.

Figure 2. The Cubieboard A20 embedded system.

connections after the conversion mode of operation, like the new SEL relays.

1.2 *SEL have Ethernet ports support only Telnet ASCII command*

Can via TCP/IP 23 ports from the remote logon, login to get the first layer set state.

1.3 *SEL have Ethernet ports, supports FTP command*

Open FTP connection service SEL relays can use SHELL SCRIPT, using FTP commands, crawl relay profiles CFG.XML, SET_61850.CID, ERR.TXT.

1.4 *GE are supported Ethernet*

GE's protective relays must use MODBUS protocol, grabbing the corresponding values of all settings.

2 METHODOLOGY

2.1 *Hardware architecture*

For the desired hardware management platform, and considering the need to add an embedded system, the system uses the open platform of Cubiboard A20 board, as shown in Figure 2.
Specifications are as follows:

- AllWinnerTech SOC A20, ARM® Cortex™-A7 Dual-Core ARM® Mali400 MP2 Complies with OpenGL ES 2.0/1.1
- 1GB DDR3 @480M
- 3.4GB internal NAND flash, up to 64GB on SD slot, up to 2T on 2.5 SATA disk
- 5VDC input 2 A or USB otg input
- 1x 10/100 Ethernet, support usb wifi
- 2x USB 2.0 HOST, 1x mini USB 2.0 OTG, 1x micro sd
- 1x HDMI 1080P display output

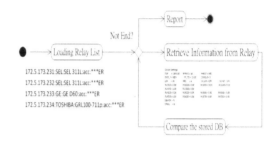

Figure 3. The state diagram for the management platform.

- 1x IR, 1x line in, 1x line out
- 96 extend pin interface, including I2C, SPI, RGB/LVDS, CSI/TS, FM-IN, ADC, CVBS, VGA, SPDIF-OUT, R-TP, and more system installation used Lubuntu, and the system has been tested and found to be stable and able to provide a good service environment and support for Shell script, Perl, SQLite 3, and so on, i.e., for all programming languages and databases that are needed for the management platform.

2.2 *Software architecture*

According to managers of the relay, there is no immediate need for information comparisons, that is, periodically you can grab one. The system will be Linux system with Cron command schedule fixed in 1:00 am, polling all protective relays to conduct the operation, and, using the process described in Figure 3, to fetch data and conduct the comparison.

Figure 4 shows how to execute the Cron command in the Linux system setting.

The above Cron command is specified to occur daily at 1:00 am sharp; starting with/usr/bin/chkrelay.sh, this SHELL SCRIPT conducts polling of the substations for all relays, and also conducts a data comparison.

Figure 4. The Cron command in Linux system.

Table 1. Cron command.

Min	Hour	day	month	Week	command
0	1	*	*	*	/usr/bin/chkrelay.sh

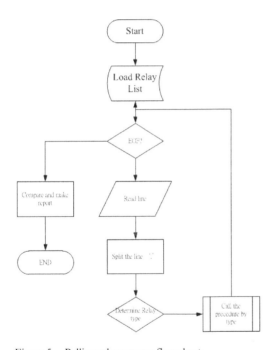

Figure 5. Polling relay process flow chart.

Because that EHV substation has only 61 protective relays, we put the IP address list into a text file archive, which can be beneficial for editing. The format used is shown in Table 2.

In Linux, you can use the Perl language to conduct simple and fast data analysis. For the machines that are listed above, you can use the following program to conduct an analysis easily (Figure 5).

By using powerful Perl commands, with just a few simple instructions you can do the following: load the file and conduct a line-by-line analysis.

In the disassembled process, can be based on the type of relay, calling the corresponding program to crawl the relay for data, and stored in SQLite3 database, and finally making compared pair and generate reports to E-MAIL sent to the administrator. When an administrator goes to work the next morning, that administrator can check the reports, and relay setting changes to master the situation.

For example, setting values obtained from the SEL relay include approximately 500 items, according to different models, and there will be some differences in the fields obtained, as follows:

Global Settings:
TGR = 1800.00   NFREQ = 60 PHROT = ABC
DATE_F = MDY FP_TO = 15.00 SCROLD = 5
LER = 30   PRE = 6   DCLOP = OFF
DCHIP = OFF
IN101D = 0.00   IN102D = 0.00   IN103D =
0.00   IN104D = 0.00
IN105D = 0.00   IN106D = 0.00
IN301D = 0.00   IN302D = 0.00   IN303D =
0.00   IN304D = 0.00
IN305D = 0.00   IN306D = 0.00   IN307D =
0.00   IN308D = 0.00
EBMON = N
EPMU = N

This format can also be resolves by the Perl language. The program flow chart is shown in Figure 6.

After parsing, the output is as follows:
IN304D = > 0.00
IN106D = > 0.00
SCROLD = > 5
IN102D = > 0.00
PHROT = > ABC

Table 2. The relay list.

IP address	Type	Brand	Model	Username	Password
172.5.173.231	1	SEL	SEL 311 L	acc	***ER
172.5.173.232	3	SEL	SEL 311 L	acc	***ER
172.5.173.233	2	GE	GE D60	acc	***ER
172.5.173.234	4	TOSHIBA	GRL100–711p	acc	***ER

IN302D = > 0.00
IN301D = > 0.00
DCLOP = > OFF
TGR = > 1800.00
EPMU = > N
DATE_F = > MDY
DCHIP = > OFF

Finally, we can use these data stuffed into a Sqlite3 database, with instructions like the following:

sqlite3 "insert into (relayid, key, value, datetime) values ....."

And then using SQL command analyzes the database data than can be done.

In addition, for GE's protective relays, the Modbus RTU mode must be used to grab the set values. The Modbus protocol, despite the passage of time (the Modicon company developed it in 1980), is still the most standard for industrial controllers (PLC) and is widely used in industrial automation systems. Modbus can operate in two modes, ASCII and RTU, and uses a RS232 interface.

In recent years, it has also gained in popularity based on Modbus TCP TCP/IP Ethernet network/IP-Modbus protocol. With the significant decline in microcontroller interface RS232 and Ethernet network prices, MODBUS recently began to be frequently used in embedded systems. We can use the ready-made MODBUS RTU developed in the Perl modules. And by referencing SERIAL and XML, CRC and other modules, you can mimic the MODBUS server to retrieve settings for the protective relays made by GE.

3 CONCLUSIONS

As usual, maintaining the correct flag settings depends on the maintenance staff avoiding setting errors. Actually, it is impossible to completely avoid typographical errors. In the current system, maintenance staff cannot find a settings error until a line failure occurs or while changing the settings. This is highly problematic in developing a smart substation. Any trip mistake at the EHV substation will lead to a large regional power outage. Most of the industries today, meanwhile, are heavily dependent on electric power, such that the economic losses that might result from such a power failure are difficult to estimate. So, a reliable intelligence management platform will assist managers so that they do not have to face the unknowable consequences of such errors.

4 FUTURE DIRECTIONS AND RECOMMENDATIONS

The direction of current smart grid development is mainly focused on a common IEC61850-based protocol. Some older and already installed devices, however, cannot all support the IEC61850 protocol, such that the underlying device operating status cannot be well monitored. In setting up a smart management platform, a low cost will paid, but big rewards may result from using it as a base for future connections. When it is necessary to implement an IEC61850 protocol converter, you can walk a little injustice to jump the road.

Figure 6. Type 1, Flow chart for getting data from an SEL only legacy serial port connection.

REFERENCES

1. Jiirgen Heckel," Smart Substation And Feeder Automation for a Smart Distribution Grid", 20th International Conference on Electricity Distribution, Prague, 8–11 June 2009.
2. John McDonald, P.E., "Substation Automation Basics—The Next Generation", Electric Energy T&D Magazine, May-June 2007 Issue.
3. Markets and Markets, "Global Substation Automation Market (2013–2018)".

Power and Energy – Kong (Ed.)
© 2015 Taylor & Francis Group, London, ISBN 978-1-138-02782-4

Review of grounding grids corrosion diagnosis methods using electromagnetic and ultrasonic guided wave

W.R. Si, C.Z. Fu & Q.Y. Lu
State Grid Shanghai Electric Power Research Institute, Shanghai, China

X. Guo & Z.B. Xu
Xiamen Red Phase Instruments Inc., Fujian, China

J.H. Li
Xi'an Jiaotong University, Shanxi, China

ABSTRACT: This article introduces several existing diagnosis methods of grounding grids corrosion detection with pointing out the shortcoming of the method based on electric network theory, which has so much research results. Subsequently, this article provides the principles, experimental research and practical applications of electromagnetic and ultrasonic guided wave methods for ground grids corrosion diagnosis. Finally, the problems currently existed and development direction is proposed.

Keywords: grounding grids corrosion; electric network theory; electromagnetic field-based analysis; ultrasonic guided wave

1 INTRODUCTION

The grounding grids of power station and substation are directly related to ensure the electrical equipment and personal safety. Because devices of grounding grids are running in harsh conditions, especially in humid places and harmful gases, or in acidic soil, corrosion happens easily (see Fig. 1). As a result, the grounding grids corrosion will change electrical parameters, and even cause conductor parted in the grounding network.

In the past, ground grids corrosion diagnosis is mainly depended on grounding resistance detection and digging the related areas to find the corrosion. This method leads to blindness, heavy work, low efficiency, and a great waste of human, financial, material. Electric network theory regards the whole grids as an equivalent of pure resistive network. By measuring resistance increment of each conductor of the grounding grids (or the node voltage change) as the fault parameters, the fault diagnosis equating is established. Through analyzing, the location where corrosion or conductor disconnection happened can be determined. However, this method is difficult to guarantee the precision in calculation and practical measurement. Besides, it is limited by the specific location and the number of the down-lead conductors. An accurate ground network drawing of design is required to obtain the network diagnosis.

In order to find out status of grounding grids and avoid a terrible accident, developing new diagnostic methods of detecting grounding grids corrosion to realize targeted partial digging or maintaining without digging is needed.

(a)

(b)

Figure 1. Grounding grids corrosion in substation.

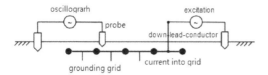

Figure 2. Principle chart of testing system.

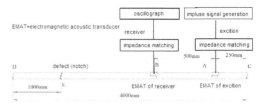

Figure 3. Principle chart of testing system.

2 ELECTROMAGNETIC FIELD-BASED ANALYSIS

The main principle of the electromagnetic field-based analysis is to detect the magnetic intensities, the current has excited on the surface by using magnetic induction coil probe after injecting different frequency sine wave current into grounding grid through two down-lead-conductors, which bases on the principle of electromagnetic induction. Then the change of the distribution of characteristic of the magnetic induction will provide a reference for the diagnosis to show the status of grounding grids currently which may has serious corrosion or even breakpoint fault.

Yang Liu from North China Electric Power University conducted a research based on this method. First, through simulation, the feasibility and correctness of the idea of measuring ground surface magnetic flux density distribution to determine the corrosion or breakpoints by injecting different excitation frequency current into grounding grid has been tested. Then, the characteristic of complex electromagnetic environments is given in transformer substations of different voltage levels, including the power frequency magnetic field and transient electromagnetic interference, and a diagnosis system and device is designed. Finally, through experiments in a testing grounding grid and applications of actual substation, they verify the validity of this method and practicality of this technique.

(a)

(b)

Figure 4. Testing system applied in actual substation.

3 ULTRASONIC GUIDED WAVE METHOD

Ultrasonic guided wave has the advantage of non-contact, no need of coupling agent and less demanding of specimen surface in detection corrosion happening on the specimen with rust complex geometry and paint or rust stained on its surface. By using an electromagnetic acoustic transducer, the ultrasonic waves are generated. Ultrasonic wave emitted by an excitation probe will reflect in the discontinuous area of specimen and form an ultrasonic guide wave which will be received by another probe. By multiplying the time difference of the received reflected wave through the two probes and speed of ultrasonic guided waves, the

actual position of the defect can be obtained. By analyzing the amplitude of received signal data, the degree of corrosion or defect of the flat steel can be estimated.

Due to the different modes of guided waves, the structure and dispersion characteristics are not the same. As a result, determining the type of guided wave is the key. Literature studied dispersion characteristics, multi-modal phenomena, wave structure and distribution of energy flow of the guided wave. They got the conclusion, the most suitable wave was SH0 wave in nondestructive detection of buried flat steel theoretically. Through experimental testing in defective flat steel with different thickness which is buried underground or exposed in the air, the sensitivity and practicality of SHO wave is confirmed in defecting the corrosion or defect happening in the steel.

Theoretically in detecting grounding grid, only transducer needs to consider to be installed outside, and the specific location of corrosion can be detected without digging a wide range of grounding grid.

4 ANALYSIS AND CONCLUSION

Detecting the grounding grids corrosion includes identifying the current status, monitoring process of ground grids corrosion on-line, and forecasting the development trend of the ground network state possibly. Besides, the joint use of various diagnostic methods will be very useful while we need to select different methods for a variety of different situation. For example, without explicit ground network drawings, electromagnetic field-based method will be the first choice. And in a poor surrounding electromagnetic environment, the application of ultrasonic guided wave method looks like to be more suitable. At the same time, through modern computer technology and sensor technology, real-time online monitoring of ground grid can be established to provide information for a prediction system which can provide a reference advice when bad situation happens.

In short, the grounding grids play a key role in protecting electrical device and staff of the substation. Early detection of grounding grids corrosion or even breakpoints has very important significance.

REFERENCES

1. Zhang, X.L. and Chen, X.L., "The technique of the optimization applied in the grounding grids failure diagnosis of the power plant and substation," High Voltage Engineering, Vol. 36, No. 4, 2000, pp. 64–66.
2. Liu, Y.G. and Teng, Y.X., "A method for corrosion diagnosis of grounding grid," High Voltage Engineering, Vol. 30, No. 6, 2004, pp. 19–21.
3. Liu, Y.G., Wang, S. and Tian, J.H., "A corrosion diagnosis method for optimized measurement of grounding grids," Journal of Chongqing University, Vol. 31, No. 11, 2008, pp. 1303–1306.
4. Hu, J., Sun, W.M. and Yao, J.X., "Novel method of corrosion diagnosis for grounding grid," In Proc. of Power Conference, 2000, pp. 1365–1370.
5. Liu, Y., Cui, X. and Zhao, Z.B., "Method for diagnosing the conductor, broken point of grounding grids in substation Based on measuring the magnetic induction intensity," High Voltage Engineering, Vol. 34, No. 7, 2008, pp. 1389–1394.
6. Liu, Y., Cui, X. and Zhao, Z.B., "Method of structure estimation and fault diagnosis of substations grounding grids," In Proc. of the CSEE, Vol. 30, No. 24, 2010, pp. 113–117.
7. Liu, Y. 2008. Research of method and technology on the defect diagnosis of grounding grids in substations. North China Electric Power University.
8. Wu, B., Li, L.T. and Wang, X.Y., "Non-destructive test of a surface defect on a steel bar based on ultrasonic guided wave techniques," Engineering Mechanics, Vol. 20, No. 5, 2003, pp. 149–154.
9. He, C.F., Sun, Y.-X. and Wu, B., "Application of ultrasonic guided waves technology to inspection of bolt embedded in soils," Chinese Journal of Geotechnical Engineering, Vol. 28, No. 9, 2006, pp. 1144–1147.
10. Beard, M.D., "Guided wave inspection of embedded cylindrical structures," London: Imperial College, 2002.
11. Cui, J.Y., Sun, Y.X. and He, C.F., "Application of low frequency ultrasonic guided waves on inspection of full-length-bonding resin bolt," Engineering Mechanics, Vol. 27, No. 3, 2010, pp. 240–245.
12. Zhong, X., "Research on non-destructive testing of grounding grids in substations using ultrasonic guided waves," Beijing University of Technology, 2012.

Power and Energy – Kong (Ed.)
© 2015 Taylor & Francis Group, London, ISBN 978-1-138-02782-4

Wide-area signals time-delay STATCOM additional damping control for power system

F. Liu, W. Yan & M. Wu

Institute of Advanced Control and Intelligent Automation, School of Information Science and Engineering, Central South University, Changsha, China

ABSTRACT: This paper proposes a customized new design method for STATCOM additional damping controller. Wide-area signals introduced through Wide Area Measurement System (WAMS) result in time delay problems in the process of transmission and processing, which will inevitably deteriorate system performances or even destroy the stability of the system. Design of STATCOM additional damping controller can be expressed as control problem of delay output feedback. Then, based on the direct feedback linearization theory, robust control theory and the design of state observer, the controller parameters are set through free weighing matrix method and pole placement method. Finally, a typical two-area and four-machine power system is used for system simulation, the results of which show that the designed controller is not only inhibits inter-area low-frequency oscillation obviously, but also is robust to various operating conditions and time delay.

Keywords: STATCOM; WADC; inter-area oscillations; time delay; free weighting matrices method

1   INTRODUCTION

The application of Flexible AC Transmission Systems (FACTS) improves the stability of power system. Static synchronous compensator (STATCOM) is a kind of typical FACTS devices. With the development of power electronics technology, the STATCOM has been progressing continuously. And the application of STATCOM to power system is increasing prevalent[1].

In recent years, we tend to improve power system stability and inter-area low-frequency oscillation of damping systems by means of designing a variety of STATCOM Additional Damping Controllers (ADCs). However, since traditional ADCs employ local signals such as the angle and speed of a generator as control signals, there are some limitations when considering the inter-area oscillations of the power system[2]. The development and application of Wide Area Measurement System (WAMS) provide successful solutions to these limitations for us. WAMS can synchronize the collection and monitoring of the many key variables (such as voltage, rotor angle, frequency, and so on) in power grid operation, providing first-hand information for power system analysis and dynamic monitoring of the system. However, the introduction of WAMS into stability control of wide-area power system brings an unavoidable problem, that is, time delay. Therefore, in the design of damping controllers, it is extremely significant to take into account how to

solve the problems brought about by time delay of wide-area signal transmission in power system[3-6].

Considering the feedback time delays brought about by WAMS system and the application prospects of STATCOM controllers in China, it is worthy of intensive studies on design of high-performance wide-area STATCOM controllers for interconnected power system in order to damp low-frequency oscillations which may exist in power system. This paper is based on STATCOM Wide-Are additional Damping Controller (WADC) of Wide-Area Control Systems (WACSs), expressed as control problem of time delay output feedback. Then, Direct Feedback Linearization (DFL) theory, robust control theory and the state observer are employed for the design of setting controller parameters through free weighting matrices method[7] and pole placement method[8]. Finally, a typical two-area and four-machine power system is used for system simulation, the results of which show that the designed controller is useful for the damping of low-frequency oscillations of the power system and meanwhile respond to various situations and achieve good stability margin of time delays.

2   DESCRIPTION OF STATCOM WADC

Figure 1 is the control block diagram of WADC proposed in this paper to equip STATCOM in the power system.

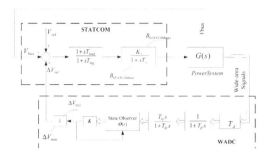

Figure 1. Control structure diagram of time delay output feedback of STATCOM WADC.

Dynamic model of STATCOM in the power system is resembled with linearization method. The closed-loop feedback control produced by STATCOM WADC can be used to damp the potential inter-area power system oscillations. Based on the small-signal analysis methods, STATCOM WADC chooses the operating variables of remote generators systems as control input signals. By following the basic rules of selecting controller signals, oscillations between the generators in some operating conditions can be directly reflected by the controller input signals. The use of wide-area measurement system to transmit the wide-area signal will inevitably bring about time delay problem, so in essence, the application of WAMS has made power system a typical delay system. In order to represent the delay characteristics of the control input signal, model $e^{-s\tau}$ is used to indicate delay τ, as shown in Figure 1. Meanwhile, to achieve the purpose of filtering the remaining relational oscillation frequency information in the process of WADC control input signals, the High-Pass Filters and Low-Pass Filters (HPF and LPF) are of great necessity.

For WADC, the state feedback gain matrix K is an integral part of the controller. The design of K can be optimized through free weighting matrices method, which will be further elaborated in the third section followed. Furthermore, as state variables K both obtained through design and operated in the actual system cannot be fully observed, the state observer $O(s)$ is employed to observe the state variables. In this paper, the state observer $O(s)$ is designed through the classical pole placement method, which will also be discussed in details in the next section 3.

3 DESIGN OF STATCOM WADC

WAMS can make real-time and simultaneous measurement of the operating variables of electric power network, STATCOM controllers and

generators. These synchronization signals transmitted through WAMS may be used to control the operation of the power system. Therefore, in the wide-area control system, a large number of signals transmitted and fed back through WAMS form a complex Network Control Systems (NCS). In the NCS, time delay of signals is a serious problem worthy of consideration.

According to linear feedback linearization theory, the linear state equations of wide-area power system with STATCOM WADC can be obtained as follows:

$$\begin{cases} \Delta\dot{x}(t) = A\Delta x(t) + B\Delta u(t) \\ \Delta y(t) = C\Delta x(t) + D\Delta u(t) \end{cases} \tag{1}$$

considering delay τ brought about by controlling input signal in NFS, when Δ is omitted and $D = 0$, equation (1) can be abbreviated as equation (2):

$$\begin{cases} \dot{x}(t) = Ax(t) + Bu(t - \tau) \\ y(t) = Cx(t) \end{cases} \tag{2}$$

where A, B and C are state variable matrix, input variable matrix and output variable matrix, respectively.

And then, through the free weighting matrices method, a memory state feedback controller is designed for STATCOM WADC gain K in time-delay power systems seen in equation (2) to improve the stability of the power system in which wide-area signals result in time delays.

$$u(t - \tau) = Kx(t - \tau) \tag{3}$$

According to the given control rules, when put into equation (4), the state feedback control system can be expressed as follows:

$$\begin{cases} \dot{x}(t) = Ax(t) + BKx(t - \tau) \\ y(t) = Cx(t) \end{cases} \tag{4}$$

Therefore, the objective of this section is to develop a new method for delay-dependent stability determination to obtain controller gain K and Delay τ, in which case the closed-loop system seen in equation (4) is asymptotically stable.

Theorem: Scalar h is given (suppose h = max (τ)), then there exist control rules in equation (3) which make the closed-loop system in equation (4) asymptotically stable. If there exist $L = L^T > 0$, $Q = Q_1^T > 0$, $R = R^T > 0$, $Y = \begin{bmatrix} Y_{11} & Y_{12} \\ Y_{12}^T & Y_{22} \end{bmatrix} L^T > 0$, and any suitable dimension of matrix M_1, M_2, V, then LMI below is established. Besides, feedback control gain $K = VL^{-1}$[2].

$$\bar{\Phi} = \begin{bmatrix} \bar{\Phi}_{11} & \bar{\Phi}_{12} & hLA^T \\ \bar{\Phi}_{12}^T & \bar{\Phi}_{22} & hV^T B^T \\ hAL^T & hBV & -hR \end{bmatrix} < 0 \qquad (5)$$

$$\bar{\Psi} = \begin{bmatrix} Y_{11} & Y_{12} & M_1 \\ Y_{12}^T & Y_{22} & M_2 \\ M_1^T & M_2^T & LR^{-1}L \end{bmatrix} > 0 \qquad (6)$$

$$\bar{\Phi}_{11} = AL + LA^T + M_1 + M_1^T + Q_1 + hY_{11}$$
where: $\bar{\Phi}_{12} = BV - M_1 + M_2^T + hY_{12}$
$$\bar{\Phi}_{22} = -M_2 - M_2^T - Q_1 + hY_{22}$$

As the nonlinear element $LR^{-1}L$ exists in equation (6), conditions of the theorem will no longer be LMIs, and we cannot calculate the minimum through convex optimization algorithm. However, as mentioned in equation (7), we can use Cone Complementarity Linearization algorithm (CCL algorithm) based on LMIs to convert this problem.

Minimize $tr\{FF_1 + LL_1 + RR_1\}$

According to equation (5) and

$$\begin{bmatrix} Y_{11} & Y_{12} & M_1 \\ Y_{12}^T & Y_{22} & M_2 \\ M_1^T & M_2^T & F \end{bmatrix} > 0 \qquad (7)$$

$$\begin{cases} \begin{bmatrix} F & I \\ I & F_1 \end{bmatrix} > 0, \begin{bmatrix} F_1 & L_1 \\ L_1 & R_1 \end{bmatrix} > 0, \\ L > 0, F > 0, R > 0, \\ \begin{bmatrix} F & I \\ I & F_1 \end{bmatrix} > 0, \begin{bmatrix} F_1 & L_1 \\ L_1 & R_1 \end{bmatrix} > 0, \end{cases} \qquad (8)$$

Then by calculating inequalities (5), (7) and (8) the maximum of Delay $h = \max(\tau)$ can be obtained.

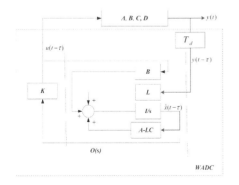

Figure 2. Structure diagram of the designed STATCOM WADC based on the state observer.

Therefore, according to Theorem we can get the system maximum delay h and the feedback controller gain K. But in the actual power system state variables cannot be observed fully, so feedback control is employed for measuring the system state in the actual system. Figure 1 describes how system state variables are monitored by the state observer $O(s)$. As the impacts of time delays have to be considered in the WAMS, Figure 2 indicates the design of STATCOM WADC based on the state observer $O(s)$. In Figure 2, L is the control gain of the state observer, and it can be obtained through the traditional and reliable pole placement method.

4 SYSTEM SIMULATION

Two-area and four-machine power system is built by Ontario Hydro in order to study the basic characteristics of low-frequency oscillations of the power system[8], and its structure is shown in Figure 3. Four generators divided into two groups are located in Area 1 and Area 2 respectively, which are connected by a long double circuit line. During normal operations, Area 1 transport 400 MW active power to Area 2. In this paper, the two-area and four-machine standard system intended for resembling STATCOM WADC is used to verify the effectiveness of the proposed design of STATCOM WADC in damping low-frequency oscillations of power system. As shown in Figure 3, the STATCOM devices are resembled at bus 8 between the two areas of power systems, while WADC is resembled in the whole interconnected system for damping oscillations. Considering that the system oscillations are caused mainly by G1 in Area-1 and G3 in Area-2, the angles and speed differences of remote G1 and G3 measured through WAMS are chosen as the input signals of the designed WADC. Time delays of wide-area measurement signals have the characteristics of randomness and boundedness[10]. This paper uses the maximum time delay as time delay values for design of the controller and the constant delay for simulation of time-domain.

Line fault is a severe test for the control performance of WADC, because when the line fault occurs,

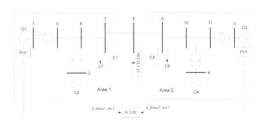

Figure 3. Two-area and four-machine benchmark system with STATCOM WADC.

373

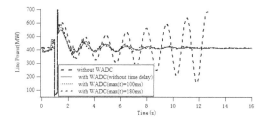

Figure 4. Linear power flow response of the system from Area 1 to Area 2.

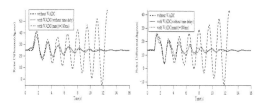

Figure 5. Response of angle differences, Diagram (a) is that of G1 and G3, Diagram (b) is that of G2 and G3.

the network structure of the system may change accordingly. Suppose that the nonlinear dynamic systems simulation is made at bus 8 by using a group of three-phase grounding oscillations lasting 12 cycles. Figure 4 indicates the linear power flow response of the system from Area 1 to Area 2, from which we can see when the whole interconnected system receive such great interferences system oscillations will occur and the system even breakdown around 13s. However, when the system is resembled with the designed WADC, it can effectively damp the system oscillations; even when wide-area control signals generating time delays of 180 ms are introduced, the designed controller can still maintain the good characteristics of system oscillations damping. Figure 5 shows the angle differences between two generators in two areas, from which we can see when oscillations occur in the entire interconnected system generators in one area will oppose against generators in another area, and that this kind of oscillation is the direct cause of power oscillations in the system. The simulation results also showed good properties of WADC can damp oscillations of the system generators, which further confirms the authenticity of our designed WADC.

5 CONCLUSION

This paper presents a new design method of STATCOM WADC. The main feature of this WADC is the use of remote electric motor wide-area signals to improve the damping of system inter-area oscillations through WAMS. Moreover, a set of delay dependent robust control standards which help the design of controllers is obtained based on free weighing matrix method. Finally, in a typical two-area and four-machine standard power system, the transient characteristics of the power system resembled with STATCOM WADC are verified. Simulation results show that the controller designed in this paper can fully damp low-frequency oscillations of the power system and that the superiority of STATCOM can be reflected in its strain capacity to respond to multiple situations and its delay insensitivity to well-functioned systems.

ACKNOWLEDGEMENT

This work was supported in part by the National Natural Science Foundation of China (No. 61304092), in part by the Research Fund for the Doctoral Program of Higher Education of China (No. 20130162120022) and in part by the Hunan Provincial Natural Science Foundation of China (No. 13JJ6004).

REFERENCES

1. K. Wang. The research of power system low frequency oscillation suppression with FACTS devices [M]. 2013, 14–17.
2. F. Liu, R. Yokoyama, Y. Li, et al. SVC robust additional damping controller design for power system with considering time-delay of wide-area signals[C]. Environment and electrical engineering (EEEIC), 2010, 313–316.
3. H. Wu. The impact of time delay on robust control design in power systems [C]. IEEE Power Engineering Society Winter Meeting, 2002, 2: 27–31.
4. J. Zhang, Y.Z. Sun. Effect of delayed input on oscillation damping using wide area power system stabilizer [C]. IEEE/PES Transmission and Distribution Conference and Exhibition: Asia and Pacific, 2005: 1–4.
   A. Snyder, M. Alali, N. Hadjsaid. A robust damping controller for power system using linear matrix inequalities [C]. IEEE Power Engineering Society Winter Meeting, 1999, 1: 519–524.
5. Y. Yuan, L. Cheng, Y. Z. Sun, et al. Effect of delayed input on wide-area damping control and design of compensation[J]. Automation of electric power systems, 2006, 30(14): 6–9.
6. M. Wu, F. Liu, P. Shi, et al. Exponential stability analysis for neural networks with time-varying delay. IEEE Trans on Systems, Man and Cybernetics, Part B, 38(4), August 2008, 1152–1156.
7. P. Kundur. Power system stability and control [M], McGraw-Hill, New York, 1994.
8. H. Gao, J. Lam, C. Wang, et al, Delay-dependent output-feedback stabilization of discrete-time systems with time-varying state delay [C], IEE Proc. Control Theory Appl, 2004, 691–698.
9. Z. X. Hu, X. R. Xie, L. Y. Tong. Characteristic analysis and polynomial fitting based compensation of the time delays in wide-area damping control system [J]. Automation of Electric Power Systems, 2005, 29(20): 29–34.

Power and Energy – Kong (Ed.)
© 2015 Taylor & Francis Group, London, ISBN 978-1-138-02782-4

Effect of hydrocyclone cylinder length on the separation performance

Meiru Liu, Peikun Liu, Xinghua Yang & Yuekan Zhang
Shandong University of Science and Technology, College of Mechanical and Electric Engineering, Qingdao, China

ABSTRACT: The cylinder length is seldom considered in present research on the separation performance and production capacity of a hydrocyclone. In order to study the effect of cylinder length variation on the separation performance, a series of experiments have been carried out. It is found that with the increase of the cylinder length, both the classification efficiency and the production capacity of the hydrocyclone are significantly improved. Moreover, according to the experimental data, a new production capacity formula considering the cylinder length is put forward, which can enrich the rotational flow separation theory further.

Keywords: hydrocyclone; cylinder length; classification efficiency; production capacity calculation

1 INTRODUCTION

The hydrocyclone has been widely used as classification equipment because of its simple structure, no moving parts, low cost, large capacity, convenient maintenance and low operating cost (B.A. Wills, 2006). It consists of a cylinder part and a conical part (Luiz G.M. Vieira, 2010). Material into the hydrocyclone should first go through the cylinder section, which has the entered material flow, circumferential material flow, and return overflow material flow inside simultaneously. Centrifugal sedimentation in conical part causes solid particles to enter conical part orderly, and expedites the process of fractionation in conical part (Chu L.Y., 2002). Many researches consider that cylinder part deals with pre-separation (Pei-kun Liu, 2008). Therefore, the parameter of the cylindrical part has certain influence on the separation process.

However, there are few references concerned with the research on cylinder length for the effect of the hydrocyclone separation performance. And another question is that, the production capacity is an important indicator of the hydrocyclone (Feridun Boylu, 2010; Lei Wu, 2012), but its representative calculation formula, Pang Xueshi and А.И.Пваровн formula, does not reflect the influence of the cylinder length's either.

This paper aims to study the impact of cylinder length on hydrocyclone separation performance. Different kinds of cylinder lengths are used for comparison in the experiments, then, according to the experiment data, a new production capacity formula considering the cylinder length is put forward.

2 THE SEPARATION PERFORMANCE EXPERIMENTS

2.1 Testing program

In order to find out the optimum height of the cylindrical section, the separation performance test is conducted. The geometrical parameters of the hydrocyclone are displayed in Figure 1.

The material concentration is 16.37%, and the inlet pressure is 1.00 kgf/cm². The test bench is shown as Figure 2.

The cylinder length H is variable. When other structural and operating parameters are fixed, the influence of cylinder length H on separation performance can be studied by changing its values. The different values of H are listed in Table 1.

2.2 Main performance index

1. Comprehensive classification efficiency

This experiment adopts the Hancock formula to evaluate the comprehensive classification efficiency. The formula 1 can be written as:

$$E_{\text{效}} = \frac{(\alpha - \vartheta)(\beta - \alpha)}{\alpha(\beta - \vartheta)(100 - \alpha)} \times 10000\% \quad (1)$$

where
$E_{\text{效}}$—classification efficiency (such as desliming), %;
α—400 mesh particle content in the mine, %;
β—400 mesh particle content of overflow, %;
ϑ—400 mesh particle content of underflow, %;

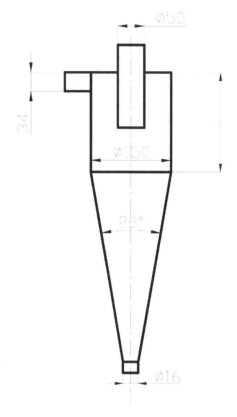

Figure 1. Geometry of the hydrocyclone in experiment.

Figure 2. Photo of the test bench.

Table 1. Cylinder length and length diameter ratio.

	1	2	3	4	5	6	7
$H(mm)$	180	245	310	375	440	505	570
H/D	1.2	1.63	2.07	2.5	2.93	3.37	3.8

2. Productive rate

According to the principle of mass balance, the underflow productive rate can be calculated by:

$$\gamma_u = \frac{C_u\left(C_i - C_o\right)}{C_i\left(C_u - C_o\right)} \times 100\% \qquad (2)$$

where

γ_u—underflow solid phase production rate, %;
C_i—feeding weight concentration, %;
C_o—overflow weight concentration, %;
C_u—underflow concentration in weight, %.

3 RESULT ANALYSIS AND DISCUSSION

3.1 The classification efficiency

The impact of length diameter ratio on the representative size fraction (overflow −400 mesh, which also means the size under 38 μm) and the comprehensive classification efficiency are shown as Figure 3 and Figure 4 respectively. It is found that, with the increase of H, the content of overflow −400 mesh increases 14.41%, which means that more fine particles (size under 38 μm) running into the overflow. Furthermore, the comprehensive classification efficiency increases 16.43% and reaches the maximum of 75.80%, which means that the increase of the cylinder length plays a significant role in promoting fine particles classification. Actually, it can be attributed to the enlarged separating space and extended separating time when the cylindrical length is increased.

3.2 The overflow concentration and the underflow concentration

The impact of cylinder length diameter ratio on the overflow concentration is shown in Figure 5. With the growing of cylinder length, overflow concentration see a decline trend, and underflow concentration just fluctuates in Figure 6, which means cylinder length does not have obvious effect on underflow concentration. The reducing of the overflow concentration indicates solid particle content decrease in the overflow, and more solid particles separated from the underflow, making most of the solid particles separate from hydrocyclone. In view of separation effect, for the seven cylinder lengths, the longer the cylinder length, the

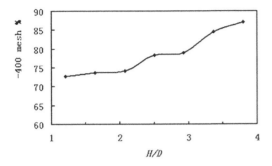

Figure 3. The impact of cylinder length on overflow-400 mesh.

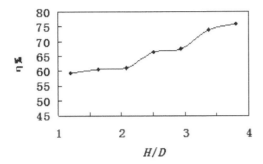

Figure 4. The impact of cylinder length on comprehensive classification efficiency.

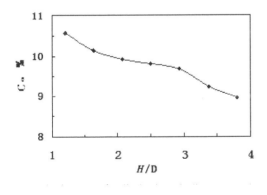

Figure 5. The impact of cylinder length diameter ratio on the overflow concentration.

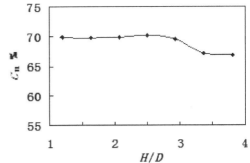

Figure 6. The impact of hydrocyclones cylinder length diameter ratio on the underflow concentration.

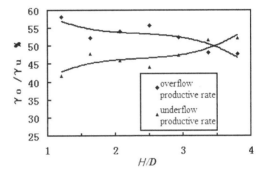

Figure 7. The impact of cylinder length diameter ratio on the overflow and underflow production rates.

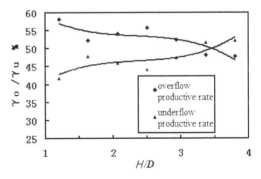

Figure 8. The impact of cylinder length diameter ratio on feed flow.

lower the overflow concentration, and the better the hydrocyclone separation effect.

3.3 Productive rate

The variation of the productive rate of the overflow and underflow is shown as Figure 7. It reveals that with the increase of length diameter ratio,

the overflow productive rate appears a downward trend, while the underflow productive rate presents an upward trend. The reason is that more solid particles are separated from the underflow, and then the separation efficiency is improved to some extent. Thus, it is clearly that the increase of the cylinder length is good for separation.

3.4 Handling capacity

The handling capacity means the volume of slurry flowing in the feed pipe per unit time. The experimental results verify that the cylinder length has great influence on the feed port flow, which can be observed in Figure 8. The handling capacity reaches its top value of 23.41 m³/h when the cylinder length diameter ratio is 3.8. It is also clear that, the longer the cylinder length is, the larger the handling capacity is.

4 THE MODIFIED А.И.ПВАРОВ PRODUCTION CAPACITY FORMULA

The most widely used production capacity formula is Pang Xueshi and А.И.ПВАРОВ formula. They believe that, in the separation process, the energy loss is the total pressure drop (or called the total head loss) from the surrounding to the air cylinder interface. The reference plane, which is used to calculate pressure drop, should be the air cylinder interface. When ignore the fluid resistance loss, after revising, the cylinder of air interface production capacity formula is acquired.[4]

$$q_m = 3K_D K_\theta d_i d_o \sqrt{\Delta P} \qquad (3)$$

where:
q_m—the hydrocyclones production capacity, m³/h
K_D—the hydrocyclone diameter correction factor
$K_D = 0.8 + (1.2/(1+0.1D))$;
K_θ—the hydrocyclone taper angle correction factor, $K_\theta = 0.79 + (0.044/(0.0379 + \tan\theta/2))$;
θ—taper angle, (°);
d_i—Inlet diameter, cm;
d_o—the overflow pipe diameter, cm;
ΔP—the hydrocyclone inlet pressure, MPa.

Obviously, the above formula does not consider the influence of cylinder length. But from the experimental data, the production capacity shows major difference, as shown in Table 2.

By use of the least squares principle, the abundant experimental data are fitted and the revised equation is proposed as follows:

$$q_m = 2.3K_D K_\theta d_i d_o \sqrt{\Delta P} \left(H/D\right)^{0.29}$$

where, H/D is the hydrocyclone cylinder length diameter ratio. This equation gives full consideration to the impact of cylinder length on the production capacity and thus can provide more accurate results during production practice.

Table 2. The experimental data.

	1	2	3	4	5	6	7
H/D	1.2	1.63	2.07	2.50	2.93	3.37	3.80
q_m	16.44	18.45	19.7	20.16	21.15	22.23	23.41

5 CONCLUSIONS

The cylinder length has certain impact on the separation performance. It plays a promoting role for the advancement of the classification efficiency and for solid-liquid separation. With the increase of cylinder length, the production capacity can be improved obviously.

The revised А.И.ПВАРОВ production capacity formula provides a reference basis to explore the influence of cylinder length on the separation performance. And it can provide more accurate results during production practice.

ACKNOWLEDGEMENT

The work is supported by the National Natural Science Foundation of China (No. 21276145) and Graduate Innovation Project of Shandong University of Science and Technology of China (YC130315).

REFERENCES

1. B.A. Wills, T. Napier-Munn, Wills, "Mineral Processing Technology," Seventh ed. Butterworth-Heinemann, Boston, 2006.
2. Chu L.Y., Chen W.M., Lee X.Z., "Enhancement of hydrocyclone performance by controlling the inside turbulence structure," Chem Eng Sci. 2002. 57: 207–212.
3. Feridun Boylu, Kenan Çinku, Fahri Esenli, "The separation efficiency of Na-bentonite by hydrocyclone and characterization of hydrocyclone products," International Journal of Mineral Processing. 94 (2010) 196–202.
4. Lei Wu, Tianyu Long and Xuping Lu, "Improvement of separation efficiency and production capacity of a hydrocyclone," Water Science & Technology: Water Supply. 2012.12:281–299.
5. Luiz G.M. Vieira, João J.R. Damasceno, Marcos A.S. Barrozo, "Improvement of hydrocyclone separation performance by incorporating a conical filtering wall," Chemical Engineering and Processing. 49 (2010) 460–467.
6. Pei-kun Liu, Liang-Yin Chu, "Enhancement of Hydrocyclone Classification Effect for Fine Particles by Introducing a Volute Chamber with a Pre-Sedimentation Function," Chem. Eng. Technol. 2008, 31, No. 3, 474–478.

Power and Energy – Kong (Ed.)
© *2015 Taylor & Francis Group, London, ISBN 978-1-138-02782-4*

The research on transformer vibration model in winding insulation weakening fault diagnosis

P.F. Shen, C.N. Wang, K. Li & Y. Li
Jiangsu Nanjing Power Supply Company, Nanjing, China

H.Z. Ma & Z.D. Zhang
College of Energy and Electrical Engineering, Hohai University, Nanjing, China

ABSTRACT: With the running time increased, the winding insulation weakening (winding nonmetallic inter-turn short circuit, the same below) is the most common phenomenon of power transformer, and are the main factors to cause the transformer winding short circuit. This paper proposes a new fault diagnosis method for transformer insulation weakening based on vibration signal. The relationship model of oil tank surface vibration and winding insulation weakening of transformer is established, and carries on the analysis; by monitoring tank vibration signal, combined with the transformer load current, both normal and fault states, the fundamental frequency signal of the transformer oil tank surface vibration were calculated; analysis shows that, when the transformer winding insulation weakening, the local asymmetric short-circuit current (circulation form short circuit within the coil) will make the fundamental frequency component of vibration signal changes; according to the results, to diagnose the winding insulation weakening the of transformer. The normal and fault experiments for the 10/0.4 kV 200 kVA transformer, the experimental results indicate that the model can diagnose of winding effective insulation weakening effectively.

Keywords: transformer; vibration; model; winding insulation weakening; diagnosis

1 INTRODUCTION

It is known that power transformer is an essential element in the power system. Transformer reliability is directly related to the safe operation of the power system, transformer failure will result in electricity supply interruption, and cause great economic losses. Condition monitoring and fault diagnosis technology in recent years has been applied in monitoring transformer operating condition, which not only reduce the transformer failures, but also can replace regular maintenance and reduce economic losses [1].

In Garcia's paper, the use of a model to monitor tank vibrations is proposed. The strategy of model-based transformer monitoring systems was proposed by MIT researchers with the aim of allowing the early detection of failures in a transformer [2]. For this purpose, some transformer key variables are calculated by models from some input variables. A great difference between calculated and measured values is an indication as to a structural change in the transformer, and an alarm is then emitted.

Dr. Wang Hongfang from EPRI study the viscoelastic nonlinear mechanical properties of the insulation pad of windings, propose the mathematical model of the nonlinear vibration of the axial winding, study the dynamic characteristics of the winding axial vibration under short-circuit conditions, analysis the influence on the winding axial stability caused by axial preload and other factors [3][4].

Winding inter-turn short circuit is caused by insulation damage as a result of long-period operation. When inter-turn short circuit occurs, the loop current in windings will seriously increase. This paper proposes a baseband time domain waveform model for transformer tank vibration signal, and calculates the time domain waveform of vibration signal fundamental frequency component by monitoring the load voltage and load current. When winding inter-turn short circuit fault occurs, the short-circuit circulation will cause an increase in vibration signal, and the model results differ.

Supported by Jiangsu Power Supply Company, China (J2014055).

2 TRANSFORMER VIBRATION

Inside a transformer, the magnetic and electric forces act on the core and windings, causing them to vibrate [5]. The vibration from windings and core penetrates into transformer oil, travels through it, and reaches the tank walls, exciting their oscillations. When there are loosened separate turns in windings or separate sheets in the core, the fundamental frequency component of the vibration signal will get larger in magnitude and high harmonic components will appear.

Steady-state vibration measurement on the tank surface under online conditions will then provide essential information about fixation conditions for both windings and the magnetic core.

Therefore, vibration measurement is an effective method for online transformer monitoring.

3 TRANSFORMER TANK VIBRATION MODEL BASED ON THE DYNAMICS

3.1 Model of transformer winding

Windings of large power transformers is the superposition of a series of copper cake, defined as a winding unit, separated by insulating pad between the adjacent winding unit [6][7]. A yoke of iron is fixed on the top and bottom of the winding to clamp it. Transformer winding current and the leakage magnetic flux interacts with each other, producing electromagnetic force, which applies to the winding unit. With the excitation of the electromagnetic force, a single line of a winding unit can be simplified as a mechanics of simple harmonic motion. Insulating oil plays the role of damping of winding vibration. Figure 1 is a winding kinetic

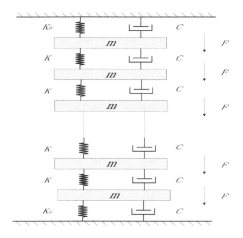

Figure 1.   Winding kinetic model.

model, and the motion of the winding model can be described as equation (1):

$$
\begin{cases}
m\dfrac{d^2z_1}{dt^2} + C\dfrac{dz_1}{dt^2} + K_B z_1 + K(z_1 - z_2) = F_1 + mg \\[2mm]
m\dfrac{d^2z_2}{dt^2} + C\dfrac{dz_2}{dt^2} - K(z_1 - z_2) + K(z_2 - z_3) = F_2 + mg \\[2mm]
m\dfrac{d^2z_3}{dt^2} + C\dfrac{dz_3}{dt^2} - K(z_2 - z_3) + K(z_3 - z_4) = F_3 + mg \\[2mm]
\cdots\cdots\cdots \\[2mm]
m\dfrac{d^2z_n}{dt^2} + C\dfrac{dz_n}{dt^2} - K(z_{n-1} - z_n) + K_H z_n = F_n + mg
\end{cases}
$$

(1)

In which: z_n—displacement of the n-th copper cake; C—damping coefficient of transformer insulating oil; K_B—stiffness coefficient of the segment between the iron yoke and the upper end of the winding; K—stiffness coefficient of the segment between adjacent winding unit; K_H—stiffness coefficient of the segment between the iron yoke and the bottom of the winding; F_n—electromagnetic force exerted on the n-th winding unit; mg—the quality of a single winding unit.

The electromagnetic force acting on a single unit is proportional to the square of the load current:

$$
F = bi^2 = b(I_m \cos \omega t)^2 = bI_m^2\left(\frac{1}{2} + \frac{1}{2}\cos 2\omega t\right)
$$

(2)

In which: ω—Grid frequency, namely 50 Hz.

Assume the displacement of each winding unit is identical, the equation can be simplified as follow:

$$
M\frac{d^2z}{dt^2} + C'\frac{dz}{dt} + K'z = bI_m^2\left(\frac{1}{2} + \frac{1}{2}\cos 2\omega t\right) + Mg
$$

(3)

In which, z is the displacement. Calculate second derivative of time, we get the acceleration of winding unit a:

$$
a = -\omega_0 A e^{\frac{C't}{2M}} \sin(\omega_0 t + \theta) - 4\omega^2 G \sin(2\omega t + \Psi)
$$

(4)

In which: ω_0—the natural frequency of the system, determined by the winding's essential attribute; A, θ—integral constant, determined by the initial conditions of the system.

3.2 Model of transformer tank vibration

The vibration signal from the transformer windings and core passes through the transformer oil to the tank surface [8][9]. Assume that the transfer function is linear time-invariant, so the signal will attenuate in amplitude and change in phase. Vibration signals from windings and core transfer to the tank via different path, so the change of the two signals is different. Therefore the vibration signal passes to the tank surface can be expressed as:

$$a_{\text{tank}} = a_{winding} + a_{core}$$
$$= K_w I^2 \cos(2\omega t + \theta_w) + K_C U^2 \cos(2\omega t + \theta_C) \quad (5)$$

In which: K_w, K_c—amplitude attenuation coefficient in the transfer process of winding and core vibration signal, respectively; θ_w, θ_c—phase of the winding and core vibration signal after they transferred to the tank surface, respectively; I, U—load current and load voltage of the transformer.

4 LABORATORY TESTS

The vibration signals of a three-phase testing transformer were measured. The parameters of this transformer are as follows: Rated power: 100 kVA; Secondary voltage: 0.40 kV; Primary voltage: 10 kV; Windings: Series and concentric style.

The vibration acceleration sensors, CA-YD-103, are attached on the transformer tank via the magnetic pedestal and its position is also fixed. During the measurement, voltage is exerted on the low voltage side, while the high voltage side is no-load. Use a voltage regulator to adjust the voltage, to make it change within a spectrum near rated voltage. Use a current sensor to monitor load current signal. Use a data collection apparatus and a PC to obtain the vibration signal and current signal, as in Figure 2. To simulate a winding inter-turn short-circuit fault, weld a wire between the coil of the A-phase winding, as in Figure 3.

Figure 4 and Figure 5 are frequency spectrum for normal condition and fault condition

Figure 3. Simulate a winding inter-turn short-circuit fault.

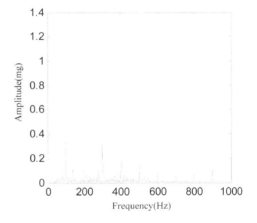

Figure 4. Vibration frequency spectrum under normal condition.

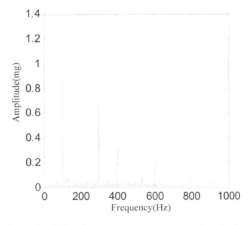

Figure 5. Vibration frequency spectrum under winding inter-turn short circuit fault.

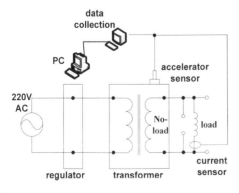

Figure 2. Test system.

respectively when load voltage is 360 V and load current is 60 A. It can be seen that the vibration signal in both cases are mainly consist of the fundamental frequency of 100 Hz, and there exists the 100 Hz integer harmonic. When the winding inter-turn short-circuit fault occurs, the 100 Hz and some harmonic amplitude significantly increased, consistent with the foregoing theoretical analysis.

Since the signal generated by the winding and the core consists mainly of 100 Hz base frequency signal, which theoretically is possible to characterize the state of the winding. The baseband signal integer harmonics is not taken into account. So filter the collected vibration signals, leaving only the fundamental frequency component and components near fundamental frequency. Figure 6 is time-domain waveform comparison before and after filtering the vibration signal. Use fundamental frequency transformer tank vibration signal under different voltage and current to polynomial fitting, then get the model parameters, and then get the transformer tank vibration model under the normal conditions.

Contrast the time-domain waveform of the vibration signal before and after the filtering, both under the same load voltage and load current, as shown in Figure 7. It can be seen that the phase of the two waveform are consistent, with some errors in the maximum part, but not more than 2%, within an acceptable range. So the simulation waveforms can perfectly present the baseband part of vibration signal.

Figure 8 shows the contrast of measured tank vibration signal and simulation signal under winding inter-turn short-circuit fault. Table 1 is the ratio of max amplitude of measured signal and simulation signal under winding inter-turn short-circuit fault. It can be seen from Figure 8 and Table 1, when transformer has inter-turn short circuit fault, the magnitude of the baseband component of the measured vibration signal is 1.5 times higher than the calculated value of the normal transformer vibration fundamental frequency component in

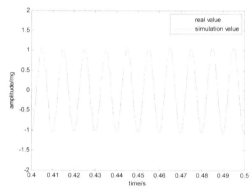

Figure 7. Contrast of measured tank baseband vibration signal and simulation signal under normal condition.

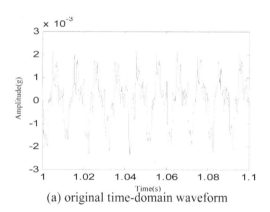

(a) original time-domain waveform

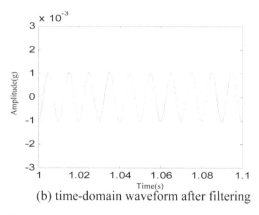

(b) time-domain waveform after filtering

Figure 6. Contract of tank baseband vibration signal before and after filtering.

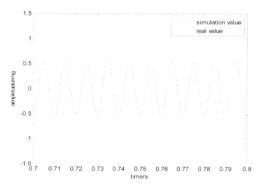

Figure 8. Contrast of measured vibration signal and simulation signal under winding inter-turn short-circuit fault.

Table 1. Ratio of max amplitude of measured signal and simulation signal under winding inter-turn short-circuit fault.

Load current (A)	70	80	90	100	110
Measurement/ calculation	1.52	1.65	1.67	1.68	1.71

the same input conditions, and with the increase of load current, the ratio increases, which means the fault characteristics is obvious. This conclusion is consistent with Figure 5 and Figure 6.

5 CONCLUSIONS

This paper proposes a transformer tank vibration model based on baseband vibration signal. The model can calculate the tank vibration signal by monitoring load voltage and load current, then compare the result with measured signals to see if the vibration is normal. If not, it's considered a winding inter-turn short-circuit fault. From the result of simulation and experiment, the model is proved effective.

REFERENCES

1. Ma Hongzhong, Electric Machinery, Bei Jing: Higher Education Press. 2005.
2. B. Garcia, J.C. Burgos, and A.M. Alonso, "Transformer tank vibration modeling as a method of detecting winding deformations-part II: experimental verification," IEEE Trans. Power Delivery, vol. 21, pp. 164–169, Jan. 2006.
3. Wang Hongfang, Wang Naiqing, and Li Tongsheng, "Axial nonlinear vibration of large power transformer winding," Power System Technology, vol. 24, pp. 42–45, Mar. 2000.
4. Wang Hongfang, Wang Naiqing, and Li Tongsheng, "Influence of axial precompression level on axial vibrations in transformer windings," Power System Technology, vol. 23, pp. 8–11, Sept. 1999.
5. Yao Zhisong, Yao Lei, The structure, principle and application of new type transformers, Bei Jing: Mechanical Industry Press, 2006.
6. Wu Shuyou, "Research on condition monitoring and fault diagnosis of power transformer based on vibration signal analysis," Ph.D. dissertation, University of Science and Technology of China An Hui: 2009.
7. Xie Po'an, Rao Zhushi, and Zhu Zishu, "Finite element modeling and analysis on transformer windings," Journal of Vibration and Shock, vol. 25, pp. 134–137, 2006.
8. Xie Po'an, "Study on application of vibration analysis to the condition monitoring of power transformers windings," Ph.D. dissertation, Dept.electrical engineering, Shanghai Jiaotong University, 2008.
9. Ji Shengchang, Li Yanming, and Fu Chenzhao, "Application of on-load current method in monitoring the condition of transformer's core based on the vibration analysis method," Proceeding of the CSEE, vol. 23, pp. 154–158, 2003.

Power and Energy – Kong (Ed.)

Lightning zones map drawing based on statistic of lightning parameters for Shanghai power grid

W.R. Si
State Grid Shanghai Electric Power Research Institute, Shanghai, China

H. Ni
State Grid Shanghai Procurement Company, Shanghai, China

C.Z. Fu
State Grid Shanghai Electric Power Research Institute, Shanghai, China

Y.Q. Wang
State Grid Changxing Power Supply Company, Shanghai, China

A.F. Jiang & Z.C. Fu
Shanghai Jiaotong University, Shanghai, China

ABSTRACT: Scientific lightning zones map drawing can improve the pertinence and effectiveness of lightning protection measures for power grids and realize the differentiation technology and strategy of lightning protection for transmission lines. In this paper, the statistics of Ground Flash Density (GFD) and Lightning Current Distribution (LCD) from the year 2006 to 2010 of Shanghai district are made, respectively. Combined with the typical lightning withstand level and lightning faults records of typical 500 kV transmission lines, the lightning zones map with $0.01° \times 0.01°$ mesh for Shanghai power grid are plotted using a statistical method, which is composed of the GFD map and 500 kV shielding failure and back flashover lightning hazard map of Shanghai power grid. The results show that the power grid lightning hazard distribution is correlated well with the historical lightning faults of 500 kV transmission lines.

Keywords: ground flash density; lightning current distribution; lightning hazard; lightning zones map

1 INTRODUCTION

Lightning flashover of Chinese power grid is rising in recent years, and the reasons could be attributed to: ① The increase of lightning activities. According to the data of lightning monitoring network, lightning ground flash density increased significantly from 2005 in many regions of China. ② The development of new power grid, the increased length of transmission lines, the increased height of towers and easily be stroke by lightning. ③ Due to lack of understanding of the activities and distribution of lightning, the lightning protection measures are less effective. Therefore, with the abnormal climate change and the rapid development of today's power grid, it's very urgent to fully grasp the distribution characteristics and regular patterns of lightning.

Currently, large amounts of long-term lightning data accumulated by lightning monitoring system have become the basis of lightning parameters statistics. State Grid Electric Power Research Institute started carrying out research on the basic parameters of lightning massive automatic monitoring data from the end of 2004, and had analyzed Fujian, Shaanxi, Zhejiang and Jiangsu Power Grid and achieved good results. This paper is based on lightning parameters accumulated by Shanghai Power Grid's lightning location system. According to the 'lightning zone level standards and rules of drawing lightning zone distribution map' [3] and related research [4], the synthesis of typical tower insulation level of 500 kV transmission line in Shanghai, and years of operating experience as well as factors such as topography features, this paper tries to draw the grid of lightning zone maps and provides reference and basis for the Shanghai power grid to differentiate lightning protection.

2 LIGHTNING LOCATION SYSTEM OF SHANGHAI POWER GRID

Shanghai Power Grid started constructed the lightning location system in 1996 [5], and expanded the detection stations from the original 4 (Jinshan, Nanhui, Chongming Island, Changxing Island) to 8 (Zhu Jing, Changxing Island, Nanhui and Chongming Island, Electric Power Research Institute, Jinshan, new Qingpu and Sui Gang) by the project of 2005, which increased the redundancy of lightning detection stations, greatly improved the reliability of lightning detection efficiency and accuracy of the system, and laid the foundation for the lightning applied of entire Shanghai EHV transmission and distribution system. Eight detection stations basically realized the real-time monitoring of lightning activities of the Shanghai power grid. Based on the web pages and lightning information systems, the real-time display and inquiries of lightning activity can be achieved. After years of operation, a massive lightning activity data have been accumulated, scientific basis of differentiated design, reform, assessment of transmission lines and finding and processing the equipment lightning failures of Shanghai power grid transmission lines has been provided.

3 THE STATISTICS OF GROUND FLASH DENSITY DISTRIBUTION

Based on the years of statistical data of lightning location system, the grid method for calculating the ground flash density of Shanghai area is adopted, and thus the distribution map of ground flash density is drawn. The so-called ground flash density is the number of ground lightning per square kilometer in statistical regions, and the corresponding lightning network method is to divide the number of lightning in each grid by the area of the grid. Generally the annual average ground flash density is used to represent by (time/km^2/a), which is an accurate reflection of the characteristics of lightning activity parameters.

The average annual spatial distribution of ground flash density in Shanghai area from 2006 to 2010 is shown in Figure 2. There are differences in different regions of lightning activity levels. Mean annual data of several years can reflect the nature of distribution better. According to five-year (from 2006 to 2010) average annual ground flash density spatial distribution, lightning activity in the southeast Shanghai area (including Nanhui, Pudong, Fengxian and other regions) is relatively weak while strong lightning activity mainly concentrated in the north-central region (including Baoshan, Jiading and Shanghai area) and Jinshan District in the north, where most of the land area of Baoshan, Jiading District, and downtown of Shanghai, the ground flash and density values were more than 7 times/km^2/a.

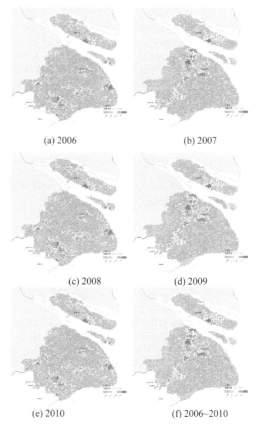

(a) 2006   (b) 2007

(c) 2008   (d) 2009

(e) 2010   (f) 2006~2010

Figure 2. Ground flash density distribution in Shanghai from 2006 to 2010 (natural hierarchical segmentation method, the grid resolution of 0.01° × 0.01°).

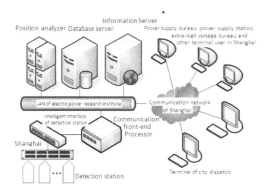

Figure 1. The schematic of lightning location system of Shanghai power grid.

4 CLOUD-TO-GROUND LIGHTNING DENSITY DISTRIBUTION DIAGRAM

The ArcGIS9.3 platform is used as the ground flash density distribution map, the involved layers are as follows: a) Base map: Shanghai geographic information map (marked each administrative domain), 80 national coordinate system is used, the map scale at all levels of the electronic is 1:150000, b) Main Layer 1: Draw connection geographic diagram of power grid with voltage level 220 kV and above, symbols and names of the transmission lines, substations (converter stations) and power plants are marked; c) Main Layer 2: value of ground flash density Ng distribution layer based on the lightning location system from 2006 to 2010 in the Shanghai area. Data are from the statistics division of the grid method and the resolution is selected as $0.01° \times 0.01°$. Based on Ng value of ground

Table 1. CG lightning density (Ng) graded.

Grade	Corresponding values	Grade	Corresponding values
A	Ng < 0.78/km^2/a	C2	5.00/km^2/a ≤ Ng ≤ 7.98/km^2/a
B1	0.78/km^2/a ≤ Ng ≤ 2.00/km^2/a	D1	7.98/km^2/a ≤ Ng ≤ 11.00/km^2/a
B2	2.00/km^2/a ≤ Ng ≤ 2.78/km^2/a	D2	Ng ≥ 11.00/ km^2/a
C1	2.78/km^2/a ≤ Ng ≤ 5.00/km^2/a		

Figure 3. Ground flash density distribution of Shanghai power grid counterattack.

flash density, the frequency of lightning activity is divided into four levels, seven layers from weak to strong, as shown in Table 1.

In Table 1, A-level is less-lightning area and the corresponding average annual thunderstorm days are not more than 10. B1 and B2 level are moderate-lightning area and the corresponding average annual number of thunderstorm days is more than 10 but less than 30. C1 and C2 level are more-lightning area and the corresponding average annual number of thunderstorm days is more than 30 but less than 70. D1 and D2 level are strong-lightning area and the corresponding average annual number of thunderstorm days is over 70. Ground flash density distribution of Shanghai power grid is shown in Figure 3.

5 LIGHTNING DAMAGE RISK MAPS

5.1 Drawing procedures

The drawing program of the distribution diagram of the lightning damage risk is shown as in the Figure 4. The typical tower structure is from the design data provided by the design institute and the insulation level is from the "Typical tower type of transmission line with voltage level 220 kV and above" edited by the Shanghai Power Company. The history record of the accident caused by the lightning is from the ultrahigh voltage company of the Shanghai Power Company (in the period from 2005 to 2011, the lightning trip-out of 500 kV transmission lines is two). Statistical sample database is from the lightning activity data measured by the lightning location system in the Shanghai power grid from 2006 to 2010. The cumulative probability distribution of lightning current amplitude is

Figure 4. Flow chart of lightning damage risk maps drawing procedure.

Table 2. Lightning risk level division.

Level	Descriptions	Level	Descriptions
I level	Risk lightning density is smaller, line lightning tripping is lower	III level	Risk lightning density is large, line lightning tripping is large
II level	Risk lightning density is small, line lightning tripping is low	IV level	Risk lightning density is large, line lightning tripping is larger

shown as formula (2) and the density distribution of ground flash density is shown as Figure 3.

5.2 *Dangerous lightning density distribution*

The stage division method of the dangerous lightning density distribution bases on the stage division of the ground flash density. According to the value of the ground flash density corresponding to the three break point in the stage division of the ground flash density (the break point between stage A and stage B1 is 0.78 time/km^2/a, the break point between stage B2 and stage C1 is 2.78 time/km^2/a and the break point between stage C2 and stage D1 is 7.98 time/km^2/a) and the cumulative probability distribution P(>I) of the lightning current amplitude in the locality. By formula (3) to obtain the three corresponding division point value from the danger of shielding failure density distribution and by the formula (4) to get the three respective break point value from risky lightning counterstrike density distribution. Based on these values, lightning shielding failure risk ground flash density distribution and dangerous lightning back flashover risk ground flash density distribution are divided into four levels respectively, as lightning damage risk maps I, II, III and IV.

5.3 *Maps drawing*

The distribution diagram showing the damage risk caused by lightning is also drawn by the ArcGIS9.3 platform. The related drawing and the main layer 1 is same with the section 4.2. The main layer 2 is the data layer of the 500 kV dangerous lightning density distribution. According to the ground flash density, the dangerous lightning density distribution of the 500 kV voltage (back flashover failure and shielding failure), the operating experience and the overview of terrain geology, the risk of lightning on the transmission lines of the 500 kV voltage level could be divided into four grades as shown in the Table 2.

The risk map of the damage caused by the shielding failure in Shanghai power grid is shown in the Figure 5. The risk map of the damage

Figure 5. Lightning shielding failure risk maps of Shanghai power grid.

caused by the back flashover of the lightning in Shanghai power grid belongs to the Class I, which corresponds to that there is none lightning back flashover of the transmission lines of the 500 kV voltage classes in Shanghai power grid.

6 CONCLUSION

Through the statistics of the ground flash density distribution and the lightning current amplitude probability distribution of Shanghai from 2006 to 2010, and by combining several years of operating experience and topography and other factors of 500 kV lines, the lightning distribution map is drawn based on meshing method (resolution ratio is 0.01° × 0.01°). The results show that: the lightning density of Shanghai area is in the C-class, Baoshan, Jiading and the downtown and the junction of Jinshan and Songjiang are in D-class. The distribution of lightning shielding failure risk of

500 kV transmission lines is mainly Class II, and a small part of the region (mainly the downtown, combining Baoshan and Jiading) is Class III, the distribution of lightning back-flashover risk of 500 kV transmission lines is Class I. In summary, the Shanghai power grid lightning zone maps are very important for guidance of differentiate lightning protection work of Shanghai area in the future.

REFERENCES

1. Chen, J.H. & Wang, H.T. 2007. Statistical method of lightning parameters. High Voltage Engineering. 33(10): 6–10.
2. Electric Power Research Institute of State Grid. 2011. Transmission lines differential lightning protection technology. The New Technology for Overhead Transmission Lines Operation and Maintenance of The State Grid. (in Chinese).
3. State grid corporation. 2010. Minefields classification standard and mapping rules of lightning distribution map.
4. Chen, J.H. & Wang, H.T. 2008. Research on power grid lightning hazard maps. High Voltage Engineering. 34(10): 2016–2020.
5. Gu, C.Y. 2004. The application of lightning location technology in the east china power grid. East China electric power. 32(1):32–35. (in Chinese)
6. DL/T 620–1997. Overvoltage protection and insulation coordination for AC electrical installations. 1997.
7. IEEE Std 1243–1997. IEEE guide for improving the lightning performance of transmission lines. 1997.

Power and Energy – Kong (Ed.)
© *2015 Taylor & Francis Group, London, ISBN 978-1-138-02782-4*

A new dynamic electricity purchasing strategy for Load Serving Entity

Yanzhou Chen
Maintenance and Test Center, CSG EHV Power Transmission Company, Guangzhou, China

Fushuan Wen
Electrical and Electronic Engineering, Institut Teknologi Brunei, Bandar Seri Begawan, Brunei

Yong Yan
Customer Service Center, State Grid Zhejiang Electric Power Company, Hangzhou, China

Lili Wu
Foshan Power Supply Bureau, Guangdong Power Grid Corporation, Foshan, China

ABSTRACT: To maximize the profits and minimize the risks, the Load Serving Entity (LSE) should properly allocate the amount of electricity purchased in various markets. The purchasing process is dynamic and repeated. A new dynamic model is proposed for purchasing electricity in multiple markets by quantizing the risk and expected revenue rate of the LSE employing the semi-absolute deviation. An average reverse model is employed for load forecasting. Case studies are used to demonstrate the feasibility and efficiency of the developed model and method.

1 INTRODUCTION

Power industry restructuring is gradually evolving in China. A potential reform in the near future is to open the retail markets [1–2]. By then, the Load Serving Entity (LSE) will no longer be a monopoly, but an independent market participant instead. From the trading point of view, the risk faced by a LSE is mainly from electricity price volatility of the generation side and the uncertainty of load demand. Future load can be estimated using load forecasting methods [3–4], and electricity price volatility can be forecasted by methods such as the wavelet transform and artificial neural networks [5–6], but forecasting error is inevitable. The LSE should properly and dynamically allocate the amount of electricity to be purchased among various markets so as to maximize the profits and minimize the risks.

Much research work on optimal power purchasing and selling strategy has been done [2][7–14], but most of the existing studies is only concerned with single-stage purchasing strategies. However, the purchasing process of the LSE is dynamic, the purchasing strategy in the current stage will be affected by that of the previous phase. Thus, a multi-stage dynamic purchasing strategy is demanding. A multi-period combined bidding model for power suppliers based on weighted CVaR is introduced in [15]; [16] proves that the actual generation price

does not obey the normal distribution and present a dynamic power purchasing model for power supply company based on fractal CVaR; in [17], a semi-absolute deviation based single-stage power purchasing strategy for LSE is proposed. Theoretically, the effect of a CVaR based risk diversification is better than that of a semi-absolute deviation based one when the loss distribution is continuous. Nevertheless, the CVaR measurement does not satisfy the consistency axiom when the loss distribution is discontinuous [18]. Furthermore, the CVaR optimization model is over-reliant on a given confidence level and the calculating procedure for evaluating the VaR is very complicated.

A large number of empirical analysis in electricity markets indicate that that the normal distribution assumption of electricity price as well as return series is questionable [19–21]. The semi-absolute deviation model does not depend on the existence of variance and co-variance of yield, compared with other models such as mean-variance and mean-semivariance model. The semi-absolute deviation model can easily be converted to a linear problem to solve and effectively control the risk of non-normal distribution assets without demanding the loss distribution strictly like the CVaR model does [22]. Meanwhile, the semi-absolute deviation measurement is a good method of risk measuring which meets the theorem of positive homogeneity, subadditivity and coordination [23] and has been successfully exploited in financial application [24–27]. A semi-absolute deviation based dynamical power purchasing strategy is proposed and investigated thereby under such background.

2 SYNOPOSIS OF SEMI-ABSOLUTE DEVIATION

The semi-absolute deviation refers to the difference of the return rates which is lower than the expected average yield and the expected average yield. Return rates which are higher than expectation do not constitute a loss to investors and are recorded as '0' accordingly under the proposed measurement. The semi-absolute deviation is defined as:

$$E|p - \mu|_- = \int |p - \mu|_- f(p) dp \tag{1}$$

where:

$$|p - \mu|_- = \begin{cases} 0 & p \geq \mu \\ \mu - p & p \prec \mu \end{cases} \tag{2}$$

p is the yield; μ represents the expected average return; $f(p)$ is probability density function.

Thus, the risk of a portfolio can be modeled as

$$\min E \left| X^T (\vec{p} - \vec{\mu}) \right|_- \tag{3}$$

s.t.

$$X^T \cdot \vec{\mu} \geq a \tag{4}$$

$$X^T \cdot L = 1 \tag{5}$$

where: $\vec{p} = [p_1, p_2, ..., p_N]^T$ represents the yield vector; $\vec{\mu} = [\mu_1, \mu_2, ..., \mu_N]^T$ is the expected average return vector; $X = [x_1, x_2, ..., x_N]^T$ is the investment ratio vector; $\sum_{i=1}^{N} x_i = 1$; $L = [1, 1, ..., 1]^T$; a is the lower limit of the investor's expected return; N is the number of the invested markets.

3 SEMI-ABSOLUTE DEVIATION BASED ELECTRICITY PURCHASING MODEL

The electricity purchasing process of a LSE is typically dynamic and therefore by purchasing dynamically with a multi-stage strategy, the LSE can reduce the risk rather than using the single-stage strategy.

The research premises in this paper are as below:

– The LSE purchases electricity from the generation company and retails to users dynamically.

– The LSE holds a certain capacity of LSE-owned thermal plants whose generation price is the average operating cost.

– In order to guarantee the system's reliability, the LSE keeps a certain capacity of spinning reserve which can be either obtained from the LSE-owned thermal plants or the spinning reserve capacity market.

Suppose the LSE purchases from N markets in I stages (where the LSE-owned thermal plants is No. N-1 and the spinning reserve capacity market is No. N). Although the semi-absolute deviation method does not require loss distribution of assets to be normal, the clearing price in each market is assumed to obey a normal distribution as shown below:

$$\lambda_{i,j} \sim (\mu_{\lambda_{i,j}}, \sigma_{\lambda_{i,j}}) \tag{6}$$

Then, the LSE sells the energy to the users and gets profits:

$$p_{i,j} \sim (\mu_{p_{i,j}}, \sigma_{p_{i,j}}) \tag{7}$$

$$\lambda_s = \frac{1}{N \times I} \sum_{j=1}^{N} \sum_{i=1}^{I} \lambda_{i,j} (1 + b) \tag{8}$$

where λ_s is the reference retail price; b is the expected return rate after selling the electricity according to relevant provisions and market quotation.

The relationship between parameters in (6)~(8) is:

$$\mu_{p_{i,j}} = \frac{\lambda_s - \mu_{\lambda_{i,j}}}{\lambda_s} \quad i = 1, 2, ..., N-1 \tag{9}$$

$$\mu_{p_{i,j}} = \frac{0 - \mu_{\lambda_{i,j}}}{\lambda_s} \quad i = N \tag{10}$$

$$\sigma_{p_{i,j}} = \frac{\sigma_{\lambda_{i,j}}}{\lambda_s} \tag{11}$$

The spinning reserve capacity is not to be retailed to users, hence its retailed price is recorded as '0'.

The load is periodic because it has a lot to do with the seasons and climate and therefore it can be estimated with the average reserve model (AR(1)) [28]:

$$\begin{cases} Q_i = \bar{Q}_w + \Delta Q_i \\ \Delta Q_i = \hat{\phi} Q_{i-1} + \hat{\sigma} \varepsilon_i \end{cases} \tag{12}$$

where Q_i is the load sequence of each stage; $\bar{Q}_w$ is the average load of the past stages; ΔQ_i is the residual sequence of Q_i; $\hat{\phi}$ is the regression coefficient of ΔQ_i; $\hat{\sigma}$ is the standard variance of the load sequence; $\varepsilon_i \sim N(0,1)$ is white noise which can be generated through standard normal distribution in the simulation. The effectiveness of the model has been verified in [16].

The dynamic energy purchasing and retailing proceeds of the LSE can be described as:

$$\sum_{i=1}^{I}\sum_{j=1}^{N-1} x_{i,j}\left(\lambda_s - \lambda_{i,j}\right) + \sum_{i=1}^{I} x_{i,N}\left(0 - \lambda_{i,N}\right) \quad (13)$$

where $\lambda_i = [\lambda_{i,1}, \lambda_{i,2}, ..., \lambda_{i,N}]^T$ is the clearing price vector of stage i; $x_i = [x_{i,1}, x_{i,2}, ..., x_{i,N}]^T$ is the amount of electricity purchased in stage i and subject to $x_i \in X = \{x_i \mid \sum_{j=1}^{N-1} x_{i,j} = Q_i, x_{i,j} \geq 0, j = 1, 2, ..., N\}$; λ_s is the retailed price.

It can be figured out that the proceeds of the LSE are dynamic because the load of one stage will be affected by the previous one as could be seen in (12).

The dynamic energy purchasing and retailing risk of the LSE can be expressed as:

$$\sum_{i=1}^{I} E\left|X_i^T\left(\vec{p}_i - \vec{\mu}_i\right)\right|_- \quad (14)$$

M sample values of $\lambda_{i,j}$ and $p_{i,j}$ are taken as $\lambda_{i,j}^1, \lambda_{i,j}^2, ..., \lambda_{i,j}^M$ and $p_{i,j}^1, p_{i,j}^2, ..., p_{i,j}^M$ respectively and (13) and (14) can be rewritten as:

$$\frac{1}{M}\sum_{k=1}^{M}\sum_{i=1}^{I}\left[\sum_{j=1}^{N-1} x_{i,j}\left(\lambda_s - \lambda_{i,j}^k\right) + x_{i,N}\left(0 - \lambda_{i,N}^k\right)\right] \quad (15)$$

$$\frac{1}{M}\sum_{k=1}^{M}\sum_{i=1}^{I}\sum_{j=1}^{N}\left|x_{i,j}\left(p_{i,j}^k - \mu_{i,j}\right)\right|_- \quad (16)$$

Suppose that $u_{i,j}^k (k = 1, 2, ..., M)$ is a dummy variable which meets $u_{i,j}^k = |x_{i,j}(p_{i,j}^k - \mu_{i,j})|_-$, $u_k \geq 0$ and $u_{i,j}^k + x_{i,j}(p_{i,j}^k - \mu_{i,j}) \geq 0$ can be easily concluded. And (16) can be updated as:

$$\frac{1}{M}\sum_{k=1}^{M}\sum_{i=1}^{I}\sum_{j=1}^{N} u_{i,j}^k \quad (17)$$

The LSE is pursuing the highest profits for the lowest risk and thus, the Model(y) of the energy purchasing process can be described as:

$$\max\frac{1}{M}\sum_{k=1}^{M}\sum_{i=1}^{I}\left[\sum_{j=1}^{N-1} x_{i,j}\left(\lambda_s - \lambda_{i,j}^k\right) + x_{i,N}\left(0 - \lambda_{i,N}^k\right)\right] \quad (18)$$

$$\min\frac{1}{M}\sum_{k=1}^{M}\sum_{i=1}^{I}\sum_{j=1}^{N} u_k \quad (19)$$

s.t.

$$Q_i = \bar{Q}_w + \Delta Q_i \quad (20)$$

$$\Delta Q_i = \hat{\phi}Q_{i-1} + \sigma\varepsilon_i \quad (21)$$

$$u_{i,j}^k \geq 0 \quad (22)$$

$$u_{i,j}^k + x_{i,j}\left(p_{i,j}^k - \mu_{i,j}\right) \geq 0 \quad (23)$$

$$\sum_{j=1}^{N-1} x_{i,j} = Q_i \quad (24)$$

$$x_{i,1} \leq c \cdot Q_i \quad (25)$$

$$Q_{min} \leq x_{i,N-1} \leq Q_{max} \quad (26)$$

$$Q_{max} - x_{i,N-1} + x_{i,N} = r \cdot Q_i \quad (27)$$

$$x_{i,j} \geq 0, j = 1, 2, ..., N \quad (28)$$

where $i = 1, 2, ..., I$; $k = 1, 2, ..., M$; c is the upper proportion limit of the long-term contract; Q_{min}, Q_{max} is the minimum and maximum generating capacity of the LSE-owned thermal plants in one stage; r is the proportion of spinning reserve capacity ratio of the load estimated.

4 CASE STUDIES

4.1 Solution of the model

The Model(y) is a bi-objective dynamic programming model and, according to solute method of the LINGO software: If the LSE is risk averse, the risk object is to be solved preferentially and the result will be taken as a constraint in solving the profits object; if the LSE is risk-loving, contrary steps are followed in the model solving; and if the LSE is risk neutral, risk or profits value of the first step should be adjusted properly and be taken as a constraint in the second step. Case studies in this paper assume that the LSE is risk adverse.

Suppose the LSE is allocating the electricity of one year in 4 markets (long-term, day ahead, LSE-owned thermal plants and spinning reserve capacity market) by 4 stages. The parameters of each market's clearing price probability distribution are shown in Table 1. The parameters of each market's profit margin probability distribution is provided in Table 2 assuming that the retailed price is 45 U.S. Dollar/(MW·h) according to (9)–(11).

By analyzing the monthly historical statistics of the electricity purchased by a provincial LSE from September 2007 to December 2009 (28 months) with the average reserve model

(AR(1)), the values of the parameters in AR(1) can be acquired as: $\bar{Q}_w = 71.05 \times 10^5$ MWh, $\hat{\sigma} = 13.86 \times 10^5$ MWh and $\hat{\phi} = 2.20 \times 10^{-2}$ MWh. Then, each month's load in 2010 can be estimated and the seasonal load of the year is as follow: $\{Q_1, Q_2, Q_3, Q_4\} = \{2004.7, 2253.5, 2675.9, 2132.1\} \times 10^4$ MW·h.

Moreover, assume that the generating limits of the LSE-owned thermal plants is $Q_{min} = 30 \times 10^4$ MW·h, $Q_{max} = 200 \times 10^4$ MW·h.

Simulated data for case study is shown in Table 1 and Table 2.

4.2 Comparative analysis of dynamic model and static model

The static model can be solved by converting the seasonal parameters in Table 1 and Table 2 to annual parameters and repeating the calculation. The results are shown in Table 3.

The results in Table 3 indicate that by using the dynamic purchasing model, the LSE obtain higher profits under lower risk than using the static one and therefore the dynamic portfolio strategy is more robust. Furthermore, the risk adverse LSE always allocates as much purchasing energy in less-fluctuating long term market before allotting the rest to the other markets (that is, more from the LSE-owned thermal plants when price of the day ahead market is relatively high and fluctuant and less when opponent; this also shows that the owning of thermal plants of the LSE makes the purchasing strategy more flexible).

Table 1. The parameters of each electricity price probability distribution.

Market	Season	Mean/$·(MW·h)	Standard Variance/$·(MW·h)
Long term	1	33.2	7.93
	2	33.5	8.59
	3	34.3	8.67
	4	32.2	7.24
Day ahead	1	38.4	15.67
	2	42.1	19.23
	3	44.6	20.34
	4	37.5	16.11
LSE-owned thermal plants	1	35	0
	2	35	0
	3	35	0
	4	35	0
Spinning reserved capacity	1	7.63	1.95
	2	8.12	2.09
	3	8.83	2.42
	4	7.64	1.99

Table 2. The parameters of each profit margin probability distribution.

Market	Season	Mean	Standard variance
Long term	1	0.2622	0.1762
	2	0.2556	0.1909
	3	0.2378	0.1927
	4	0.2844	0.1609
Day ahead	1	0.1467	0.3482
	2	0.0644	0.4273
	3	0.0089	0.4520
	4	0.1667	0.3580
LSE-owned thermal plants	1	0.2222	0
	2	0.2222	0
	3	0.2222	0
	4	0.2222	0
Spinning reserved capacity	1	−0.1696	0.0433
	2	−0.1804	0.0464
	3	−0.1962	0.0538
	4	−0.1698	0.0442

4.3 Comparative analysis with the traditional variance-revenue model

The risk of the variance-revenue model can be expressed as:

$$E(x \cdot p - x \cdot \mu)^2 = x^T G x = \sum_{j=1}^{N} x_j^2 \sigma_j^2 \qquad (29)$$

Extended to a dynamic situation, the optimize object of risk can be described as:

$$\min \sum_{i=1}^{I} \sum_{j=1}^{N} x_{i,j}^2 \sigma_{i,j}^2 \qquad (30)$$

The revenue object and constraints are identical with (18) and (20)–(28).

The results after solving the model is shown in Table 4.

As is shown in Table 4, the expected revenue is less than that of the Model(y) in part A while the LSE being risk adverse as the same. And this means that the strategy proposed in this paper is more effective.

4.4 The impact of retail price fluctuation on risk and revenue

Repeat the calculation in part A after changing the retail price λ_s and the alteration of risk and expected revenue is shown in Figure 1.

Figure 1 states clearly that the benefits increase constantly while the risk drops with the ascending

Table 3. Allocation of power purchases under dynamical model and static model.

Purchasing strategy	Stage	Allocation of power purchases/(10^4 MW·h)				Risk and Profits/(10^6 U.S. Dollar)	
		Long term	Day ahead	LSE-owned thermal plants	Spinning reserve capacity	Risks	Profits
Dynamic	1	1303.05	501.65	200.00	200.47	7.89	833.33
	2	1464.77	588.73	200.00	225.35		
	3	1739.33	736.57	200.00	267.59		
	4	1385.86	716.24	30.000	43.21		
Static	1	5893.01	2373.19	800.00	906.62	8.01	829.42

Table 4. Allocation of power purchases under variance-revenue model.

Model	Stage	Allocation of power purchases/(10^4 MW·h)				Risk and Profits/(10^6 U.S. Dollar)	
		Long term	Day ahead	LSE-owned thermal plants	Spinning reserve capacity	Risks	Profits
Variance-revenue	1	1303.06	501.64	200.00	200.47	1086.01	830.01
	2	1464.78	588.72	200.00	225.35		
	3	1739.34	736.56	200.00	267.59		
	4	1385.87	546.23	200.00	213.21		

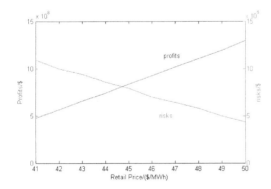

Figure 1. Relationship between retail price and LSE's benefits and risks.

of the retail price. This is in line with real market situations and the profits and risk of the LSE have so much to do with the retail price that a fair one is important to an equitable electricity market.

5 CONCLUSION

The clearing price of each market in each stage is fluctuant and the load is uncertain as well, and both of them should be considered in the modeling of electricity purchasing of the LSE. A semi-absolute deviation based electricity purchasing strategy for LSE is proposed in this paper and several case studies indicate that the proposed model reflects the essential characteristics of risk faced by the LSE. Therefore, a new approach for risk assessment and risk control of the LSE is provided.

ACKNOWLEDGEMENTS

This work is supported by National Natural Science Foundation of China (No. 51361130152).

REFERENCES

1. D. Kirschen, "Demand-side view of electricity markets," IEEE Trans. Power Syst., vol. 18, no. 2, pp. 520–527, May 2003.
2. X. Zhang and X. Wang, "Survey of financial markets for electricity," Automation of Electric Power Systems, vol. 29, no. 20, pp. 1–10, October 2005.
3. M. Ghiassi, D.K. Zimbra and H. Saidane, "Medium-term system load forecasting with a dynamic artificial neural network model," Electric Power Systems Research, vol. 76, no. 5, pp. 302–316, March 2006.
4. B. Chen, M. Chang and C. Lin, "Load forecasting using support vector machines: a study on EUNITE competition 2001," IEEE Trans. Power Syst., vol. 19, no. 4, pp. 1821–1830, November 2004.

5. A.J. Conejo, M.A. Plazas, R. Espinola and A.B. Molina, "Day ahead electricity price forecasting using the wavelet transform and ARIMA models," IEEE Trans. Power Syst., vol. 20, no. 2, pp. 1035–1042, May 2005.
6. N. Amjady, "Day-ahead price forecasting of electricity markets by a new fuzzy neural network," IEEE Trans. Power Syst., vol. 21, no. 2, pp. 887–896, May 2006.
7. Y. Liu and X. Guan, "Purchase allocation and demand bidding in electric power markets," IEEE Trans. Power Syst., vol. 18, no. 1, pp. 106–112, February 2006.
8. J. Xu, P.B. Luh, F.B. White, E. Ni and K. Kasiviswanathan, "Power portfolio optimization in deregulated electricity markets with risk management," IEEE Trans. Power Syst., vol. 21, no. 4, pp. 1653–1662, November 2006.
9. M. Zhou, Y. Nie, G. Li and Y. Ni, "Long-term electricity purchase scheme and risk assessment in power markets," Proceedings of the CSEE, vol. 26, no. 3, pp. 116–122, January 2006.
10. X. Zhang, X. Wang, J. Wang and Z. Hu, "A long-term allocating strategy of power generators," Proceedings of the CSEE, vol. 25, no. 1, pp. 6–12, January 2005.
11. R. Wang, J. Shang, Y. Feng, X. Zhou, Y. Zhang and Y. You, "Combined bidding strategy and model for power supplies based on CVaR risk measurement techniques," Automation of Electric Power Systems, vol. 29, no. 14, pp. 5–9, July 2005.
12. R. Wang, J. Shang, X. Zhou, Y. Zhang, and S. Zhang, "Conditional value at risk based optimization of power purchase portfolio in multiple electricity markets and risk management," Power System Technology, vol. 30, no. 20, pp. 72–76, October 2006.
13. R. Dahlgren, C.C. Liu and J. Lawarree, "Risk assessment in energy trading," IEEE Trans. Power Syst., vol. 18, no. 3, pp. 503–511, August 2003.
14. N.A. Iliadis, M.V.F. Pereira, S. Granville, R.M. Chabar and L. -A. Barroso, "Portfolio optimization of hydroelectric assets subject to financial indicators," Proceedings of IEEE Power Engineering Society General Meeting, vol. 1, pp. 1–8, June 2007.
15. X. Zhang, L. Chen and R. Wu, "Analysis of multiperiod combined bidding of power suppliers based on weighted CVaR," Proceedings of the CSEE, vol. 28, no. 16, pp. 79–83, June 2008.
16. M. Wang, Z. Tan, Y. Guan, P. Xie and X. Li, "Dynamical power purchasing model for power supply company based on fractal conditional value at risk," Automation of Electric Power Systems, vol. 33, no. 16, pp. 50–54, August 2009.
17. R. Liu, J. Liu, M. He, M. Wang and Y. Chen, "Power purchasing portfolio optimization and risk measurement based on semi-absolute diviation. Automation of Electric Power Systems," vol. 32, no. 23, pp. 9–13, December 2008.
18. J. Liu, "Comparative studies of VAR and CVAR based on the extreme value theory," The Journal of Quantitative & Technical Economics, vol. 24, no. 3, pp. 125–133, March 2007.
19. F.A. Longstaff and A.W. Wang, "Electricity forward prices: A high-frequency empirical analysis," The Journal of Finance, vol. 59, no. 4, pp. 1877–1900, August 2004. Eydeland and K. Wolyniec, Energy and Power Risk Management, Hoboken, New Jersey: Wiley, 2003.
20. E. Eberlein and G. Stahl, "Both sides of the fence: a statistical and regulatory view of electricity risk," Energy Power Risk Management, vol. 8, no. 6, pp. 34–38, September 2003.
21. H. Konno, H. Waki and A. Yuuki, "Portfolio optimization under lower partial risk measures," Asia-Pacific Financial Markets, vol. 9, no. 2, pp. 127–140, June 2002.
22. P. Wen and X. Da, "The risk measures based on Minkowski gauge," Journals of Systems and Management, vol. 17, no. 2, pp. 177–180, April 2008.
23. F. Guo and F. Deng, "A dynamic semi-absolute deviation portfolio selection model," Systems Engineering, vol. 24, no. 9, pp. 68–73, September 2006.
24. F. Guo and F. Deng, "Extended semi A.D and dynamic portfolio selection," Mathematica Applicata, vol. 20, no. 3, pp. 446–451, March 2007.
25. X. Xu, X. Yang and Y. Chen, "Portfolio model with semi-deviation risk measure," Journal of Wuhan University (Natural Science Edition), vol. 48, no. 3, pp. 297–300, March 2002.
26. Z. Hu, F. Peng and D. Huang, "An extended with portfolio optimization model mean-semi deviation," Systems Engineering, vol. 22, no. 3, pp. 57–61, March 2004.
27. X. Zhang, X. Wang and Y.H. Song, "Modeling and pricing of block flexible electricity contracts," IEEE Trans. Power Syst., vol. 18, no. 4, pp. 1382–1388, November 2004.

Power and Energy – Kong (Ed.)
© 2015 Taylor & Francis Group, London, ISBN 978-1-138-02782-4

A gravure drying device of intelligent control system

Bin Zhang, Peng Cao & Zhong-ming Luo
Beijing Institute of Graphic Communication, Beijing, China

ABSTRACT: In view of the problem that the low temperature control precision and the poor stability of the gravure drying equipment, which make it difficult to achieve coordinated control between the temperature and the wind speed, an intelligent temperature tracking control system is designed. The system kernels with ARM microprocessor, uses thermocouple to access the temperature signal and finally achieve the closed-loop tracking control of the temperature through the way of PWM control which controls the average power of the heater. We describe the construction principle and implementation method of this system and focus on the fuzzy PID subsection control principle and implementation of the algorithm.

1 INTRODUCTION

As one important class of printing technology, gravure printing occupies very important position in the field of printing and packaging and graphic publishing with its thick ink layer in printing products, bright color, high saturation, stable printing quality and fast printing speed. In our country, gravure printing is mainly used for paper packaging, plastic flexible packaging, wood grain decoration, leather materials, pharmaceutical packaging, etc. There are strict requirements needed for the oven temperature and wind speed if we want to achieve high gravure printing quality. Traditional gravure dryers make use of the electric heater which is controlled by the temperature controller to dry the printing products, but it is not convenient because the heat energy and the quantity of wind both need to be adjusted by hand.

This paper introduces a temperature control system which applies an improved fuzzy PID control algorithm. The system kernels with ARM microprocessor use thermocouple to access the temperature signal and acquire real-time control parameters which fit the system best through way of self-adaptive algorithm to achieve the accurate temperature control. The system includes driver, controller and temperature sensor, which constitutes a closed loop control circuit. In this way, we can achieve accurate temperature control as well as avoid tedious manual operation.

2 DESIGN OF OVERALL SYSTEM

2.1 Framework

This system consists of main control computer, intelligent controller and industrial Ethernet. The system block diagram was shown in Figure 1.

The goal of this system is to control the oven temperature in real-time. After the operator uses the main control computer to set the start or stop instruction, the target temperature and wind speed, the intelligent controller accepts the instruction via the industrial Ethernet. The intelligent controller can closed-loop control the temperature automatically. At the same time, the intelligent controller can feedback their own running state to main control computer through industrial Ethernet. And the system has fault detection device, if there is a fault, the system will display the error information. When situation means danger, the system will stop automatically.

2.2 Design in intelligent controller

The intelligent control system was designed by combining PID control system with fuzzy control system, based on actual requirements of this temperature control system. Each intelligent control system was constructed based on the single-chip microcomputer with ARM kernel. The temperature

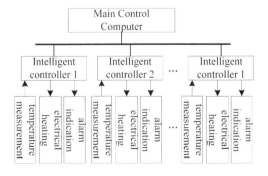

Figure 1. The system block diagram of overall system.

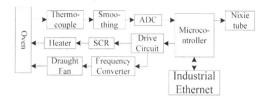

Figure 2. The block diagram of intelligent control system.

control of the gravure drying device always works in the closed-loop state. The intelligent control system constitutes of the temperature sensing circuit, control output circuit, electrical heater and industrial Ethernet. The design diagram is shown in Figure 2.

When the system starts to work, intelligent controller will always wait for the commands from the master control computer. After the intelligent controller receives data from the main control computer via the industrial Ethernet, the intelligent controller will change the inverter output frequency through the driver to adjust the fan speed. And intelligent controller will use thermocouples to detect the actual temperature of oven. After filtering and amplifying circuit, the temperature signal is sent to A/D port of STM32. The intelligent controller will calculate the actual temperature according to the formula and display this temperature by digital tube. At the same time, the intelligent controller will send temperature data to the main control computer.

The intelligent controller compares the actual temperature with the setting temperature and the result will determine the output PWM duty ratio in the next control period according to the fuzzy PID control algorithm. After that, the intelligent controller can control the opening time of the solid state relay. The average power of electric heating will change along with the change of the solid state relay, so that we can achieve the goal of temperature control.

3 FUZZY PID CONTROL ALGORITHMS

PID (Proportional, Integral and Derivative) is a linear controller with the advantages of simple algorithm, good robustness and high reliability etc. So it is widely applied in process control and motion control. But the temperature control system has its own special qualities that the heat transfers in the form of thermal field, so it has obvious nonlinear nature, time-variation, distribution, and hysteresis. However, conventional PID cannot self-tune parameter according to the on-line situation,

so it is necessary to improve the conventional PID algorithm.

3.1 Structure of fuzzy self-adaptive PID controller

Fuzzy self-adaptive PID control is a computer mathematics method which is based on fuzzy set theory, fuzzy linguistic variables and fuzzy logic. According to the system error absolute value $|e|$ and the error rate absolute value $|ec|$, the system self-tunes the parameters Kp, Ki and Kd online. Fuzzy self-adaptive PID controller is shown in Figure 3.

Fuzzy PID adjustment process: the system samples, quantifies and fuzzes the signal and achieves the fuzzy control relation according to the fuzzy control rules. Here we use Mamdani fuzzy reasoning algorithm to synthesis. Mamdani fuzzy reasoning algorithm adopts fuzzy implication operator:

$$\mu_c : a \to b = a \wedge b$$
$$A \to B = A \times B \tag{1}$$

The formula for the fuzzy output reasoning:

$$C' = A' \circ (A \times C) \cap B' \circ (B \times C) \tag{2}$$

In multi-rule, the fuzzy relation can be shown as follows:

$$C' == A' \circ R_1 \cap B' \circ R_2$$
$$= A' \circ \bigcup_{j=1}^{n} (A_i \times C_i) \cap B' \circ \bigcup_{j=1}^{n} (B_j \times C_j) \tag{3}$$

We use the weighted average method to defuzzify, the calculating formula can be expressed as:

$$z^* = \frac{\sum \mu_{cj}(w_j) \cdot w_j}{\sum \mu_{cj}(w_j)} \tag{4}$$

where w_j, $\mu_{cj}(w_j)$ is respectively the centroid and grade value.

At the beginning, we set the three parameters K_{p0} K_{i0} and K_{d0} for the PID controller, and then the system will calculate parameter of PID correction

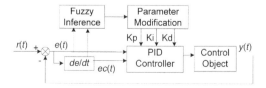

Figure 3. Block diagram of fuzzy PID controller.

ΔK_p ΔK_i and ΔK_d through fuzzy reasoning, and it will adjustment PID parameter according to the formula (5).

$$\begin{cases} K_p = K_{p0} + \Delta K_p \\ K_i = K_{i0} + \Delta K_i \\ K_d = K_{d0} + \Delta K_d \end{cases} \tag{5}$$

3.2 Control algorithm is improved

Parameter self-tuning fuzzy PID control has the advantage of accuracy high precision, and it has a good dynamic performance and robustness. All parameters of the PID need to be online setting, while fuzzy inference will occupy part of the processing time. When the deviation between the set value and current value is larger, the time needed for the system output value tend to set point of system will become bigger. To improve the control system, By subsection control algorithm is a good solution.

In this paper, we use the fuzzy PID subsection control strategy. Namely, we set a threshold of EP, when the system error |e| ≥ EP, the system will choose the proportional control mode, which makes the system response quickly; and when the deviation |e| < EP, it will choose the parameter self-tuning fuzzy PID control method described above. So that we can make the system response has no overshoot and static deviation. Of course, we must choose a proper threshold. If it is chosen too large, the system will earlier into fuzzy PID control, the system response speed will then be slower; if it is chosen too small, the system switch to the fuzzy PID control when the system response closes to the setting point, that may cause overshoot.

3.3 Algorithm implementation

Fuzzy self-tuning PID subsection control algorithm is used in the control system. The process block diagram is shown as Figure 4.

We must pay attention to several questions in the programming:

1. After detecting temperature, intelligent control system will filter the data and to judge whether e and ec are higher than the setting threshold. If higher than the threshold, the system sets them to be the setting upper or the setting lower limit value. Otherwise, the system will make the input data multiplied by the quantitative factor.
2. According to the input value of fuzzy variables, the output value can be achieved by looking up the table. The PID parameters of adjustment Kp, Ki and Kd can be got when the output value of fuzzy reasoning is multiplied with the quantitative factor.

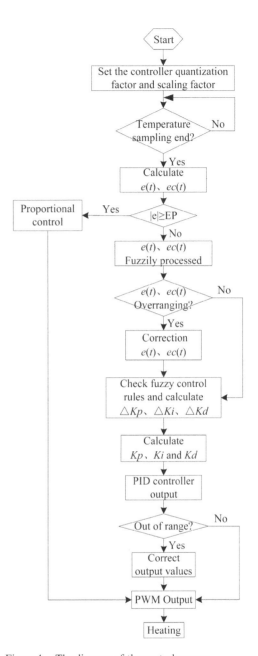

Figure 4. The diagram of the control process.

Before the next cycle, the system uses this value to calculate the output of the PID.

4 RESUALTS OF EXPERIMENT

The conventional PID algorithm and fuzzy PID subsection control algorithm are used in gravure

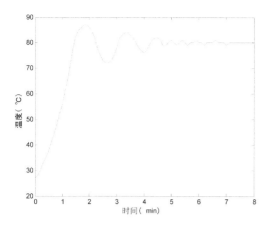

Figure 5.   Conventional PID control chart.

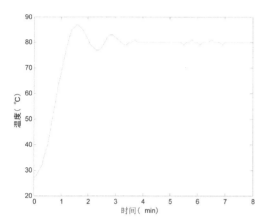

Figure 6.   Fuzzy PID subsection control chart.

drying control system respectively, and their temperature curve are shown in Figures 5 and 6.

The results show that the method we proposed has higher accuracy, faster response, smaller overshoot, and better performance.

5   CONCLUSION

The designed gravure dryers temperature tracking control system in this paper, which builds the reasonable ARM embedded platform and uses the improved fuzzy PID subsection control strategy, can make the controlled object's temperature fluctuations decrease significantly. In our experiments, the control system works stable and reliable, and meets the requirement of system temperature control precision. It is of highly practical value.

REFERENCES

1. Ling, Lu & Yan, Gao. 2006. Temperature Control System based on S3C44B0X. *Microcomputer Information*: 5–2:113–115.
2. Lin, Lei & Hou-jun, Wang. 2006. Adaptive Fuzzy PID Control Method Based on Identification Structure. *Microcomputer Information*: 01:15–21.
3. Zhi-bao, Tang. & Xing-wang, Guo. 2008. A design of temperature control system based on ARM. *Microcomputer Information*, 1–2:144–146.
4. Xiao-yang, Xiao. & Ling, Wang. 2010. Study on Application of Self-adaptive Fuzzy PID Controller in Central Air-Conditioning System. *Microcomputer Information*: 11–1:31–33.
5. Liang, Yin. & Wen-ning, Gong. 2011. Analysis of Fuzzy Adaptive PID Control Algorithm. *The World of Inverters*: 9:86–88.
6. Guang-qi, Zeng. & Jun-an, Hu. 2006. Fuzzy control theory and engineering application. Wuhan: Wuhan: Huazhong Science and Technology Press.
7. Jing, Zhu. 2005. Fuzzy control principle and application. Wuhan: Beijing: China Machine Press.

Power and Energy – Kong (Ed.)
© 2015 Taylor & Francis Group, London, ISBN 978-1-138-02782-4

Equivalent cross-section calculation of orthogonal magnetization reactor

Zhengrong Jiang, Shaowei Meng & Xiaodan Li
College of Electrical and Control Engineering, North China University of Technology, Beijing, China

ABSTRACT: Orthogonal magnetization reactor in this study consists of axisymmetric iron core cylinder and Rectangle iron yoke, whose three-dimensional magnetic field belongs to the unequal distribution. According to the axis of symmetry and two-dimensional field distribution characteristics, which were combined with field boundary conditions, this paper proposed a method to calculate the reactor's magnetic field distribution with two-dimensional plane field and axisymmetric cross-sectional equivalent method. By comparing the results with those that have been measured before, this method is characterized by simpler model, few calculations and higher precision, and it is suitable for the new reactor construction, which can optimize the design.

1 INTRODUCTION

1.1 Introduction of orthogonal magnetization reactor structure

Orthogonal magnetization reactors can be used for high-voltage transmission shunt reactors to adjust the terminal reactive power, or it can be used as the high-pressure filter on the primary side of the adjustable inductor trial. Experiment with orthogonal magnetization reactor consists of a semi iron core cylinder and outer surrounded iron yoke. Two coils wind on the semi iron core perpendicular with each other: 300-turn coil winds around center as the AC working winding, 150-turn coil winds inside and outside as the DC bias winding. By adjusting the DC bias wingding's current to change iron core tensor permeability, the AC working winding inductance changes, the continuously adjustable inductor can be obtained.

Such Orthogonal magnetization reactors have the feature of small harmotic, less winding and high security that non-electrical contact caused AC and DC coil. It has great value in HVDC.

1.2 Instruction of equivalent cross-section calculation

Orthogonal magnetization reactor's sectional view is in Figure 2, including iron core cylinder cross-section's symmetric circular area and iron yoke's rectangular area. AC coil wounds around the outside of the core barrel.

This method combines analysis method of symmetrical field with finite element analysis of two-dimensional plane field, to approach this complex

three-dimensional field distribution. For the iron core cylinder area a-b-c-d-a, can be approximated to axisymmetric magnetic field distribution area. For the iron yoke area d-c-e-f-d, magnetic field distribution is a typical two-dimensional field. If the two

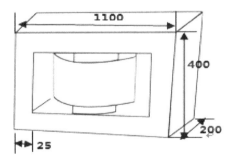

Figure 1. New reactor construction.

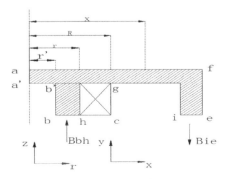

Figure 2. New reactor calculation area.

regions field distribution are only simply superimposed, the magnetic induction of two regions boundary cannot be continuous everywhere. Equivalent algorithm is to establish new magnetic vector potential field to satisfy the boundary of the two regions in a row. This method can obtain a higher accuracy in the case of very small magnetic flux leakage.

2 MAGNETIC VECTOR POTENTIAL

2.1 Magnetic vector potential definition in continuous boundary

Assuming vector potential in boundary b-a-f-e is zero, iron core cylinder's magnetic vector potential A_R is defined:

$$A_R = rA_\theta \tag{1}$$

This represents per radian magnetic flux density in cylindrical area, where A_θ is axisymmetric field tangential component, r is equivalent cross-section radius. In radius area, total flux can be expressed as:

$$\phi_{rz} = 2\pi A_R \tag{2}$$

For Two-dimensional field, magnetic vector potential A is corresponding to magnetic induction per unit length in the Z direction in the iron yoke, where the total flux in iron yoke is obtained:

$$\phi_{xy} = \omega_0 A_z \tag{3}$$

where ω_0 is the thickness of iron yoke.

From the structure of the two regions can obtain the two parts of the magnetic flux with the following relationship:

$$\phi_{rz}/2 = \phi_{xy} \tag{4}$$

Based on the equation (2)~(4), the following can be obtained:

$$\pi A_R = \omega_0 A_z \tag{5}$$

where, A_R and A_z is magnetic vector potential in their respective areas, at the interface of these two magnetic vector potential should be equal. As can be seen apparently from the equation (5), at the interface between the two areas, the boundary conditions continuity is not met the equation that A_R eruals A_z.

To satisfy the condition of magnetic vector potential equivalent in boundary, defining along rectangular region thickness direction (along z direction) of the new magnetic vector potential is:

$$B_x = \frac{1}{\omega}\frac{\partial A}{\partial y}, \quad B_y = \frac{1}{\omega}\frac{\partial A}{\partial x} \tag{6}$$

where B_x and B_y is the magnetic flux density of the rectangular area in the X and Y direction components respectively. With a new definition of the flux, the total magnetic flux on the iron yoke can be expressed as:

$$\phi_{xy} = A\omega_0/\omega \tag{7}$$

Based on equations (2), (4) and (7), the following relationship can be obtained:

$$\pi A_R = A\omega_0/\omega \tag{8}$$

where A_R, A is expressed as two regions' magnetic field, these two values should be equal at the interface. Therefore, iron yoke depth ω that meets continuous magnetic field can be introduced by the following formula:

$$\omega = \omega_0/\pi \tag{9}$$

Combined equation (7) with (9), the flux can be expressed as:

$$\phi_{xy} = \pi A \tag{10}$$

Further, on the air boundary, the flux should be continuous, where A_R and A are equal everywhere.

2.2 Magnetic vector potential equation

For axisymmetric area, Poisson equation can be expressed as:

$$\frac{\partial}{\partial r}\left(\frac{v}{r}\frac{\partial A_R}{\partial r}\right) + \frac{\partial}{\partial r}\left(\frac{v}{r}\frac{\partial A_R}{\partial z}\right) = -J_0\theta \tag{11}$$

where v and J_0 are expressed as magnetoresistive and current density toroidal component.

Using the new definition of the magnetic vector potential A, rewrite the Poisson equation:

$$\frac{\partial}{\partial x}\left(\frac{v}{d}\frac{\partial A}{\partial x}\right) + \frac{\partial}{\partial y}\left(\frac{v}{d}\frac{\partial A}{\partial y}\right) = -J_{0z} \tag{12}$$

where J_{0z} is Z-component of the current density.

Rayleigh Ritz equation can be obtained by the principle of minimum potential energy:

$$\frac{\partial E}{\partial A_i} = 0 \quad (i = 1, 2 \dots n) \tag{13}$$

where, A_i is magnetic vector potential of node of i. A_i is A_R of axisymmetric area and A of rectangular area. E is the two regions' magnetic energy:

$$E = E_{rz} + E_{xy} \tag{14}$$

For axisymmetric area, E is determined by the following equation:

$$E_{rz} = 2\pi \iint_{Srz} \left(\frac{1}{2} \int_0^{B_{rz}^2} v dB_{rz}^2 - J_{0\theta} \frac{A_R}{r} \right) r dr dz \tag{15}$$

where S_{rz} represents the entire axisymmetric region. For the rectangular region, E is determined by the following equation:

$$E_{xy} = 2\omega_0 \iint_{Sxy} \left(\frac{1}{2} \int_0^{B_{xy}^2} v dB_{xy}^2 - \frac{\pi}{\omega_0} J_{0z} A \right) dx dy \tag{16}$$

where S_{xy} represents the entire rectangular region.

3 THE IMPACT ON ACCURACY

The orthogonal magnetization reactor combines axisymmetric iron core with rectangular iron yoke, magnetic circuit performs uneven, and therefore, new calculation method's accuracy is highly affected by the equivalent boundary dividing positions.

3.1 The division of the boundary position

First, analyze the situation of AC coil group radius as border. Figure 3 shows with the changes of the boundaries of R, the changes of magnetic flux density in the iron core cylinder and iron yoke. Respectively, calculated and measured in ampere turns to be 0.25 * 850 and 1 * 850, the results show that when the size setted in the boundary surface of the yoke is much smaller than the entire yoke size, calculated results are in good agreement with the measurements.

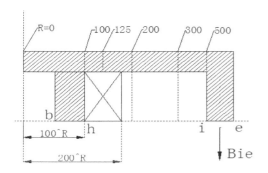

Figure 3.   The boundaries position.

Figure 4 shows the X component of the magnetic flux density in center line of iron yoke, that is B_X, when X is less than or equal to 80 mm, B_X increases gradually with the increasing of X. When X is greater than 80.16, since the increase of equivalent cross-section is great than the increase of magnetic flux, the flux will appear a peak. In this model, on the border the cross area between axisymmetric area and rectangular area reach balance at R = 127.39 (mm). Assuming cylindrical core cross-sectional at the conditions that R respectively is 95, 127 and 300 mm, shown in Figure 5. While considering the boundary surface continuity of magnetic flux density in 2.2, whereas in fact, when R exceeds 127 mm, the flux is no longer continuous. Accordingly, the conclusion can be drawn that the border interface must be set equal to cross-sectional area between two parts, magnetic flux can keep continuous.

3.2 The magnetic circuit structure

The sample's size of the magnetic circuit is determined by following parameters: cylindrical core's inner and outer diameter, length and thickness of iron yoke, and the width ω_0, the height H of the yoke. Using usual two-dimensional field to solve the magnetic circuit structure, radius of the core barrel will bring large error, which cannot

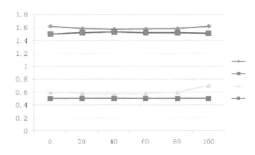

Figure 4.   The relationship between R and flux density (limb).

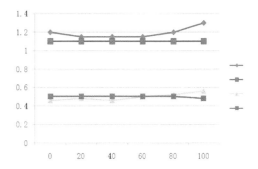

Figure 5.   The relationship between R and flux density (yoke).

be ignored. It is because two-dimensional field solving treats entire field as a rectangular area. If assuming the entire field is axisymmetric region or two-dimensional field, the calculated magnetic flux density contains large errors with real magnetic field distribution. Cross-face flux density calculation error is generated by the cylindrical core's inner and outer diameter, length and thickness of iron yoke. The following calculation errors caused by various factors were analyzed.

Figure 6, Figure 7 and Figure 8 show the influence on B_{bh} and B_{ie} is affected by cylindrical core's inner and outer diameter, length and thickness of iron yoke. By comparison three kinds of calculation methods with experimental result, the calculated value of the equivalent cross-section calculation is

better consistent with experimental test value, and the results obtained by directly using axisymmetric method and two-dimensional field distribution calculation have large deviation with the experimental results. Because the new method takes the core barrel and the iron yoke intersection equivalent cross-section problems into account.

4 NEW METHOD FOR INDUCTANCE CALCULATE

In this paper, new method for orthogonal magnetization reactors were calculated and compared to

Figure 6. The sections of the rectangular and the axisymmetric at the boundary.

Figure 9. The results for four methods (limb).

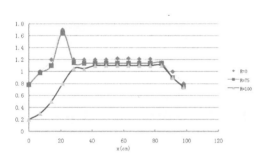

Figure 7. The relationship between R and the flux (R < 42).

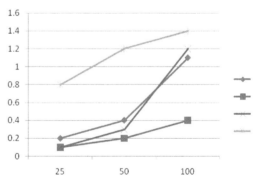

Figure 10. The results for four methods (yoke).

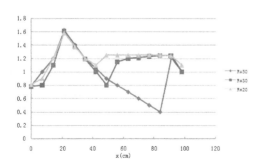

Figure 8. The relationship between R and the flux (R > 42).

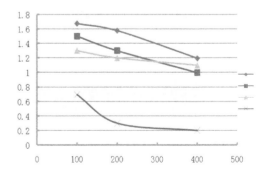

Figure 11. The relationship between length of yoke and flux (I = 2A, limb).

404

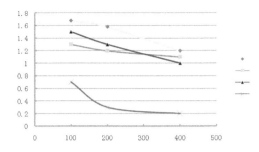

Figure 12. The relationship between length of yoke and flux (I = 2A, yoke).

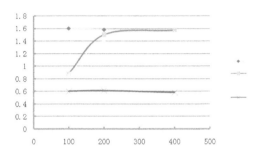

Figure 13. The relationship between yoke thickness and the flux (limb).

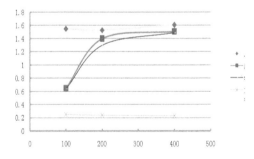

Figure 14. The relationship between yoke thickness and the flux (yoke).

the results of the use of axisymmetric two-dimensional rectangular field distribution method and the calculation method, and in each DC bias current 1 A, 2 A, and 3 A situation, the inductance of reactor were calculated.

The results show inductance, which is calculated by new method, can be matched well with actual measured inductance.

5 CONCLUSIONS

For magnetic circuit structure that combined axisymmetric structure with a rectangular structure, equivalent cross-domain method can well to ensure accuracy, and simplified calculation. The calculation accuracy depends on the equivalent interface of two regions. The method can also be applied to magnetic circuit with varied different thickness rectangular area, the analytic results have very important guiding significance to new reactor structure optimization.

REFERENCES

1. Aharoni. 2002. Principles of Ferromagnetismusm. Lanzhou: Lanzhou University Press.
2. Jianye Chen et al. 2005. Study on the Tuned Filter Based on Splitted-core Type Controllable Reactor. Automation of Electric Power System, 29(15): 53–56.
3. Shaobing Liao. 1988. Ferromagnetismusm. Beijing: Science Press.
4. LiQun Shang & Wei Shi. 2005. Study of the secondary Arc Extinction at EHV Double Circuit Transmission Lines. Automation of Electric Power Systems, 29(10): 60–63.
5. Tommy Helegren & Peter Hidman. 2002. A test installation of a self-tuned ac filter in the Konti-Skan 2 HVDC link. ABB Power Systems, 4(25): 64–66.
6. Xiangqian Tong & Junyi Xue. 2005. Control characteristics of electronic continuously tunable reactor. Electric Power Automation Equipment, 25(1): 25–26.
7. Y. Bi & D.C. Jiles. 1999. Finite element modeling of an electrically variable inductor. IEEE transactions on magnetics. Vol. 33, NO. 5. September.

Power and Energy – Kong (Ed.)
© *2015 Taylor & Francis Group, London, ISBN 978-1-138-02782-4*

Power transformer fault diagnosis based on wavelet de-noising and HHT method of vibration signal

C.N. Wang & H.H. Xu
Jiangsu Nanjing Power Supply Company, Nanjing, China

H.Z. Ma, Z.D. Zhang, H.T. Liu & Z.H. Geng
College of Energy and Electrical Engineering, Hohai University, Nanjing, China

ABSTRACT: Transformer is one of the most important equipment in power system. For the transformer inner mechanical fault diagnosis, a kind of wavelet de-noising method based on Hilbert-Huang Transform is introduced to extract characteristics of vibration signal and diagnose fault in transformer. First the noise vibration signal is eliminated by wavelet de-noising, and then separate the data to components with different time scale, that is, Intrinsic Mode Functions (IMF), using Empirical Mode Decomposition (EMD), then apply the Hilbert transformation to every IMF. After calculating the energy of the IMF, the EMD energy entropy can be obtained. The method and algorithm is explained in this paper. This paper applies the method to a testing fault vibration signal analysis of the transformer. Through analyzing the result of the testing signal, the conclusion can be obtained that we can extract the fault characteristic by contrast the normal state and fault state energy spectrum.

Keywords: transformer; vibration signal; wavelet de-noising; EMD; Hilbert transformation; spectral analysis

1 INTRODUCTION

In a variety of power system equipment, power transformers as the key equipment, will bear the responsibility of the voltage conversion, power distribution, transmission, and the provision of electric service, will be an important guarantee of the power system security, reliability, quality, economic run. Although the production quality of the transformer is improved continuously, due to the long-running of transformer, failures and accidents will not be completely avoided. According to the data from abroad, it can be found that the failure rate of transformers which have run more than four years is 2.6%. The national average statistics show that the failure rate of 500 kV transformer is 5.1% in 2001; the failure rate of 220 kV transformer is 0.4% [1]. Therefore, it is a great significance of the transformer fault detection research for safe operation of the transformer [2].

Moreover, core and winding vibration resulting from core magnetostriction and electro-dynamic forces will induce body vibration of transformer. The coupling vibration is transmitted to the tank

through the body and cooling system, will also induce body vibration of transformer. After being propagated, vibration data can be obtained by applying some vibration sensors on the transformer tank. Vibration level will increase if the transformer performance is going down. To analyze vibration signal, extract fault features is an effective monitoring and diagnosing method [3].

The earlier research based on vibration methods focused on the relationship between vibration signals and load current, which ignores the vibration characteristics of transformer. Transformer vibration signal contains a lot of transient signals, are non-stationary signals, while the traditional methods such as Fourier transform and wavelet transform are used for smooth or piecewise smooth vibration signals, their effects are not obvious. To compensate for these limitations, a time-scale-frequency analysis method based on HHT (Hilbert-Huang Transform) has been developed for extracting vibration features of power transformer.

Hilbert-Huang transform, which consists of two parts: Empirical Mode Decomposition (EMD) and Hilbert transform. Finally, the Hilbert energy spectrum can be used to compare the normal state with the winding short-circuit state of vibration signal to find fault exception and extract fault features [4].

Supported by Jiangsu Power Supply Company, China (J2014055).

2 SIGNAL PROCESSING METHODS

2.1 *Wavelet de-noising*

Wavelet de-noising process can be divided into three steps. The first step, choose a wavelet and determine the wavelet decomposition level M. Then make M layer wavelet decomposition of the noisy signal with the discrete wavelet transform. Because the noise signals are mainly located in the details section of each signal decomposition, so it can be realized de-noising by dealing with these detail parts. With different wavelet bases, the same signal will receive different noise cancellation. Therefore it is very important to find a suitable wavelet basis for signal de-noising. The second step, the high frequency coefficients of each layer from the first layer to the M-layer need threshold processing. The default threshold de-noising method is producing default threshold of the signal and then de-noising. The third step, according to wavelet decomposition of the M-layer of low-frequency coefficients and high-frequency coefficients after quantification from the first layer to layer M, the wavelet reconstruction of signals can be done [5].

2.2 *Principle*

EMD method is built on the assumption that any signal consists of different, simple intrinsic modes of oscillations. According to this, each signal could be decomposed into a number of IMFs (Intrinsic Mode Functions) and each of IMFs must satisfy the following two conditions: a) In the whole data set, the number of extreme and the number of zero-crossings must either equal or differ at most by one; b) At any point, the mean value of the envelope defined by local maxima and the envelope defined by the local minima is zero [6].

An IMF represents a simple oscillatory mode compared with the simple harmonic function. Set signal as $x(t)$. With the definition, any signal can be decomposed as follows:

Identify all the local maxima and minima of the signal $x(t)$, and then connect all the local maxima, minima by a cubic spline as the upper, lower envelopes $u(t)$ and $v(t)$. It should be noted that the upper and lower envelopes should encompass all the data between them. The mean of the two envelopes is designated as:

$$m_1(t) = \frac{u(t) + v(t)}{2} \tag{1}$$

The remaining part of the subtracted:

$$h_1(t) = x(t) - m_1(t) \tag{2}$$

Ideally, if it is an IMF, $h_1(t)$ is the first component of $x(t)$.

If $h_1(t)$ cannot satisfy the condition of IMF, see as the original data, repeat the step(I), then get the mean of the upper, lower envelopes, and denote by $m_{11}(t)$. then If $h_{11}(t) = h_1(t) - m_{11}(t)$ satisfies the above IMF conditions, that is an IMF, stop the first screening process and denote the first IMF by $c_1(t)$, otherwise, substitute $h_{1k}(t)$ for $x(t)$ and repeat the Steps (I).

Subtract IMF $c_1(t)$ from $x(t)$ to obtain the first residue $r_1(t)$

$$r_1(t) = x(t) - c_1(t) \tag{3}$$

The screen process is repeated until the last residue $r_n(t)$ becomes a monotonic function or constant. The signal $x(t)$ then can be expressed as follow:

$$x(t) = \sum_{i=1}^{n} c_i(t) + r_n(t) \tag{4}$$

The original signal $x(t)$ is broken down into intrinsic mode functions and a residual component, the component $c_i(t)$ contains signals of different frequency from high in the end, the high-frequency signal is always the first to be isolated [7].

2.3 *Hilbert-Huang transform*

To extract the instantaneous characteristic frequency from a mono-component signal through EMD, Hilbert transform can be adopted. Where $a_i(t)$ and $\theta_i(t)$ are the instantaneous amplitude and instantaneous phase:

$$a_i(t) = \sqrt{c_i^2(t) + H^2[c_i(t)]}, \ \theta_i(t) = \arctan\left(\frac{H[c_i(t)]}{c_i(t)}\right) \tag{5}$$

And the instantaneous frequency can be obtained as:

$$\omega_i(t) = \frac{d\theta}{dt} \tag{6}$$

Then, Signal $x(t)$ can be expanded into the following form:

$$x(t) = \text{Re} \sum_{i=1}^{n} a_i(t) \exp\left(i \int \omega_i(t) dt\right) \tag{7}$$

where, Re is said as the real part.

We call the right of equation (7) as the Hilbert time-frequency representation of signal $x(t)$, denoted by

$$H(\omega,t) = \mathrm{Re} \sum_{i=1}^{n} a_i(t) \exp\left(i \int \omega_i(t)\, dt\right) \qquad (8)$$

$H(\omega,t)$ Accurately describe the variation of the signal amplitude with time and frequency.

Hilbert spectrum can be further defined as the marginal spectrum:

$$h(\omega) = \int_{-\infty}^{+\infty} H(\omega,t) dt \qquad (9)$$

$h(\omega)$ Reflect the signal amplitude variation with frequency.

Above EMD and Hilbert-Huang transform signal analysis method are known as of EMD-HT method [8].

3 EXPERIMENT AND DISSCUSION

Transformer failure, the mechanism is not the same, in order to be able to accurately diagnose transformer faults, it is necessary to study various failure domain—frequency domain—the energy spectrum characteristics, and contrast, to find out their differences, and then see this as a basis of the transformer fault diagnosis.

3.1 Vibration characteristics of the normal operation transformer

The vibration signal of the normal operation transformer is collected at the scene is shown in Figure 1, the noise greatly, the signal after wavelet de-noising is shown in Figure 1, compared with the original signal. The new signal waveform has been greatly improved. The EMD decomposition of signal is shown in Figure 2, the Hilbert transform of order intrinsic mode functions, the Hilbert spectrum shown in Figure 3.

It can be seen from Figure 3, the Hilbert spectrum of the normal operation transformer has the following characteristics:

The low frequency part of the time interval the concentration of almost all the energy of signal, the irregular distribution of small amounts of energy in high-frequency part, the high-frequency part ratio of the energy is small. That frequency of the highest amplitude almost is 100–300 Hz low frequency, high-frequency part of the amplitude of the smaller, almost no. This shows that the transformer is operating normally, there is no fault.

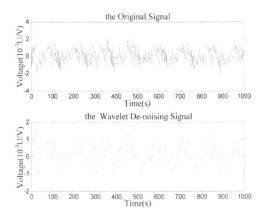

Figure 1. Normal operation of the transformer vibration signal and noise signal.

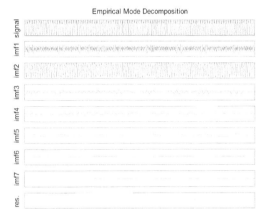

Figure 2. Normal operation of the IMF components of the transformer vibration signal.

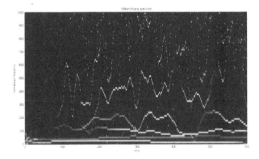

Figure 3. Normal operation of the Hilbert energy spectrum of the transformer vibration signal.

3.2 Transformer vibration signal characteristics of winding short-circuit

Through the transformer core hanging experiment, winding short circuit through two turns of copper wire, the winding short-circuit vibration signals collected at the scene are shown in Figure 4. Similarly, the de-noising vibration signal is shown in Figure 4, compared with the original signal, the new signal waveform is greatly improved, the order of intrinsic mode functions of the new signal is shown in Figure 5; The Hilbert transform of order intrinsic mode functions, the Hilbert spectrum shown in Figure 6.

3.3 Results

Comparing the results of these analyzes with the transformer vibration signal of the normal operation, we get the following conclusions:

I) when the transformer winding short-circuit occurs, the vibration amplitude of 100 Hz and

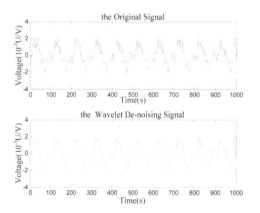

Figure 4. Winding short-circuit of the transformer vibration signal and noise signal

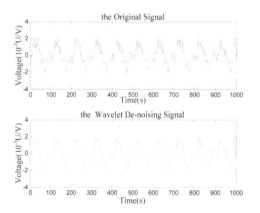

Figure 5. Winding short-circuit operation of the IMF components of the transformer vibration signal.

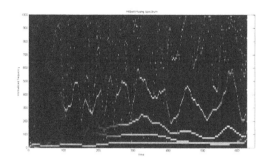

Figure 6. Winding short-circuit of the Hilbert energy spectrum of the transformer vibration signal.

Table 1. Summary of the transformer characteristics on different conditions.

Signal type	Characteristic
Normal	100 Hz–400 Hz is the main frequency
Winding short-circuit	i. The vibration amplitude of 100 Hz and 300 Hz increase
	ii. The energy ratio of high frequency significantly increases, the energy ratio of low frequency significantly reduces

Table 2. The energy ratio of the transformer low and high under different conditions.

Signal type	The energy ratio of low frequency	The energy ratio of high frequency
Normal	0.99411	0.00589
Winding short-circuit	0.96125	0.03875

300 Hz are larger than normal transformer vibration signal. Other multiples of magnitude have different degrees of attenuation increases, but not significantly.

II) From the binding energy of the Hilbert spectrum, when short-circuit fault occurs, the energy ratio of high frequency (400 Hz) significantly increases, the following of low frequency (400 Hz), although the increase in amplitude, but the share of the energy ratio is lower than normal. The above results are shown in Table 1.

Make a compare of the transformer fault signal and the normal vibration signal, can be found when the transformer failure, there will be a significant feature: the energy distribution is very different before and after the failure. We set 400 Hz as the standard to distinguish between low and high frequency, to calculate the energy ratio of high and

410

low-frequency energy in different circumstances, the result is shown in Table 2.

Through analysis, we can see that the energy ratio of high frequency energy is significantly increased after the failure. This feature can be used as an important basis for transformer failure distinguishes.

4 CONCLUSION

In this paper, a HHT-signal noise analysis method based on noise reduction is introduced into the field of transformer vibration signal processing, theory and algorithms of the method is introduced in this paper. First, after wavelet de-noising, the non-stationary simulation signal with white noise, through the EMD decomposition and transformation, the frequency amplitude spectrum and marginal spectrum can be got the results verify the correctness and practicality of this method. And then the fault vibration signals of a measured transformer are analyzed by this method. The following conclusions can be drawn from the results of the analysis:

I. There is a great impact of noise on the EMD accuracy, making the EMD method cannot be adaptive decomposition. Therefore, before using the EMD-HT method, it is sure to deal with the noise.
II. From the time-frequency analysis of EMD method, and the corresponding intrinsic mode functions, the vibration components and their vibration signal amplitude and phase can be obtained.
III. From the time-frequency analysis of EMD-HT, and the corresponding time-frequency spectrum, it can be available to get frequency components of the vibration signal, also the dynamic changes of the vibration signal amplitude and frequency over time.
IV. Through the analysis of practical examples, analysis of the diagnostic instance, the proposed diagnostic methods can accurately reflect the physical state of the transformer; it is able to diagnose the transformer failure directly. The proposed diagnostic methods can be widely applied to the transformer fault diagnosis.

The comparison results have demonstrated that the HHT analysis method based on vibration signal de-noising is a more effective for the transformer fault feature extraction from vibration signals of power transformer.

REFERENCES

1 Z. Berler, A. Golubev, et al, "Vibra-Acoustic method of transformer clamping pressure monitoring," Conference Record of the 2000 IEEE International Symposium on Electrical Insulation, pp. 263–266, 2000.
2 Ji Shengchang, Liu Weiguo and Shan Ping, et al, "The vibration measuring system for monitoring core and winding condition of power transformer," High Voltage Technology, vol. 6, pp. 1–3, 2000.
3 Xie Poan, "Study on application of vibration analysis to the condition monitoring of power transformers windings," Ph. D, Shanghai Jiao Tong University, 2008.
4 Gu Xiaoan, Shen Miqun, Zhu Zhenjiang et, al, "Test research on vibrations and noise level in transformer core," Transformer, vol. 4, pp. 1–4, 2003.
5 Gu Xiaoan, Shen Rongyin, et, al, "Model of magnetic forces in ferromagnetic materials under magnetization," Journal of Shanghai Jiaotong University, vol. 5, pp. 794–797, 2003.
6 B. Garcia, J.C. Burgos, A and M. Alonso, "Transformers tank vibration modeling as a method of detecting winding deformations-part II: experimental verification," IEEE Transaction on Power Delivery, vol. 21, pp. 164–169, 2006.
7 C. Bartoletti, "Vibrio-acoustic technique to diagnose power transformers," IEEE Trans Power Del, vol. 19, no. 1, pp. 221–229, Jan. 2004.
8 Lu Jian, "The Least Square Method and Its Application," Science and Technology of West China, vol. 19, pp. 19–21, 2007.

Power and Energy – Kong (Ed.)
© *2015 Taylor & Francis Group, London, ISBN 978-1-138-02782-4*

Study on a high efficiency coordinated control strategy for combined PV-Storage microgrid

Feijin Peng, Hongyuan Huang & Xiaoyun Huang
Foshan Power Supply Bureau, Guangdong Power Grid Co. Ltd., Foshan, Guangdong, China

Aidong Xu, Lei Yu, Zhan Shen & Jinyong Lei
Electric Power Research Institute of China Southern Power Grid, Guangzhou, Guangdong, China

ABSTRACT: PV-Storage microgrid is likely to play an important role for improving the reliability and photovoltaic permeability, specifically in remote areas and load dense regions. This paper presents a coordinated power control strategy to switch the operation mode of PV-Storage microgrid smoothly between the emergency operation mode and energy transmission operation mode. The structure of PV-Storage microgrid has a DC-DC converter as an interface between the PV and energy storage system, and the PV-storage unit also can directly connect to the gird through a DC-AC converter. Three output channels are designed for PV generation unit to increase the reliability and efficiency of the PV-Storage microgrid. Finally, the proposed coordinated control strategy is simulated by PSCAD after building the PV-storage microgrid model. The results of simulation verified that the control strategy of PV-storage system is reliable and efficiency.

1 INTRODUCTION

The power supply technologies related with distributed energy resources, such as photovoltaic and wind power, can relieve the energy shortage and environment problems caused by the excessive consumption of traditional fossil fuels. However, the output power of distributed energy resources is random and intermittent. So when these distributed generations are connected to the power grid, they will easily affect the power supply security and energy quality of the power grid (Fakham et al. 2008 & 2011). Under the background of the government promoting the popularization and application of PV energy, the price of PV generation power has fallen dramatically. Combined the solar energy technology with energy storage technology, establishing the PV-Storage microgrid is an important and promising method to improve the ability of the power grid for PV energy integration and the power supply reliability of the heavy load area (Cai et al. 2013). The PV-Storage microgrid has many characteristics, such as improving the power supply reliability, enhancing the ability of disaster resisting and realizing the power supplying in the no-electricity area.

The structure of four typical PV-Storage microgrids is shown in Figure 1 (Aharon et al. 2011). In the structure of Figures 1a, b, the PV unit and battery storage unit can connect to the DC bus

directly. The simple structure is the most obvious advantage while its disadvantage is that the PV unit and battery storage unit cannot be controlled flexibly. Take the structure of Figure 1a as example, the voltage of the DC bus need to be controlled in a relative wide range in the Maximum Power Point Tracking (MPPT) control. Besides, the bidirectional DC-AC inverter need to be equipped with the DC bus voltage control and MPPT control function. To make the PV unit and battery storage unit control more flexible and intelligent, the structure of Figure 1c is adoptable, in which the PV unit and battery storage unit are connect to the DC bus through DC-DC inverter. Generally, when the microgrid is connected to the power grid, the voltage of DC bus controlled by the DC-AC unit is constant to keep the energy balance. The

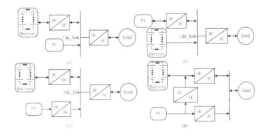

Figure 1. Typical PV-Storage microgrid structure.

battery storage unit and related DC-DC module can be controlled for charging and discharging based on the system optimization objective. In this structure, two DC-DC modules share a common DC bus and are connected to the power grid by a DC-AC bidirectional inverter. Therefore, the high quality on the reliability of the directional inverter is needed for stable operation of the system. Once the inverter or the DC bus fails to work, the PV unit cannot output power to the power grid. To solve this problem, this paper proposes a reasonable PV-Storage microgrid structure (Fig.1d). To improve the system reliability and control flexibility, the PV unit and the battery storage unit are connected with the DC-DC inverter. Compared with the other three topologies, the PV unit could output power through three ways in the proposed structure, which improves the power supply reliability of the PV energy greatly.

To meet the high utilization of the PV-Storage microgrid and grid connection requirement, the coordinated control strategy is the developed in this paper. The influence of a high-penetrated PV system on the safe operation cannot be ignored. Hence, the PV system should be able to ride through the power failures and maintain the grid connection state to supply energy during the failure period. Within the range of the grid connection capacity, it shall support and maintain the voltage of the local grid. Considering the output power of the PV system is random and not able to maintain the transient voltage of the local grid, Kong et al. (2013) proposes a coordinated control strategy to adjust the power efficiently and improve the situation of low voltage ride through. And Pedro et al. (2007) and Wang et al. (2011) propose many control objectives, such as instantaneous active and reactive power control and three-phase current balance control after considering the oscillating component of the instantaneous power and current distortion. The control objective is mainly to eliminate the active AC component of the power output by the inverter and control the voltage fluctuation of the DC bus. Under the premise of guaranteeing the voltage of the DC bus under the upper limit, it could eliminate the negative sequence component of the power output by the inverter.

In the above references, the researches stress on the quality control of the output current of the inverter and seldom involves in the control of system power balance. If the grid side suffers some severe failures or the PV inverter breaks down, the PV system will stop supplying energy to the grid, which greatly reduces the reliability and utilization of the PV system. Therefore this paper analyzes its operation mode and related energy flow firstly. And then the system coordinated control strategy is studied to improve the power supply reliability

and generating efficiency of the PV-Storage microgrid. Finally, the related simulation model is established in the PSCAD to verify the effectiveness of the proposed method.

2 COORDINATE CONTROL STRATEGY FOR PV-STORAGE MICROGRID

The main circuit structure of the PV-Storage microgrid proposed in this paper is shown in Figure 2. The grid-connecting inverters of battery storage unit and PV unit are both three-phase Voltage Source Converter (VSC), which connects to the microgrid AC bus through LCL filter and isolation transformer. DC-DC converter is boost converter, and energy flow direction is from PV side to battery side (Chen 2002, Xu 2005).

2.1 PV-storage microgrid basic control methods

In Figure 2, this simple circuit structure can not only reduce the cost, but also has the advantages of high energy conversion efficiency. The PV grid inverter *dq* decoupling current closed-loop control system is shown in Figure 3 (Zhou et al. 2008),

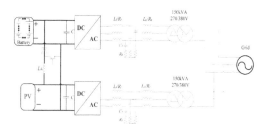

Figure 2. Main circuit diagram of PV-Storage microgrid.

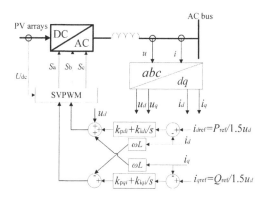

Figure 3. PQ control structure for PV grid connected inverter.

which is based on grid voltage orientation and space vector modulation (SVPWM).

In this control system, d axis of dq synchronous rotating coordinate system is oriented on the grid voltage vector direction using grid voltage as reference for voltage and frequency. Under the condition that the power grid voltage approximate constant, active power and reactive power outputted by inverter are proportional to the d and q axis component of its output current respectively. Therefore, by adjusting the d, q axis component of current i_d, i_q, active power and reactive power outputted by the inverter to the grid can be controlled. As this paper takes coordinated control strategy for PV-Storage microgrid, active power reference of PV grid inverter is not directly obtained by the MPPT control system, and this will be analyzed in the latter part.

When the PV-Storage microgrid connected to the grid, energy storage grid inverter works under constant power control mode, and the basic control structure is similar to Figure 3. When PV-Storage system operated in island mode, energy storage grid inverter needs to adopt the constant voltage/constant frequency control mode to maintain voltage and frequency stable and establish the reference of the voltage and frequency for PV grid inverter. The control system is shown in Figure 4, which adopts the voltage and current double closed-loop control structure to ensure the accuracy of the valid values of output voltage (Li et al. 2014). The voltage ring output result is operated as the current input reference for the current ring. Inverter output frequency (50 Hz) is generated by sine signal of the controller, and this signal is used as orientation reference vector and d axis orientation vector.

The control structure of DC-DC converter connected between PV arrays and energy storage system DC bus is shown in Figure 5. Current ring adopts traditional PI control structure to realize current tracking.

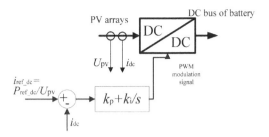

Figure 5. Power control structure for DC-DC converter.

2.2 PV-Storage microgrid coordinated control strategy

For PV-Storage system shown in Figure 2, the energy flow diagram of its grid-connected operation and its island operation is shown in Figure 6. Three output channels are designed for photovoltaic power generation unit, which are shown as follow:

1. PV → DC-AC converter → Grid;
2. PV → DC-DC converter → storage grid inverter → Grid;
3. PV → DC-DC converter →energy storage system;

When PV-Storage microgrid is operated in grid-connected and island mode, photovoltaic power generation unit directly accesses to AC bus through PV grid inverter, and PV grid inverter applies maximum power output control at this time.

When a failure occurs in the PV grid inverter, photovoltaic power generation unit output power could be directly injected into energy storage system through DC-DC converter, or output to grid through grid inverter. If failure occurs in AC side, which results the PV and energy storage grid inverter is in the current limit or stop state, PV cell can charge energy storage system through DC-DC converter till the energy storage is full. In order to ensure the stable operation of PV-Storage microgrid and switch smoothly between system operation mode and power transmission, the proposed coordinated control strategy is shown in Figure 6.

In Figure 6, the maximum output power reference P_{ref_PV} of the PV system can be obtained through MPPT control for output current I_{PV} and voltage U_{dc} of PV array. There are many MPPT control algorithm, such as disturbance observation method, which constantly adjusts the output voltage of PV arrays with a constant step and observes its output power at the same time. When the output power increases it may maintain the original direction to continue the search, and when output power decreases it may searches in the opposite

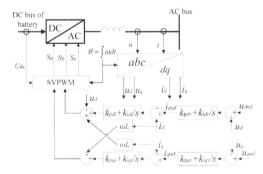

Figure 4. V/F control structure for battery grid connected inverter.

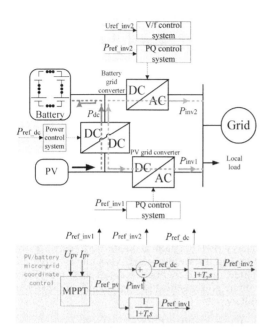

Figure 6. Energy flow diagram for the PV-Storage microgrid.

direction, finally it searches the best working voltage of photovoltaic array (Dend 2007).

As shown in Figure 6, the output power of PV grid inverter reference P_{ref_inv1} is written as:

$$P_{ref\_inv1} = \frac{1}{1 + T_1 s} P_{ref\_pv}$$ (1)

where T_1 is time constant of filter wave, in this paper its value is 1 ms.

As the reference of output power of PV grid inverter, Equation (1) could be used to smooth the fluctuation of photovoltaic array output power and reduce the impact on the local power grid. The power reference P_{ref_dc} of DC-DC converter, which connects PV array and energy storage system, is written as:

$$P_{ref\_dc} = P_{ref\_pv} - P_{inv1}$$ (2)

where P_{inv1} is actual output active power of PV grid converter.

Applying Equation (2), on the one hand it can undertake the fluctuation component of PV array output power through DC-DC converter when PV-Storage microgrid operates in normal condition; on the other hand, when failure occurs in the grid or PV grid inverter that makes the PV power cannot be sent out, the PV array output power

can be injected to the battery energy storage system through the DC-DC converter or sent to grid through energy grid inverter.

When PV-Storage microgrid operates in grid-connected mode, output reference P_{ref_inv2} of energy storage grid inverter is written as:

$$P_{ref\_inv2} = \frac{1}{1 + T_2 s} P_{ref\_dc}$$ (3)

where T_2 is time constant of filter wave, in this paper its value is 1 ms. Equation (3) can also smooth output power.

When the PV-Storage microgrid operates in island mode, energy storage grid inverter should adopt constant voltage/constant frequency (V/F) control mode, set up AC bus voltage and frequency, the control mode as shown in Figure 4.

3 SIMULATION AND RESULTS ANALYSIS

To verify the validity of the proposed control strategy, the simulation is made in PSCAD/EMTDC to verify the case system shown in Figure 2. The parameters of the system are shown in Table 1.

The rated output power of PV is 100 kW while the PV-Storage microgrid is operated in grid-connected state. Assume that there is a sudden failure of PV grid-connected converter at the time 0.5 s, the related power curve and current curve of each converter in the power system are shown in Figure 7 and Figure 8, respectively.

Table 1. PV-Storage microgrid system parameters.

Parameter name	Parameter value
The inverter side L_f/R_f	0.4 mH/0.04 Ω
The grid side Lg/Rg	0.2 mH/0.01 Ω
Filter capacitor and damping resistance	100 μF/0.2 Ω
Switching frequency	4 kHz
DC side voltage	600 V
Rated capacity	100 kVA
Filter inductance (L_{dc})	2 mH
Switching frequency of DC-DC converter	4 kHz
Rated capacity of DC-DC converter	100 kVA
Parameters of power control system (k_p/k_i)	1/200
Terminal voltage of battery	600 V
Rated capacity of battery	200 Ah
Operating voltage of MPPT for PV	450–550 V
Rated capacity of PV	100 kW
Line voltage and frequency of microgrid	380 V/50 Hz

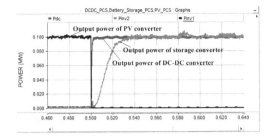

Figure 7. Power output of each converter.

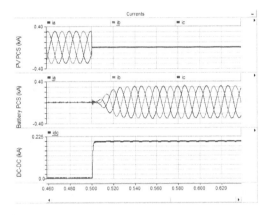

Figure 8. Current output of each converter.

It can be seen from Figure 6 and Equation (2), the reference value of the DC-DC converter output power is derived from PV array MPPT reference value minus the actual output power of PV converter. As can be seen in Figure 7 and Figure 8, the output power of PV array was transferred to DC-DC channel rapidly when PV converter stops outputting power due to the power failure. From Equation (3), it can be known that, compared to the reference value of DC-DC converter, the reference value of battery storage power delays with a constant filter time. In this paper, the value of a is 1 ms. It can be seen from Figure 8, the output power of battery storage converter delay with 20 ms compared to the output of DC-DC converter, which is decided by the dynamic response of current controller in battery storage converter. To adjust the dynamic response, parameters of current controlling can be changed on the condition that the stability is guaranteed (Wang et al. 2012, Gou et al. 2012).

4 CONCLUSIONS

The PV-Storage microgrid can greatly improve the photovoltaic permeability and enhance the power distribution reliability in the high load dense regions. According to the PV-Storage microgrid structure, this paper studied the corresponding power coordinated control strategy. Theory analysis and simulation results show that this method can effectively switch the operation mode of PV-Storage microgrid smoothly between the emergency operation mode and energy transmission operation mode. The proposed method can greatly increase the power distribution reliability and the generating efficiency of PV-Storage microgrid.

ACKNOWLEDGMENT

This paper is supported by Science and Technology Project of Guangdong Power Grid Co. Ltd (K-GD2013–044) and the National High Technology Research and Development Program of China (863 Program) (2011 AA05114).

REFERENCES

1. C. Wang, X. Li, L. Guo, "Coordinated Control of Two-stage Power Converters Based on Power Balancing and Time-delay Compensation," Proceedings of the CSEE. Vol 32, No. 25, 2012, pp. 109–117.
2. D. Xu, "Power electronics system modelling and control. China Machine Press," 2005.
3. D. zhou, Z. Zhao, L. Yuan, B. Feng, "Optimum Control and Stability Analysis for a 300 kW Photovoltaic Grid-Connected System," Transactions of China Electrotechnical Society, Vol 23, No. 11, 2008, pp. 116–120.
4. F.L. Dend, "A battery charging controller based on the self-adapted hill climbing method," Solar Energy. Vol 6, No. 2, 2007, pp. 22–23.
5. F. Wang, L.D. Jorge, et al. "Pliant Active and Reactive Power Control for Grid-Interactive Converters Under Unbalanced Voltage Dips," IEEE Transactions on Power Electronics, Vol 26, No. 5, 2011, pp. 1511–1521.
6. G. Cai, L. Kong, et al. "System Modeling of Wind-PV-ES Hybrid Power System and Its Control Strategy for Grid-Connected," Transactions of China Electrotechnical Society, Vol 28, No. 9, 2013, pp. 196–204.
7. H.L. Chan, D. Sutanto, "A new battery model for use with battery energy storage systems and electric vehicles power systems," Power engineering society winter meeting, Vol 1, No. 7, 2000, pp. 470–475.
8. H. Kakigano, Y. Miura, T. Ise, et al. "A DC microgrid for super high-quality electric power distribution," Electrical Engineering in Japan, Vol 164, No. 1, 2008, pp. 34–42.
9. H. Fakham, D Lu, B Francois, "Power control design of a battery charger in a hybrid active PV generator for load-following applications," IEEE Transactions on Industrial Electronics, Vol 58, No. 1, 2011, pp.85–94.

10. I.Aharon, A. Kuperman, "Topological overview of powertrains for battery-powered vehicles with range extenders," IEEE Transactions on Power Electronics, Vol 26, No. 3, 2011, pp. 868–876.

11. J. Chen, "Power electronics conversion and control technology," Higher Education Press. 2002.

12. L. Guo, X. Li, C. Wang, "Coordinated Control of Hybrid Power Supply Systems Considering Nonlinear Factors," Proceedings of the CSEE. Vol 32, No. 25, 2012, pp. 60–69.

13. L. Kong, G. Cai, et al. "Modeling and Coordinated Control of Grid-Connected PV Generation System With Energy Storage Devices," Power System Technology, Vol 28, No. 9, 2013, pp. 196–204.

14. Tremblay, L.A. Dessaint, A.I. Dekkiche, "A generic battery model for the dynamic simulation of hybrid electric vehicles," Vehicle power and propulsion conference, 2007, pp. 284–289.

15. R. Pedro, V. Adrian et al. "Flexible Active Power Control of Distributed Power Generation Systems During Grid Faults," IEEE Transactions on Industrial Electronics, Vol 54, No. 5, 2007, pp. 2583–2591.

16. S.K. Kim, J.H. Jeon, C.H. Cho et al. "Modeling and simulation of a grid-connected PV generation system for electromagnetic transient analysis," Solar Energy, Vol 83, No. 5, 2009, pp. 664–678.

17. X. Li, L. Guo, C. Wang, "Stability Analysis in a Master-Slave Control Based Microgrid," Transactions of China Electrotechnical Society, Vol 29, No. 2, 2014, pp. 24–34.

18. Z.M. Salameh, M.A. Casacca, W.A. Lynch, "A mathematical model for lead-acid batteries," IEEE transactions on energy conversions, Vol 7, No. 1, 1992, pp. 93–97.

Power and Energy – Kong (Ed.)
© *2015 Taylor & Francis Group, London, ISBN 978-1-138-02782-4*

Research on the sound source model of UHV shunt reactor

Yuan Ni, Bin Zhou, Chun-ming Pei & Zhen-huan Liu
China Electric Power Research Institute, Wuhan City, Hubei Province, China

ABSTRACT: The key to accurately predicting the sound field distribution is to carefully analyze the sound sources of UHV shunt reactor. Through measuring the sound fields of main body of 1,000 kV UHV shunt reactor and its surroundings, the sound power level of shunt reactor is determined in this paper, which is mainly at 100 Hz. In addition, the software SoundPLAN is used to calculate several sound source models. The contrastive analysis of theoretical and actual measured data shows that the main factor affecting the reactor noise calculation is the sound source parameters. It is suggested to use the theory of acoustic interference to analyze the sound field in the near-field region of reactors.

1 INTRODUCTION

In order to meet demands for long-distance and high-capacity power transmission, China has built up the 1,000 kV UHV AC pilot demonstration project and in recent years, China has also put into operation the Huainan-Nanjing-Shanghai UHV double-circuit AC transmission line on the same tower. With the rapid development of power grid construction, domestic resources have gradually been fully deployed and used, but at the same time the projects are running, the problem of the electromagnetic environment has become more prominent, especially the noise of the substation, which has become the focus of public concerns.

At present, a number of enterprises use Cadna/A, SoundPlan and other noise prediction software to carry out the assessment and management study of the substation noise[1], but because there isn't uniform understanding of the acoustic characteristics and the model of the noise source intensity in the substation, the prediction results by software have certain deviations[2, 3].

Based on such situations, this paper combined with the measured noise data of Jingmen 1,000 kV UHV substation, analyzed the sound power level and the surrounding sound field of 1,000 kV shunt reactor, and used software SoundPLAN to calculate different sound source models of shunt reactor. It compared the theoretically calculated sound field with the measured value, analyzed the reasons for several models' calculation error, and obtained the optimum acoustic model of the reactor. The conclusion of this paper is a practical guidance for the establishment of the noise model of the UHV power transmission project substation as well as for noise control of shunt reactors.

2 NOISE TEST OF SHUNT REACTOR

The shunt reactor of Jingmen 1,000 kV UHV substation is located in the northeast corner of the substation, 35 m away from the north wall. The main instruments used in the measurement are AWA6291 noise statistics real-time signal analyzer and LDS Photon II portable four-channel noise tester.

The measuring points are arranged in two groups[5]. One group is arranged along the central axes at the north side of phases A, B and C of the UHV shunt reactor and parallel to the firewall direction, where phases A and C axial measuring points are at a distance of 0.5–25 m from the HV shunt reactor. Within the distance of 5 m from the HV shunt reactor, a measuring point is arranged every 0.5 m; and if beyond 5 m, a measuring point is arranged every 1 m. Phase B axial measuring points are arranged at a distance of 0.5–34 m from the HV shunt reactor, the arrangement interval between measuring points is the same as phases A and C, and the measuring point is 1.5 m away from the ground, specifically shown in Figure 1. The other group is arranged around phase B of HV shunt reactor, 1 m away from the case of the reactor; there are 58 measuring points, the interval between two adjacent measuring points is 0.5 m, and the measuring point is 1.5 m above the ground. See Figure 2 for their specific locations.

In the first group, the measuring points are all axial points of the HV shunt reactor parallel to the firewall, which are respectively A-01#–A-30#, B-01#–B-39#, and C-01#–C-30#. The sound level meter is used to measure 1/3 octave spectra and LAeq of the noise at each measuring point and the duration is 5 s. While LDS is used for the sampling of the noise of the measuring points, and the sampling time is about 10 s. In the second group,

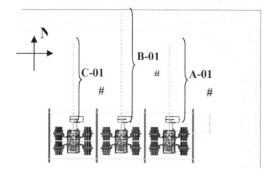

Figure 1. Schematic diagram of axial measuring points from three-phase shunt reactor.

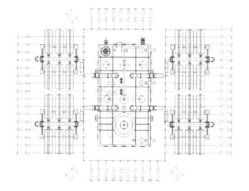

Figure 2. Schematic diagram of measuring points around phase B.

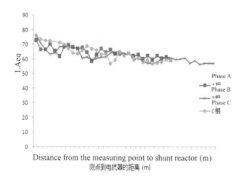

Figure 3. Data of measuring points in the first group arrangement.

the measuring points around the case of the HV shunt reactor of phase B are respectively RN-01#–RN-10#, RW-01#–RW-18#, RS-01#–RS-12#, and RE-01#–RE-18#, and the LDS Photon II portable four-channel noise tester is used to collect samples of noises at different measuring points[4].

Figure 3 shows LAeq of the measuring points with the arrangement of the first group. The analysis shows that the distribution of axial sound fields of the HV shunt reactor parallel to the fire-wall, despite some degree of attenuation, attenuate gradually with several maxima and minima in certain interval in the measurement path, which shows that the sound field in the vicinity of the shunt reactor has superimposition to some extent.

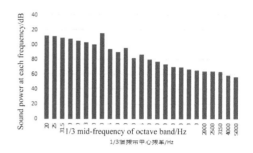

Figure 4. Sound power of 1/3 octave band about phase B of HV reactor.

Table 1. Predictive parameters of reactor noise.

Frequency (Hz)	Area sound noise	Model of volume sound noise		
		Side (7 × 4.5)	Side (9 × 4.5)	Top (9 × 7)
20	61.9	53.7	54.8	56.7
25	66.9	58.7	59.8	61.7
31.5	70.5	62.4	63.4	65.4
40	73.6	65.4	66.5	68.4
50	75.1	66.9	68.0	70.0
63	76.8	68.7	69.7	71.7
80	78.1	69.9	71.0	72.9
100	96.4	88.3	89.4	91.3
125	78.0	69.8	70.9	72.8
160	76.5	68.3	69.4	71.3
200	84.9	76.7	77.8	79.7
250	73.1	64.9	66.0	67.9
315	80.5	72.3	73.4	75.3
400	75.1	66.9	68.0	69.9
500	74.1	65.9	67.0	68.9
630	71.9	63.7	64.8	66.7
800	69.7	61.6	62.7	64.6
1000	69.4	61.3	62.3	64.3
1250	67.6	59.5	60.6	62.5
1600	66.7	58.5	59.6	61.5
2000	65.3	57.1	58.2	60.1
2500	65.4	57.2	58.3	60.2
3150	64.4	56.2	57.3	59.3
4000	59.7	51.5	52.6	54.5
5000	57.2	49.0	50.1	52.1
Total	96.8	88.6	89.7	91.6

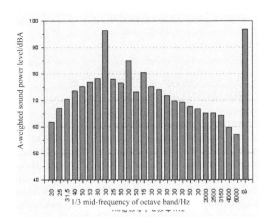

Figure 5. A-weighted sound power level of 1/3 octave band about B phase of HV reactor.

Table 2. Acoustical absorption coefficient of each side at different frequencies.

Reflecting surface	Mid-frequency of octave band (Hz)					
	125	250	500	1000	2000	4000
Ground	0.01	0.01	0.02	0.02	0.02	0.04
Firewall	0.03	0.03	0.04	0.05	0.06	0.06
Enclosure	0.02	0.02	0.02	0.03	0.03	0.04

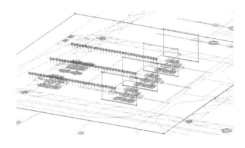

Figure 6. Model of cross-section sound noise.

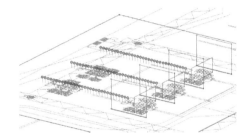

Figure 7. Model of vertical plane sound noise.

In the arrangement of the measuring points in the second group, the measurement is done around phase B of reactor to obtain L100 Hz and LAeq at each measuring point. According to conversion of Sound Pressure Level (SPL) and sound power level[6, 7], Figure 4 shows the sound power at each frequency of phase B calculated on the basis of Table 1. Figure 4 analysis shows that in the entire 1/3 octave band, the low-frequency component accounts for a larger proportion, and the sound power at 100 Hz is maximum. Taking into account the auditory perception, A-weighted conversion is made for Figure 4[8], with the results shown in Figure 5. Figure 5 is more intuitive to draw the sound power level at 100 Hz is 96.4 dBA, 11.5 dBA higher than that at 200 Hz. The A-weighted sound power of 1/3 octave band where 100 Hz is accounts for 91.2% of that of single-phase HV shunt reactor.

3 SOUND SOURCE MODEL CALCULATION OF SHUNT REACTOR

Through the measurement of sound fields around the HV shunt reactor in Jingmen UHV substation, the volume size of the reactor is determined as 9 m × 7 m × 4.5 m (L × W × H). Combined with measurement data analysis, the sound power level of the reactor in each phase at 1/3 frequency is calculated. Table 1 shows the power level on each frequency component as the reactor is regarded as an area sound noise and a volume sound source.

In the modeling process, the acoustical absorption coefficient of the obstacle around the reactor or the reflecting surface in the spreading process[9] should be set, and with review of relevant references, the specific absorption coefficient of each frequency is as shown in Table 2.

The following analysis takes Jingmen UHV substation shunt reactor as example, uses software SoundPLAN to set reactors as area sound noise and volume sound noise[10], and calculates the distribution of sound fields parallel to each phase of reactor and on the central axis in the firewall direction.
1. Model of area sound noise
   As shunt reactor is simulated as the area sound noise, it can be set as a cross-section sound noise[11, 12] and a vertical plane sound noise, both of which are on the central axis of shunt reactor. The specific model is as shown in Figure 6 and Figure 7.
   Figure 8 and Figure 9 show the distribution of sound fields around shunt reactor calculated in line with Figure 6 and Figure 7.
2. Model of volume sound noise
   As shunt reactor is simulated as the volume sound noise, it can be set as five planes, four sides and a top, and the specific model is as shown in Figure 10.

421

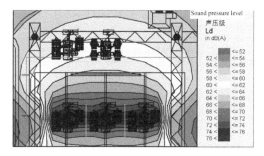

Figure 8. Calculation result of model of cross-section sound noise.

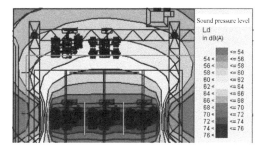

Figure 9. Calculation result of sound field of model of vertical plane sound noise.

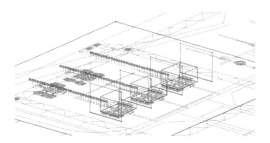

Figure 10. Model of volume sound noise.

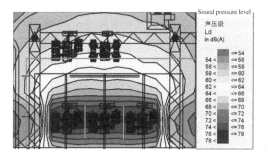

Figure 11. Calculation result of sound field of model of vertical plane sound noise.

Figure 1 shows the distribution of sound fields around shunt reactor calculated in line with Figure 11.

4 ANALYSIS OF SOUND FIELDS OF SHUNT REACTOR

From the calculation results of the noise attenuation of the above models, the reactor's noise attenuates with the increasing distance, and the attenuation trend of three phases is consistent. The calculated value of the noise that has the same distance from the vertical plane of reactor has little difference. Meanwhile, the distribution of the calculated result of the sound field shows that the reactor noise mainly spreads along the open side without the firewall, and the greater the distance from the reactor, the smaller the noise. Outside the firewall on both ends of phase A and phase C, the noise value is low, almost less than 60 dB(A).

Combined with the measured value of the noise, the calculated value of the noise of each model is compared as shown in Table 3[13].

The comparative analysis of the three models in Table 3 shows that if the sound source is an area model, within the close distance from the reactor, there is some difference between noise values of cross-section and vertical plane models, while the difference is little for the distance of 5 m or above, almost less than 1.0 dB(A).

The calculated result of the volume sound noise has no significant difference from the cross-section sound noise. The volume sound noise is modeled with five area sound noises, and the area sound noise is modeled with single area sound noise. The calculated results have little difference[14, 15], indicating that the number of area sound noises has little effect on the calculated results.

The comparison of the calculated result and the measured value of the above three models shows that within 10 m, the difference is large, even over 10 dB(A). But with the increase of the distance, the difference between calculated and measured values gets smaller and smaller. The difference over 20 m is little, generally no more than 2 dB(A). So the sound noise model built with software SoundPLAN has some difference in the calculation of the noise within the short distance from reactors (20 m), while the calculation of the noise with a long distance from reactors (over 20 m) is accurate.

In summary, the main factor affecting the noise calculation is sound noise parameters; the area, the layout form and the number of the sound noise model have less impact on the calculated result of the noise. It is feasible to use noise prediction software SoundPLAN to build the area sound noise model to calculate the noise attenuation in the far field region

Table 3. Comparison of calculated results of noises of models.

Measuring point	Cross-section model	Vertical plane model	Volume sound noise model	Measured value
1	74.2	70.6	75.6	/
2	72.3	69.7	73.4	**71.5**
3	70.8	68.9	71.9	/
4	70.8	68.9	71.9	**60.6**
5	69.7	68.4	70.7	/
6	68.8	68.1	69.8	**65.6**
7	68.1	67.5	69.0	/
8	67.4	67.0	68.3	**59.9**
9	66.9	66.5	67.6	/
10	66.4	66.1	67.0	**63.7**
11	65.9	65.6	66.5	/
12	65.5	65.2	66.0	**63.0**
13	65.0	64.8	65.5	/
14	64.6	64.5	65.0	**65.3**
15	64.2	64.1	64.5	/
16	63.8	63.7	64.0	**65.6**
17	63.4	63.1	63.6	/
18	63.0	62.7	63.2	**62.3**
19	62.7	62.3	62.8	/
20	62.3	62	62.5	**63.0**
21	62.0	61.7	62.2	/
22	61.6	61.4	61.9	**58.8**
23	61.3	61.1	61.6	/
24	61.0	60.8	61.3	**61.3**
25	60.7	60.5	61.1	/
26	60.4	60.2	60.8	**60.4**
27	60.1	59.9	60.5	/
28	59.8	59.6	60.2	**59.4**
29	59.5	59.4	60.0	/
30	59.3	59.1	59.7	**58.7**

of reactors. For the near-field region, the interference of 100 Hz sound wave needs to be considered. It is suggested to use the sound field theory to derive the acoustic interference in the near-field region.

5 CONCLUSIONS

1. The measurement of the sound field around shunt reactors shows that the sound power level at 100 Hz is a major component of total sound power level.
2. The calculated result of the models of the volume sound noise and the area sound noise indicates that the difference between the two sound noise models is small. The main factor affecting the noise calculation is sound noise parameters; the area, the layout form and the number of the sound noise model have less impact on the calculated result of the noise.

3. The analysis of the measured data shows that maxima and minima appear in the near-field region of shunt reactor and the calculated result of software SoundPLAN takes on linear attenuation. It is suggested to use the theory of acoustic interference to calculate the sound field in the near-field region.

REFERENCES

1. China Electric Power Research Institute, Zhejiang University. Research on acoustic interference characteristics in the near-field region of 1,000 kV UHV shunt reactor [R]. Wuhan, 2013. China Electric Power Research Institute.
2. Central Southern China Electric Power Design Institute, Wuhan High Voltage Research Institute of SGCC. Special study on analysis and preventive measures of noise sources in UHV AC substation [R]. Wuhan. Central Southern China Electric Power Design Institute.
3. David Snell. Measurement of noise associated with model transformer cores [J]. Journal of Magnetism and Magnetic Materials, 320 (2008):535–538.
4. Du Gonghuan, Zhu Zhemin, Gong Xiufen. Fundamentals of Acoustics [M]. Nanjing: Nanjing University Press. 2003.
5. GB 7328-1987 Determination of transformer and reactor sound levels [S]. 1987.
6. GB/T 1094.10-2003 Power transformers—Part 10: determination of sound levels [S].
7. Li Xue-liang, Xu Zhen, Zhou Ying, Wang Fei, Zhu Geng-fu, Zhao Gang. The simulation of acoustic environment impact in 1,000 kV extra-high voltage substation [J]. Journal of Environmental Engineering Technology. 2012, 2(3):264–270.
8. Mao Dong-xing, Hong Zong-hui. Environmental noise control engineering [M]. Higher Education Press. 2010.
9. Ni Yuan, Zhang Guang-zhou, Zhang Xiao-wu, et al. Assessment and treatment of noise in UHV AC test base [J], High Voltage Engineering, 2009, 35(8):1856–1861.
10. Ni Xue-feng, Lin Hao, Yan Fei, Jiang Sheng-bao. Experimental research noise characteristics of filter capacitors in UHVDC converter station [J]. High Voltage Engineering. 2010, 36(1):160–166.
11. Popeck R.A, Knapp R.F. Measurement and analysis of audible noise [J]. IEEE Transaction on Power Apparatus and Systems, 1981, 100(4):1440–1452.
12. Ruan Xue-yun, Li Zhi-yuan, Wei Hao-zheng, et al. Studies on noise prediction model and simplification for current convert transformers [J]. Applied Acoustics. 2011, 30(3):235–240.
13. Sanjay Patil, George G. Karady and Wesley Knuth. Effect of load-generated transformer noise in a substation [J]. Euro. Trans. Electr. Power, 2011, 21:596–607.
14. Zhou Bing, Pei Chun-ming, Ni Yuan, Zhang Jian-gong. Noise measurement and analysis of UHVAC substation [J]. High Voltage Engineering. 2013, 39(6):1447–1453.

Power and Energy – Kong (Ed.)
© 2015 Taylor & Francis Group, London, ISBN 978-1-138-02782-4

Development of a leakage current on-line monitoring unit for arresters

Chien-Nan Chen, Hung-Chang Hsu & Ming-Yuan Cho
Department of Electrical Engineering, National Kaohsiung University of Applied Sciences, Kaohsiung, Taiwan

ABSTRACT: Surge arresters are important pieces of over-voltage protection equipment in substations. For a Metal Oxide Surge Arrester (MOA), there will be a leakage current through the arrester. Arrester status can be estimated with such a leakage current. Leakage currents and weather conditions are closely related. The first purpose of this project was to establish a remote Supervisory Control and Data Acquisition (SCADA) system which integrates weather and arrester leakage current data. The second purpose was to perform real-time monitoring of a critical connecting station. The third was to analyze the relationship between weather parameters and arrester leakage current. The fourth was to propose a rational maintenance procedure for the arrester of a connecting station. This project collected information to test and investigate different types of arrester maintenance conducted by the world's advanced countries, and Taipower's current arrester maintenance procedure (outage and non-outage) was compared with those from overseas to propose effective arrester testing methods, maintaining cycle for charged department. Furthermore, this project sought to compare the world's arrester monitoring methods to propose an on line monitoring system for Taipower. The proposed system is integrated with 3.5G wireless communication technique to develop a real-time monitoring system to detect various types of leakage current of the third order harmonic current, resistive leakage current, and total leakage current, lightning count, temperature and humidity and to automatically send back related data via a GPRS module. In addition, an artificial intelligent technique was applied to analyze the collected data, and an alarm signal was enabled to notify engineers when the collected data violate the limit. Finally, these proposed and developed systems were installed at connecting stations and extra high voltage substations for field testing. The results show that the practicality and effectiveness of the proposed system can be justified.

1 INTRODUCTION

A surge arrester is one of the important pieces of equipment for ensuring the safe operation of a power system. Traditional maintenance requires regular testing conducted by personnel on the scene to view and record current arrester leakage current states. This method takes up a lot of manpower and resources. So, alternative methods are needed to save such manpower and material resources.

This research project aimed to develop online real-time monitoring system in order to collect the electrical characteristics of arresters and analyze their current states. Such a system could then be used to get the performance conditions of an arrester at any time.

The electrical characteristics in question include the following: power loss P, power factor p.f., voltage distribution, resistive leakage current I_R, and third harmonic currents. From the leakage current, insulation resistance and other characteristics of the material can be determined, and it is also possible to analyze resistant polymeric and ceramic arrester ability with regard to pollution.

2 SYSTEM ARCHITECTURE

The proposed system is comprised of three parts. One is the power supply system unit. One is the Front-End unit built-in microprocessor chip with a high order that serves as the computing core for data processing and sending. The other is the Back-End unit for data receiving and analysis.

The data processing unit may collect data that includes meteorological data, such as temperature and humidity parameters, and data on the arrester leakage current, including surface leakage current, surge arresters, and resistive leakage current.

The back-end data integration unit is integrated with GPRS (3.5G) wireless communications as well as the amount of the data measurement function. The embedded system links all the functions to operate above. The embedded system itself has a variety of RS-485 communications built in, and other digital input/output interfaces.

In this system, RS-485 communications are used to connect the meteorological parameters subsystem and the arrester leakage current subsystem, and the common industrial automation Modbus protocol is used to transmit information.

The system captures collected data into the unit, and then converts the analog signal into digital signals through the internal A/DC operation. The unit then sends the data to a GPRS wireless communication unit by RS-232 communications.

After receiving complete information on GPRS wireless communications, the unit will send data back to the control terminal at a fixed time through the GPRS network. The monitoring client unit receives data from a base station through an ADSL line telecommunications network. Finally, the received information is transmitted to the SQL Server Database Monitoring end for the data to be saved.

We developed man-machine interface software specially to monitor and operate the system. When the user wants to view the real-time status of personnel, the human-machine interface can be displayed from the SQL Server Database.

2.1 Arrester leakage current monitoring subsystem

The leakage current monitoring subsystem is composed of three parts, namely, the leakage current meter unit, a data processing unit and a data sending and receiving unit.

Its data processing unit includes an AFE (Analog Front-end) circuit, a comparator circuit (Comparator) and a Microprocessor Control Unit (MCU).

The leakage current signal is transmitted through the AFE circuit MCU. The signal of the arrester surface leakage current through the comparator circuit will be transmitted to the MCU.

The data is sent internally using the RS-485 protocol. The data is sent to the back-end using the GPRS module. Likewise, the transmission is sent through the RS-485 communication interface to the embedded system and then spread through the UART port GPRS module for wireless transmission.

2.2 Monitoring principle analysis

In order to analyze the fundamental components and third harmonic arrester components, the leakage current using the microprocessor to retrieve and make the Fast Fourier Transform (FFT) in order to obtain the third harmonic component of the leakage current was analyzed, but the system voltage harmonics generated by the capacitive were similar to the third harmonic components of the arrester ingredients, so an electric field sensor system was used to remove the harmonic components of the signal voltage and deducted from the total amount of current harmonics, thereby allowing on to obtain the arrester third harmonic currents.

In addition to the analysis of the third harmonic, an electric field sensor was used to capture the system voltage and the total leakage current of the phase detector.

2.2.1 Phase extraction method

The leakage current and the electric field induced voltage are converted to the square wave signals by a square-wave signal circuit. Using the microprocessor detects the rising edge of its square wave signals trigger point in order to calculate a time difference Δt between the rising edge, and then convert that value to phase angle α. The phase between voltage and current is thus calculated.

The hardware used for phase extraction is shown in Figure 1a and 1b.

2.3 Resistive leakage current calculation method

The phase α difference between the electric voltage signal and the total leakage current can be calculated by a microprocessor.

The resistive leakage currents may be calculated by the equation $I_r = I_t \times \cos\alpha$. The relationship between I_r and I_t is shown in Figure 2.

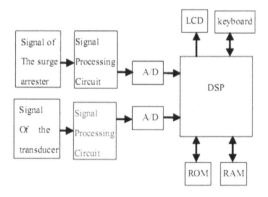

Figure 1a. Hardware of phase extraction.

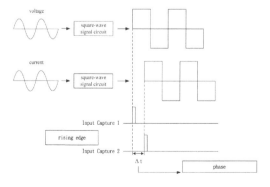

Figure 1b. Hardware of phase extraction.

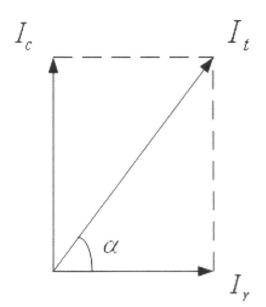

Figure 2. The relationship between I_r and I_t.

3 HARDWARE DEVELOPMENT

3.1 *ARM embedded system*

The basic functions mainly include digital signal processing, SD memory card access, USB interface identification, Shell command execution, communication, and intrusion signal detection, among other functions.

This embedded system operates in the Linux OS. The system is shown in Figure 3. It also possesses Internet communication abilities, surrounds the USB, includes an SD memory card, and can perform HSDPA wireless communications. The hardware circuit design software was made using C language written on a PC platform using a GCC (GNU Compiler Collection) compiler. That can be executed in ARM machine code.

3.1.1 *GSM/GPRS (3.5G) communication module*

GSM mobile phone functionality with the same general, GPRS is a way of transmitting data packets, billing, while the communications module provides a set of RS-232 serial interfaces for connection lines.

The GSM SMS or GPRS packet can communicate with or connect to a higher rate of 3.5G, with GSM SMS using the GSM communication network primarily in text messages transmitted between the mobile stations.

Figure 3. ARM9 embedded system at Linux OS.

Figure 4. LCM II system made by Norwegian TransiNor.

3.2 *Voltage probe*

The voltage probe is one of key parts of the system. Usually, there is no PT next to a surge arrester at a connect station or substation. It is not easy to get arrester voltage signals at same time. So, this system uses the LCM II system made by Norwegian TransiNor as a reference voltage sampling signal method. The LCM II is shown in Figure 4.

3.3 *Current transformer*

This current transformer involves the use of electromagnetic induction on the leakage current. In order to collect the total current leakage current, the current transformer responsible must be applied to measure small current (μA) requirements.

In order to obtain a higher sampling accuracy than the stream's current structure, the parameters and anti-jamming capabilities and other aspects of selection and design requirements are set relatively high. The performance of this current transformer directly affects the accuracy and reliability

Figure 5.   Current transformer.

Figure 6.   DSPIC33FJ128MC710 made by Microchip.

of the monitoring. The transformer is shown in Figure 5.

3.4   Microprocessor circuit

The microprocessor uses a DSPIC33FJ128MC710 made by Microchip. It is high-performance DSC CPU, 16-bit DSP and 40 MIPS high-speed computing microprocessor. There are 2 ADC 10-bit@1.1 Msps or 12-bit@500 Ksps conversions in main function of the internal. The sample point of a signal cycle is 512 at 60 Hz. It is very suitable as a front-end signal acquisition conversion use. The microprocessor is shown in Figure 6.

4   SYSTEM INSTALL AND TEST

The hardware circuit passed the 10kA impulse current level of tolerance repeatedly in the laboratory tests conducted during the development of this system. It have be encountered that isolating protective devices is normally working. Each phase data acquisition system is equipped in a cabinet. The leakage current data and other information are transmitted to the communication circuit

through the RS485. Meteorological parameters of temperature and humidity and arrester leakage current data are also collected.

These systems were installed in a connecting station at No.2 #19 tower Hsing-Ing Shya-Ing, Taipower and in an extra high voltage substation in Long-Chi, Taipower. The data was then collected for more than one and a half years.

4.1   Install in connect station

The system is currently installed in the No.2 #19 Hsing-Ing Shya-Ing, Taipower connect station. The field probe is shown in Figure 7. A phase data acquisition system in a cabinet is shown in Figure 8. A sensor for the meteorological parameters of temperature and humidity is shown in Figure 9.

4.2   EHV Substation #5 ATR

The system installed at an EHV subsystem is shown in Figure 10.

4.3   User interface

The function list for the user interface is shown in Figure 11. The real-time online information screen

Figure 7.   Field probe.

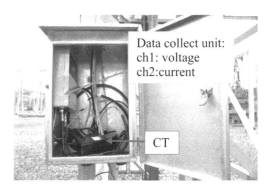

Figure 8.   A phase data acquisition system in a cabinet.

Figure 9. Meteorological parameters of temperature and humidity sensor.

Figure 10. The system installed at an EHV subsystem.

Figure 11. Function list.

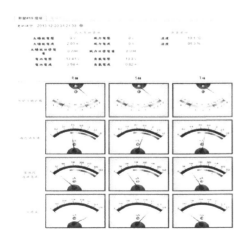

Figure 12. Real-time online information screen.

Figure 13. R-phase third harmonic leakage current.

is shown in Figure 12. The R-phase third harmonic leakage current data for August, 2013, is shown in Figure 13.

5 CONCLUSIONS

Online monitoring with the GPRS communication function can collect data on an arrester's total leakage current, resistive leakage current, third harmonic currents, surface contamination level and environmental temperature, humidity and other diverse parameters. Such monitoring thus provides important diagnosis reference information about potential deterioration in an arrester.

At any instant, the remote access operation status of the arrester can be checked with its portability. Maintenance officers can use laptops or mobile phones for online monitoring, and the latent faults of MOA can be found by online resistive leakage current monitoring.

This research developed a "lightning arrester online monitoring and maintenance system" (later referred to as this system) to achieve concrete realization of the possibility of real-time, online performance monitoring of arresters.

REFERENCES

1. H. Kado, K. Izumi, S. Shirakawa, K. Komatsu, H. Watanabe, M. Yamaguchi, M. Nakajima, M. Kobayashi, S. Nishimura, "Artificial Pollution Tests on Porcelain-housed Metal-Oxide Surge Arresters for 275kV Power Systems," IEEE Power Engineering Society Winter Meeting, Vol. 3, 2000, pp. 2087–2092.
2. Juyong Kim, Park Chulbae, Jung Yeunha, Ilkeun Song, "An investigation of aging characteristics of polymer housed distribution surge arresters by accelerated aging test," Annual Report Conference on Electrical Insulation and Dielectric Phenomena, 2008, pp. 280–283.
3. B. Richter, W. Schmidt, K. Kannus, K. Lahti, V. Hinrichsen, C. Neuman, W. Petrusch, K. Steinfeld, "Long term performance of polymer-housed MO-surge arresters", CIGRE Session 2004, pp. A3–110.

4. IEC 60099-5, Surge Arresters—Part 5: Selection and Application Recommendations, March 2000.
5. D.W. Lenk, "An Examination of the Pollution Performance of Gapped and Gapless Metal Oxide Station Class Surge Arresters," IEEE Trans. Power Apparatus and Systems, Vol. PAS-103, No. 2, 1984, pp. 337–344.
6. IEC 60099-4 2009, Surge Arresters—Part 4: Artificial pollution test with respect to the thermal stress on porcelain-housed multi-unit metal-oxide surge arresters.
7. IEC 62217 Ed.1, Surge Arresters—Part 9: Tests on shed and housing material.
8. S.C. Oliveira, E. Fontana, F.J. do Monte de Melo Cavalcanti, "Real-Time Monitoring of the Leakage Current of 230-kV Glass-Type Insulators During Washing," IEEE Trans. Power Del., Vol. 24, No. 4, pp. 2257–2260, 2009.
9. E. Fontana, S.C. Oliveira, F. Jd. Md. M. Cavalcanti, R.B. Lima, J.F. Martins-Filho, E. Meneses-Pacheco, "Novel sensor system for leakage current detection on insulator strings of overhead transmission line," IEEE Trans. on Power Delivery, Vol. 21, No. 4, 2006, pp. 2064–2070.
10. J. Lundquist, L. Stenstrom, A. Schei, B. Hansen, "New Method for Measurement of the Resistive Leakage Currents of Metal-Oxide Surge Arresters in Service," IEEE Trans. on Power Del., Vol. 5, No. 4, 1990, pp. 1811–1822.
11. G. Montoya, I. Ramirez, J. I. Montoya, "Correlation among ESDD, NSDD and leakage current in distribution insulators," IEE Proceedings—Generation, Transmission and Distribution, Vol. 151, No. 3, 2004, pp. 334–340.
12. J. Taylor, Production Guide, ABB, 2008.
13. T. Zhao, Q. Li, and J.L. Qian, "Investigation on Digital Algorithm for On-Line Monitoring and Diagnostics of Metal Oxide Surge Arrester Based on An Accurate Model," IEEE Trans. on Power Del., Vol. 20, No. 2, 2005, pp. 751–756.
14. Novizon, Zulkurnain Abdul Malek, Aulia, "A new method to separate resistive leakage current of ZnO surge arrester," Vol. 2, No. 29, 2008, pp. 67–71.
15. R.T. de Souza, E.G. da Costa, S.R. Naidu, M.J.A. Maia, "A virtual bridge to compute the resistive leakage current waveform in ZnO surge arresters," IEEE/PES Transmission and Distribution Conference and Exposition, 2004, pp. 255–259.
16. Chandana Karawita, and M. R. Raghuveer, "Onsite MOSA condition Assessment-a new approach," IEEE Trans. Power Delivery, Vol. 21, No. 3, 2006, pp. 1273–1277.
17. Huijia Liu, Hanmei Hu, "Development of Tester of the Resistive Leakage Current of MOA," Power and Energy Engineering Conference Asia-Pacific, 2010, pp. 1–4.
18. IEC 60099–5, section 6: Diagnostic indicators of metal-oxide surge arresters in service, 2000.
19. IEEE Std C62.11, IEEE Standard for Metal-Oxide Surge Arresters for AC Power Circuits (>1 kV), 2005.

Power and Energy – Kong (Ed.)
© 2015 Taylor & Francis Group, London, ISBN 978-1-138-02782-4

Research on the influence of tip clearance on the aerodynamic performance of the turbine

Shi Ma
College of Power Engineering, Naval University of Engineering, Wuhan, China

Yong-bao Liu
College of Power Engineering, Naval University of Engineering, Wuhan, China
Institute of Thermal Science and Power Engineering, Naval University of Engineering, Wuhan, China

Yu-jie Li
College of Power Engineering, Naval University of Engineering, Wuhan, China
Military Key Laboratory for Naval Ship Power Engineering, Naval University of Engineering, Wuhan, China

ABSTRACT: A 2 stages turbine was studied by using a numerical simulation program. The efficiency and flow of turbine was analyzed when tip clearance changed. The flow angle, the pressure and the entropy at the outlet of the first rotor, the static pressure and streamline of the second stator were analyzed when tip clearance changed. The results show that the flow increased linearly and the efficiency decreased linearly with the increase in tip clearance value. The 20~30% area at the top of the outlet channel of the first rotor was influenced by the tip clearance flow, and the area increased with the increase in tip clearance value. Besides, the flow inlet angle of the second stator increased too. Then it caused more loss of angle attack and reduced aerodynamic performance.

1 INTRODUCTION

Bindon[1] analyzed the flow structure of tip clearance of turbine in the earliest time. He found the blade tip separation vortex and the attachment phenomenon for the first time. He also observed that the static pressure on the pressure side of the blade tip clearance decreased by the effects of blade tip vortex separation, or even below the static pressure on the suction side. In addition, he put forward a new model of the leak-flow. Donton[2] concluded that the loss inside the tip clearance is related to the flow structure closely. As the flow structure inside the tip clearance is closely related to the factors such as height, blade shape and the rotor speed, the loss and the distribution of leak-flow is different significantly for the tip clearance of different turbine. Yamamoto[3] concluded the influence degree on the tip leakage flow by the speed of inlet flow strengthened with the increase in tip clearance value. Moore[4] has done a large number of experiments on the tip clearance flow, he found that the leakage flow entering into tip clearance with high speed and mixing at about 20% of blade chord, then it mixed before it out of tip clearance. Xiao[5-7] analyzed the flow field near the tip clearance detailed and argued that the flow field in tip clearance of turbine was more complex than compressor. NIU

Mao-sheng[8,9] concluded that the leakage vortex of tip clearance formed on suction side about 25% of the axial chord length position after the fluid through the tip clearance. The larger tip clearance value, the leakage flow has more and the leakage vortex is stronger. ZHAO Wang-dong[10] concluded that the change of tip clearance has no effect on turbine inlet flow field, but it has greater influence on the flow field of the top of outlet channel. The total pressure and total temperature reduced with the decrease in tip clearance value, and the airflow angle is closer to the area angle of mainstream. SHI Bao-long[11] concluded that the same tip clearance flow area, the step and shrink tip clearance shape perform better than line tip clearance shape, while the groove and bulge shape is worse relatively.

2 THE NUMERICAL METHOD AND THE PHYSICAL MODEL

In this paper, the commercial CFD software NUMECA FINE/Turbo module is used to calculation. The solving equations are the three dimensional unsteady Reynolds averaged Navier-Stokes equations, and the turbulence model is Spalar-Allamaras model. Using Autogrid5 of NUMECA software package to generate the grid of model,

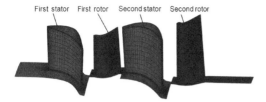

Figure 1. The computational mesh of model.

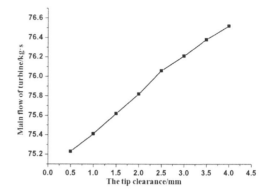

Figure 2. The flow changed with the tip clearance of the turbine.

O4H type topology structure is used in cascade passage in order to simulate the flow characteristics inside the boundary layer accurately. All the value of Y+ on the wall is less than 2. The grid number of model is about 4 million, and Figure 1 is the computational mesh of model, the date of boundary conditions according to the gas turbine laboratory test data. The inlet total pressure is 1 949 230 Pa, inlet total temperature is 1 543 K, exit static pressure is 376 438 Pa, the first rotor speed is 9 740 r/min, the second rotor speed is 7 436 r/min.

3 CALCULATION RESULT ANALYSIS

The design of tip clearance is 0.5 mm, 1.0 mm, 1.5 mm, 2.0 mm, 2.5 mm, 3.0 mm, 3.5 mm and 4.0 mm in height. Main-flow of turbine rate with the change of tip clearance as shown in Figure 2 through the numerical calculated. As can be seen from the Figure the main-flow increased with the increase in tip clearance value. Quantitative results show, when the tip clearance increased from 0.5 mm to 4.0 mm, the main flow increased by 2.55%, that is 1.29 kg/s. At the same time, the main flow increased linearly with the increase in tip clearance value.

The efficiency of turbine rate with the change of tip clearance is shown in Figure 3. As can be seen from the figure the efficiency decreased with the increase in tip clearance value. When the tip clearance increased from 0.5 mm to 4.0 mm, the efficiency decreased by 4.29%. The change of efficiency is closely related to the strength of the leakage flow. The strength of leakage vortex increased with the increase in tip clearance value, and then the leakage loss increased, the efficiency decreased. So the study of the influence of tip clearance on the turbine aerodynamic performance is important to improve the performance of the gas turbine. On the other hand, it can be concluded from the Figure that the efficiency decreased linearly with the increase in tip clearance value.

Based on the numerical results of the turbine, compared the calculation results when the tip clearance is 1.0 mm, 2.0 mm, 3.0 mm and 4.0 mm in height. The rule of turbine aerodynamic parameters of mainstream was analyzed. Comparing the

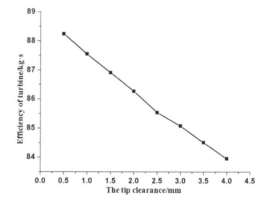

Figure 3. The efficiency changed with the tip clearance of the turbine.

flow angle, the total pressure and the entropy of the outlet of the first rotor and the static pressure distribution of the second stator under different tip clearance is the main content.

Figure 4 shows the distribution of the relative flow angle at the outlet of the first rotor along the blade under different tip clearance. As shown in Figure 4, under 70% blade height, the relative flow angle is basically the same under different tip clearance. Starting from 70% blade height, the relative flow angle increased with the increase in tip clearance value, and it began to increase sharply from 80% blade height. From 80% blade height to the top, the relative flow Angle increased from 20° to about 100°. On the other hand, above 80% blade height, the bigger the tip clearance, the bigger the relative flow angle. The loss of attack angle of the second stator caused by the deflection insufficient of the relative flow angle, and then the aerodynamic performance of the turbine reduced.

432

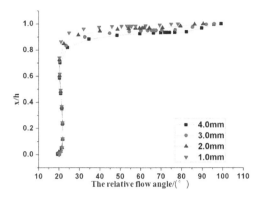

Figure 4. The relative flow angle at the outlet of the first rotor.

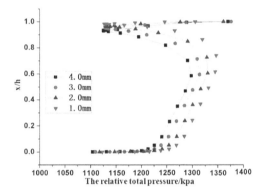

Figure 5. The relative total pressure at the outlet of the first rotor.

Figure 5 shows the distribution of the relative total pressure at the outlet of the first rotor along the blade under different tip clearance. As shown in Figure 5, from blade root to the top, the relative total pressure increased first, then began to decrease rapidly when the height came to 70% blade height, and then increased rapidly from 95% blade height to the top. Under 95% blade height, relative total pressure will be reduced about 20 kPa when tip clearance increased 1 mm in height. From the distribution of relative total pressure of the whole channel can be obtained, the lager the tip clearance, the loss of relative total pressure has more.

Figure 6 shows the cloud picture of the distribution of the relative pressure at the outlet of the first rotor along the blade under different tip clearance. As shown in the figure the annular low pressure area appeared at the top of the channel, the causes of low pressure area is the effect of the leakage vortex which formed by the leakage flow. On the

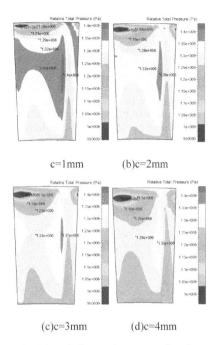

c=1mm   (b)c=2mm

(c)c=3mm   (d)c=4mm

Figure 6. The relative total pressure direction at the outlet of the first rotor.

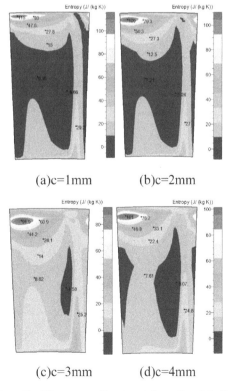

(a)c=1mm   (b)c=2mm

(c)c=3mm   (d)c=4mm

Figure 7. The entropy direction at the outlet of the first rotor.

433

other hand, the center of low-pressure area downward from the top and the area of low pressure increased with the increase in tip clearance value. Because the tip clearance leakage flow increased with the increase in tip clearance value, then the intensity of leakage vortex became stronger, and then the low-pressure area becomes larger. At the same time, the low-pressure distribution of the whole channel was affected by the leakage vortex, the total pressure decreased with the increase in tip

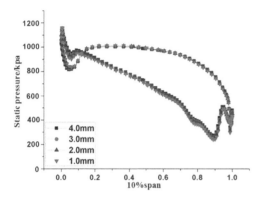

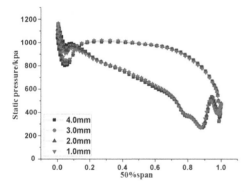

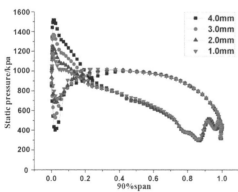

Figure 8. Static pressure distributions along the blade surface at the 2nd stator.

clearance value. The results of Figure 6 are consistent with the results of Figure 5.

Figure 7 shows the cloud picture of the distribution of the entropy at the outlet of the first rotor along the blade under different tip clearance. As shown in the figure, the maximum value of entropy appeared in the center of leakage vortex of the channel, and the maximum value of entropy decreased with the increase in tip clearance value, but the center of the leakage vortex moved down, and the area of leakage vortex increased. So the loss of entropy increased and the efficiency decreased with the increase in tip clearance value.

Figure 8 shows the distributions of static pressure along the blade surface at the 2nd stator under different tip clearance. The distribution of static pressure with a small difference in the 10% blade height section, The distribution of static pressure on the front edge of 20% suction side and 10% press side with a small difference under different tip clearance, and the difference between the pressure of suction surface and press surface increased with the increase in tip clearance value. The difference between the pressure on the front edge of 35% of suction surface and press surface becomes greater in the 90% blade height section. The difference between the pressures increased with the increase in tip clearance value. Because the intensity of leakage vortex increased with the increase in tip clearance value, and the loss of attack angle got serious. So the static pressure of press surface decreased and the static pressure of suction surface increased.

4 CONCLUSIONS

1. The main flow of turbine increased linearly with the increase in tip clearance value, at the same time, the efficiency of turbine decreased linearly with the increase in tip clearance value.
2. Due to the existence of tip clearance, the flow field structure at the top of 20~30% channel was changed. The intensity of leakage vortex became stronger with the increase in tip clearance value, then the flow angle at the outlet of the first rotor increased and leading to the insufficient deflection of the flow into the second stator became more serious, and then the loss increased.

REFERENCES

1. Bindon J.P. The measurement and formation of tip clearance loss [J]. ASME Journal of Turbo-machinery, 1989, 111(3):257–263.
2. Denton J.D. Loss mechanisms in turbine machines [C], ASME Paper, 1993-GT-415. 1993.

3. Yamamoto A. End-wall Flow/Loss Mechanisms in a linear turbine cascade with blade tip clearance [J]. Journal of Turbo machinery, 1989, 111(2): 246–275.

4. Moore J, Tilton J.S. Tip leakage flow in a linear turbine cascade [J]. Journal of Turbo machinery, 1988, 110(1):18–26.

5. Xin-wen Xiao, McCarter Andrew A. Tip clearance effects in a turbine rotor: Part I-Pressure field and loss [R]. ASME 2001-GT-0476.

6. McCarter Andrew A, Xin-wen Xiao. Tip clearance effects in a turbine rotor flow part II-Velocity field and physics [R]. ASME 2001-GT-0477.

7. Xinwen Xiao. Investigation of Tip clearance flow physics in axial flow turbine Rotors [D]. Pennsylvania: The Pennsylvania State University, 2000.

8. Niu Mao-sheng, Zang Shu-sheng, Huang Ming-hai. Effect of tip clearance height on tip clearance flow in turbine rotors [J]. Journal of Engineering Thermo-physics, 2008, 29(6):935–939.

9. Niu Maosheng, Zang Shusheng, Huang Minghai. Effect of tip clearance height on tip clearance flow in turbine rotors [J]. Gas Tuebine Technology, 2008, 21(4):26–31.

10. Zhao Wangdong, Zhou Yubin, Yang Rui. Effects of Turbine Tip Clearance on Gas Turbine Performance [J]. Gas Turbine Experiment and Research, 2009, 22(3):19–22.

11. Shi Baolong, Qi Xingming, Jiao Jinyi, et al. Numerical analysis of the secondary flow and tip clearance leakage loss [J]. Journal of Aerospace Power, 2009, 24(5):1096–1100.

Power and Energy – Kong (Ed.)
© *2015 Taylor & Francis Group, London, ISBN 978-1-138-02782-4*

Research and application of the 750 kV AC voltage vehicle test platform

Min Chen, Jun Chen, Tao Wang & Yao Bai
Electric Power Research Institute of Hubei Electric Power Company, Wuhan, Hubei, China

Hong Zhou & Qinqin Zhou
Suzhou Huadian Electric Limited by Share Ltd., Shuzhou, Jiangshu, China

ABSTRACT: In order to arrange boost resonant reactor and measuring capacitive voltage divider with their grading ring in a limited space on the 750 kV AC voltage vehicle test platform through optimizing design, the structure of the reactor sharing the grading ring with the voltage divider is designed, and the electric field distribution of the platform is simulated and calculated. Next, the test is carried out to ensure that the distributed capacitance of the test platform doesn't bring obvious influence on the measurement accuracy and the influence can be negligible. At last, the measurement system of the platform is calibrated at the national high voltage metering station. Currently, the platform has been successfully applied in a ±800 kV UHV converter station.

1 INTRODUCTION

AC voltage withstand test is the most direct identification method for insulation intensity of electrical equipment, which is the decisive significance to determine whether the electrical equipment is put into operation, this is because the AC voltage withstand test can effectively detect the insulation defects (including the damage in the process of packaging, transportation, storage, installation and commissioning and the existence of the foreign body, etc.) [1–10]. At present, the variable frequency series resonance test method widely used in the field has become the most effective method of AC voltage withstand test, this method as the advantages of small volume, light weight, small power supply capacity and many other advantages. In addition, once the sample breakdown occurs, as the circuit lose resonant conditions, the output current of the power supply automatically reduces, and the voltage across sample plummets, and over voltage won't be produced, and discharge energy is small, therefore, the influence and the damage to the experimental article is very small [11–19].

In order to carry out the AC withstand voltage test of high voltage electrical equipment, such as power transformer, circuit breaker, isolating switch, GIS, insulator, bushing, generatrix, mutual inductor and so on, the corresponding test equipment and instruments (mainly for the series resonant test equipment) should be transported to the site. Due to the initial vehicle layout, structure and equipment configuration etc. Without the corresponding planning, the overall coordination is poor, the

equipment is scattered, the vehicle loading rate is low, the field test site requires large space; equipment lifting and installation and combination is very cumbersome, workload is heavy and mobility is poor, shift after installing is very trouble; wiring is complex, reliability is poor, work efficiency is low [20].

In order to solve many problems completely in the process of the field AC withstand voltage test, the 750 kV AC withstand voltage vehicle test platform with motorization, integration, automation and information is successfully developed.

2 REALIZATION WAY

In the traditional high voltage variable frequency test device, the boost resonant reactor and the measuring capacitor voltage divider respectively need the support base and the voltage equalizing system, and also need the anti corona connection between the reactor and the voltage divider, which is clearly contrary to the vehicle test platform with very limited space. Therefore, the structure of the reactor sharing the grading ring with the voltage divider is designed for the platform.

The 750 kV AC withstand voltage vehicle test platform is shown in Figure 1.

The upper end of the upper segment reactor and voltage divider keep the same potential through sharing of grading ring, and the lower end is fixed through an epoxy plate (the potential is not clamped together). In order to prevent the possible potential difference between the reactor

Figure 1. Test state diagram of the 750 kV AC withstand voltage vehicle test platform.

Figure 3. Schematic diagram of the 750 kV AC withstand voltage vehicle test platform.

Figure 2. Integrated design physical map of the reactor sharing the grading ring with the voltage divider.

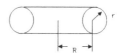

Figure 4. Grading ring structure.

and the voltage divider and the leakage magnetic interference of reactor, a certain gap is maintained between the reactor and the voltage divider. The specific is shown in Figure 2.

3 THE SIMULATION CALCULATION OF ELECTRIC FIELD DISTRIBUTION

The main electrical equipment layout on the test platform after the preliminary design is shown in Figure 3.

In the figure, the installation height of the big grading ring (i.e. the grading ring shared by the upper segment reactor and voltage divider) is 3.47 meters, the big grading ring size is shown in Figure 4.

The size of the big grading ring can be changed by changing the values of R and r. The different dimensions are changed as follows: R = 1000 mm, r = 400 mm; R = 1000 mm, r = 200 mm; R = 800 mm, r = 200 mm.

Electric field distribution with three kinds of different sizes is simulated. Among them, the cascade reactors uniform boost, the small grading ring (i.e. the lower segment grading ring) boosts to 290 kV, and the big grading ring boosts to 750 kV, with the metal shell grounding. Electric field distribution of

boosting uniformly in three kinds of different sizes is shown in Figures 5–7.

The results shows that, the maximum electric field intensity of the platform in the above three kinds of different sizes are located in the edge of the large grading ring, the value are respectively 11 kV/cm, 15 kV/cm, 16 kV/cm.

According to the electric field empirical calculation formula of of grading ring Emax = U[1 + (r/2R) × ln(8R/r)]/[rln(8R/r)], the maximum electric field intensity of the isolated grading ring is calculated. And the maximum electric field intensity of the grading ring with three kinds of different sizes are calculated and compared by using ansys and the empirical formula as shown in Table 1.

It can be seen from Table 1 that, the calculated values in three kinds of different sizes by using the empirical formula are slightly lower than the ANSYS finite element method.

In general, the larger the grading ring is designed, the lower the electric field intensity will be, and the better the equalizing effect will be. However, to the vehicle test platform, large size may cause many problems on the layout in the limited space of the platform. Therefore, the size of the grading ring shared by the reactor and the voltage divider should be designed as small as possible in the premise of the equalizing effect meeting the actual needs.

The experience shows that, it can be guaranteed not to appear the corona discharge in the condition that the electric field intensity is less than 15 kV/cm. It can be seen from Table 1, when the size of the large grading ring is designed as R = 1000 mm, r = 200 mm, the equalizing effect can meet the actual needs, but the margin of the maximum electric field intensity is too large, so the size is designed not small enough; when the size of the large grading ring is designed as R = 800 mm, r = 200 mm, the size is

438

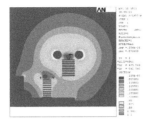

Figure 5. Field distribution when R = 1000 mm, r = 400 mm.

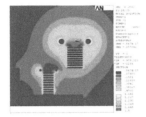

Figure 6. Field distribution when R = 1000 mm, r = 200 mm.

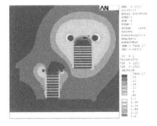

Figure 7. Field distribution when R = 800 mm, r = 200 mm.

Table 1. The maximum electric field intensity at grading ring (kV/cm).

Calculation method	R = 1000 mm r = 200 mm	R = 800 mm r = 200 mm	R = 1000 mm r = 400 mm
Ansys finite element	15	16	11
Empirical calculation	13.91	15.5	10.01

small, but the maximum electric field intensity at the grading ring (i.e. the maximum electric field intensity on the platform) is more than 15 kV/cm, and it doesn't meet the actual needs; when the size of the large grading ring is designed as R = 1000, mm r = 200 mm, the maximum electric field intensity at the grading ring (i.e. the maximum electric

field intensity on the platform) is just 15 kV/cm, not only the equalizing effect can meet the actual needs, but also the size has been designed to be as small as possible on this premise.

Therefore, the size of the grading ring on the vehicle experiment platform is confirmed as R = 1000 mm, r = 200 mm.

In order to meet the national standard 《Transfinite transport vehicle highway management regulations》, 《Road vehicle dimensions, axle load and quality limit》 and traffic conditions on the substation standards way of 4 meters main road, 3.5 meters fire road, 3 meters maintenance road, the size of the whole vehicle test platform is designed as 8.91 meters long, 2.48 meters wide, 3.95 meters high.

4 EFFECT OF THE TEST PLATFORM DISTRIBUTED CAPACITANCE TO THE MEASUREMENT ACCURACY

The distributed capacitance of the test platform is shown in Figure 8.

C1 is the distribution capacitance between the reactor and the voltage divider capacitor. It has been fixed, and the parameter is considered and already exists when correction table in the vehicle type series resonant test equipment factory test, and the parameter has a certain effect on the voltage linearity under different voltages, but from the correction table situation, this impact is small and can be neglected.

C2 is the distribution capacitance between the grading ring and the voltage divider capacitor. The parameter is fixed and exists in traditional voltage divider, and it has little effect on accuracy and can be negligible.

C3 is the distribution capacitance between the high voltage wire and the upper end of the lower segment voltage divider. The parameter changes as the high voltage lead changes, and it has some influence on precision. If the high voltage wire is just above the lower section voltage divider, the capacitance is large, and the capacitor current flows into the lower section voltage divider, and the current flowing into the lower section voltage

Figure 8. The distributed capacitance of the test platform.

divider increases, therefore, sampling voltage of the sampling capacitor and the displayed value is high; If the high voltage wire is drawn out from the other side, there are less influence on precision.

C4 is the distribution capacitance between the upper end and middle of the lower segment voltage divider and the ground. The current flows directly into the ground through the capacitor, so the parameter is fixed and it has no effect on measuring accuracy.

C5 is the distribution capacitance between the high voltage wire and the ground. The current flows directly into the ground through the capacitor, and it doesn't flow through the low voltage arm of the voltage divider, therefore, the parameter has no effect on measuring accuracy, it just adds a little bit of capacitive load.

C6 is load capacitance; the current flowing through the capacitor doesn't flow through the low voltage arm of the voltage divider, so the parameter has no effect on measuring accuracy.

However, the influence that the distributed capacitance of the test platform bring son the measurement accuracy should be determined by testing.

The location of all equipment and the connected line on the vehicle test platform are fixed. Through the above analysis, the distributed capacitance that may change in test mainly are C3, C5, C6, while C5 and C6 have no effect on measuring accuracy, therefore, the following mainly carries out the voltage test by changing the thickness and connection direction of the high voltage wire, namely changing C3, to determine the influence that the change of the distributed capacitance bring on the measurement accuracy.

4.1 Changing the connection direction of the high voltage wire

The voltage tests are carried out by changing three different directions of the high voltage wire between the series resonant device and the standard voltage divider source which is also the sample. The three different directions are that, the high voltage wire is just above the lower segment voltage divider, the high voltage wire is perpendicular to above the lower segment voltage divider, and the high voltage wire is contralateral to above the lower segment voltage divider.

The test site photo (the high voltage wire is contralateral to above the lower segment voltage divider) is shown in Figure 9.

The voltage value shown on the standard voltage divider source is actual measured value; the voltage value shown on the frequency conversion cabinet is the displayed detected value.

Figure 9. The test site photo (the high voltage wire is contralateral to above the lower segment voltage divider).

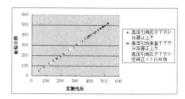

Figure 10. The curve comparison chart of the displayed detected value to the actual measured value in three ways.

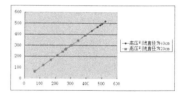

Figure 11. The curve comparison chart of the displayed detected value to the actual measured value with two different diameters of the corona proof corrugated pipe.

According to test results, the curve comparison chart of the displayed detected value to the actual measured value in three ways can be obtained as shown in Figure 10.

It can be seen from Figure 10, the voltage tests are carried out by changing three different directions of the high voltage wire between the series resonant device and the standard voltage divider source, and the measurement error compared to the relative standard source doesn't change significantly. Therefore, the connection direction of the high voltage wire doesn't bring obvious influence on the measurement accuracy, and the influence can be negligible.

4.2 Changing the thickness of the high voltage wire

When the high voltage wire between the series resonant device and the standard voltage divider source, which is also the sample is respectively corona proof corrugated pipe with the diameter of 40 cm or 20 cm, and just above the lower segment voltage divider, the voltage tests are carried out. The voltage value shown on the standard voltage divider source is actual measured value. The voltage value shown on the frequency conversion cabinet is displayed detected value.

According to test results, the curve comparison chart of the displayed detected value to the actual measured value with two different diameters of the corona proof corrugated pipe can be obtained as shown in Figure 11.

It can be seen from Figure 11, the voltage tests are carried out when the high voltage wire between the series resonant device and the standard voltage divider source is respectively corona proof corrugated pipe with two different diameters, and the measurement error compared to the relative standard source doesn't change significantly. Therefore, the thickness of the high voltage wire doesn't bring obvious influence on the measurement accuracy, and the influence can be negligible.

In summary, the distributed capacitance of the test platform doesn't bring obvious influence on the measurement accuracy and the influence can be negligible.

5 MEASUREMENT SYSTEM CALIBRATION TEST

The measurement system of the reactor sharing the grading ring with the voltage divider is calibrated at the national high voltage metering station, and the calibration data is shown in Table 2.

Table 2. Measurement system calibration data of the reactor sharing the grading ring with the voltage divider.

Actual value (kV)	Displayed value (kV)	Error ((displayed value-actual value)/ actual value)
69.4	68.9	−0.72%
125.8	124.8	−0.79%
179.7	179.2	−0.28%
231.7	230.5	−0.52%
298.9	297.8	−0.37%
364.7	362.8	−0.52%
414.3	410.4	−0.94%
480.0	479.0	−0.21%
533.0	529.8	−0.60%
593.7	592.4	−0.22%
665.6	659.6	−0.90%
700.5	694.1	−0.91%
743.1	737.8	−0.71%

Among them, the test frequency is 157.4 Hz. Extended uncertainty of calibration result is

$$U_{rel} = 1.8 \times 10^{-2}, \ k = 2.$$

It can be seen that, to thirteen measuring points almost uniformly distributed in the range of voltage values 0~750 kV, the error between the measurement system displayed value and the actual voltage value of the reactor sharing the grading ring with the voltage divider is kept throughout in the range of −1%~−0.21%, fully meet the field engineering application.

6 THE FIELD APPLICATION

The 750 kV AC withstand voltage vehicle test platform with specially modificative vehicle as the carrier, adopt integration design of the reactor sharing the grading ring with the voltage divider, and integrate the 750 kV complete sets of series resonant frequency test device and auxiliary systems in the vehicle, and has automatic deployment test platform without relying on external lifting equipment, and wiring is convenient, thereby completing the field AC withstand voltage test of 750 kV tank type circuit breaker in the vehicle is realized.

Compared to completing the field AC withstand voltage test of a 750 kV tank type circuit breaker with the conventional lifting mode for two days, the 750 kV AC withstand voltage vehicle test platform with the highest voltage level at home and abroad is applied in a ±800 kV UHV converter station, can complete the field AC withstand voltage test of three 750 kV tank type circuit breakers in one day, and complete a total of seven 750 kV tank type circuit breakers, greatly improves work efficiency and decreases security risk.

7 CONCLUSION

In order to arrange boost resonant reactor and measuring capacitive voltage divider with their grading ring in a limited space on the 750 kV AC voltage vehicle test platform through optimizing design, the structure of the reactor sharing the grading ring with the voltage divider is designed, and the electric field distribution of the platform is simulated and calculated, then the size of the grading ring on the vehicle experiment platform is confirmed. Next, the test is carried out to ensure that the distributed capacitance of the test platform doesn't bring obvious influence on the measurement accuracy and the influence can be negligible.

At last, the measurement system of the platform is calibrated at the national high voltage metering station; the result fully meets the field engineering application. Currently The platform has been successfully applied in a ±800 kV UHV converter station.

REFERENCES

1. CIGRE WG 33/23-12. Insulation Coordination of GIS: Return of Experience, on Site Tests and Diagnostic Techniqures [R]. Electro, No. 176, 1998.
2. A. Diessner, G.F. Luxa, W. Mosca, A. Pigini. High voltage Testing of GIS Insulated Substation on Site [R]. CIGRE session 1986 Report 33–06.
3. CIGRE WG 33/23-12. Insulation Coordination of GIS: Return of Experience, on Site Tests and Diagnostic Techniqures [R]. August, 1996.
4. Shi Hai-Jun, Wang Guang-Qian, Zhang Shao-Yan. GIS site Test [J]. Voltage electrical, 2005, 41(1):55–58.
5. Chen Hua-Gang. Electrical equipment preventive test methods [M]. Beijing: China Water Power Press, 1999.
6. Lin Meifen. The application of series resonant frequency in Transformer AC voltage test [J]. Hydro-electric energy, 2010, 28(2), 155–157.
7. Shi Zhixia. The situation and development of field test to GIS [J]. Voltage electrical, 2001, 37 (4):42–45.
8. Wang Yi-Ping. GIS site AC voltage test techniques [J]. East China Electric Power, 1998(4):38–41.
9. Electrical equipment installation engineering electrical equipment over test. GB50150-2006 [S]. Ministry of Construction, China State Administration of Quality Supervision, Inspection and Quarantine jointly issued. China Planning Press.
10. Li Chen. AC voltage test research of 500 kV substation the GIS system [J]. Power system technology, 2012(2), 14–16.
11. Luo Zhuowei. Intelligent UHV inverter resonant test power development and engineering application [D]. Hunan University.
12. Shi Feng, Zhang Dalin, Han Tao. Resonant frequency AC voltage test application in GIS fault detection [J]. Northeast Electric Power Technology, 2010(10), 41–43.
13. Zhang Ren Yu, Chang Chen Yue, Wang Long. High Voltage Engineering (the 2nd edition) [M]. Beijing: Tsinghua University Press, 2003.
14. Sun Jianjun. FM series resonant AC high voltage test equipment development [D]. Dalian University of Technology, 2002.
15. Quan Chao-Chun, Gan Gen-Zhi, Yuan Yong-Wu etc. series resonant frequency dielectric tests in the GIS application [J]. Voltage electrical, 2006, 42(3), 199–200.
16. Shen Cong Shu. GIS equipment series resonant AC voltage test probe [J]. Power System Technology, 2011(10), 32–34.
17. Chen Zhong. Series resonant voltage withstand test site problems and solutions [J]. Power System Technology, 2006, 30, 205–207.
18. Pan Hua. UHV AC withstand and partial discharge electrical test equipment research and development [D]. Hunan University, 2008.
19. Ding Gan, Xu Zhi-Xin, Chen Fang-Liang, etc. The reaseach of series resonant voltage withstand test system automatic frequency power [J]. Power Electronics, 2006, 30(6), 71–73.
20. Wang Ya-Zhou, Jiang Jian-wu, Zhong Jian-Ling, etc. power integrated test vehicle usage theoretical analysis and experimental summary [J]. Central power, 2010, 23(5):59–62.

Power and Energy – Kong (Ed.)
© 2015 Taylor & Francis Group, London, ISBN 978-1-138-02782-4

Characteristics analysis of switching transient disturbance to secondary equipment port of 1,000 kV substation

Jiangong Zhang
Huazhong University of Science and Technology, Wuhan, Hubei Province, China
China Electric Power Research Institute, Wuhan, Hubei Province, China

Junjia He
Huazhong University of Science and Technology, Wuhan, Hubei Province, China

Jun Zhao & Yemao Zhang
China Electric Power Research Institute, Wuhan, Hubei Province, China

Weidong Zhang
North China Electric Power University, Beijing, China

ABSTRACT: In order to obtain characteristics of switching transient disturbance to the secondary equipment port of 1,000 kV substation, regulate the level of the Electromagnetic Compatibility (EMC) test of secondary equipment and ensure safe and stable operation of secondary equipment of 1,000 kV substation, a transient electromagnetic disturbance measuring system was developed in this paper for the field research of voltage, current and nearby space magnetic field of the secondary equipment port as the Gas Insulated Switchgear (GIS) was operated. The results showed that the waveform of transient disturbance to the secondary equipment port generated by the operation of the GIS was approximate to the damped oscillatory wave, the maximum amplitude of the common-mode voltage reached 1,071 V, and that of the space magnetic field reached 70.2 A/m. It is recommended that as EMC is tested, the test standard of the immunity to the damped oscillatory wave and the damped oscillatory magnetic field is respectively not less than Level-3 and Level-5.

1 INTRODUCTION

As the GIS of 1,000 kV substation switches on/off the un-loaded busbar, gas breakdown and restrike generally occur between contactors due to the slow movement of contactors and the weak arc extinguishing capacity of the switch. Every breakdown will have a very steep traveling wave, and refraction and reflection will occur as the traveling wave encounters the mismatched point of wave impedance, resulting in Very Fast Transient Overvoltage (VFTO) with high-frequency oscillation. As the traveling wave arrives at the outlet bushing from GIS, electromagnetic leaks will occur and induce Transient Earth Voltage (TEV) at GIS enclosure. The space electromagnetic field and TEV will be coupled into the secondary equipment port by means of induction and conduction, and the traveling wave on the busbar will also pass directly to the secondary side via the potential transformer and the current transformer. During the commissioning of the expansion project of the UHVAC

pilot and demonstration project, as the GIS was operated, the false alarm of monitoring data of the air chamber pressure occurred, and such EMC problem will affect the safe operation of the UHV substation and the power grid.

The electromagnetic disturbance to the secondary equipment caused by switching transient of the substation is always a research hotspot at home and abroad. American Electric Power Research Institute (EPRI) carried out a lot of tests to obtain characteristics of transient disturbance of the substation, and combined with test data to establish a more accurate predictive model of the transient electromagnetic environment of the substation. P. H. Pertorus from South Africa gave a preliminary result of the switching transient electromagnetic disturbance measured at 132 kV, 275 kV and 400 kV substations, and scientists from Japan and Italy also carried out researches in this field. But there is no research on transient disturbance of 1,000 kV substations conducted at home and abroad. Because the amplitude of the switch arc current is equal to the ratio of the breakdown voltage and GIS wave impedance, the amplitude of switching transient of 1,000 kV substation is higher. In addition, because on-line monitoring devices scattered in the switch yard and some protection and control equipment increase and get closer to the source of disturbance, the EMC problem tends to occur.

Firstly, a transient electromagnetic disturbance measuring system was developed in this paper; secondly, the actual measurement was carried out in several 1,000 kV substations for the switching operation's electromagnetic disturbance to secondary equipment to obtain characteristics of transient electromagnetic disturbance; finally, the contrast was conducted with the EMC immunity test standard to put forward requirements of EMC immunity of equivalent switching disturbance to secondary equipment of 1,000 kV substation.

2 ELECTROMAGNETIC DISTURBANCE MEASURING SYSTEM

The substation is a typical representative of the complex electromagnetic environment, its transient electromagnetic disturbance measuring system will be disturbed by a variety of outside factors, and at the same time the measured transient disturbance has high frequency and high amplitude; therefore, the substation transient measuring system should meet the measurement requirements in the aspects of shield, bandwidth, measurement range, etc.

In terms of anti-electromagnetic disturbance, the following measures were taken for the measuring system: Firstly, with respect to 50 MHz and higher bandwidth of the disturbance signal to be measured, voltage and current sensors was placed in the merging unit shielding box, and the wiring length was controlled within 0.5 m. Secondly, for the sensor's secondary cable is susceptible to the disturbance, the oscilloscope was placed nearby the substation's secondary equipment and in the shielding box, and powered by the inverted lithium battery. The BNC bi-pass was embedded in the wall of the box to connect the sensor's secondary cable to ensure reliable lap joint between BNC bi-pass's outer wall and shielding box. The sensor's secondary cable was wrapped in the copper shielding net, which was lapped with the box wall at both ends. Thirdly, the magnetic field sensor was placed on the ground near the merging unit, and its secondary cable's copper shield was reliably lapped with the measurement shielding box. Fourthly, the background computer was connected with the oscilloscope via the optical fiber transmission system to prevent electromagnetic disturbance.

Figure 1. Site layout of measuring system.

Specific technical parameters of the measuring system are as follows:

Maximum DC-AC voltage peak of the voltage sensor: 40 kV; maximum effective value of AC voltage: 28 kV; rise time: 2.4 ns, bandwidth: DC – 150 MHz. Maximum peak current of the current sensor: 600 A; rise time: 5 ns; bandwidth: 25 Hz – 60 MHz. The magnetic field sensor consists of the magnetic field probe and the integrator. Time constant of the integrator: 1.2 us; cutoff frequency of overall-3dB: 1 GHz; no-load voltage ratio: 1.1 A/m/mV; maximum output voltage: 1 kV. The digital storage oscilloscope has four channels for collection. Bandwidth: DC – 500 MHz; maximum storage depth: 125 Mpts; maximum sampling rate: 2.5 GS/s.

The developed substation transient disturbance measuring system is mainly comprised of sensor, equipment shielding box, instrument shielding box, digital storage oscilloscope, inverter, battery, optical fiber transmission control system, trigger system, optical fiber, storage media, etc. The digital storage oscilloscope records the voltage waveform and supplies power after the battery is inverted by the inverter. The external trigger system is used to control the measuring system via the optical fiber transmission control system, and the entire measuring system is in electromagnetic shielding. The site layout of the measuring system is shown in Figure 1.

3 ACTUAL MEASUREMENT OF SUBSTATION

3.1 Measured contents

The measurement of switching transient disturbance to secondary equipment was carried out in Changzhi Station, Nanyang Station and

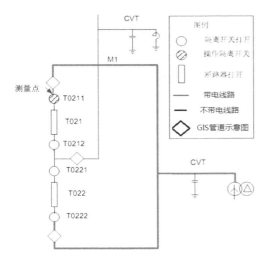

Figure 2. GIS operation schematic.

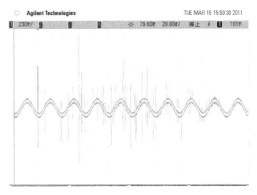

Figure 3. Common-mode voltage of CVT secondary port in the switching-on process.

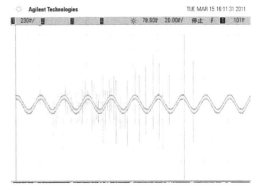

Figure 4. Common-mode voltage of CVT secondary port in the switching-off process.

Jingmen Station of Southeastern Shanxi—Nanyang—Jingmen 1,000 kV UHVAC pilot and demonstration project which had been put into operation, as shown in Figure 2. As the GIS was in hot-line work, the measured physical quantities included site control cabinet's CVT secondary side common-mode voltage, control signal port's common-mode voltage, CT secondary side disturbing current as well as the space magnetic field near the GIS, and each operation was conducted three times.

3.2 Measurement results

The GIS operation produced dozens of pulses at the secondary equipment port. As the GIS was switched on, the disturbing pulse changed from sparse to dense, and the amplitude got smaller. It was opposite as the GIS was switched off. Because the switching operation was slow, multiple arc breakdowns and restrikes were produced. As the distance between open contacts of the switch in the switching-on process became smaller, the breakdown got easier, and the breakdown voltage became low. The initial amplitude of the arc current depended on the ratio of the breakdown voltage and the impedance of the busbar. Then overlap might occur due to the multiple reflections; therefore the current wave with the maximum amplitude appeared first as the GIS was switched on, and the disturbance coupled to the secondary equipment circuit was also the maximum.

Figure 3 shows the common-mode voltage waveform of local cabinet's CVT secondary port as the GIS is switched on. The duration of pulse train was 150 ms, the maximum peak voltage was 828 V, and the maximum value of voltage peak was 1,500 V. Figure 4 is the expanded view of local cabinet's CVT secondary transient voltage as the GIS was switched on and shows its spectrum. The duration of pulse train was 130 ms, the maximum peak voltage was 920 V, and the maximum value of voltage peak was 1,700 V. It can be seen from common mode disturbance to the CVT secondary port that the transient waveform had about 30 pulses as the GIS was switched on and off. The pulse amplitude got smaller as the GIS was switched on. At the beginning of switching-on process, the distance between open contacts was longer, the breakdown voltage was higher, but the pulse interval was large. As the distance between open contacts became smaller, the pulse interval decreased. It was opposite in the switching-off process. If the electromagnetic disturbance is taken into account, generally the largest transient waveform is taken for analysis. It can be found after the waveform expansion that the waveform is damped oscillatory wave, and its spectral characteristics can be obtained after Fourier transform.

The common-mode voltage at the control signal port was measured respectively in relay room and site control cabinet, and the maximum transient waveform was taken for analysis.

Figures 5 and 6 show the typical waveforms in the switching-on and switching-off process measured at the relay room. For the switching-on, the maximum peak voltage was 48 V, the maximum peak-to-peak voltage was 94 V, the longest duration of pulse train was 146.6 ms, the longest duration of single pulse was 3.46 μs, and the maximum basic frequency was 12.02 MHz. For the switching-off, the maximum peak voltage was 68 V, the maximum peak-to-peak voltage was 112 V, the longest duration of pulse train was 114.8 ms, the longest duration of single pulse was 4.6 μs, and the maximum basic frequency was 12.02 MHz. Figures 7 and 8 show the typical waveforms in switching-on and switching-off process measured by the site control cabinet. For the switching-on, the maximum peak voltage was 245.06 V, the maximum peak-to-peak voltage was 440.84 V, the longest duration of pulse train was 144.36 ms, the longest duration of single pulse was 2.99 μs, and the maximum basic frequency was 12.10 MHz. For the switching-off, the maximum peak voltage

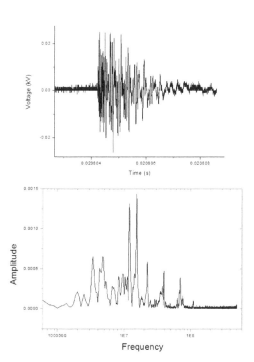

Figure 6. Waveform and spectrum analysis of single pulse in relay room in the switching-off process.

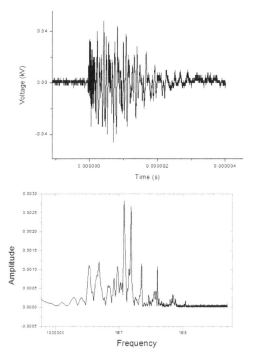

Figure 5. Waveform and spectrum analysis of single pulse in relay room in the switching-on process.

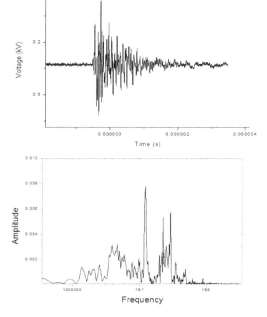

Figure 7. Waveform and spectrum analysis of single pulse in site control cabinet in the switching-on process.

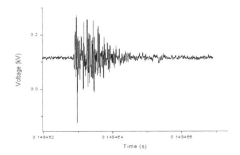

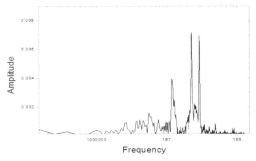

Figure 8. Waveform and spectrum analysis of single pulse in site control cabinet in the switching-off process.

Table 1. Maximum electromagnetic disturbance generated by GIS operation.

Voltage level and type	CVT (V)	Control signal (V)	Space magnetic field (A/m)	CT (A)
Changzhi Station	163	358	13.6	24
Nanyang Station	216	311	35.8	7.5
Jingmen Station	1046	498	70.2	4.5

electromagnetic disturbances generated by the GIS operation are summarized in Table 1. In the table, the maximum common-mode disturbing voltage of CVT secondary side reaches 1,071 V, the maximum common-mode voltage of the control signal is up to 665 V, the maximum amplitude of the space magnetic field reaches up to 70.2 A/m, the maximum transient current amplitude of CT secondary side up to 24 A, and TEV of GIS enclosure up to 577 V. The range of basic frequencies of these disturbances is 3–30 MHz, largely concentrated around 10 MHz, similar to the VFTO waveform.

was 239.44 V, the maximum peak-to-peak voltage was 397.18 V, the longest duration of pulse train was 142.29 ms, the longest duration of single pulse was 2.76 μs, and the maximum basic frequency was 11.81 MHz.

Both amplitude and frequency measured in the control cabinet were higher than those measured in the relay room, indicating that the switch yard's electromagnetic environment was more severe than the relay room, which was because the relay room had better shielding and grounding and was farther from the source of disturbance.

Characteristics of the switching transient are greatly affected by the structure and layout of the circuit. In the same circuit, repetitive operations can get consistent basic waveforms; on the contrary, in different circuits, there is a great difference between the transient waveforms generated in the operation. The actual measurement was carried out in different test circuits and test points selected in six 1,000 kV substations, so the secondary equipment port's disturbing waveform had some differences in both the amplitude and the frequency. From the perspective of EMC, what is concerned is the biggest electromagnetic disturbance that may arise.

In this paper, space magnetic field, CT secondary side current, TEV of GIS enclosure and other signals were also measured, and due to their similar waveforms to the damped oscillatory wave, no further illustrated. The maximum

4 ACTUAL MEASUREMENT OF SUBSTATION

The waveform of the disturbance to the substation's secondary equipment is dispersible, thus its EMC assessment can not accurately reproduce the actual waveform of the disturbance at the port. The waveform measured shows that the conductive voltage disturbance generated by the GIS operation is similar to the damped oscillatory wave and the magnetic field is similar to the damped oscillatory magnetic field.

1. Damped oscillatory wave immunity test
   As specified in IEC61000-4-18, the damped oscillatory wave has two kinds, one is slow damped oscillation, including oscillation frequencies of 100 kHz and 1 MHz, and the other is fast damped oscillation, including 3 MHz, 10 MHz and 30 MHz. The fast damped oscillatory wave mainly simulates the transient voltage of the secondary equipment port as the GIS is operated. From the measured disturbance of 1,000 kV substation, the secondary port's maximum amplitude is 1,071 V, thus the test level not less than Level-3 can meet the requirements.

2. Damped oscillatory magnetic field immunity test
   The magnetic field, to which the equipment exposes, may affect the reliable operation of the equipment and system. The damped oscillatory

Table 2. Fast damped oscillatory wave test level (3 MHz, 10 MHz, 30 MHz).

Level	Common-mode voltage (kV)
1	0.5
2	1
3	2
4	4
X[a]	X

X is an open level, which can be given in the product specification.

Table 3. Damped oscillatory magnetic field test level.

Level	Damped oscillatory magnetic field intensity (A/m (peak))
1	–
2	–
3	10
4	30
5	100
X	

X is an open level, which can be given in the product specification.

magnetic field immunity test is used to assess the ability of the electronic equipment installed in the vicinity of the high-voltage device to withstand the damped oscillatory magnetic disturbance generated by the GIS operation. Seen from the measured disturbance, the space magnetic field's maximum amplitude is 70.2 A/m; therefore the test level not less than Level-5 can meet the requirements.

5 CONCLUSIONS

As the GIS of 1,000 kV substation is operated, the VFTO is generated on one side, which at the same time can also be coupled to the secondary equipment port by means of conduction and induction; therefore it is necessary to strengthen the EMC assessment of secondary equipment. The analysis in this paper recommends that during the damped oscillatory immunity test, the test level is not less than Level-3 and during damped oscillatory magnetic field immunity test, the test level is not less than Level-5.

In order to ensure the normal operation of secondary equipment, the equipment's capacity of resisting disturbance shall be also improved, such as better shielding, reasonable arrangement for grounding, and optimization of electronic circuit design.

REFERENCES

1. Bajramovic Z, Turkovic I, Mujezinovic A, et al. Overvoltages in secondary circuits of air-insulated substation due to GIS switching [C]//Power System Technology (POWERCON), 2012 IEEE International Conference on. IEEE, 2012: 1–4.
2. B.D. Russell, S.M. Harvey, S.L. Nisssol. Substation electromagnetic disturbance: Part 2 susceptibility testing and EMI simulation in high voltage laboratories, IEEE Transactions on PAS, 1984, 103(7): 1871~1879.
3. CIGRE W G. 36. 04: Guide on EMC in Power Plants and Substations [J]. Techn. Rep, 1997 (124).
4. Cui Xiang. A Review of CIGRE'2002 on Power System Electric-magnetic Compatibility[J]. Automation of Electric Power System, 2003, 27(4): 1–5.
5. IEC. 61000-4-18. Electromagnetic Compatibility (EMC)–Part 4–18: Testing and measurement techniques-Damped oscillatory wave immunity test[S]. International Electrotechnical Commission, 2006.
6. IEC. 61000-4-10. Electromagnetic Compatibility (EMC)–Part 4–10: Testing and measurement techniques-Damped oscillatory magnetic field immunity test[S]. International Electrotechnical Commission, 2001.
7. P.H. Pertorus, A.C. Britten, Jpreynders. Radiated transient measurements with substation operation[R]. CIGRE 1996, 36–203: 49–53.
8. S. Okabe, M. Kan, T. Kouno. Analysis of surges measured at 550 kV substations[J], IEEE Transactions on Power Delivery, October 1991 Vol.6(4):1462–1468.
9. Wiggins C.M., Thomas D.E., Nickel F.S., et al. Transient electromagnetic disturbance in substations[J]. Power Delivery, IEEE Transactions on, 1994, 9(4): 1869–1884.

Power and Energy – Kong (Ed.)
© 2015 Taylor & Francis Group, London, ISBN 978-1-138-02782-4

Augmented node branch incidence matrix based back/forward sweep flow calculation

Wanyue Zhang & Xingying Chen
College of Energy and Electrical Engineering, Hohai University, Nanjing, China

Hong Zhu
Nanjing Power Supply Company, Nanjing, Jiangsu, China

Kun Yu & Yingchen Liao
Nanjing Engineering Research Center of Smart Distribution Grid and Utilization, Nanjing, China

ABSTRACT: Topology changes in Distribution Network Reconfiguration, a topology analysis method based on Augmented Node Branch Incidence Matrix is proposed, combined with back/forward sweep power flow algorithm, using plus-minus sign represents the actual flow direction, without renumbering the network after topology changes. This algorithm can also check Loops and Islands in network structure, results of IEEE-69 nodes system verify the correctness of the proposed method.

1 INTRODUCTION

Distribution network closed-loop design, open-loop operation, changing the state of switch can isolate fault, restore power and reconfigure network, so distribution network is a variable structure dissipative network actually. The other characteristic of distribution network is its structure is different from transmission network, destruct the diagonal dominance in Jacobi matrix elements, and likely to cause pathological network, classic flow algorithm may not converge. The high ration of R/X in distribution network make P-Q decoupled load flow algorithm does not converge, although some improvement methods are put forward, but convergence is still poor. Therefore, Back/Forward Sweep (B/FS) power flow fit for the special network in distribution systems has been widely used in the power distribution network. But the back/forward sweep power flow algorithm has strict order when calculating the electricity, traditional method cannot meet the network topology changes.

To solve these problems, solutions have been proposed in the literature, such as hierarchical node based back/forward sweep power flow, using arc matrix structure to determine the change of the flow, using the tree node class to reflect the link relations between nodes dynamically, using digraph matrix C to recognition topology change, using node-layered incidence matrix, those above methods do not introduce how to check Loop and Islands, and need too much auxiliary matrix,

programming complex. The proposed method in this paper, using Augmented Node Branch Incidence Matrix (ANBIM), can adapt the topology change, check loops and islands, can also combined with back/forward sweep power flow algorithm, results of IEEE-69 nodes example verify the correctness of the proposed method.

2 TOPOLOGY ANALYSIS

2.1 *Branch layer analysis*

Without loss of generality, description of the method uses the diagram (Fig. 1).

First, form ANBIM using network topology information, denoted as A. There are x nodes, y branches, z Tie Switches (TS), we can get $x \times (y + z)$ ranks ANBIM. Keeping TS in this matrix, we can analyze topology flexibly when network changes. If the element in row i and column j in the matrix is "1", then node i and branch j directly connected.

Where the rows represent node 1 to 8 and the columns represent branch 1 to 9.

Secondly let column i be zeros if branch i break. Although many simplified methods were proposed, there are also non-feasible solutions in the optimization process of network reconfiguration, such as loops, islands, which cannot meet the operational requirements.

A modified topology identification method is proposed in this paper, which can deal with the above situation, flow chart is shown in Figure 3.

BLM in Figure 3 stand for branch-level matrix. The original network status as Figure 1, let the 8-th, 9-th column be zeros, search form the root node 1, results are shown below:

$$L = \begin{bmatrix} 1 & 0 & 0 \\ 2 & 3 & 0 \\ 4 & 5 & 6 \\ 7 & 0 & 0 \end{bmatrix}$$

Which is BLM, the nonzero elements in i-th row represent the i-th level branches.

2.2 Topology layer analysis with islands

If branch 2, 4 were break in Figure 1, let the 2-th, 4-th columns be zeros, denoted as A', analyze A' using the flowchart in Figure 3, results are shown below:

$$L = \begin{bmatrix} 1 & 0 \\ 3 & 0 \end{bmatrix}$$

If the number of the nonzero elements in L less-than the number of supply branches under the initial state, then there are isolated branches.

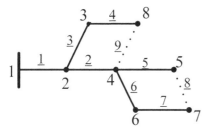

Figure 1. 8 node distribution network.

	Initial state branches						branches with TS	
$A=$	1	0	0	0	0	0	0	0	0
	1	1	1	0	0	0	0	0	0
	0	0	1	1	0	0	0	0	0
	0	1	0	0	1	1	0	0	1
	0	0	0	0	1	0	0	1	0
	0	0	0	0	0	1	1	0	0
	0	0	0	0	0	0	1	1	0
	0	0	0	1	0	0	0	0	1

Figure 2. Augmented node branch incidence matrix.

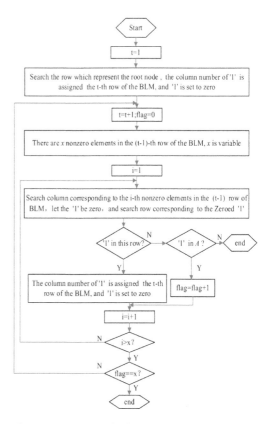

Figure 3. Flowchart for finding the branch-level.

2.3 Topology layer analysis with loops

There may be loops in the optimization process of network reconfiguration, such as branch 3, 4 were break in Figure 1, there will be a loop in the network, then let the 3-th, 4-th columns be zeros, denoted as A'', analyze A'' using the flowchart in Figure 3, results are shown below:

$$L = \begin{bmatrix} 1 & 0 & 0 \\ 2 & 0 & 0 \\ 5 & 6 & 9 \\ 8 & 7 & 0 \\ 7 & 0 & 0 \end{bmatrix}$$

Branch 7 appears twice, then there is one loop in the network, and a rule exists here:
There will be m loops in the network if there are m branches appear twice in the BLM.

2.4 Topology analysis with distributed generation

When the main power grid or feeder fault, part of the load can be supplied from Distributed

450

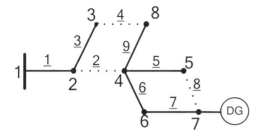

Figure 4.　Network with DG.

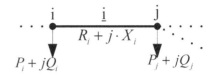

Figure 5.　Schematic diagram of the network parameters.

Generation (DG), the proposed method can also analysis topology with DG.

Assume operating condition as Figure 4, load point 2, 3 are supplied by main power, the other loads are supplied by DG.

Branch 2, 4, 8 are break, then let the corresponding columns in A be zero. First, let node 1 be root node, search the BLM, we can get L_M, then let node 7 be root node, search the BLM, we can get L_{DG}:

$$L_M = \begin{bmatrix} 1 & 0 \\ 3 & 0 \end{bmatrix} \quad L_{DG} = \begin{bmatrix} 7 & 0 \\ 6 & 0 \\ 5 & 9 \end{bmatrix}$$

Analysis is correct.

3　COMBING WITH B/FS

When network changes, the direction of the flow may change. To identify the real direction, the following conventions were made:

The positive direction of the flow is promised as head node to end node, if the flow is from the end to the head, then a minus sign was used to distinct.

To avoid too many auxiliary matrixes, structure is used to store branch and node parameter, the parameters of the following network is stored as Figure 6 and Figure 7.

3.1　Set the flag of branch structure

The level of the branch lies in BLM, now we will set the flag of branch structure in order to get the correct direction of power flow.

Branches in two layers may connect with each other only when the layers are adjacent, traversal search from the first layer to the last layer, we will meet four conditions as follows:

1. $branch(i).mo = branch(j).shou$
   Set flags of the two branches be 0;

branch(i)	← Structure name
R_i	← Resistance of branch i
X_i	← Reactance of branch i
i	← Head of branch i
j	← End of branch i
p	← Active power of branch i
q	← Reactive power of branch i
flag	← Flag of branch i

Figure 6.　Branch structure for Figure 5.

node(i)	← Structure name
P_i	← Active power of node i
Q_i	← Reactive power of node i
u	←Voltage magnitude of node i
δ	←Voltage phase of node i

Figure 7.　Node structure for Figure 5.

2. $branch(i).mo = branch(j).mo$
   Set flag of branch i be 0, flag of branch j be 1;
3. $branch(i).shou = branch(j).mo$
   Set flag of branch i be 1, flag of branch j be 1;
4. $branch(i).shou = branch(j).shou$
   Set flag of branch i be 1, flag of branch j be 0;

The four cases above, i stand for any branch except the last layer, j stand for branch in adjacent lower layer corresponding to i.

3.2　Forward sweep calculating power

Branches in the last layer, process is as follows:
Judge the flag,
If $flag = 0$, flow direction is positive, calculate using the following formula:

$$branch(i).p = P_m + \frac{P_m^2 + Q_m^2}{U_m^2} r_i \qquad (1)$$

$$branch(i).q = Q_m + \frac{P_m^2 + Q_m^2}{U_m^2} x_i \qquad (2)$$

If *flag* = 1, flow direction is negative, calculate using the following formula:

$$branch(i).p = P_s + \frac{P_s^2 + Q_s^2}{U_s^2} r_i \tag{3}$$

$$branch(i).q = Q_s + \frac{P_s^2 + Q_s^2}{U_s^2} x_i \tag{4}$$

where *branch(i).p, branch(i).q*, stand for power flow of the head of branch under the real flow direction; P_m, Q_m, stand for power of end node and P_s, Q_s, stand for power of head node of branch *i*; U_m, U_s stand for voltage magnitude of end node and head node of branch *i*, the first calculating adopt rated value, then adopt last calculated value.

Branches not in the last layer, process is as follows:

Judge the flag,

If *flag* = 0, *flow* direction is positive, calculate using the following formula:

$$branch(i).p = \sum P + \frac{\left(\sum P\right)^2 + \left(\sum Q\right)^2}{U_m^2} r_i \tag{5}$$

$$branch(i).q = \sum Q + \frac{\left(\sum P\right)^2 + \left(\sum Q\right)^2}{U_m^2} x_i \tag{6}$$

$$\sum P = P_m + NLP \tag{7}$$

$$\sum Q = Q_m + NLQ \tag{8}$$

If *flag* = 1, flow direction is negative, calculate using the following formula:

$$branch(i).p = \sum P + \frac{\left(\sum P\right)^2 + \left(\sum Q\right)^2}{U_s^2} r_i \tag{9}$$

$$branch(i).q = \sum Q + \frac{\left(\sum P\right)^2 + \left(\sum Q\right)^2}{U_s^2} x_i \tag{10}$$

$$\sum P = P_s + NLP \tag{11}$$

$$\sum Q = Q_s + NLQ \tag{12}$$

where ΣP, ΣQ stand for power flow of the end of branch under the real flow direction; *NLP, NLQ* stand for total power of branches in the next layer which can connect to branch *i*, illustrate the calculating method using the case of *NLP*:

$$NLP = \sum_{j=1}^{a} branch(j).p \tag{13}$$

where, *a* stand for number of branches in the next layer, which can connect to branch *i*, any connection type in Section 3.1 should be counted.

3.3 *Back sweep calculating voltage*

It is simple when calculating voltage, calculate from the first layer to the last layer in the BLM:

Judge the flag, if *flag* = 0

$$n(br(i).mo).u = \sqrt{(U_s - \Delta U)^2 + \delta U^2} \tag{14}$$

$$n(br(i).mo).\delta = n(br(i).s).\delta - arc \frac{\delta U}{U_s - \Delta U} \tag{15}$$

If *flag* = 1

$$n(br(i).s).u = \sqrt{(U_m - \Delta U)^2 + \delta U^2} \tag{16}$$

$$n(br(i).s).\delta = n(br(i).mo).\delta - arc \frac{\delta U}{U_m - \Delta U} \tag{17}$$

where $n(br(i).mo).u$, $n(br(i).s).u$ stand for voltage magnitude of the end node, head node of branch *i*; $n(br(i).mo).\delta$, $n(br(i).s).\delta$ stand for voltage phase of the end node, head node of branch *i*; ΔU, δU stand for vertical component, horizontal component of voltage drop, which is well known.

Iteration calculate over and over using the above formula, until reach convergence precision:

$$\max |U_i(k) - U_i(k+1)| \leq \varepsilon \tag{18}$$

where *k* stand for iteration times, ε stand for convergence precision.

When complete calculating, judge the flag, if *flag* = 0, branch power will be positive, if *flag* = 1, branch power will be negative.

4 CASE STUDY

Figure 8 is a 12.66kV distribution system, which has 69 nodes and 5 base loops, 68 section switches and five Tie Switches, total load of the system is 3802 kW + j2694 kvar. The initial state network is shown in Figure 8 (a), switches {69, 70, 71, 72, 73} are break, let corresponding columns be zeros, then calculate the power flow. Figure 8 (b) is an alternative solution, switches {18, 13, 56, 8, 61} are break, let corresponding columns be zeros, then calculate the power flow. Results are shown as Table 1.

Result shown in Table 1 is per unit value, the reference value is 12.66 kV, 10 MVA. As we can see in Table 1, there is negative branch power, because the flow is reverse direction corresponding to reference direction.

452

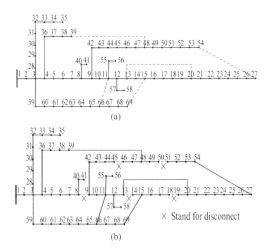

(a)

× Stand for disconnect

(b)

Figure 8. 69-nodes system.

Table 1. Power flow of two state.

Voltage magnitude			Active power of branch		
Node	(a)	(b)	Branch	(a)	(b)
1	1.0000	1.0000	1	0.4027	0.3914
2	1.0000	1.0000	2	0.4027	0.3914
3	0.9999	0.9999	3	0.3750	0.2428
4	0.9998	0.9999	4	0.2899	0.0162
5	0.9990	0.9998	~	~	~
~	~	~	10	0.0748	−0.0113
51	0.9120	0.9492	11	0.0566	0.0716
52	0.9117	0.9493	12	0.0362	0.0511
53	0.9098	0.9495	13	0.0353	0.0000
54	0.9092	0.9519	14	0.0344	−0.0008
55	0.9713	0.9728	~	~	~
56	0.9713	0.9728	68	0.0028	0.0028
~	~	~	69	0	−0.1015
66	0.9983	0.9790	70	0	0.0500
67	0.9983	0.9784	71	0	−0.0174
68	0.9982	0.9784	72	0	0.0320
69	0.9982	0.9781	73	0	0.1397

5 CONCLUSION

A flexible identify topology method is proposed, which base on augmented node branch incidence matrix with TS. Let the columns corresponding to disconnect branch be zeros, which can represent the topology changes, search from the root node we can get branch-level matrix. This algorithm can also check Loops and Islands in network, combined with back/forward sweep power flow algorithm, can calculate the result of the flow conveniently. This method doesn't need too many auxiliary matrix, and easy to program, the example proved the correctness of this method.

ACKNOWLEDGEMENTS

This research is supported by the National High Technology Research and Development of China 863 Program (2012AA050214), National Natural Science Foundation of China (51207047) and State Grid Corporation of China (key technology research, equipment development and demonstration of demand side's optimal operation, 2013-1209-6998).

REFERENCES

1. D. Rajicic and A. Bose, "A Modification to the Fast Decoupled Power Flow for Entworks with High R/X Rations," IEEE Transactions on Power system, Vol. 3, No. 2, 1988, pp. 743–746.
2. F Baran-M-E, Wu-F, "Optimal capacitor placement on radial distribution system," IEEE Transactions on Power Systems, Vol. 4, No. 1, 1989, pp. 725–734.
3. H.B. Sun, B.M. Zhang, N.D. Xiang, "Study on Convergence of Back/Forward Sweep Distribution Power Flow," Proceedings of the CSEE, Vol. 19, No. 7, 1999, pp. 27–30.
4. J.J. Wang, L. Lu, J.Y. Liu, "Reconfiguration of Distribution Network Containing Distribution Generation Units Based on Improved Layered Forward-Backward Sweep Method," Power System Technology, Vol. 34, No. 9, 2010, pp. 60–64.
5. J. Wang, P.J. Zong, "Power Flow Calculation for Radial Distribution Network Based on Forward and Backward Substitution Method," Northeast Electric Power Technology, Vol. 34, No. 2, 2008, pp. 7–10.
6. L. Liu, Y.B. Yao, X.Y. Chen, X.P. Sun, "Topology structure identification and flow calculation of 10 kV distribution network," RELAY, Vol. 28, No. 2, 2000, pp. 19–21.
7. L. Liu, X. Zhao, X.L. Jiang, "Distribution network topology analysis and flow calculation based on layer matrix," Power System Protection and Control, Vol. 40, No. 18, 2012, pp. 91–94,100.
8. P. Sun, "Research on Method of Urban Power Network Renovation Based on the Fuzzy Multi-objective Decision Theory," North China Electric Power University, 2013.
9. P.X. Bi, J. Liu, W.Y. Zhang, "A Refined Branch-Exchange Algorithm For Distribution Networks Reconfiguration," Proceedings of the CSEE, Vol. 21, No. 8, 2001, pp. 99–104.
10. R.Q. Li, L.F. Xie, Z.Y. Wang, J.S. Zhe, "Back/forward substitution method for radial distribution load flow based on node-layer," Power System Protection and Control, Vol. 38, No. 14, 2010, pp. 63–66.
11. X.F. Gu, C.Y. Guan, "Analysis and study on topology for power flow of distribution network," Journal of North China Electric Power University, Vol. 35, No. 2, 2008, pp. 47–50.

12. X.S. Zhang, Z. Liu, E.K. Yu, J.C. Chen, "A Comparison on Power Flow Calculationmethods For Distribution Network," Power System Technology, Vol. 22, No. 4, 1998, pp. 47–51.
13. Y. Zhang, Q. Wang, W.N. Song, "A Load flow Algorithm for Radial Distribution Power Networks," Proceedings of the CSEE, Vol. 18, No. 3, 1998, pp. 217–220.
14. Z.K. Li, X.Y. Chen, K. Yu, "Hybrid Particle Swarm Optimization for Distribution Network Reconfiguration," Proceedings of the CSEE, Vol. 28, No. 31, 2008, pp. 35–41.

Power and Energy – Kong (Ed.)
© 2015 Taylor & Francis Group, London, ISBN 978-1-138-02782-4

Multi-thread parallel data integration for large-scale power grids

Yifeng Dong & Sanen Du
China Electric Power Research Institute, Beijing, China

Tao Mu
State Grid Tianjin Electric Power Company, Tianjin, China

Yi Wang & Junxian Hou
China Electric Power Research Institute, Beijing, China

E. Zhijun & Weihong Zheng
State Grid Tianjin Electric Power Company, Tianjin, China

ABSTRACT: Data integration is the basis of online Dynamic Security Assessment system (online DSA), Strong and Smart Grid Dispatching Supporting System (SG-OSS) as well as other online simulation analysis systems rely on EMS data. There are two main objectives of data integration: accuracy and swiftness. This paper is to present a quick method of multi-thread parallel data integration for large-scale power grid. This paper uses the method mentioned above to integrate the online data of North China Grid, South China Grid as well as Central China Grid and then generate the data of PSD-BPA format to prove that this method is much swifter than the traditional data integration method. In practical operation, online data availability evaluation index is verified to be reasonable.

1 INTRODUCTION

On August 14th, 2003 the large-scale blackout in North America power grid affected more than 50 million people and resulted in the loss of 61.8 GW load [1]–[3]. On November 10th, 2009 and February 4th, 2011 two large-scale blackouts in Brail resulted in the loss of 28.83 GW load and 8 GW load respectively and caused a great economic loss [4][5]. These disturbances raise an alarm to the safe operation of China Power Grid. With the development of economy and the implementation of "West-to-East Power Transmission, North-south power exchange", China will form four synchronous power grids: North China—Central China—East China Grid (referred to as "Sanhua" Grid), Northeast Grid, Northwest Grid and South Grid. "Sanhua" Grid is UHV large synchronous interconnected power grid, which connects large coal base, hydropower base and load centers. Against this background, trans-regional power grids are very complicated, in order to ensure their security and stability and establish corresponding defense system, all provincial power grids or above establish online Dynamic Security Assessment (DSA) system [6]–[11].

Data integration is the basis of online DSA system. By making full use of the existing data resources, its goal is to provide quality data which can show the grid operation state and meet requirements of calculation and analysis, so as to improve the application of advanced grid calculation and analysis program.

In the literature [13] basic methods of online data integration are introduced. However, these methods cannot fully meet the requirements of the current online DSA System for following two reasons: Firstly, there are huge amounts of data under the background of trans-regional power grids. Therefore, the speed of online data integration is an important index to reflect its performance, then how to accomplish online data integration more quickly is an urgent problem. Secondly, with its developments in recent years, online DSA System has been transitioned from research to practical application and it can instruct production to a certain extent. But its analysis results, especially the data source, are widely doubted. Therefore, to improve the accuracy of online data integration and evaluate the credibility of data source are essential way to enhance the credibility of DSA analysis results.

This paper is to present Parallel Multi-thread data integration based on online E data [13] and offline PSD-BPA data and its implementation scheme. Power grids are naturally divided.

According to this characteristic, an overall design to multi-thread data integration is completed. It mainly solves the problems about the speed of online data integration as well as the modification and availability evaluation of online data. This paper offers the synchronous data as an example, which verifies effectiveness of multi-thread parallel data integration.

2 MAIN FUNCTION MODULES

Online data integration includes four parts: Data reading and parsing; topological connection and analysis; Data error handling and availability evaluation; calculated data generation. The procedure is shown in Figure 1 (The yellow ones are parallel processing steps.)

3 DATA READING AND ANALYZING

3.1 *Data reading*

For large-scale power grid, online data has a huge amount of information, and the file containing these data is relatively large. The best way to read

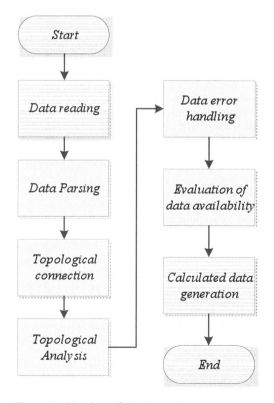

Figure 1.   Flowchart of data integration.

large files is using Memory-Mapped Files. Memory-Mapped files are mapped by a file into the process address space, which are similar to virtual memory. Compared with traditional file reading method, files can be read more quickly by Memory-Mapped Files. Therefore, Memory-Mapped Files play a very important part in processing files which contain large amount of data.

Memory-Mapped Files are created by calling function of CreateFileMapping() in windows system, while in Linux system by calling function of mmap().

3.2 *Data parsing*

Online data includes information about substations, buses, transformers, AC lines, generators, loads, parallel compensators, series compensators, rectifiers and inverters, DC lines, topological nodes and so on.

After the grid data is mapped to memory through Memory-Mapped Files, the memory will be divided into blocks according to the corresponding data from different kinds of equipment, and then each block will be parsed. The OpenMP [15]–[17] parallel multi-thread technology processes each block so as to realize the fast parsing of data file.

4 TOPOLOGICAL CONNECTION AND ANALYSIS

4.1 *Topological connection*

The purpose of topological connection is to connect various equipment and topological node, which can calculate the exertion, load, compensation and fixed voltage, etc.

For single-terminal equipment, connect the single-terminal equipment with topological node; for double-terminals or multi-terminals equipment, after connecting the double-terminals or multi-terminals equipment with topological nodes, connective topological node shall be generated for each topological node. Single-terminal equipment includes buses, generators, loads, etc. Double-terminals or multi-terminals equipment include AC lines, transformers, etc.

Topological connection could use OpenMP multi-thread parallel technology based on different electric equipment and establish connective topological node diagram.

4.2 *Topological analysis*

The purpose of topological analysis is to conduct topological division for the whole power grid, and to determine the topological node and the effective output range of different equipment.

Topological analysis conducts traversal analysis to connect topological node diagram, and it can use depth-breadth prioritized searching methods. For multi-area interconnected network, it can use OpenMP multi-threading parallel technology based on network division.

Each parallel conducts topological search and encode for each division, and then encoding and combination are done by main thread. So it can accelerate the speed of topological analysis.

5 DATA ERROR HANDLING AND AVAILABILITY EVALUATION

5.1 Data error handling

Data Error Handling is mainly used to solve the problems of initial voltage over-limit of generators, the ratio-missing of transformers and the node-unbalanced power. The initial voltage over-limit of generators will set limit to a predetermined value (for example 1.0 pu); transformer ratio missing can choose the rated transformation ratio. The result can also be calculated by the voltage on both sides as well as the power, but the transformer corresponding gear ratio of non-continuity should be taken into consideration.

Node imbalance of power is usually due to some of the estimated equivalents (such as low voltage level line). For the imbalance of power exceeding a predetermined limit, the unbalanced node load should be supplemented. Node unbalanced power can be calculated by the formula as follows:

$$\Delta P_i = \sum_{m=1}^{n_G} P_{Gm} + \sum_{m=1}^{n_L} (-P_{Lm}) + \sum_{m=1}^{n_B} P_{BINm} \qquad (1)$$

$$\Delta Q_i = \sum_{m=1}^{n_G} Q_{Gm} + \sum_{m=1}^{n_L} (-Q_{Lm}) + \sum_{m=1}^{n_B} Q_{BINm} + \sum_{m=1}^{n_{Sh}} Q_{SHm} \qquad (2)$$

where ΔP_i = the imbalance of active power of node i; ΔQ_i = the imbalance of reactive power of node i; P_{Gm} = the active power of the m-th generator on node i; Q_{Gm} = the reactive power of the m-th generator on node i; P_{Lm} = the active power of the m-th load on node i; Q_{Lm} = the reactive power of the m-th load on node i; P_{BINm} = the active power injection of the m-th branch connecting with node i; Q_{BINm} = the reactive power injection of the m-th branch connecting with node i; Q_{Lm} = the reactive power of the m-th compensator on node i.

5.2 Evaluation of data availability

Evaluation of data availability should be considered comprehensively depending on whether each state variable surpasses the limits, whether or not there is a big imbalance of power, etc. This article tries to evaluate by verifying whether the data source meets the flow equations.

Construct the two-node model line and evaluate whether the value of the line meets the trend by verifying the difference between the calculated value and the actually measured value as well as the difference between the output power of the two terminals of the line.

The trend lines are verified through formula (3) and (4).

$$\Delta \dot{S}_{ni} = \dot{S}_{ni} - \dot{V}_i \overset{*}{I}_n - j\frac{1}{2}V_i^2 B_n \qquad (3)$$

$$\Delta \dot{S}_{nj} = \dot{S}_{nj} + \dot{V}_j \overset{*}{I}_n - j\frac{1}{2}V_j^2 B_n \qquad (4)$$

$$\dot{I}_n = \frac{\dot{V}_i - \dot{V}_j}{R_n + jX_n} \qquad (5)$$

where ΔS_{ni} = the power output error of n-th branch's i-terminal; ΔS_{nj} = the power output error of n-th branch's j-terminal; S_{ni} = the power output of n-th branch's i-terminal; S_{nj} = the power output of n-th branch's j-terminal; V_i = the voltage of n-th branch's i-terminal or the voltage of i-th node; V_j = the voltage of n-th branch's j-terminal or the voltage of j-th node; I_n = the current from i-terminal to j-terminal of n-th branch. R_n = the resistance of the n-th branch; X_n = the reactance of the n-th branch; B_n = the susceptance of the n-th branch. All the values are per unit values. The value marked with a dot is phasor. The value marked with an asterisk is conjugate phasor.

Finally, calculate the errors of all the nodes and conduct normalization processing, and then get the evaluation index of source data availability.

6 CALCULATED DATA GENERATION

Output the information about topological nodes, AC lines and transformers in the concerned area into power flow file in the format of **PSD-BPA**. If generator dynamic parameter is prepared, transient stability file in **PSD-BPA** format can also be generated.

In the **PSD-BPA** format node name has 8 bytes length limit, but most of the node names in online data are over 8 bytes, so it is necessary to generate node name of **PSD-BPA** format, and generate the corresponding name mapping table.

7 EXAMPLE

The example uses source data from State Grid on April 10th 2013 at 8:36 and 2013 annual operation

Table 1. Comparison between serial program time and parallel program time.

Program	Time (second)
Serial	15.64
Parallel (with 4 threads)	9.93

mode. Data include 15752 substations, 19476 buses, 12087 AC lines, 4023 generators, 7777 transformers, 30643 loads, 8114 compensators.

The program is running in the Linux system, and the configuration of experimental computer is 2.5 GHz quad-core CPU, 4 GB of memory. Results are shown in Table 1.

By calculating the results in the above table, the acceleration ratio is 1.575. At the same time, the generated PSD-BPA data pass the tests of power flow calculation, transient stability calculation, short circuit calculation and small disturbance calculation, so it can meet the requirements of online DSA.

8 CONCLUSION

A fast approach to Parallel Multi-thread data integration based on online E data and offline PSD-BPA data is proposed in this paper. By applying OpenMP parallel multi-thread technology, this approach can guarantee fast calculation. It also put an end to conflicts between threads because it employs different parallelization schemes for different steps. The example has verified the correctness of this approach.

REFERENCES

1. Yin Yonghua, Guo Jianbo, Zhao Jianjun, et al. "Preliminary analysis of large scale blackout in interconnected North America power grid on August 14 and lessons to be drawn", Power System Technology, Vol. 27, No. 10, 2003, pp. 1–5.
2. Hu Xuehao, "Rethinking and enlightenment of large scope blackout interconnected North America power grid", Power System Technology, Vol. 27, No. 9, 2003, pp. 2–6.
3. Liu Yongqi, Xie Kai, "Analysis on blackout of interconnected North America power grid occurred on Aug 14, 2003 from the viewpoint of power system dispatching", Power System Technology, Vol. 28, No. 8, 2004, pp. 10–15.
4. Lin Weifang, Sun Huadong, Tang Yong, et al. "Analysis and Lessons of the Blackout in Brazil Power Grid on November 10, 2009", Automation of Electric Power Systems, Vol. 34, No. 7, 2010, pp. 1–5.
5. Lin Weifang, Tang Yong, Sun Huadong, et al. "Blackout in Brazil Power Grid on February 4, 2011 and Inspirations for Stable Operation of Power Grid", Automation of Electric Power Systems, Vol. 35, No. 9, 2011, pp. 1–5.
6. K. Morison, L. Wang and P. Kundur, "Power system security assessment", IEEE Power and Energy Magazine, Vol. 2, No. 5, 2004, pp. 30–39.
7. P. Kundur, G.K. Morison and L. Wang, "Techniques for on-line transient stability assessment and control", IEEE Power Engineering Society Winter Meeting, 2000.
8. Yan Jian-feng, Yu Zhi-hong, Tian Fang, Zhou Xiao-xin, "Dynamic Security Assessment & Early Warning System of Power System", Proceedings of the CSEE, Vol. 28, No. 34, 2008, pp. 87–92.
9. Zheng Chao, Hou Jun-xian, Yan Jian-feng, et al. "Functional Design and Implementation of Online Dynamic Security Assessment and Early Warning System", Power System Technology, Vol. 34, No. 3, 2010, pp. 55–60.
10. Shi Li-bao, Zhou Hua-feng, Peter T.C. Tam, et al. "Implementation of Power System Real Time Dynamic Security Assessment and Early Warning System", Journal of System Simulation, Vol. 19, No. 23, 2007, pp. 5401–5405.
11. Yuan Zeng, Pei Zhang, Meihong Wang, et al. "Development of a new tool for dynamic security assessment using dynamic security region", International Conference on 2006 PowerCon, 2006.
12. Yan Yaqin, Tao Hongzhu, Li Yalou, et al. "Research and practice of integration of information in power dispatching center", Relay, Vol. 35, No. 17, 2007, pp. 37–40.
13. Xin Yao-zhong, Tao Hong-zhu, Li Yi-song and Shi Jun-jie, "E Language for Electric Power System Model Description", Automation of Electric Power Systems, Vol. 35, No. 10, 2006, pp. 48–51.
14. Yang Yihan, Zhang Dongying, "Study on the architecture of security and defense system of large-scale power grid", Power System Technology, Vol. 28, No. 9, 2004, pp. 23–27.
15. OpenMP Architecture Review Board, http://www.openmp.org.
16. OpenMP: a proposed industry standard API for shared memory programming. http://www.openmp.org. 1997.
17. Jiang QuanYuan, Jiang Han, "OpenMP-based parallel transient stability simulation for large-scale power systems", Vol. 55, No. 10, 2012, pp. 2837–2846.

Power and Energy – Kong (Ed.)
© 2015 Taylor & Francis Group, London, ISBN 978-1-138-02782-4

Status and expectation of control & protection for Energy Internet

Qingping Wang
ABB Corporate Research, Beijing, China

Jingyi Wang
Beihang University, Beijing, China

ABSTRACT: The control and protection of Power Energy Internet is one of the most important research topics. The research status of Energy Internet is introduced including principle research, technical development and pilots. According to new requirements of Energy Internet, the challenges of control, protection and energy management technology are analyzed. The new technical trends are presented as follow: architecture of P&C, local real-time control, autonomous management, protection & FLISR, wide-area control & management and communication interface standard.

1 INTRODUCTION

Issues of fossil energy and environmental deterioration are more and more severe these years with global incremental consumption of energy. It has been the new global energy strategy that development and utilization of energy is based on electric power technology. The traditional fossil energy will be replaced totally by electric power produced by renewable energy. But the traditional power grid is not suitable to connection of a mount of distribution generations anymore because of intermittence, variability and randomness of renewable energy. In order to meet requirements of huge distributed generation and bidirectional power flow, Power Energy Internet has been direction of smart grid technology in the future.

Energy Internet is presented by Jeremy Rifkin in "The Third Industrial Revolution". Because the electric power is easy for collection, transmission and distribution, power grid is one of the best choices to realize Energy Internet. Power Energy Internet is defined as a power grid which can support bidirectional exchange by integration of distributed renewable generation, based on electronic and information technology.

The control and protection of Power Energy Internet is one of the most important parts of research area for Power Energy Internet as one of the five pillars of Energy Internet, which is based on information network, wide area measurement, high-speed sensor, high efficient calculation and intelligent control. Both research status and technical forecast of control & protection for Power Energy Internet are presented in this paper.

2 CONCEPT OF ENERGY INTERNET

2.1 Concept of Energy Internet

The concept of Energy Internet is firstly presented by American economic and social theorist Jeremy Rifkin in his famous book called "The Third Industrial Revolution". Millions of people would produce their own renewable energy in their homes, offices, and factories and share green electricity with each other in an "Energy Internet" just like we now generate and share information online.

Rifkin describes how the five pillars of the Third Industrial Revolution will create thousands of businesses and millions of jobs, and usher in a fundamental reordering of human relationships, from hierarchical to lateral power, that will impact the way we conduct business, govern society, educate our children, and engage in civic life.

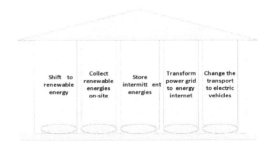

Figure 1. The pillars of the Third Industrial Revolution.

The five pillars of the Third Industrial Revolution are (1) shifting to renewable energy; (2) transforming the building stock of every continent into green micro–power plants to collect renewable energies on-site; (3) deploying hydrogen and other storage technologies in every building and throughout the infrastructure to store intermittent energies; (4) using Internet technology to transform the power grid of every continent into an energy internet that acts just like the; and (5) transitioning the transport fleet to electric plug-in and fuel cell vehicles that can buy and sell green electricity on a smart, continental, interactive power grid.

2.2 *Architecture of Energy Internet*

The energy grid of the future will not appear to be very different from today's infrastructure. Figure 2 shows this schematically—just like today, there will be large-scale power plants (1) that transport energy to the consumers (3) via transmission grids (6) and distribution grids (7). In order to make greater use of regenerative energy sources, an even larger number of decentralized generators (2) will be installed than currently exist today, which—just like the large-scale power plants—will contribute to meeting the demand for energy. Since energy feed-in will be increasingly decentralized and consumers will react more flexibly and intelligently (4), more situations will occur in which load flows (8) in sub-grids will be reversed. If this dynamic infrastructure is to be operated and coordinated efficiently, all individual components must be integrated (10) into a uniform communication infrastructure—the Energy Internet (9), which will serve to map all producers and consumers of the energy grid onto one virtual level.

3 RESEARCH STATUS AND PROGRESS

3.1 *New technologies of power Energy Internet*

CIMEG (Consortium for Intelligent Management of Electric-power Grid) advanced an anticipatory control paradigm named Local Area Grid. It uses a bottom-up approach to circumvent the technical difficulty of defining the global health of a power system at the top level.

The concept of intelligent Power Routing (PR) is proposed as a new function in power delivery systems. The energy router is actually formed by an SST (Solid state transformer) device that performs control and voltage step-down function.

A single-phase SST circuit topology is illustrated in Figure 3. The SST is a power electronic device that does much more than just a voltage step-down function. The power frequency input signal of the primary side of the transformer turns into the high frequency signal. As the volume of the core type transformer is inversely proportional to frequency, so the volume of the SST is much smaller than the legacy transformer.

The SST can realize the control of power frequency, voltage and waveform. It can send messages to communicate with the same level SSTs through the optical fiber network. Besides, the SST can control the power flow real time; minimize the network loss; control voltage and reactive power optimization of the distribution system.

As the generation and storage subsystems components are added to (or removed from) the distribution network, which may have different interface types (ac or dc), different voltage levels, different power levels and different power quality requirements. The communication interface acts as a smart plug-and-play interface and any device coupled to the grid can be instantly recognized as soon as it is connected.

The energy router support the functions can be categorized into the following seven domains: bulk generation, transmission, distribution, operation, market, customer, and service provider. Therefore the energy router is an enabling technological component in the energy internet operations.

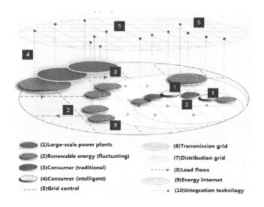

Figure 2.   The infrastructure of future power grid.

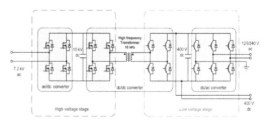

Figure 3.   Single-phase SST circuit topology.

The control of the energy internet is inherently much more complicated than the traditional power distribution system. The multi layers control system in the FREEDM (Future renewable electric energy delivery and management) system is classified in Figure 4. In the control schematic, these controls are further classified into two groups: IEM (Intelligent energy management) control and IFM (Intelligent fault management) control. This architecture allows the FREEDM system to be scalable and the number of IEM & IFM nodes can be increased or decreased without any major change to the operating system.

The intelligent terminal system of the energy internet is composed of the smart meter, the intelligent electrical switch and control system.

Smart meter is a product based on the combination of modern computer technology and modern measurement technology. It has capacities of automatic correction; automatic compensation; automatic storage, computing and logic judgment of data; automatic operation and network transmission. One could generate periodic reports of the electric power drained by home appliances, in order to come up with policies. Smart meters within a house may be in charge of detecting anomalous situations locally, and report only the most relevant sets of samples for external analysis.

As the use of large amounts of DESDs and DRERs in the energy internet, a great quantity of the control data and flexible control modes were created. This makes difficult for the centralized control mode that unified judgment and scheduling by the scheduling center to control in an effective way. The control will be distributed to network elements; these elements change their running states according to the grid scheduling. This distributed coordination control pattern will effectively solve these problems. Multi-agent control System is

divided into control layer, application layer and system layer.

Some new intelligent control methods were raised in the MAS architecture. A new method for reliability evaluation of active distribution systems with multiple micro-grids based on a Monte Carlo simulation was proposed. Based on optimal power flow solutions, Dynamic optimal power flow was put forward which is an extension of OPF to cover multiple time periods.

3.2 Pilots and applications of Energy Internet

1. Europe: ARTEMIS-IoE Project

The 3 year ARTEMIS-project (Advanced Research & Technology of Embedded Intelligence and Systems) is focusing on opportunities from smart-grid developments to enable and support the large-scale uptake of electric mobility. The Internet will be connected with the energy grids to enable intelligent control of energy production, storage and distribution; these are all key infrastructure enablers for the widespread use of electric vehicles. The project is proposing distributed systems to implement the real-time interface between the energy grid and a cloud of devices.

A bi-directional power/energy transport on the power transmission lines provided by a new smart electric grid relying on data/information transferred across both power lines and data links; greater consumer participation thanks to capillary spread information and economic benefits on the energy production, use, storage and transportation; significant levels of energy storage capabilities and generation that will greatly help balance the disparities between peak supply and demand.

2. France: Nice Grid

The Nice Grid project is about demonstrating the management of distributed energy resources located in a solar district. Three innovative operation modes of the distribution grid have been separately tested in three nested areas. The biggest one, the Load Management Area, will demonstrate how demand side management can benefit the Transmission System Operator and Distribution System Operator, and how these two entities could coordinate when their needs are incompatible. The medium size area, the Solar District, is known to expect voltage issues in the future, mainly during the summer due to photovoltaic power production that cannot be locally absorbed by any loads. The smallest area, the Micro-Grid, will be used to show how planned islanding can be realized by the coordinated management of various active devices.

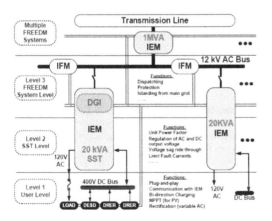

Figure 4. Control schematic of the FREEDM system.

3. Germany: E-Energy

E-Energy: The future energy system based on Information & Communications Technology is a technology innovation promotion plan, which was sponsored by the German Federal Ministry of economics and technology. It's a part of the German "green IT pioneer" action plan. E-Energy has six demonstration projects.

4 NEW CHALLENGE TO CONTROL & PROTECTION OF ENERGY INTERNET

With the development of Energy Internet, the existing control and protection system of power grid cannot meet the requirements of Energy internet. The corresponding control & protection system is facing new challenge for technology reform.

4.1 *Diverse ports for DERs and loads*

More and more distributed generators and DC loads are connected to distribution network. The different grid-connected ports including DC & AC interface are necessary for different demands. Only AC port based on AC transformer is provided in traditional power grid. More types of ports can be provide to DC source and DC load by taking advantage of Solid-State Transformer (SST). The current control system is definitely not suitable to the diverse grid-connected interface.

4.2 *Flexible power flow control*

The power flow control is very important to minimize loss or eliminate overload in distribution network. It is realized by switch operation to change network topology in current power grid, which cannot optimize system accurately and flexibly. But many power electronic devices can be used for active power control in energy internet, such as grid-connected inverter, SST and Unified Power Flow Controller (UPFC). The coordinative control of active power is one of the most important functional requirements for Energy Internet.

4.3 *Voltage and reactive power optimization*

The traditional solutions for eliminating voltage violation are tap changing of OLTC (On Load Tap Changer) and shunt capacitor switching. But there are other methods for precise voltage regulation in Energy Internet, for example SST, DSTATCOM and APF (Active Power Filter). The control algorithm of power electronic elements in Energy Internet for voltage regulation and reactive power optimization needs to be researched in P&C system.

4.4 *Custom power (high-quality power supply)*

In order to improve the quality of power supply, many devices are used in Energy Internet, which include DVR (Dynamic Voltage Regulator), APF, UPQC (Unified Power Quality Conditioner), etc. According to the demand of custom, different quality is guaranteed correspondingly. The control system of custom power is also one of the important issues of which needs to be solved in Energy Internet.

4.5 *Fault handling with distributed source*

There are different kinds of DGs and power electronic devices in Energy Internet. Therefore Fault handling in Energy Internet will be quite different to traditional distribution network. The Fault Location, Isolation and System Restoration (FLISR) should be developed for Energy Internet.

4.6 *Demand response with DGs*

Demand response is usually used for balance of power generation and load. Only active loads are concerned in current distribution network instead of distributed generation in demand side. New functionality of demand response should realize coordinative management of DGs and active loads.

4.7 *Real-time control for mass devices*

Most of DGs and power electronic devices in Energy Internet can be controlled for energy management. But in traditional distribution network, only switch and OLTC need to be controlled. The current distribution management system is composed by master station and remote terminals, and the coordinative control is realized by master station. It is impossible to implement the real-time control of Energy Internet based on master station. The distributed real-time control is new requirement of Energy Internet.

5 FURTHER PROSPECT OF CONTROL & PROTECTION FOR ENERGY INTERNET

In order to meet the challenge of Energy Internet, the control and protection technology for Power Energy Internet becomes one of the important trends in future power system.

5.1 *Architecture of P&C for energy internet*

The architecture of traditional Distribution Automation (DA) system is not suitable to Energy Internet in the future anymore. The new architecture for

Energy Internet needs to be researched to support real-time control and management of massive DGs and active loads.

The architecture of P&C system for Energy Internet should be divided into different levels, shown in Figure 5. The basic level is composed by sensors, meters and measuring devices for measurement and control, which are combined with energy routers or power electronic devices. Intelligent Terminal Unit (ITU) is the local control unit for coordinative control with other ITUs, which can be configured by master station and control energy routers and power electronic devices. The wide-area control and energy management system is used for optimization of power grid and configuration of local ITUs, which is installed in master station.

5.2 The real-time control of energy router with diverse ports

The energy router provides different types of ports for AC/DC source and AC/DC load. The interface of energy router based on SST is shown in Figure 6.

The energy router based on SST has 4 kinds of voltage ports: low-voltage AC, low-voltage DC,

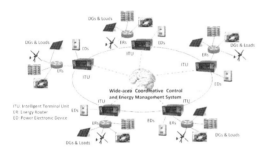

Figure 5. Architecture of P&C system for Energy Internet.

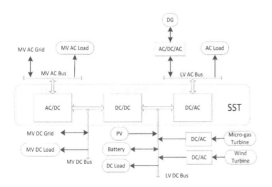

Figure 6. Energy router with diverse ports.

mid-voltage AC and mid-voltage DC. The low-voltage ac and dc ports are used to connect DGs and loads. Some DC loads and DGs with DC output (such as PV and battery) can be connected to energy router directly, and other DGs with AC output (such as micro-gas turbine and wind turbine) can be connected to DC bus by AC/DC converter. The mid-voltage AC port is connected into the three-phase power distribution line. It can also be connected to other energy routers by mid-voltage DC port. The real-time control is expected to support the diverse ports of energy router. It is also the basis of the coordinative control and autonomous management for multi energy routers.

5.3 The autonomous management based on distributed multi-agent

The nodes of Energy Internet are sub-grid of distribution network. Each of them should be autonomous to realize distributed control and management. Based on multi-agent technology, ITU can implement fast network analysis and real-time control by communication with corresponding ITUs. The configurations of local control are set by master station.

The functionalities of coordinative control and autonomous management are shown in Figure 7. The basic functionalities are included but not limited as follow:

– Local SCADA
– Distributed power flow calculation & state estimation
– Local dynamic topology analysis
– Communication with other ITUs
– Power flow optimization and loss minimization
– Reactive power control and voltage regulation
– Power quality optimization and custom power
– Load & DG forecast and demand response
– State monitoring and dynamic rating.

5.4 FLISR for power energy internet

The typical protection and feeder automation for current distribution network are developed without concerning distribution power source in consumer side. The DGs and power electronic devices

Loss Minimization	Voltage Regulation	Custom Power	Demand Response with DG	Dynamic Rating
Power Flow Optimization	Reactive Power Control	Power Quality Optimization	Load & DG Forecast	State Monitoring
Distributed Power Flow Calculation & State Estimation		Local Dynamic Topology Analysis		Communication with other ITUS
Local SCADA				

Figure 7. Coordinative control and autonomous management.

will influence protection and fault location greatly. It is necessary to research new algorithm of fault location for Energy Internet.

Many new power electronic technologies can be used to implement fault isolation and service restoration. SFCL (Superconducting Fault Current Limiter) can limit short current when fault occurs. The SST and grid-connected inverter can isolate fault feeder or limit short current. SSTS (Solid-state Transfer Switch) can switch load to other power source within 10 ms to restore power supply.

5.5 The wide-area coordinative control and energy management for energy internet

The network analysis and optimization are mainly implemented by master station in current distribution network. There are many big differences between Energy Internet and traditional distribution network. The distributed analysis and real-time control are executed by ITUs instead of master station because of the huge amount of information and the real-time requirement.

The wide-area coordinative control and energy management system in master station is a brain of Energy Internet, which can set control logic and configuration to ITUs according network analysis results. Based on coordinative configuration, ITUs can work individually without master station.

5.6 Communication standard interface

Communication protocols already exist worldwide for individual sections of integrated communication. These protocols are mainly limited to one section, meaning that an integrated communication system is still a long way off. A key task to the drive towards realizing the Energy Internet will be the need to establish an integrated, bi-directional communication system from generation to the end consumer.

Furthermore, the importance of virtual power plants will continue to grow, and these will be dependent on up-to-date data. If an integrated communication system exists, new products and services can be developed. The communication system must also be available for new products. The communication standards must be open to enable other applications to be integrated.

6 CONCLUSION

The progress of Energy Internet is introduced in this paper, especially control and protection of Energy Internet. Based on analysis of new challenges of control and protection in Energy Internet, some technical trends have been presented, such as:

1. Architecture of P&C system for Energy Internet;
2. Real-time control of energy router with diverse ports;
3. The autonomous management based on distributed multi-agent;
4. Fault location, isolation and system restoration for Power Energy Internet;
5. The wide-area coordinative control and energy management system;
6. Communication standard interface.

REFERENCES

1. Amin, S.M.; Wollenberg, B.F., "Toward a smart grid: power delivery for the 21st century," Power and Energy Magazine, IEEE, vol.3, no.5, pp.34,41, Sept.-Oct. 2005.
2. Belitz, H.-J.; Winter, S.; Rehtanz, C., "Load shifting of the households in the E-Energy project E-DeMa," PowerTech (POWERTECH), 2013 IEEE Grenoble, vol., no., pp.1,6, 16–20 June 2013.
3. Bui, N.; Castellani, A.P.; Casari, P.; Zorzi, M., "The internet of energy: a web-enabled smart grid system," Network, IEEE International Conference, vol.26, no.4, pp.39,45, July–August 2012.
4. Gill, S.; Kockar, I.; Ault, G.W., "Dynamic Optimal Power Flow for Active Distribution Networks," Power Systems, IEEE Transactions, vol.29, no.1, pp.121,131, Jan. 2014.
5. Huang, A.Q.; Crow, M.L.; Heydt, G.T.; Zheng, J.P.; Dale, S.J., "The Future Renewable Electric Energy Delivery and Management (FREEDM) System: The Energy Internet," Proceedings of the IEEE, vol.99, no.1, pp.133,148, Jan. 2011.
6. Hollinger, R.; Erge, T., "Integrative energy market as system integrator of decentralized generators," European Energy Market (EEM), 2012 9th International Conference on the, vol., no., pp.1,6, 10–12 May 2012.
7. Jeremy Rifkin. The Third Industrial Revolution[M]. Washington: AEI Press, 2011, 9.
8. Lannez, S.; Foggia, G.; Muscholl, M., "Nice Grid: Smart grid pilot demonstrating innovative distribution network operation," PowerTech (POWERTECH), 2013 IEEE Grenoble, vol., no., pp.1,5, 16–20 June 2013.
9. Ling Chen, Ge Baoming, Bi Daqiang. Research on power electronic transformer in distribution network [J].Power system protection and control, 2012, 40(2):34–39.
10. Luo Kaiming, Li Xingyuan. Study on multi-agent systembased uninterrupted power substation control system [J]. PowerSystem Technology, 2004, 28(22): 1–5.
11. Mao Chengxiong, Fan Shu, Wang Dan. Theory and application of power electronic transformer [J]. High voltage technology, 2003, 29(10):4–6.

12. Mou Longhua, Zhu Guofeng, Zhu Jiran. Design of intelligent terminal based on Smart Grid[J]. Power system protection and control, 2010,38(21):53–56.

13. Michiorri, A.; Girard, R.;, "A local energy management system for solar integration and improved security of supply: The Nice Grid project," Innovative Smart Grid Technologies (ISGT Europe), 2012 3rd IEEE PES International Conference and Exhibition on, vol., no., pp.1,6, 14–17 Oct. 2012.

14. Nguyen, P.H.; Kling, W.L.; Ribeiro, P.F., "Smart Power Router: A Flexible Agent-Based Converter Interface in Active Distribution Networks," Smart Grid, IEEE Transactions on Power System, vol.2, no.3, pp.487,495, Sept. 2011.

15. Orestis Terzidis. Internet of Energy: ICT for Energy Markets of the Future[M]. Berlin: Industrie-Förderung Gesellschaft mbH, 2010,2.

16. Rong Mingzhe, Wang Xiaohua, Wang Jianhua. Investigation on the new development of the intelligent switch electric appliance intension [J].High voltage electrical appliance, 2010,46(5):1–3.

17. Tsoukalas, L.H.; Gao, R., "From smart grids to an energy internet: Assumptions, architectures and requirements"[C], Electric Utility Deregulation and Restructuring and Power Technologies, 2008. DRPT 2008. Third International Conference, vol., no., pp.94,98, 6–9 April 2008.

18. Wang Chengshan, YUXuyang. Distributed coordinative emergency control based on multi-agent system [J]. Power System Technology, 2004, 28(3): 1–5.

19. Wang Yezi, Wang Xiwen. The German version of the smart grid "E-Energy" [J]. The Internet of things technology, 2011,05:3–5.

20. Yang Xusheng, Sheng Wanxing, Wang Sun'an. Study on multi-agent architecture based decision support system for power system operation [J]. Automation of Electric Power Systems, 2002, 26(18): 45–49.

21. Yi Xu; Jianhua Zhang; Wenye Wang; Juneja, A.; Bhattacharya, S., "Energy router: Architectures and functionalities toward Energy Internet," Smart Grid Communications (Smart Grid Comm), 2011 IEEE International Conference, vol., no., pp.31,36, 17–20 Oct. 2011.

22. Zhang Mingguang, LU yunyun. Research on multi-agent layered self-healing control of smart distribution network [J]. The Application of Electronic Technology, 2012,38(11):77–79,83.

23. Zhaohong Bie; Peng Zhang; Gengfeng Li; Bowen Hua; Meehan, M., "Reliability Evaluation of Active Distribution Systems Including Microgrids," Power Systems, IEEE Transactions on, vol.27, no.4, pp.2342,2350, Nov. 2012.

Power and Energy – Kong (Ed.)
© 2015 Taylor & Francis Group, London, ISBN 978-1-138-02782-4

Lightning Detection and Location System for Shanghai Power Grid

Z. Yin
State Grid Shanghai Jiading Electric Power Supply Company, Shanghai, China

M.X. Lu
State Grid Shanghai Shibei Electric Power Supply Company, Shanghai, China

W.R. Si, C.Z. Fu & Q.Y. Lu
State Grid Shanghai Electric Power Research Institute, Shanghai, China

A.F. Jiang & Z.C. Fu
Shanghai Jiaotong University, Shanghai, China

ABSTRACT: Lightning Detection and Location System (LDLS) is the most widely used in field of lightning engineering. Shanghai Power Grid starts the cooperation with Wuhan-Nanrui Electric Power Research Institute since 1996, and has built the LDLS with 12 lightning detection stations, which cover the whole region of Shanghai and share flash data with four nearby provinces of eastern China. This paper introduces this LDLS and its location principle, and expounds the monitoring network construction. The important roles the system playing in the operation of Shanghai Power Grid from 4 aspects are also discussed. At last, further application based on the LDLS in the future is also proposed.

Keywords: Shanghai Power Grid; lightning detection; lightning location; lightning protection; lightning parameters; further application

1 INTRODUCTION

Lightning is a natural wonder, and the enormous energy it contains can destroy structures, devices, et al, and also endanger personal safety. Since the wide geographic distribution of power grid and overhead transmission lines are generally taller than nearby structure, the power grid are more vulnerable to lightning strokes. During the 2010 to 2013, the total number of overhead distribution line trip outs due to lightning strokes is more than 5000 times, the trip outs of 10kV distribution line could occupy 94.7% of total trip outs and 35kV distribution line is 5.3%. The lightning strokes become the prime impact on the safe operation of the Shanghai Power Grid [1].

Lightning detection and location is the basis for lightning protection. Lightning Detection and Location System (LDLS) is the most widely used lightning monitoring techniques in the field of lightning engineering in China during the last 20 years. In the late 1980s, the LDLS for lightning activity monitoring has been firstly adopted by power grid in China. From the first set of domestic LDLS used in Anhui Power Grid in 1993, after more than ten years of constant construction, the LDLS a national network project has been built in 28 provinces and autonomous regions of power grid in China. A national lightning monitoring network cover the power grid and the vast majority land of China had been completed in 2006.

Shanghai Power Grid start the cooperation with Wuhan Nanrui lightning protection department since 1996, and have built a new generation of lightning detection and location system until 2013, which can cover the whole region of Shanghai, with 12 lightning detection stations (including 4 digital), new and old server running "1 + 1" mode, and sharing flash data with four provinces of eastern China (Jiangsu, Zhejiang, Anhui and Fujian). This paper will provide detailed description of the new generation of digital detector station and its locating principle, lightning monitoring network of Shanghai Power Grid, the role of new generation LDLS of Shanghai Power Grid, and give a prospect on future work to the further application based on LDLS.

2 A NEW GENERATION OF DIGITAL DETECTION STATION AND LOCATION PRINCIPLE

The new generation LDLS of Shanghai Power Grid contains four digital lightning detection stations, as shown in Figure 1. In digital lightning detection stations, GPS is used to achieve clock synchronization, and stable clock is achieved through high-precision temperature-compensated crystal. Its operating indicators are as follows: 1) the storage capacity of the ground flash signal waveform data and ground flash signal characteristic data > 60000, 2) the record length of ground flash waveform signal: the time before the peak point > 100µs, the time after the peak point > 200µs, 3) the Mean Time Between Failures (MTBF) > 20000h, 4) data transfer velocity: 1200 ~ 115200bps, 5) power dissipation < 8W, 6) 30km < effective detection radius < 300km, 7) the correct rate of lightning polarity identification in effective detection range > 99.5%, 8) the detection data content: detection station should be able to accurately detect the peak value, polarity, time, location, and waveform data of each return stroke of ground flash activity, 9) the detection station can call-back the historical data stored in detection station side, including ground flash signal characteristic data and ground flash signal waveform data.

3 THE LIGHTNING MONITORING NETWORK OF SHANGHAI POWER GRID

The current ground flash lightning observation data communication process of the new generation LDLS in Shanghai Power Grid (as shown in Fig. 2) is: the ground flash lightning observation data of the 8 old ground lightning detection stations is uploaded to the front-end processor of the old central station system (including the data base), the front-end processor of new central station system receives not only the data from four digital observation station, but also the ground flash lightning observation data of the 8 old ground lightning detection stations from the front-end processor of the old central station system, Meanwhile, the front-end processor of the new central station shares grounding data with east China four provinces (Jiangsu, Zhejiang, Anhui and Fujian) power grid through a new generation front-end processor of east China branch of State Grid. In addition, Figure 2 shows the current application page of Shanghai Power Grid Lightning Monitoring System.

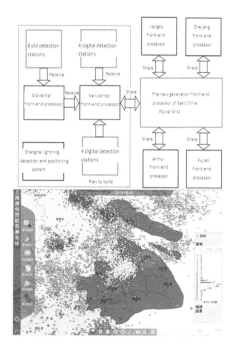

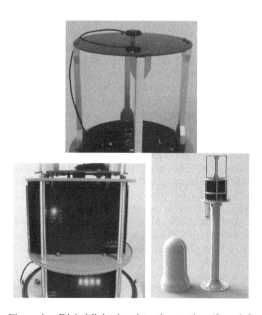

Figure 1. Digital lightning detection station (from left to right, GPS antenna, electronic control box, whole structure).

Figure 2. Ground flash detection data communication flow of the new generation LDLS of Shanghai Power Grid (left) and application page of Shanghai Power Grid Lightning Monitoring System (data in August 1, 2013) (right).

4 THE FUNCTION OF SHANGHAI POWER GRID LIGHTNING MONITORING AND LOCATION SYSTEM

4.1 *Rapid location of lightning failures point*

Compared with the traditional method, LDLS can locate the fault tower struck by lightning or lightning stroke point within seconds, the labor productivity of the patrol line workers have been greatly increased. A typical application cases was the 220kV line lightning trip in September 2012, as follows: Feng Long 4145 lines tripped at 16:15, with strong wind, heavy rain and strong thunderstorms. The fast query line in PMS lightning management module: Feng Long 4145, Relevant time: 2012-07-12 16:15:00, Time float: 60 seconds, distance float: 1000 m. As show in Figure 3, five lightning-related distributions are obtained through query. In conjunction with the lightning current

value-17.8kA and-41.9kA of the two time points in Figure 3, within the current amplitude range of shielding failure on 220kV typical tower, and each has one return stroke, consider the lightning location system positioning error, the fault lines Patrol area can range from the No. 17th tower to 25th tower. Based on these results above, the transmission line lightning stroke fault location is the upper phase double string insulators on the 17th tower close to the Tang Lu highway.

4.2 *Lightning parametric statistics*

Thunderstorm day, thunder hours, the number of ground flash, ground flash density and lightning current amplitude and other lightning parameters are very important but short of basic data for lightning protection engineering design. LDLS keep monitoring the time, location, magnitude and polarity and other parameters of the ground flash, the statistical analysis methods are as follows:

1. Grid analysis of the thunderstorm days
   The obtained thunderstorm days through grid average method can be converted to traditional weather thunderstorm days. According to the thunderstorm data from meteorological department, using the grid average method to conduct a statistical comparison for thunderstorm days in Shanghai area, results show that the 0.15° grid statistical average value can be annual average thunderstorm day, the maximum value of 0.275° grid statistics can be the maximum of annual thunderstorm days (as shown in Fig. 4).
2. Time theme analysis
   Analyzing the data samples of multiple administrative regions or power supply areas of Shanghai, important line corridors and grid, and by comparing thunderstorm days, thunderstorm hours and ground flash density and other parameters of multiple regions, the time

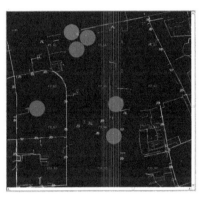

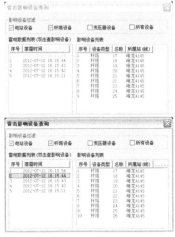

Figure 3. Lightning management module query results of PMS (from left to right the query results of PMS, the stroke time 16:15:54 and 16:15:44).

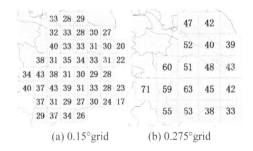

(a) 0.15°grid   (b) 0.275°grid

Figure 4. Thunderstorm grid analysis results of Shanghai in 2013.

distribution of lightning activity in multiple regions is obtained. The statistical layers in lightning parameter analysis system are shown in Figure 5, the following space theme analysis, electricity thematic analysis can be based on these layers statistics. Figure 5 shows the lightning strike tripping correlation analysis results of the 11 power company of Shanghai in June 2013 based on the system function.

3. Space theme analysis

Analyzing the overall area of Shanghai, one administrative region or power supply area, one certain types of data samples within important line corridor, comparison of thunderstorm day, thunderstorms hour and ground flash density in terms of mouth or year, time distribution lightning activity in these areas can be obtained [7]. Figure 6 shows the ground flash density

distribution obtained based on this function and the distribution of thunder day and number of ground flash in each power supply region in Shanghai.

4. Current theme analysis and buffer zones division

Analyzing the overall area of Shanghai, one administrative region or power supply area, one certain type of data samples within important line corridor, lightning current amplitude probability curve [8] and mid-value of the current can be calculated based on current amplitude. An example of current theme analysis application is shown in Figure 7. In order to study the lightning activity rule along the transmission line, dividing the transmission line by the length of the line buffer zone, we can study the lightning attractive width, ground flash density of the transmission line and lightning current amplitude probability curve of these buffer zones. An example of the buffer zone dividing application is shown in Figure 7.

4.3 *Lightning protection performance evaluation for transmission line*

In order to analysis the lightning protection performance of the transmission line with different

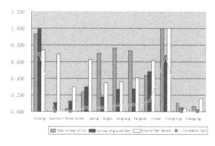

Figure 5. Distribution line lightning tripping correlation comparison of each power supply company in June 2013.

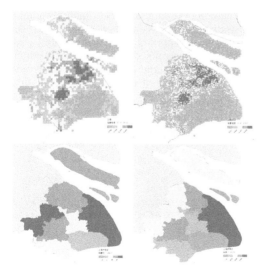

Figure 6. Ground flash density distribution map and thunder day of Shanghai area in 2013.

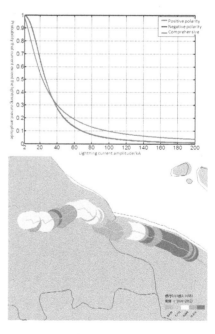

Figure 7. Lightning current amplitude cumulative probability distribution of Shanghai during 2001–2010 (left) and annual average ground flash density distribution of 500kV on Qiaoxing 5110 line (right).

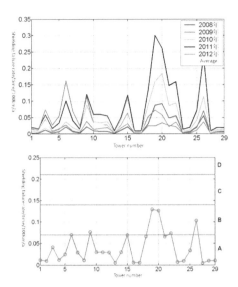

Figure 8. Shielding failure tripping rate distribution (left) and average shielding failure lightning protection performance evaluation results of different segments of Waigu 5119 line under actual lightning parameters.

temporal and zones more precisely, we use lightning parameter grid statistical method to deal with the observation data of the lightning location system to obtain the ground flash density and lightning current amplitude probability distribution in the corridor of transmission line. Total lightning tripping rate is calculated, and by using specified value and empirical value of the transmission line lightning protection design as the reference of the division of the valuation level, the assessment of transmission line lightning protection performance is obtained finally. The application examples are shown in Figure 8.

5 CONCLUSIONS

The lightning location system of Shanghai Power Grid has accumulated 10 years of ground flash data so far, based on these data and the safe operation demand of the power grid, four aspects further application can be carried out:

1. Update drawn work of the lightning stroke maps [9]. Share the lightning stroke maps to the designing institute, maintenance companies, power supply companies or other companies, to provide reference for work related to lightning protection.
2. Enhance the lightning failure investigation and lightning damage statistical analysis [10].

Summarize lightning trip rules, to provide the basis for the choice of lightning protection devices or lightning protection measures for transmission and distribution lines.
3. Carry out lightning damage risk assessment for important transmission line. Find out the easy trip out segments and towers of the lines, provide the basis for differentiation lightning protection and improvement of 110kV and above important load power lines, strategic transmission channel and 500kV and above core grid.
4. Carry out evaluation work after lightning protection improvement. Base on the ground flash data in line corridor of lightning location system and actual operation conditions, evaluate and analyze the actual effect of lightning protection improvement project, analyze the effect of lightning protection improvement, evaluate the effectiveness of the reforming scheme, and guide the line lightning protection in subsequent work.

REFERENCES

1. Yanf, L.H. & Chen, H.G. & Si, W.R. 2012. The Analysis of Lightning Activities and Distribution Network Transmission Line Lightning Tripping Of Shanghai In 2012. East China electric power, Vol. 40, pp. 153–156. (in Chinese)
2. Chen, H.J. & Zhang, Q. & Feng, W.X. 2008. Lightning Location System and Lightning Detection Network of China Grid. High Voltage Engineering, Vol. 34, No. 3, pp. 425–431.
3. Gu, C.Y. 2004. The Application of Lightning Location Technology in the East China Power Grid. East China electric power, Vol. 32, No. 1, pp. 32–35. (in Chinese)
4. Gu, C.Y. & Zhang, J.M. 2003. The Supplementary Role of Shanghai Lightning Location System Playing in Lightning Protection Safety Work. Shanghai electric power, No. 1, pp. 60–62. (in Chinese)
5. Ji, H. 2009. Lightning Protection Research of Power Network based on Lightning Location System. East China electric power, Vol. 37, No. 6, pp. 976–981.
6. Si, W.R. & Zhang, J.X. & Fu, C.Z. 2012. Lightning Activity and Parameters Analysis in Shanghai Area from the Year 2003 to 2011. East China electric power, Vol. 40, No. 10, pp. 1734–1738.
7. Si, W.R. & Zhang, J.X. & Xiao, R. 2012. Lightning Zone Map Drawing based on Lightning Parameters Statistics for Shanghai Power Grid. East China electric power, Vol. 40, No. 6, pp. 1038–1041.
8. Si, W.R. & Zhang, J.X. & Zheng, X. 2012. Analysis of characteristics of lightning current amplitude distribution in Shanghai. Power and energy, Vol. 33, No. 2, pp. 116–119.
9. State grid corporation, 2010. Minefields classification standard and Mapping rules of lightning distribution map.
10. Chen, J.H. & Wang, H.T. 2008. Research on Power Grid Lightning Hazard Maps. High Voltage Engineering, Vol. 34, No. 10, pp. 2016–2020.

Power and Energy – Kong (Ed.)
© 2015 Taylor & Francis Group, London, ISBN 978-1-138-02782-4

Persistence and development: China's solar power generation for fifty years (from year 1958 to now)

L. Xia, S.B.Y. Huang & J. Hou
School of Humanities and Social Science, North China Electric Power University, Baoding, China

ABSTRACT: As an ideal and clean energy resource, solar power has many unparalleled characters compared with other conventional energy. Solar power generation in our country began in the 1950s, and it has been more than 50 years until now. In actual, the history of Solar power generation in our country can be divided into three parts: the experimental and study period; the development and utilization period; the difficult development period and the vigorous development period. All in all, China's solar power generation not also achieved much in the last 50 years, but struggled difficultly. Nowadays, with the great help of the National Energy Administration and the National Development and Reform Commission, China's solar power will continue to persist after the "cold winter"!

Keywords: solar power generation; solar-thermal power generation; photovoltaic power generation; China

1 INTRODUCTION

Solar power generation is a huge industry including solar-thermal power generation and photovoltaic power (PV) generation. The former takes advantages of the optical system collecting radiant energy to heat working substances in order to produce high-temperature steam to drive the steam turbine to generate electricity; the latter changes light power into electric power directly through the photoelectric conversion. With the investigation and development of several decades, a considerable number of developed countries in Europe and the United States have initially achieved industrialization in solar power generation industry, it has always attracted the attention all around the globe when the world's first silicon solar cell was used and the first atomic power plant was built in 1954 in America. Solar photovoltaic generation mainly refers to solar battery generation, which transforms light energy into electrical energy directly through the photoelectric effect or photochemical effect. The Sputnik satellite the U.S. launched in 1958 used solar cells as power sources. Similarly to many other developed counties, the Chinese government has also pay much attention to the solar power generation and has successfully developed the first practical solar cells in 1959. At present, China has come into being a solar power generation industry pattern which based on solar-thermal power and supplemented by photovoltaic power. Though many difficulties we faced, solar power generation in China has made much innovation and breakthrough in the past fifty years and has been seeking a way not only meets the needs of all social but adapts to the surroundings of China.

2 THE EXPERIMENTAL AND STUDY PERIOD OF CHINA'S SOLAR POWER (YEAR 1958-YEAR 1970)

Silicon is the main material for solar cells. The United States developed the world's first silicon in 1957. China developed the country's first monocrystalline silicon in 1958 with the tireless efforts of lots of experts. In 1958, China's first film photovoltaic cells was produced in Tianjin Institute of Power Sources, which was regarded as the beginning of China's solar power generation; in 1959, the Institute of Semiconductors newly established invented China's first solar cells with practical value successfully. From 1958 to 1965, the efficiency of PN junction cells invented by the Institute of Semiconductors increased by leaps and bounds and the efficiency of 10X20MM cells were steady at 15%, similarly to international level. We can say that the researches at the initial period of solar energy generation in China keep the pace with international advanced level and have achieved a lot.

The first solar cell in our country was used in space in the same with many other countries that also research and invent solar cells. From year 1968 to the end of year 1969, the Institute of

Semiconductors took the task of developing and producing solar panel for the "Shijian-8 satellite", and researchers found that P+/N monolithic silicon solar cells would encounter electronic radiation when running in the space, which caused the energy of solar cells to decline and made it that solar cells couldn't work for a long time. In 1969, the Institute of Semiconductors stopped studying Silicon solar cells and turned to research satellite irradiation effects created by silicon solar, and they discovered that NP junction silicon solar cell was dozens of times stronger than PN junction silicon cells in the ability of anti-electron irradiation,which was one of the greatest innovations in the technology of solar cells in our country.

3 THE DEVELOPMENT AND UTILIZATION PERIOD OF CHINA'S SOLAR POWER (YEAR 1970-YEAR 1980)

China's solar power stepped into its second period-the development and utilization period on the basis of lots of innovations achieved in the first stage. In 1970, after they decided to replace PN junction silicon cells by NP junction silicon to finalize the design production, the Institute of Semiconductors manufactured 5690 pieces of NP junction silicon solar cells, among which 3350 pieces of NP junction silicon solar cells reached space application standards. In 1971, the "Shijian-8 satellite" run by the NP junction silicon solar cells launched successfully and the power of solar cells declined less than 15% in its eight-year lifetime, which proved that the Institute of Semiconductors completed the developmental and producing task successfully in virtue of Solar panels. At the same time, the solar cell array invented by the Tianjin Institute of Power Sources was applied to a series of geosynchronous satellites successfully, such as "dongfanghong-2 satellites", "dongfanghong-3 satellites" and "dongfanghong-4 satellites".

In 1975, Ningbo and Kaifeng set up plants to produce solar cells one after another, manufacturing technique of solar cells imitated the technology of the early space cells, which marked that the solar cells in China started to apply on the ground besides space and created conditions for solar-photovoltaic power's popularity and universality. In the same year, the first national solar energy utilization conference to exchange work experience held in Anyang, Henan Province. Solar power's research and extension work was included in the government programs of our country, and solar energy won the support of special funds and material meanwhile. The meeting not only furthered the development of technologies of solar energy, but accelerated the development of solar power

industry including solar power generation industry on the basis of the conversion and utilization of solar power technology.

4 THE DIFFICULT PERIOD OF CHINA'S SOLAR POWER (YEAR 1981-YEAR 1996)

The solar power industry in China was in embryonic form before the 1980s, the annual output of solar cells has been hovering at 10 kilowatts or less and its price was also very expensive. The solar cell market evaluated slowly on account of its limited price and production.

Affected by the international environment, in the "Seventh Five-Year Plan" and "Sixth Five-Year Plan" period,China also started to give support to photovoltaic power industry, photovoltaic power market and photovoltaic power technology. For instance, State Development Planning Commission listed and emphasized the dissertation of solar cells in the rural energy strategy of "SeventhFive-Year Plan" (from 1986 to 1990) "dongfanghong-2 satellites", "dongfanghong-3 satellites" and "dongfanghong-4 satellites", subsequently there were six universities and six research institutes beginning to study crystalline silicon cells all over the country. More importantly, the central and local governments also invested a certain amount of money to accelerate the development of PV industry and making it that the very weak solar cell industry in our country got consolidated and many demonstrations have also been built, such as the microwave relay stations, the communications systems for military "dongfanghong-2 satellites", "dongfanghong-3 satellites" and "dongfanghong-4 satellites", the cathodic protection system for Sluices and oil pipeline, carrier telephone systems in rural areas, the system for small users and the power systems, and so on. However, the development of solar power generation industry in China struggled not only because the solar power itself is scattered, random and intermittent, but also there were difficulties in the theory of solar photovoltaic cells, materials and device researches. As a whole, the development of solar power generation industry in our country evolves slowly.

5 THE VIGOROUS DEVELOPMENT PERIOD OF CHINA'S SOLAR POWER (YEAR 1997-YEAR 2010)

In September 1996, the Global Solar Energy Summit was held in Zimbabwe. "Photoelectric Engineering" was put forward for the first time, and it practiced in the place where without electricity. Later, the Chinese government proposed and implemented the "Chinese bright project" plan. According to the plan, by 2010, the use of wind power and photovoltaic power generation technology would solve the power problems of the 23 million population in rural areas, making it that they can reach the level of per capita power generation capacity of 100 w, which is equivalent to the national per capita as a third generation capacity level. Since 1997, China's solar power enters a period of vigorous development of "golden age".

China's solar power business has reached the "climax" during this period. It mainly manifested in those aspects.

5.1 The development of solar-thermal power generation industry had a fierce momentum

In the past 30 years, China's solar power has mainly focused on the photovoltaic power, but there was almost no development in thermal power generation which was an important form of solar power. China also attached great importance to the solar-thermal power generation, one case in point is that according to the "Renewable energy medium and long-term development plan" during the period of "11th five-year plan", in Dun Huang province,Gansu province and Tibet Lhasa demonstration project, the construction of large grid solar photovoltaic power station, in Inner Mongolia, Gansu, Xinjiang, construction of solar thermal power generation demonstration project. By 2020, the national solar photovoltaic power station the total capacity of 2000 mw of solar thermal power generation capacity will reach 2000 mw 973 vice President of China association of electrical Huang Xiang estimates that by 2020, Chinese thermal power generation market will reach 22.5 trillion RMB to 30 trillion RMB, the total thermal power generation can be accounted for 30%-40% of the total generating capacity. Since the Chinese government pays great attention to solar-thermal power generation, it made breakthrough progress in this period. In 2005, Nanjing built the first domestic 70 mw tower solar thermal power generation system and it passed the identification of acceptance in 2007; In 2010, the Asian first tower solar-thermal power station was started to build in Yan Qing, Beijing. Compared with photovoltaic power generation, Solar-thermal power generation is more convenient in the power grid and the current domestic docking, and there is a big problem for photovoltaic power generation and other new energy power generation that how to realize the power grid.

5.2 The solar photovoltaic industry developed rapidly

In 1998, the Chinese government proposed to build the first set of 3 mw Crystalline silicon cells and application demonstration project and open it to the public market. In 2009, China began to the implementation of support the demonstration project "golden sun" that to promote the development of photovoltaic industry progress and scale in domestic. After that, the relevant state departments launched two successive fiscal subsidy policies to promote the application of photovoltaic in less than half a year's time. Under the strong support of the Chinese policies, the development of the domestic photovoltaic industry has entered a fast lane, and has created the world photovoltaic the "China miracle" of the market. By 2010, In mainland China photovoltaic cells output of 10 million kilowatts, accounting for more than 50% of global market share, five of which the photovoltaic cell production in the world top ten enterprises. Chinese photovoltaic cell technology have a top quality in the world, Chinese have mastered the scale thousand tons of production technology of polycrystalline silicon. The comprehensive utilization level of silicon material production by products obviously improved, advanced enterprise energy consumption index is close to the international advanced level.

5.3 The solar power market is increasing

Rapid advances in the solar industry are based on the growing solar market. In the beginning, solar photovoltaic power generation technology was applied in the field of aviation not just in China even all the world, the market is very narrow. After the seventy's, as the solar cell by space down to the ground, the popularization of the solar cell applications make the solar power generation market is becoming more and more diversified. In 2002, the former state development planning commission launched the "sending electric to the countryside" project, in Inner Mongolia, Qinghai, Sichuan, Xinjiang and other places built several photovoltaic power stations, not only solved a lot of farmers and herdsmen's life that without electricity area electricity problems, but also enlarged the application of photovoltaic market in China, the rapid advances in domestic PV industry. In 2009 and 2010, the total of two consecutive years of new photovoltaic power generation capacity is the sum of the past few years. Promulgated in 2010 issued by the state council that "about to speed up the decision of the cultivating and developing strategic emerging industries" explicitly proposed to "develop diversified photovoltaic solar-thermal power generation market", how to expand the market and realize the diversity of the market are the important issues in the deepening development of solar photovoltaic solar-thermal power generation.

5.4 *The technology made much breakthrough*

The polycrystalline silicon is the "lifeline" for the whole solar cell industry, its shortages of raw materials make the solar cell keeping high cost, which seriously restrict the development of solar energy industry and the market In 2005, the national development and reform commission organized and implemented the "high purity silicon high technology industrialization of major projects", around the major technical problems of each polycrystalline silicon production link, the implementation of key research, made a series of research and industrialization progress, has the independent intellectual property rights technology system. In 2004, Luoyang monocrystalline silicon factory and China non-ferrous & design institute jointly established the silicon in high-tech 12 is developed by ourselves for energy-saving polycrystalline silicon reduction furnace, on this basis, in 2005, the domestic first 300 tons of polycrystalline silicon production projects completed and put into operation, thus opened the prelude of China's polycrystalline silicon big development.[1]

6 THE STUMBLING PERIOD OF CHINA'S SOLAR POWER (YEAR 2011-PRESENT)

In October 2011, Solar World, the American Solar photovoltaic enterprises demand for China PV enterprises launched the "double reverse" survey; The United States began to initiate an investigation on China's 75 PV enterprises. After experienced the "golden decade", Chinese photovoltaic industry began to enter the "winter". In September 2012, the EU formally issued a circular, will start the PV solar products anti-dumping investigation on China, for China PV industry in the "winter" is definitely a big blow, Even a fatal blow. Because in the past decade the miracle of Chinese photovoltaic owes its European and American markets, especially in Europe, in 2011, China's solar photovoltaic cells components more than 80% of exports abroad, including Europe accounted for 60%[2]. But 2011 years later, due to the influence of the debt crisis in Europe, including Germany, Italy, Europe's main photovoltaic installed photovoltaic application market began to slash subsidies, which make the global photovoltaic products

prices fell quickly, photovoltaic modules prices fell more than 50%, China's photovoltaic module exports collapsed. But, fundamentally, the plight of China's photovoltaic industry is mainly due to the rapid photovoltaic industry development which led to severe overcapacity in the past decade, the domestic market narrowly and relies on foreign markets heavily. In photovoltaic industry progresses day by day, the government leading established numerous "sun city" and "photovoltaic industrial park", to attract more PV on the enterprise and project though the investment promotion and capital introduction. By the end of 2009, our country has built only 300 mw photovoltaic power generation project, The PV modules over 90% exported to foreign countries, relies on foreign markets heavily, intensifies the competition in the market and industry risks. Once through the EU anti-dumping case, not only means China PV enterprises will be lost, will directly lead to more than 350 billion RMB output value of the loss, more than 200 billion RMB of nonperforming loans risk, more than 300000 to 500000 people at the same time are under unemployment.[3]

Announced on August 2, 2013, the EU has officially approved the china- EU PV trade disputes "Price Undertaking Agreement", to accept Chinese solar cell exporters to submit the scheme of price undertakings. The Chinese enterprises which participate in the scheme shall be exempted from the provisional anti-dumping duty, the European Union to China has the largest trade actions to a win-win situation, which also brings hope to China as in "cliff" at the edge of PV enterprises, and it is expected to be out of the "winter" enter the development of another "spring".

For China, according to the national energy bureau issued "solar" twelfth five-year "development planning", by the end of 2015, China solar power installed capacity will reach 21 million KW or more, generating 25 billion kilowatt-hours. Among them, the photovoltaic power station installed capacity of 10 million kilowatts, distributed photovoltaic system accounts for about 10 million kilowatts, solar-thermal power installed capacity of 1 million kilowatts. It is obvious that Chinese government will promote the development of Chinese solar power from two aspects of market and technology, not only by increasing the solar power installed capacity to expand the domestic market, but also through the photovoltaic industry marketization photovoltaic enterprise merger and reorganization, promote evolution; More important

[1]Chinese solar cells are hard and glorious history [N] Science and Technology Journal, 2010, 01 13.
[2]Zhang Fang. Journey to the Photovoltaic Market [J]. Solar Power Generation, 2012, 17:13–15.

[3]Suffer from Europe and the United States double reverse unbreakable frequently. [N]. The Counterattack of China PV Enterprises. China Joint Business in 2012-08-27D03.

is to grasp the core of the solar power technology, reverse the awkward position of "world factory", not only it can improve the international competitiveness of China's solar power industry, also can promote further expand the domestic market.

In July 2013, the national development and reform commission issued the "state council on promoting the healthy development of photovoltaic industry several opinions". Which puts forward in view of the development of photovoltaic industry to adjust the energy structure, promote the revolution of energy production and consumption, to promote ecological civilization health has an important significance. It is necessary for China that take various measures to regulate and promote the healthy development of photovoltaic industry, which also put forward the photovoltaic (PV) power generation with a total installed capacity in 2015 to reach 35 million KW or more, more than the planning of installed capacity is expected to increase by 3.5 times, China's solar power will significantly expand domestic market, we believe that after the "cold winter", the Chinese solar power will also continue to persistent forward with the strong support of national policy.

REFERENCES

1. National energy administration solar power development twelfth five-year plan, [J]. Solar Energy 2012, 18:6–13.
2. Chinese solar cells are hard and glorious history [N] Science and Technology Journal, 2010.01 13.
3. Zhang Fang. Journey to the Photovoltaic Market [J]. Solar Power Generation, 2012, 17:13–15.
4. The Investigation of Chinese Photovoltaic Industry [N]. China Business News, 2012-10-29A10.
5. Suffer from Europe and the United States double reverse unbreakable frequently. [N]. The Counterattack of China PV Enterprises. China Joint Business in 2012-08-27D03.

Power and Energy – Kong (Ed.)
© 2015 Taylor & Francis Group, London, ISBN 978-1-138-02782-4

A hybrid fault processing scheme for active distribution network

Q.Z. Zhao, Y. Li & Y.J. Cao
College of Electrical and Information Engineering, Hunan University, Changsha, Hunan, China

B.R. Pan, J.B. Xin & R.X. Fan
Electric Power Research Institute of Jiang Xi Electric Power Corporation, Nanchang, Jiangxi, China

ABSTRACT: Since the large-scale connection of distribution generations to distribution network, the network faces bidirectional power flows during normal operation, which leads to great difficulties in fault processing of distribution network. In order to solve this problem, this paper proposes a hybrid fault processing method that combines the advantages of distributed and centralized protection. When a fault occurs, the distributed protection plays a role in the location and isolation of fault area, judges and processes the instantaneous fault. When determining that the fault is permanent, centralized protection is adopted to implement the optimization of power recovery in the non-responsible fault zones. Compared to current methods, the proposed method has advantages of significant adaptive ability, small impacts on network operation, and intelligent back-up protection.

Keywords: distributed generation; combination of distributed protection and centralized protection; instantaneous fault; intelligent back-up protection

1 INTRODUCTION

In recently years, more and more distributed generations are being integrated to distribution networks [1], which not only changes the network topology and leads to bi-directional power flow during normal operation, but also changes the characteristics of fault currents [2]. At present, several protection schemes have been proposed in [3]-[6] for the reliability and security of active distribution network with high penetration of distributed generations. Especially in the field of the distributed protection and the centralized protection, some important research results have been given in [7]-[9].

In [7], an overcurrent protection method was presented for smart radial distribution systems containing lots of distributed generations. By collecting the real-time current and comparing with the offline data sheet, the fault type can be determined and the fault area can be identified. However, this method needs to process large amount of information, which spends a lot of time. In [8], a multi-agent technique based distributed protection method was proposed, but the protection structure is too complex, and when the network topology is changed, all agents in the changed region has to be reset. In [9], a protection idea, which combines distributed intelligence protection with centralized intelligence protection, was proposed, but the detailed processing scheme is not presented.

Based on the aforementioned literatures, this paper will propose a hybrid fault processing method, which effectively integrates distributed protection with centralized protection and has the advantages of these two protections at the same time.

The organization of this paper is as follows: Section II presents the new protection scheme and explains the basic operating principles. Furthermore, Section III presents the working modes of the Intelligent Equipment (IE). Section IV gives a case study to demonstrate the processing flow of the proposed method. Conclusion is given in Section V.

2 PROTECTION SCHEME AND PRINCIPLES

2.1 *Fault processing strategy*

The proposed hybrid fault processing method combines the technical features of distributed and centralized protection, which makes full use of the advantages of these two protections, such as fast fault location and isolation as well as flexible recovery of power supply. The basic processing flow of this method is described as follows: when a fault occurs in active distribution network, at the first, the distributed protection is in charge of locating the fault zone and identifying the disposing the instantaneous fault. When judging the fault is permanent, the centralized protection is active to carry out optimal recovery of power supply.

2.2 *Fault processing principle*

The hybrid fault processing method focuses on the accurately locating and rapidly isolating of fault zones. These functions are based on the principles of fault processing such as appropriate dividing zones, correctly defining the positive direction and locating the fault zones. These principles have a significant impact on the performance of the hybrid fault processing method. Therefore, we should follow the below planned principles when dividing the zones, defining the positive direction and locating the fault zones.

- Principle-1: The principle of dividing zone. Taking the IE that locates in the top of the feeder as a starting point, the nearby IEs belongs to the same zone, and the IE that locates in the end of the feeder is separately divided a zone.
- Principle-2: The definition of the positive direction. For an assigned zone, define the active power that flows into the zone is regarded as the positive direction.
- Principle-3: The principle of locating fault zone. During a fault, if identifying that the zone has both positive and negative power flows, we can judge that the fault happens outside this zone; if a zone only has the positive power flows, we can judge the fault happens in this zone.

2.3 *Fault processing flow*

Figure 1 shows the overall workflow of the proposed hybrid fault processing method. According to Figure 1, the fault processing scheme includes the following five steps:

- Step-1: Zone division. According to the IE location, the IEs belonging the same zone are incorporated into the same communication group. At beginning, IEs in the same communication group interchange the information about power flow, and IEs self-select working mode by comparing the magnitude of the normal power.
- Step-2: Fault detection. The IEs installed on the feeder monitor the real-time status of the distribution lines. When the IEs detect the fault signal, they will communicate each other in the same communicate group and interchange fault information.
- Step-3: Fault dealing. After interchanging fault information in the same zone, the IEs can distinguish the fault type by using the fault location criterion. If the fault locates in this zone, namely responsible fault zone, all IEs installed in this zone will disconnect the breaker to isolate the zone. If the fault is not in this zone, namely no responsible fault zone, all IEs installed in this zone will shift to status of intelligent backup protection. Since the priority of tripping command is higher than the command of shifting to intelligent backup protection, so the IE located in the border of responsible zone and no responsible zone can be

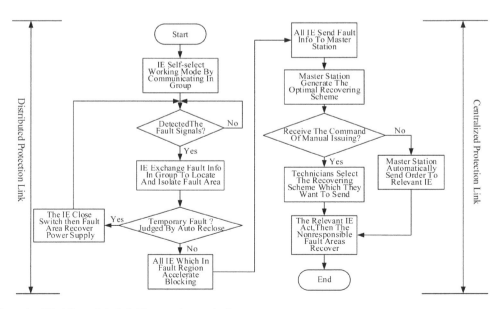

Figure 1. Workflow of the hybrid protection method.

- Step-4: Judgment of instantaneous fault. After the responsible fault zone was isolated, the IEs located in the isolated zone will attempt automatic reclosure after delay. If the fault is an instantaneous fault, the automatic reclosure can be success. After the success of automatic reclosure, the IE which works in the working mode 1 will send close command to the IEs, which works in the working mode 2, located in the isolated zone. The IEs, which works in the mode 2, will close the switch after receiving the close command and then the power supply can be recovered. If the fault is permanent, the automatic reclosure will be failed. After the failure of automatic reclosure, the IE, which works in the working mode 1, will accelerate to block and send blocking command to the IEs located, which works in the working mode 2, in the isolated zone. The IEs, which works in the mode 2, will block the switch after receiving the blocking command. After blocking, all IEs will send fault information to the master station.

- Step-5: Optimal restoration. If the fault is a permanent fault, the automatic reclosure will be failed. After the failure of automatic reclosure, all IEs will upload the fault information to the master station of active distribution network. The master station will generate the optimal restoration scheme by making full use of fault information. After then, the master station will send the action commands to the suitable IEs. The power supply of the no responsible zones will be recovered after a series of tripping or closing operation.

3 WORKING MODE OF THE IE

The performance of the proposed hybrid fault processing method depends on the cooperation between IEs. Considering the different location and different tasks, IE has two working modes. Mode 1 is an active switching mode, which is characterized as delayed automatic reclosure after the responsible zone was isolated. Mode 2 is a passive switching mode, which is characterized as the cooperation with the IE located in the same zone and operating in mode 1. In this way, the instantaneous fault can be effectively handled by the cooperation between IEs operating in mode 1 and IEs operating in mode 2.

3.1 Working mode 1

Figure 2 illustrates the workflow of this mode. The active switching function can be realized by the following four criterions. The first criterion is about fault location, which is used to determine

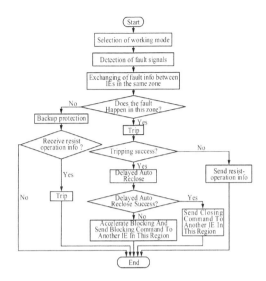

Figure 2. The working mode 1 of IE.

the fault zone. If the fault is an inner-zone fault, IE will shift to the tripping process. Otherwise it will shift to the backup protection. The second criterion is about refusing action, which is used to judge if IE tripping is successful. If tripping successfully, IE will shift to the delayed automatic reclosure. Otherwise it will send the refusing action message to the adjacent IE. The third criterion is used to judge the success of delayed automatic reclosure. If automatic reclosure is successful, the fault is an instantaneous fault, and the IE sends switching command to other IEs located in the same zone. If the automatic reclosure fails, the fault is a permanent fault, and the IE accelerates blocking and sends command of blocking to other IEs located in the same zone. The forth criterion is a detection criterion used to determine if the IE operating in the status of intelligent backup protection needs to be tripped. If the IE receives the refusing action message from adjacent IEs after delayed, it will turn to trip process, otherwise it will end the process.

3.2 Working mode 2

According to the workflow shown in Figure 3, it can be seen that mode 2 also has 4 criterions to realize passive switching function. Compared to mode 1, the first, the second and the forth criterions have the same function. But for the third criterion, it is used to determine if IE receives a switching command from the other IEs located in the same zone and operating in mode 1. If IE receives the closing command, it will turn to the closing process, otherwise it will end the process.

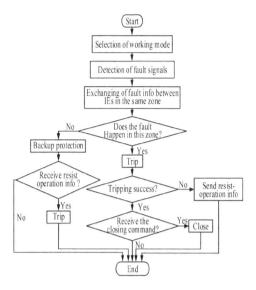

Figure 3. The working mode 2 of IE.

4 CASE STUDY

The active distribution network of Gongqing city in China is used as a case to demonstrate the proposed hybrid fault processing method.

Figure 4(a) shows one part of the network, where 17 IEs are installed. The black one represents the IE is on the closing status, and the white one represents the IE is on the disconnection status. According to the principle of dividing zone, this network is divided into 13 zones, as shown in this figure. The IEs within the same zone are incorporated into the same communication group. At the beginning of protection initialization, IEs self-select working mode by comparing the magnitude of the normal power. IE is on the disconnection status. According to the principle of dividing zone, this network is divided into 13 zones, as shown in this figure. The IEs within the same zone are incorporated into the same communication group. At the beginning of protection initialization, IEs self-select working mode by comparing the magnitude of the normal power.

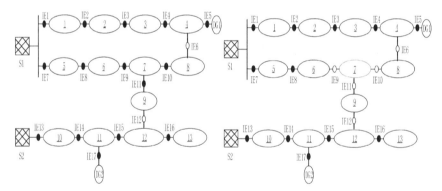

(a) One part of the active distribution network   (b) The location and isolation of fault zone

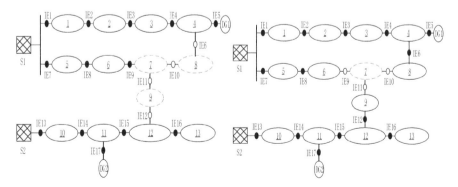

(c) Judgment of the fault type by automatic reclosure (d) Optimal restoration by centralized protection

Figure 4. Demonstration of the hybrid protection processing method.

As shown in Figure 4(b), when a permanent three-phase short circuit fault occurs in zone 7, the IE7-9 can detect the fault signal. Then the IEs located in zone 5-7 communicate with each other. Based on the principle of locating the fault zone, the IEs can judge the fault is located in the zone 7, hence zone 7 is a responsible fault zone, and the IE 9-11 in zone 7 will trip to isolate the fault.

As shown in Figure 4(c), IE 9 located in the responsible fault zone operate in the working mode 1, so it will delayed automatically reclose. Due to the fault is a permanent fault, automatic reclosure will be failure. Then IE 9 will accelerate to block and send blocking command to the IE 10-11, which work in mode 2, located in the same zone.

As shown in Figure 4(d), after IE9-11 blocking, all IEs send fault message to the master station of active distribution network. The master station will generate the optimal restoration scheme and send the closing command to IE 6 and IE 12, then the non-responsibility fault zones 8 and 9 will recovery power supply.

5 CONCLUSION

This paper proposes a hybrid fault processing method with the combination of distributed and centralized protection. Due to such a combination, its capacity of fault-tolerant gets a significant improvement. This method can accurately and fast complete the fault location and isolation. More importantly, this method has an effective ability of dealing with instantaneous fault. Since the automatic reclosure equipment is located in the isolated zone, it can decrease the impact areas of automatic reclosure, resulting in less impact on the network. Furthermore, this method is suitable for intelligent backup protection, and is able to deal with the impact of bi-directional power flow problems caused by distribution generations on the protection action.

REFERENCES

1. H.A. Gil and G. Joos, "Models for quantifying the economic benefits of distributed generation," IEEE Trans. Power Syst., vol. 23, no. 2, pp. 327–335, May 2008.
2. Liang Caihao, Duan Xianzhong, "Distributed generation and its impact on power system," Automation of Electric Power Systems, vol. 25, no. 12, pp. 53–56, Jun 2001.
3. Liang You Wei, Hu Zhi Jian, Chen Yun Ping, "A survey of distributed generation and its application in power system," Power system Technology, vol. 27, no. 12, pp. 71–76, Dec 2003.
4. S.M. Brahm and A.A. Girgis, "Development of adaptive protection scheme for distribution systems with high penetration of distributed generation," IEEE Trans. Power Del., vol. 19, no. 1, pp. 56–63, Jan 2004.
5. Zhang Chao, Ji Jianren, Xia Xiang, et al, "Effect of distributed generation on the feeder protection in distribution network," Relay, vol. 34, no. 13, pp. 9–12, Jul 2006.
6. Zhao Shanglin, Wu Zaijun, Hu Minqiang, et al, "Thought about protection of distributed generation and micro grid," Automation of Electric Power Systems, vol. 34, no. 1, pp. 73–77, Jan 2010.
7. Fred A. Ituzaro, Richard H. Douglin, Buttler-Purry K.L., "Zonal overcurrent protection for smart radial distribution systems with distributed generation," in Proc. 2013 ISGT conference, pp. 1–6, Feb 2013.
8. Wang Zhao, Ma Wen Xiao, Gao Fei, "A multi-agency based approach to the distributed feeder automation system," Automation of Electric Power Systems, vol. 34, no. 6, pp. 54–58, Mar 2010.
9. Liu Jian, Zhang Xiaoqing, Chen Xingying, et al, "Fault location and service restoration for distribution networks based on coordination of centralized intelligence and distributed intelligence," Power System Tecnology, vol. 37, no. 9, Sep 2013.

Power and Energy – Kong (Ed.)
© 2015 Taylor & Francis Group, London, ISBN 978-1-138-02782-4

Research of lighting trip-out rate model of transmission line

Shi-yang Zhu & Le Wang
Guangxi Power Grid Electric Power Research Institute, Nanning, Guangxi Zhuang Autonomous Region, China

Can Wang
*State Key Laboratory of Power Transmission Equipment & System Security and New Technology,
Chongqing University, Chongqing, China*

ABSTRACT: Failure rate of transmission line and reliability of power system are significantly affected by Lightning meteorological factor. In view of the complexity and variability of Lightning meteorological factors, this paper presents lightning trip-out rate model of transmission line in considering distribution of ground flash density and lightning day hours. Meanwhile, presents a failure rate model of transmission line in different condition, and a risk analysis method for transmission line considering multiple risk factors based on risk quantification. This method takes Lightning meteorological factor as the main evaluation standard, and establishes risk degree evaluation system for transmission line including another five evaluation standard. We put forward the risk indicators by quantifying the risk factors based on experience date of transmission line in service. Based on the risk indexes comprehensive evaluation is conducted, and the evaluation result closer to practice is achieved, providing basis for transmission line risk warning and maintenance strategy.

1 INTRODUCTION

The failure probability of overhead transmission lines under inclement weather conditions is much higher than under normal climatic conditions. Lightning is one of the harsh meteorological factors, which has significant impact on overhead transmission lines in service. Thus, the study of risk analysis methods on overhead transmission lines would contribute to the achievement of early risk warning under lightning weather conditions. Guarantee the safe and reliable of overhead transmission lines in service by taking security precautions in advance.

Currently, scholars have study on effects of meteorological factors on the reliability of power system. Literature [5] points out climate conditions can be divided into normal climate conditions and adverse weather conditions, the method of evaluating the reliability by simulating piecewise line under different climate conditions has being put forward. In literature [8], Monte-Carlo method is adopted to transmission line and climate area sampling, then determined the state of power transmission components and power grid reliability calculation. In [11] the reliability evaluation model of transmission lines in different weather influence monthly is established. The research in the past mainly considered the off-line reliability assessment of power grid, meanwhile, established the reliability

evaluation model in considering the meteorological effects. However, there is less research on risk analysis and early warning methods of transmission line on-line assessment.

Overhead transmission lines reliability is closely related to meteorological factors, which is not only affected by line construction ways, but also by line corridors region, such as landform and physiognomy. These factors are often determines the form of a lightning striking on the transmission line. We built overhead transmission line risk assessment index system on the bases of analysis of lightning disaster main factors. Based on the meteorological conditions, transmission line and the external environment factors, the level of risk analysis and weight calculation, finally make a comprehensive evaluation. We proposed multifactor risk analysis method of overhead transmission lines under the thunder and lightning disasters.

2 THE DETERMINATION OF THE TRANSMISSION LINE LIGHTNING DISASTERS RISK FACTORS

Transmission line lightning trip-out fault has its unique characteristics. There are two kinds of lightning overvoltage: one is the induction lightning overvoltage, the other is direct lightning overvoltage. The hazard of induction lightning overvoltage is small due to the high insulation level of transmission lines, with a focus on the direct lightning protection. From the part of the line lightning hit by we divided the direct lightning overvoltage into three kinds: lightning strike on tower, namely inverse flashover; lightning strike pass the wire, namely detour lightning thunder; lightning strike to the central of ground wire. Lightning strike on the middle of ground wire span and the conductor flashover causing tripping is rare. Therefore, direct lightning protection is mainly against inverse flashover overvoltage and detour lightning overvoltage.

Detour lightning thunder tripping rate and inverse flashover tripping rate which related to the influence of factors such as landform, grid insulation level, the structure of the tower, the ground tilt Angle and grounding resistance. In brief, ground flash in certain conditions can cause lightning accident. According to the cause of the breakdown of the transmission line under lightning disasters, combine the experience and the actual situation to determine the related risk factors. This article selects the six risk factors such as ground lightning density, time distribution, the regional average soil resistivity of transmission line, transmission line elevation, transmission line grounding resistance and transmission line tower height.

2.1 Ground lightning average density

The main damage on transmission line caused by lightning is unscheduled shutdown due to trip-out. Ground lightning density is the most important risk factor affecting lightning trip-out rate, determination of it directly affects the accuracy of the assessment. Lightning distribution is seasonal, most areas in China mainly happened in summer thunderstorm season. According to the monthly statistics based on the ground lightning number per day from 2000 to 2010, we get monthly average ground lightning density, which is shown in Figure 1. From June to September the ground lightning density was significantly higher than other months.

2.2 Lightning day time distribution

The region's lightning day time distribution is shown in Figure 2. The lightning is double peak distribution in 24 hours, lightning mainly concentrated in 0~3a.m, 14~23p.m. If transmission line fault occurs in 17~23p.m during electricity peak, that will cause great influence to power supply. Therefore, different distribution would affect the fault risk of overhead line.

2.3 Soil resistance rate

The soil resistivity also has bigger difference in different regions which is relevant to the probability distribution of lightning current amplitude [10]. In fact, the probability of higher lightening current amplitude will increase in the lower soil resistivity area to a great extent [1]. Lightning current amplitude distribution directly affects the detour lightning tripping rate. Thus detour lightning tripping is much easier happened in higher soil resistivity area.

2.4 Altitude

Transmission line lightning strike is also different in different topography. Altitude is another important factor affecting the lightning current distribution which affects the transmission line detour lightning tripping rate. The height of thundercloud in mountain area is lower than in plain area, they reached critical breakdown strength of the air before a large number of charge formed, so there is less chance of the higher lightening current amplitude appearing in mountains area than in plain area [2]. In a word,

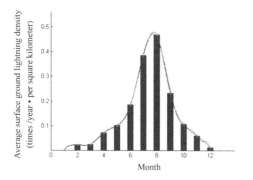

Figure 1. Monthly statistics of average ground lightning density.

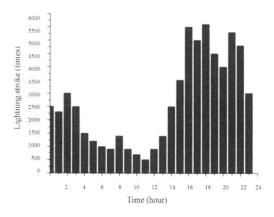

Figure 2. Hourly statistics of transmission line lightning strike.

486

lightning current amplitude decreases with increasing of altitude. Detour lightning tripping rate is more likely to happen in mountain area. Therefore, altitude should be taken into consideration in transmission line lightning risk assessment.

2.5 *Transmission line ground resistance*

Tower grounding resistance is the most important factor directly impacting on the withstand lever of inverse flashover and inverse flashover trip-out rate. When calculating a lightning withstand lever of inverse flashover, tower impulse grounding resistance is an important parameter to calculate tower impulse potential which got from measured power frequency grounding resistance multiplying by the impulse factor [3].

Smaller transmission line grounding resistance can effectively reduce overvoltage due to circulation through thunder lightning rod (wire) or lightning arrester which can protect the personal safety and insulation of the electrical equipment. But when the transmission environment conditions are severe and the grounding body is not buried deep enough, especially in the mountains, rock region which is high oxygen content. Thus, the soil corrosion of ground would be faster which directly lead to the grounding resistance value greater than the design value. Pose a threat to transmission line operation.

In conclusion, consideration should be given to the influence of environment on ground resistance and use the actual measured value to instead the design value of ground resistance when take a risk assessment for transmission line.

2.6 *Transmission line tower height*

The voltage level of transmission line affects the tower height, as shown in Figure 3, shielding

failure trip-out rate will increase with the increase of tower height [4]. Tower height has an impact on the lightning strike risk of transmission line.

3 FMEA MODEL OF TRANSMISSION LINES UNDER LIGHTNING DISASTERS

From the analysis of above factors affecting the risk of transmission line, we consider that the risk factors are independent of each other as shown in Table 1. According to the theory of fault tree analysis [7], FMEA model is set up as shown in Figure 4.

3.1 *Risk factors quantify*

Determining the probability of each risk events U_i is the key to build FMEA model. Transmission lines have different failure rate under different environmental conditions. The failure rate even varied with different levels of risk under the same condition.

Table 1. Transmission line lightning risk factors.

Risk factor	Sign
The ground lightning average density	U_1
Lightning day time distribution	U_2
Landscape (vegetation)	U_3
Altitude	U_4
Ground resistance	U_5
Tower height	U_6

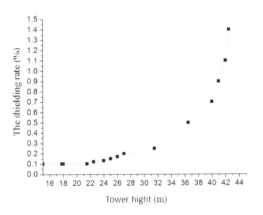

Figure 3. The relationship between shielding failure trip-out rate and tower high degree.

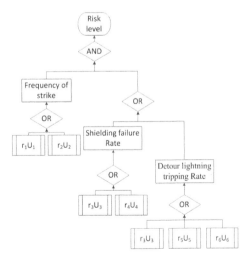

Figure 4. FMEA model of transmission line risk level.

For example, transmission line failure rate is obviously higher under red level of the lightning weather than general lightning weather. Therefore, we should select targeted indexes when assess the risk of transmission lines.

Risk assessment indicators can be obtained from building failure rate model of single external risk factors under different risk level based on transmission lines operating experience statistics [9].

Suppose ai the risk index transformed into risk level is each of the six quantitative applied risk factors such as ground flash density, lightning day time distribution, landform, altitude, ground resistance and tower height, which can be calculated by formula as follow:

$$\lambda_{ia_i} = \frac{N_{a_i}}{N_{ia_i}}, i = 1, 2, 3, ..., n \tag{1}$$

In the formula, λ_{ia} is the transmission line failure rate of risk factor i under risk level α_i. λ_{ia} is the function of the risk level α_i. N_{α} is the failure number of risk factor i under the risk level α_i. N_{α} is the total failure number of risk level α_i. Under risk factor i.

3.2 The determination of weights

Due to the incidence of various factors on the transmission line fault rate is different, this paper used Improved Analytic Hierarchy Process (IAHP) [9] to deal with the weight of each risk factor, that is to divide the problem to into two stages: first, use familiar three scale method (0,1,2) to do a comparison between any two elements, then establish a comparison matrix and calculate sorting index for each element; second, transform comparison matrix into judgment matrix and prove that it can completely satisfy the consistency. Suppose ri is the weight of six risk factors weight sets.

Establish comparative matrix

$$A = (b_{ij}) \tag{2}$$

In the formula, $b_{ij} = \begin{bmatrix} 2 & b_i \text{ is important than } b_j \\ 1 & b_i \text{ is as important as } b_j \\ 0 & b_j \text{ is important than } b_i \end{bmatrix}$,

$(i, j = 1, 2, 3, 4, 5, 6)$.

$$r_i = \sum_{j=1}^{6} b_{ij} \tag{3}$$

1. Construct judgment matrix

$$B(b_{ij}) = \begin{bmatrix} b & b_1 & b_2 & b_3 & b_4 & b_5 & b_6 & M_i & W_i & w_i \\ b_1 & 1 & & & & & & & & \\ b_2 & & 1 & & & & & & & \\ b_3 & & & 1 & & & & & & \\ b_4 & & & & 1 & & & & & \\ b_5 & & & & & 1 & & & & \\ b_6 & & & & & & 1 & & & \end{bmatrix} \tag{4}$$

In the formula, $M_i = \prod_{j=1}^{6} b_{ij}$, $w_i = \frac{W_i}{\sum_{j=1}^{6} W_i}$, $W_i = \sqrt[6]{M_i}$, $\sum_{j=1}^{6} w_i = 1$.

2. Consistency check

$$B = (b_{ij}), C = (c_i) = B \cdot [w_i]^T \tag{5}$$

The biggest characteristic value:

$$\lambda_{\max} = \frac{1}{n} \sum_{i=1}^{6} \frac{c_i}{w_i} \tag{6}$$

$$P_{B.I.} = \frac{\lambda_{\max} - n}{n - 1} \leq 0.001 = \varepsilon \tag{7}$$

3.3 Risk level rating

From six risk factors above, we can determine the risk probability index and weight index based on the possibility of harm transform into risk and weight analysis. The probability of basic risk event U_i is as follows:

$$U_i = r_i \lambda_i \tag{8}$$

$$\begin{aligned} V_1 &= U_1 + U_2 \\ V_2 &= U_3 + U_4 + U_5 + U_6 \end{aligned} \tag{9}$$

The risk level of transmission lines under the lightning disasters are for:

$$R = CV_1 \cdot V_2 \tag{10}$$

In the formula, C is a constant which is the magnification times for more expediently judging the risk level of transmission line, usually take 10.

Comprehensively analysis the effects and weights of the six risk factors on transmission lines, we can take the risk level as shown in Table 2.

High risk index greater than 1 indicates the highest risk level. That is transmission lines malfunction easily under lightning disasters, the security of power system face great risks. We need to strengthen and improve the existing lightning protection measures. Risk index greater than 0.8 shows that risk level is

Table 2. Risk analysis table of Transmission lines under the thunder and lightning disasters.

Comprehensive index R	Risk level
>1	Highest
>0.8	Higher
>0.5	Medium
>0.3	Low

higher which may lead to transmission line tripping, conductor wire breakdown and Insulator damage etc. The risk index is less than or equal to 0.5 shows the low risk level, indicates that the existing lightning protection measures can effectively ensure the safety of the transmission line and the stability of power grid. We do not need to take additional measures.

Therefore, we need to focus on the situation of larger risk index. Analysis and calculation the relevant parameters of risk factors which may cause transmission line failure, there by select suitable protection measures.

4 CONCLUSION

Transmission line risk early warning is one of the main contents of the power grid security early warning. A Gordian knot lies in how to build early warning methods and model [17]. In this paper, the key is to build a fault model considered the thunder day distribution, ground lightning density and other risk factors affecting the safety of transmission line, the actual lightning activity impact on transmission line is described more accurately. Thus get more data of risk factors which is more in line with actual situation, and then get more accurate risk rating criteria of transmission lines under lightning disasters which can provide evidence for security plans and accident treatment measures.

REFERENCES

1. Armstrong, H.R., Whitehead, E.R., 1968. Field and analytical studies of transmission line shielding. IEEE Transactions on Power Apparatus and Systems, IEEE, 87:270–281. DOI: 10.1109/TPAS.1968.291999.
2. Brown, G.W., Whitehead, E.R., 1969. Field and analytical studies of transmission line shielding: Part II. IEEE Transactions on Power Apparatus and System, 88617–626. DOI: 10.1109/TPAS.1969.292350.
3. Bali, T.G., 2003. An extreme value approach to estimating volatility and value at risk. The Journal of Business, 76:83–108. DOI: 10.1086/344669.
4. Cebrian, A.C., Denuit, M. and Lambert, P., 2003. Generalized Pareto Fit to the Society of Actuaries' Large Claims Database. North American Actuarial Journal, 20:18–36. DOI: 10.1080/10920277.2003.10596098.
5. Dai, M., Zhou, P.H. and Gu, D.Y., 2007. Studies on Lightening Protection of 750 kV Double-circuit Transmission Line. Advances of Power System & Hydroelectric Engineering, 23:1–5. DOI: 10.3969/j.issn.1674–3814.2007.10.001.
6. Du, Y., Chen, S.M. and John B., 2001. Experimental and Numerical Evaluation of Surge Current Distribution in Building During a Direct Lightning Stroke. Transactions Hongkong Institution of Engineers, 23:43–49. DOI: 10.1080/1023697X.2001.1066783.
7. Embrechts, P., Resnick, S., 1999. Extreme value theory as a risk management tool. North American Actuarial Journal, 3:30–41. DOI: 10.1080/10920277.1999.10595797.
8. Hosking, J.R.M., Wallis, J.R., 1987. Parameter and quantile estimation for the Generalized Pareto distribution. Technometrics, 29:339–349. DOI:10.1080/00401706.1987.10488243.
9. He, J.L., Zhang, D. and Liu, Y.G., 2011. Lightning Protection Analysis of the 10 kV Overhead Insulation Line Based on Real Lightning Parameters. Journal of Chongqing ElectricPowerCollege, 16:68–72. DOI: 10.3969/j.issn.1008-8032.2011.02.024.
10. IEEE Working group on estimating lighting performance of transmission lines. A simplified method for estimating lightning performance of transmission lines[J]. IEEE Transactions on Power Delivery, 1985, 104(4):919–932.
11. IEEE Working Group, 1993. Estimating lightning performance of transmission lines II-Updates to analytical models. IEEE transactions on Power Delivery, IEEE/PES Summer Meeting, Seattle WA, ETATS-UNIS (12/07/1992), 8:1254–1267.
12. McDermott, T.E. and Longo, V.J., 2000. Advanced computational methods in lightning performance-The EPRI lightning protection design workstation. In: IEEE Power Engineering Society, Power Engineering Society Winter Meeting, New York, USA. IEEE, 33(1):17–21. DOI: 10.1109/PESW.2000.847189.
13. Rochafeller, R.T., Uryasev, S., 2000. Optimization of conditional Value-at-Risk. In: Computational Intelligence for Financial Engineering, 2000.(CIFEr) Proceedings of the IEEE/IAFE/INFORMS 2000 Conference on, New York, NY, 2:14–19. DOI: 10.1109/CIFER.2000.844598.
14. Sakae, T., Toshihiro, T. and Shigemitsu, O., 2010. Method of calculating the lightning outage rate of large-sized Transmission Lines. Transactions on Dielectrics and Electrical Insulation, 117:12–18. DOI: 10.1109/TDEI.2010.5539700.
15. Sheng, H.B., 2011. Simulation Tests on Lightning Stroke-Caused Wire-Breakage of 10 kV Overhead Transmission Line. Power System Technology, 35:117–121. DOI: 1000.367.2011.01.0117.05.
16. Zhang, Z.J., Liao, R.J., Sima, W.X., Calculation and analysis of the lightning protection performance of back striking for 500 kV double-circuit transmission Lines. High voltage apparatus, 39:20–28. DOI: 10.3969/j.issn.1001-1609.2003.05.007.
17. Zhang, Z.J., Sima, W.X. and Jiang, X.L., 2005. Study on the lightning protection performance of shielding failure for UHV&EHV transmission lines. Proceedings of the CSEE, 25:1–6. DOI: 10.3321/j.issn:0258-8013.2005.10.001.

Author index

T - #0275 - 101024 - C0 - 246/174/27 [29] - CB - 9781138027824 - Gloss Lamination